Low-Dimensional Semiconductor Structures

MATERIALS RESEARCH SOCIETY
SYMPOSIUM PROCEEDINGS VOLUME 1617

Low-Dimensional Semiconductor Structures

Symposium held August 11–15, 2013, Cancún, México

EDITORS

Tetyana V. Torchynska
ESFM-Instituto Politécnico Nacional
México City, México

Georgiy Polupan
ESIME-Instituto Politécnico Nacional
México City, México

Larysa Khomenkova
V. Lashkaryov Institute of Semiconductor Physics
National Academy of Sciences of Ukraine Kyiv, Ukraine

Gennadiy Burlak
CIICAp - Universidad Autónoma del Estado de Morelos
Morelos, México

APPENDIX EDITORS

Tetyana V. Torchynska
ESFM-Instituto Politécnico Nacional
México City, México

Yuri V. Vorobiev
CINVESTAV-IPN Unidad Querétaro
Queretaro, México

Zsolt J. Horvath
Obuda University Budapest, Hungary

Materials Research Society
Warrendale, Pennsylvania

CAMBRIDGE
UNIVERSITY PRESS

University Printing House, Cambridge CB2 8BS, United Kingdom

One Liberty Plaza, 20th Floor, New York, NY 10006, USA

477 Williamstown Road, Port Melbourne, VIC 3207, Australia

314-321, 3rd Floor, Plot 3, Splendor Forum, Jasola District Centre, New Delhi - 110025, India

103 Penang Road, #05-06/07, Visioncrest Commercial, Singapore 238467

Cambridge University Press is part of the University of Cambridge.

It furthers the University's mission by disseminating knowledge in the pursuit of education, learning and research at the highest international levels of excellence.

www.cambridge.org
Information on this title: www.cambridge.org/9781605115948

Materials Research Society
506 Keystone Drive, Warrendale, PA 15086
http://www.mrs.org

First published 2013

CODEN: MRSPDH

A catalogue record for this publication is available from the British Library

ISBN 978-1-605-11594-8 Hardback

CONTENTS

SI BASED MATERIALS

ZnO NANOCRYSTALS

II-VI MATERIALS

METAMATERIALS AND OTHERS

APPENDIX

Papers in the Appendix were published in electronic format as Volume 1534.

IV GROUP SEMICONDUCTORS

III-V GROUP SEMICONDUCTORS

II-VI GROUP SEMICONDUCTORS

PREFACE

This MRS Proceedings book contains the papers presented in Symposium 7E: "Low-Dimensional Semiconductor Structures" at XXII International Material Research Congress, IMRC 2013, and in Symposium 6B: "Low-Dimensional Semiconductor Structures" at XXI International Material Research Congress, IMRC 2012, which were held in Cancun, Mexico on August 11-15, 2013 and August 12-16, 2012, respectively.

"Low-Dimensional Semiconductor Structures" represent one of the most intensively developing areas of modern semiconductor physics. It is well known that low-dimensional systems, such as quantum dots, quantum wells, nanocrystals, nanosheets, nanorods and nanowires, as well as metamaterials constitute a firm basis for advanced physics with applications in optoelectronics, microelectronics, photonics and spintronics, as well as biology and medicine. The "Low-Dimensional Semiconductor Structures" symposia are annual forums addressed to the fundamental and applied aspects of physics of low-dimensional systems, including the theory, modeling, preparation, characterization and simulation of electronic low-dimensional semiconductor structures with the whole range of their possible future applications. Both theoretical and experimental contributions have been chosen for the publication in this MRS Proceedings volume.

The grand part of papers presented in this MRS Proceedings book belongs to the symposia's speakers—the leading specialists defining the scientific and technological progress in nanoscience. Novel physical phenomena in nano-scale structures have been revealed and are applied in lasers and optical amplifiers, light-emitting diodes, photodiodes and solar cells, memory devices, biological luminescence markers, etc. The papers included in the volume conform to the subject "Low-Dimensional Nanostructures for Optoelectronics, Biology and Memory Devices."

We would like to thank all of the contributors to this MRS Proceedings volume for their excellent presentations and papers, as well as the reviewers for their important and careful job to ensure the publication of high quality.

On behalf of both Symposium's organizers

Tetyana V. Torchynska

October 2013

Materials Research Society Symposium Proceedings

Volume 1607E — Nanotechnology-Enhanced Coatings, 2014, A. Taylor, N. Ludford, C. Avila-Orta, C. Becker-Willinger, ISBN 978-1-60511-584-9
Volume 1608 — Nanostructured Materials and Nanotechnology, 2014, J.L. Rodríguez-López, O. Graeve, A.G. Palestino-Escobedo, M. Muñoz-Navia, ISBN 978-1-60511-585-6
Volume 1609E — Biomaterials for Medical Applications, 2014, S. E. Rodil, A. Almaguer-Flores, K. Anselme, J. Castro, ISBN 978-1-60511-586-3
Volume 1610E — Advanced Materials and Technologies for Energy Storage Devices, 2014, I. Belharouak, J. Xiao, P. Balaya, D. Carlier, A. Cuentas-Gallegos, ISBN 978-1-60511-587-0
Volume 1611 — Advanced Structural Materials, 2014, H. Calderon, H. Balmori Ramirez, A. Salinas Rodriguez, ISBN 978-1-60511-588-7
Volume 1612E — Concrete and Durability of Concrete Structures, 2013, L. E. Rendon Diaz Miron, B. Martínez Sánchez, N. Ramirez Salinas, ISBN 978-1-60511-589-4
Volume 1613 — New Trends in Polymer Chemistry and Characterization, 2014, L. Fomina, G. Cedillo Valverde, M. del Pilar Carreón Castro, J. Godínez Sánchez, ISBN 978-1-60511-590-0
Volume 1614E — Advances in Computational Materials Science, 2014, E. Martínez Guerra, J.U. Reveles, O. de la Peña Seaman, ISBN 978-1-60511-591-7
Volume 1615E — Electron Microscopy of Materials, 2014, H. Calderon, C. Kisielowski, L. Francis, P. Ferreira, A. Mayoral, ISBN 978-1-60511-592-4
Volume 1616 — Structural and Chemical Characterization of Metals, Alloys and Compounds, 2014, A. Contreras Cuevas, R. Pérez Campos, R. Esparza Muñoz, ISBN 978-1-60511-593-1
Volume 1617 — Low-Dimensional Semiconductor Structures, 2013, T.V. Torchynska, L. Khomenkova, G. Polupan, G. Burlak, ISBN 978-1-60511-594-8
Volume 1618 — Cultural Heritage and Archaeological Issues in Materials Science (CHARIMSc), 2014, J.L. Ruvalcaba Sil, J. Reyes Trujeque, A. Velázquez Castro, M. Espinosa Pesqueira, ISBN 978-1-60511-595-5
Volume 1619E — Modeling and Theory-Driven Design of Soft Materials, 2014, M. Rodger, M. Dutt, Y. Yingling, V. Ginzburg, ISBN 978-1-60511-596-2
Volume 1620E — Point-and-Click Synthesis—Implementations of Click Chemistry in Polymers, 2014, C.J. Kloxin, R.K. O'Reilly, T.F. Scott, J. Woods, ISBN 978-1-60511-597-9
Volume 1621 — Advances in Structures, Properties and Applications of Biological and Bioinspired Materials, 2014, T. Deng, H. Liu, S.P. Nukavarapu, M. Oyen, C. Tamerler, ISBN 978-1-60511-598-6
Volume 1622 — Fundamentals of Gels and Self-Assembled Polymer Systems, 2014, F. Horkay, N. Langrana, M. Shibayama, S. Basu, ISBN 978-1-60511-599-3
Volume 1623E — Synthetic Tools for Understanding Biological Phenomena, 2014, D. Benoit, A. Kloxin, C-C. Lin, V. Sée, ISBN 978-1-60511-600-6
Volume 1624E — Integration of Biomaterials with Organic Electronics, 2014, M.R. Abidian, M. Irimia-Vladu, R. Owens, M. Rolandi, ISBN 978-1-60511-601-3
Volume 1625E — Multiscale Materials in the Study and Treatment of Cancer, 2014, N. Moore, S. Peyton, J. Snedeker, C. Williams, ISBN 978-1-60511-602-0
Volume 1626 — Micro- and Nanoscale Processing of Materials for Biomedical Devices, 2014, R. Narayan, V. Davé, S. Jayasinghe, M. Reiterer, ISBN 978-1-60511-603-7
Volume 1627E — Photonic and Plasmonic Materials for Enhanced Optoelectronic Performance, 2014, C. Battaglia, ISBN 978-1-60511-604-4
Volume 1628E — Large-Area Processing and Patterning for Active Optical and Electronic Devices, 2014, T.D. Anthopoulos, I. Kymissis, B. O'Connor, M. Panzer, ISBN 978-1-60511-605-1
Volume 1629E — Functional Aspects of Luminescent and Photoactive Organic and Soft Materials, 2014, M.C. Gather, S. Reineke, ISBN 978-1-60511-606-8
Volume 1630E — Solution Processing of Inorganic and Hybrid Materials for Electronics and Photonics, 2014, P. Smith, M. van Hest, H. Hillhouse, ISBN 978-1-60511-607-5
Volume 1631E — Emergent Electron Transport Properties at Complex Oxide Interfaces, 2014, K.H. Bevan, S. Ismail-Beigi, T.Z. Ward, Z. Zhang, ISBN 978-1-60511-608-2
Volume 1632E — Organic Microlasers—From Fundamentals to Device Application, 2014, R. Brückner, ISBN 978-1-60511-609-9

MATERIALS RESEARCH SOCIETY SYMPOSIUM PROCEEDINGS

Volume 1633 — Oxide Semiconductors, 2014, S. Durbin, M. Grundmann, A. Janotti, T. Veal, ISBN 978-1-60511-610-5
Volume 1634E — Diamond Electronics and Biotechnology—Fundamentals to Applications VII, 2014, J. C. Arnault, C.L. Cheng, M. Nesladek, G.M. Swain, O.A. Williams, ISBN 978-1-60511-611-2
Volume 1635 — Compound Semiconductor Materials and Devices, 2014, F. Shahedipour-Sandvik, L.D. Bell, K.A. Jones, A. Clark, K. Ohmori, ISBN 978-1-60511-612-9
Volume 1636E — Magnetic Nanostructures and Spin-Electron-Lattice Phenomena in Functional Materials, 2014, A. Petford-Long, ISBN 978-1-60511-613-6
Volume 1637E — Enabling Metamaterials—From Science to Innovation, 2014, J.A. Dionne, P. J. Schuck, V.J. Sorger, L.A. Sweatlock, ISBN 978-1-60511-614-3
Volume 1638E — Next-Generation Inorganic Thin-Film Photovoltaics, 2014, C. Kim, C. Giebink, B. Rand, A. Boukai, ISBN 978-1-60511-615-0
Volume 1639E — Physics of Organic and Hybrid Organic-Inorganic Solar Cells, 2014, P. Ho, M. Niggemann, G. Rumbles, L. Schmidt-Mende, C. Silva, ISBN 978-1-60511-616-7
Volume 1640E — Sustainable Solar-Energy Conversion Using Earth-Abundant Materials, 2014, S. Jin, K. Sivula, J. Stevens, G. Zheng, ISBN 978-1-60511-617-4
Volume 1641E — Catalytic Nanomaterials for Energy and Environment, 2014, J. Erlebacher, D. Jiang, V. Stamenkovic, S. Sun, J. Waldecker, ISBN 978-1-60511-618-1
Volume 1642E — Thermoelectric Materials—From Basic Science to Applications, 2014, Q. Li, W. Zhang, I. Terasaki, A. Maignan, ISBN 978-1-60511-619-8
Volume 1643E — Advanced Materials for Rechargeable Batteries, 2014, T. Aselage, J. Cho, B. Deveney, K.S. Jones, A. Manthiram, C. Wang, ISBN 978-1-60511-620-4
Volume 1644E — Materials and Technologies for Grid-Scale Energy Storage, 2014, B. Chalamala, J. Lemmon, V. Subramanian, Z. Wen, ISBN 978-1-60511-621-1
Volume 1645 — Advanced Materials in Extreme Environments, 2014, M. Bertolus, H.M. Chichester, P. Edmondson, F. Gao, M. Posselt, C. Stanek, P. Trocellier, X. Zhang, ISBN 978-1-60511-622-8
Volume 1646E — Characterization of Energy Materials *In-Situ* and *Operando*, 2014, I. Arslan, Y. Gogotsi, L. Mai, E. Stach, ISBN 978-1-60511-623-5
Volume 1647E — Surface/Interface Characterization and Renewable Energy, 2014, R. Opila, F. Rosei, P. Sheldon, ISBN 978-1-60511-624-2
Volume 1648E — Functional Surfaces/Interfaces for Controlling Wetting and Adhesion, 2014, D. Beysens, ISBN 978-1-60511-625-9
Volume 1649E — Bulk Metallic Glasses, 2014, S. Mukherjee, ISBN 978-1-60511-626-6
Volume 1650E — Materials Fundamentals of Fatigue and Fracture, 2014, A.A. Benzerga, E.P. Busso, D.L. McDowell, T. Pardoen, ISBN 978-1-60511-627-3
Volume 1651E — Dislocation Plasticity, 2014, J. El-Awady, T. Hochrainer, G. Po, S. Sandfeld, ISBN 978-1-60511-628-0
Volume 1652E — Advances in Scanning Probe Microscopy, 2014, T. Mueller, ISBN 978-1-60511-629-7
Volume 1653E — Neutron Scattering Studies of Advanced Materials, 2014, J. Lynn, ISBN 978-1-60511-630-3
Volume 1654E — Strategies and Techniques to Accelerate Inorganic Materials Innovation, 2014, S. Curtarolo, J. Hattrick-Simpers, J. Perkins, I. Tanaka, ISBN 978-1-60511-631-0
Volume 1655E — Solid-State Chemistry of Inorganic Materials, 2014, S. Banerjee, M.C. Beard, C. Felser, A. Prieto, ISBN 978-1-60511-632-7
Volume 1656 — Materials Issues in Art and Archaeology X, 2014, P. Vandiver, W. Li, C. Maines, P. Sciau, ISBN 978-1-60511-633-4
Volume 1657E — Advances in Materials Science and Engineering Education and Outreach, 2014, P. Dickrell, K. Dilley, N. Rutter, C. Stone, ISBN 978-1-60511-634-1
Volume 1658E — Large-Area Graphene and Other 2D-Layered Materials—Synthesis, Properties and Applications, 2014, editor TBD, ISBN 978-1-60511-635-8
Volume 1659 — Micro- and Nanoscale Systems — Novel Materials, Structures and Devices, 2014, J.J. Boeckl, R.N. Candler, F.W. DelRio, A. Fontcuberta i Morral, C. Jagadish, C. Keimel, H. Silva, T. Voss, Q. H. Xiong, ISBN 978-1-60511-636-5
Volume 1660E — Transport Properties in Nanocomposites, 2014, H. Garmestani, H. Ardebili, ISBN 978-1-60511-637-2

Volume 1661E — Phonon-Interaction-Based Materials Design—Theory, Experiments and Applications, 2014, D.H. Hurley, S.L. Shinde, G.P. Srivastava, M. Yamaguchi, ISBN 978-1-60511-638-9
Volume 1662E — Designed Cellular Materials—Synthesis, Modeling, Analysis and Applications, 2014, K. Bertoldi, ISBN 978-1-60511-639-6
Volume 1663E — Self-Organization and Nanoscale Pattern Formation, 2014, M.P. Brenner, P. Bellon, F. Frost, S. Glotzer, ISBN 978-1-60511-640-2
Volume 1664E — Elastic Strain Engineering for Unprecedented Materials Properties, 2014, J. Li, E. Ma, Z. W. Shan, O.L. Warren, ISBN 978-1-60511-641-9

Prior Materials Research Symposium Proceedings available by contacting Materials Research Society

III-V materials

Mater. Res. Soc. Symp. Proc. Vol. 1617 © 2013 Materials Research Society
DOI: 10.1557/opl.2013.1156

Electrical Properties of Quantum Wells in III-NITRIDE Alloys and the Role of Defects

Daniela Cavalcoli, Albert Minj, Saurabh Pandey, Beatrice Fraboni and Anna Cavallini
Physics and Astronomy Department, University of Bologna, Viale C Berti Pichat 6/II, I-40127 Bologna, Italy

ABSTRACT

III-nitrides (III-Ns) semiconductors and their alloys have shown in the last few years high potential for interesting applications in photonics and electronics. III-Ns based heterostructures (HS) have been under wide investigation for different applications such as high frequency transistors, ultraviolet photodetector, light emitters etc. In the present contribution a III-Ns based heterostructure, in particular the nearly lattice matched $Al_{1-x}In_xN$/AlN/GaN HS will be discussed. The formation of the two dimensional electron gas (2DEG), its origin, its electrical and optical properties, the confined subband states in the well and its effect on the conduction mechanisms have been studied. Moreover, extended defects and their effect on the degradation phenomena of the 2DEG have been analyzed.

INTRODUCTION

Notwithstanding the wide application range and interest in III-Nitride based heterostructures (HS), many material-related fundamental issues of these materials are still under investigation: interdiffusion of mobile species, In or Al incorporation or segregation [1,2] and the role of extended defects on the electrical properties [3-6] are still debated. As a case study of such heterostructures, the nearly lattice matched $Al_{1-x}In_xN$/AlN/GaN system will be here analyzed. This system has been studied only quite recently, as before 2005 very few data were available on these alloys, and high electron mobilities transistor were basically developed on AlGaN/GaN HS [7]. Later on, it was found that lattice-matched AlInN/GaN heterostructures exhibit more than twice the amount of electrons confined at the AlGaN/GaN heterointerface, and hence a serious interest arose in those HS for applications in high power electronics.

In $Al_{1-x}In_xN$/GaN HS, the strong band gap difference between AlN and $Al_{1-x}In_xN$ (with low In content, x around 0.13-0.14) and the very high polarization- induced electric field create a triangular potential well at the heterointerface, which is able to confine electrons, thus a two dimensional electron gas (2DEG) forms in the well. The electrons forming the 2DEG may suffer from poor in-plane transport properties due to alloy disorder induced scattering. To overcome this difficulty, the insertion of an AlN interlayer, an approach already explored within the AlGaN/GaN system, has been used. This helps to keep the electrons better confined in the GaN channel and prevents 2DEG electrons from alloy scattering. High carrier density (around $2\times10^{13}cm^{-2}$) and high mobility (around 2×10^{3} cm^2/Vs) [7] are usually obtained in these structures. An example of such a well is sketched in fig1.

Notwithstanding the quite good structural and electrical properties of these structures, we have to remind here that III-Ns and their alloys suffer for the absence of suitable growth substrate. Sapphire is typically used, but thermal and lattice mismatch are responsible for the formation of threading dislocations (TDs) which start forming at the sapphire /GaN interface and propagate through the whole structure. Therefore, the study of the role of these defects on the transport properties of the HS is of the major importance. While structural properties of TDs have been widely investigated, electrical properties of TDs are still debated. As an example, III-Ns based HS show a quite intriguing "defect insensitivity", which is still debated. III-Ns based HS are able to support, without strong degradation, a TD density considerably higher than III-Phosphorous based HS. The reason seems to be related to the strong ionicity of III-Ns, which places the surface states (as well as defect states) very close to the band edges, and to the existence of potential fluctuations in the band structures due to stacking faults, alloying phenomena and phase separation [8,9]. Both these phenomena would help preventing trap assisted electron-hole recombination. However, notwithstanding III-Ns based devices can tolerate a quite large defect density, the knowledge of the defect electrical characteristics and of their effect on the macroscopical electrical properties of the heterostructure is a fundamental issue that should be still widely explored.

The present contribution deals with a summary and a review of the results that we recently obtained on nearly lattice matched $Al_{1-x}In_xN$/AlN/GaN heterostructures with In content around 13%. These structures have been investigated by electrical macroscopical methods (current-Voltage, Hall effect), microscopical methods (Atomic Force Microscopy, in current and phase contrast mode) and spectroscopy techniques (Photovoltage, Photocurrent and Photoluminescence). The formation of the 2DEG, its electrical and optical properties, its origin and its role in the electrical conduction will be discussed. In addition, possible defect related degradation effects on the 2DEG properties will be discussed.

EXPERIMENT

Nearly lattice matched $Al_{1-x}In_xN$/AlN/GaN heterostructures were grown by AIXTRON and III-V lab by metal–organic chemical vapor deposition (MOCVD) on c-plane sapphire substrates. The nominal thickness of AlInN was varied as 15 and 30 nm, while the thickness of the AlN layer was varied between 0 and 2.5 nm. Indium content varies from 12.5% to 14.5% as assessed by high resolution X-ray diffraction (HR-XRD) [10]. Relaxed HS (with AlN thickness of 7.5 nm, above the critical thickness) were also studied. Further details on the sample growth can be found elsewhere [4,11].

Macroscopical electrical characterization has been carried out by Current-Voltage measurements either on back-to-back Schottky contacts in a planar configuration directly formed by In-Ga alloy with a spacing of 2mm, either on Ni-Au Schottky contacts (dots of 1 mm diameter) and Ti/Al/Ni/Au Ohmic contacts (dots of 0.6 mm diameter). Hall effect measurements were carried out at 77 and 300 K on Van der Pauw structures to obtain the 2DEG density and carrier mobility. Further details on these experimental procedures can be found in ref [4,12]. Spectroscopical analyses were carried out by Surface photo voltage spectroscopy (SPS) analyses were performed at room temperature by means of a custom-made apparatus based on Xenon lamp source and a SPEX 500M monochromator [13,14]. Photoluminescence measurements have been performed by exciting the carriers with 193 nm ArF excimer lasers with at 5 K. Photocurrent analyses have been performed in an Ohmic-Ohmic configuration on SiNx passivated samples.

Microscopic morphological and electrical analyses were obtained by AFM (NT MDT Solver PRO47) by using conductive probes (Pt-Ir, diamond, nanoneedles (Ag_2Ga-Nauganeedles LLC). Conductive AFM mapping and IV curves were obtained with the sample grounded [15]). The contact was made with conductive silver on one side of the sample piece.

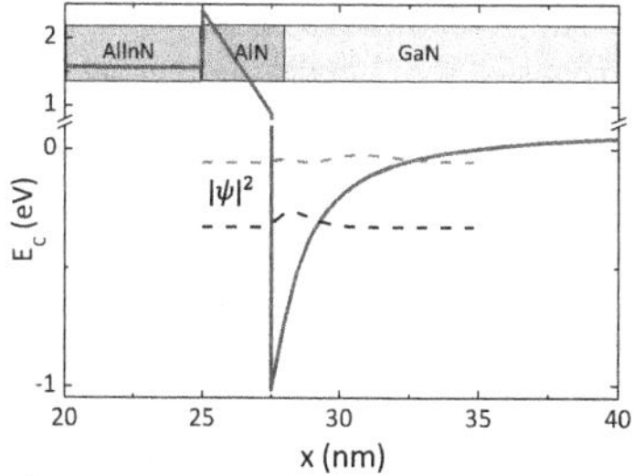

Figure 1 Triangular potential well at the GaN/AlN interface. The well shape and the first two subband energy levels and wave functions are calculated from Schrödinger-Poisson solver for an AlN interlayer thickness of 2.5 nm. A sketch of the heterostructure is also shown.

RESULTS

2DEG in $Al_{1-x}In_xN$/AlN/GaN heterostructures. Origin, electrical and optical properties.

AlN, GaN and AlInN possess polarized Wurtzite crystal structures, having dipoles across the crystal in the [0001] direction. In absence of external fields, spontaneous (pyroelectric) and strain induced (piezoelectric) polarizations contribute to the total macroscopic polarization [16]. In the AlInN/GaN system, this polarization induces an interface charge due to abrupt divergence in the polarization at the heterointerface. Even if the polarization is basically a volume effect, since vicinal dipole sheets in the z-direction cancel out each other, the polarization manifests itself mainly at interface as the difference of the total polarization of neighboring layers. Thin pseudomorphic AlN layers grown on GaN exhibit therefore a higher total polarization charge density due to the significant contribution of the piezoelectric polarization. In the case of lattice matched AlInN, this contribution vanishes completely and only the spontaneous polarization is present. If the polarization charge density is positive, electrons are attracted forming a 2DEG [7].

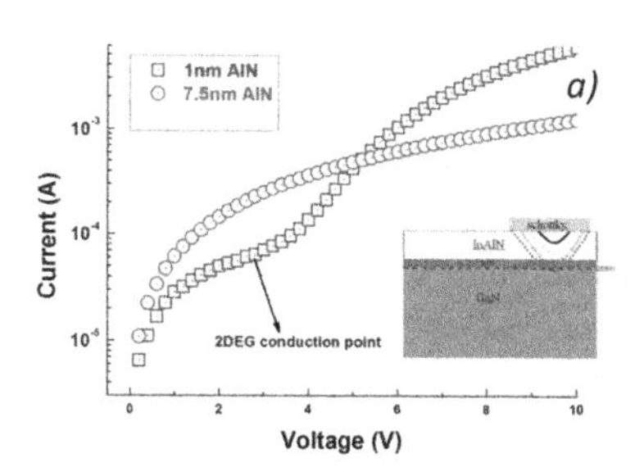

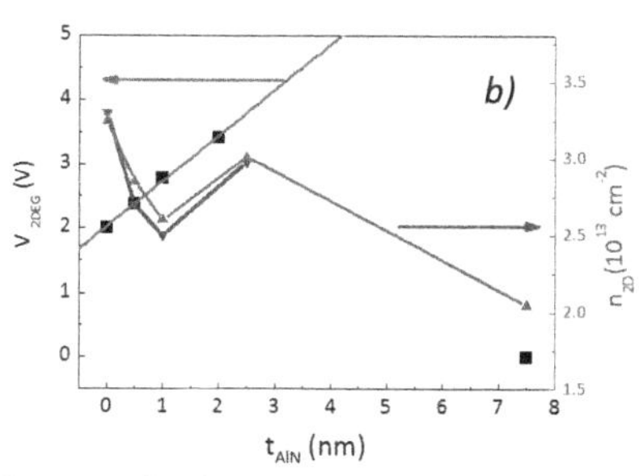

Figure 2 (a) Current vs. Voltage plot showing slope change activating 2DEG conduction in samples with different AlN thickness. The depletion of reverse-biased Schottky with increasing bias is also shown (inset). (b) The bias activating 2DEG conduction V_{2DEG} (left axis) and the 2DEG density n_{2D} (right axis) are plotted as a function of the AlN interlayer thickness. Literature data have been plotted for comparison [12]. The last point (7.5nm) of our I-V curve has been extrapolated. The lines are a guide for the eye (reprinted from [12]).

Figure 1 shows the band diagram structure of a nearly lattice matched AlInN/AlN/GaN HS and the position of the two subband states within the 2DEG channel, obtained by solving the 1-D Schroedinger-Poisson equation. Transport properties of the 2DEG in the HS are shown in figure 2. IV curves obtained on $Al_{1-x}In_xN$/AlN/GaN HS with AlN thickness (t_{AlN}) of 1 nm (below the critical thickness) and 7.5 nm (above the critical thickness, which is around 6.5 nm [17]) by back-to-back Schottky contacts in a planar configuration are shown in fig 2a. A clear change in the curve slope is visible in 1nm thick AlN samples, but not visible in 7.5nm AlN samples. The slope change occurs at an applied bias voltages which scale with the AlN thickness in different samples, except for the samples with a 7.5nm thick AlN interlayer (fig.2b). For low bias voltages the transport is limited to the top $Al_{1-x}In_xN$ barrier layer, as the applied bias voltage increases (fig. 2a inset), the depletion region extends further through the AlInN and AlN layers, allowing the current flow to reach the interface with the GaN substrate, where the 2DEG is located. The current increases when the conduction through the 2DEG is activated, apart from the sample with t_{AlN}=7.5 nm. In this sample large currents at very low bias voltages were observed, related to the presence of many TDs and nanocracks observed by AFM [3 and fig 5 below]. The V_{2DEG} values were used to calculate the total effective polarization charge density, and by applying an electrostatic model [7], the 2DEG concentration, which is plotted in fig 2b as a function of t_{AlN}, and compared with literature data [12]. It can be noted that the transport properties of the 2DEG have been reliably investigated by macroscopic IV measurements.

The 2DEG density was evaluated also by microscopical IV analyses. The effect of the 2DEG on the local transport properties of the HS has also been investigated by C-AFM. We have demonstrated that current–voltage measurements performed with a conductive AFM tip on $Al_{1-x}In_xN$/AlN/GaN can be modeled by the thermionic emission from the 2DEG assisted by image-charge-induced barrier lowering. Other transport mechanisms that could be active in these heterostructures, such as tunneling or dislocation-assisted conduction mechanisms, have been ruled out due to the contact dimension (of the order of nanometers). The barrier lowering is caused by the image charge induced by the 2DEG and depends on the 2DEG characteristics. The 2DEG density can be thus be obtained by fitting the experimental data. The so obtained 2DEG density values have been found to be in very good agreement with the Hall measurements [15].

The optical properties of the 2DEG are of fundamental importance for optoelectronic device applications. Optically induced electronic transitions in AlInN/AlN/GaN HS with different In contents have been investigated by absorption (SPV and Photocurrent, PC) and emission spectroscopy (Photo Luminescence, PL). Figure 3a shows the possible electronic transitions in the HS as evaluated by solving Schrodinger-Pöisson equation, while figure 3b shows the direct measurement of these transitions. Energy level values of the subband states within the well at the GaN / AlN heterointerface have been measured by PL and PC as well, and the obtained results well compare with the simulated ones. Moreover, a strong enhancement of the Photoluminescence intensity due to holes recombining with electrons at the Fermi Energy, known as Fermi Energy Singularity (FES), has also been observed [14].

Another interesting point still under debate is the origin of the 2DEG. Surface donor states have been proposed to be the underlying cause of the 2DEG formation in the AlGaN layer [18]. Either a single surface state energy or a Fermi level pinning at the surface are able to explain the measured variation of the 2DEG concentration with the barrier thickness in AlGaN. By determining the Schottky barrier height in (Ni-Au)/ on $Al_{1-x}In_xN$/AlN/GaN Schottky diodes, measured by current-voltage (I-V) characteristics, we obtained the density and energy distribution of surface donor states.

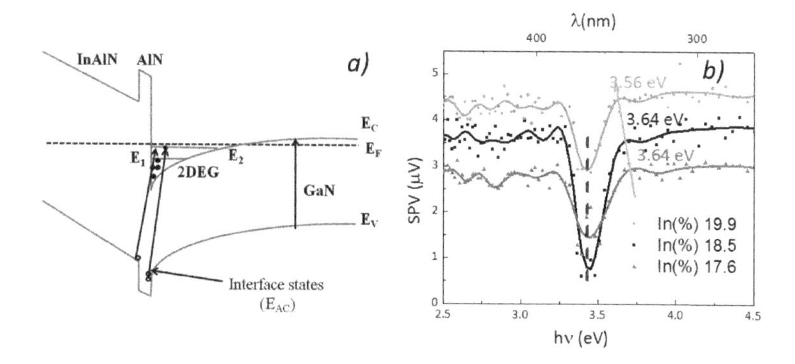

Figure. 3 (a) Schematic of the band structure for AlInN/AlN/GaN heterostructures calculated at 300 K from Schrodinger-Pöisson equation to show the possible photoexcited electronic transitions. **(b)** Surface Photovoltage signal plotted against photon energy for $Al_{1-x}In_xN/AlN/GaN$ HS with different In content.

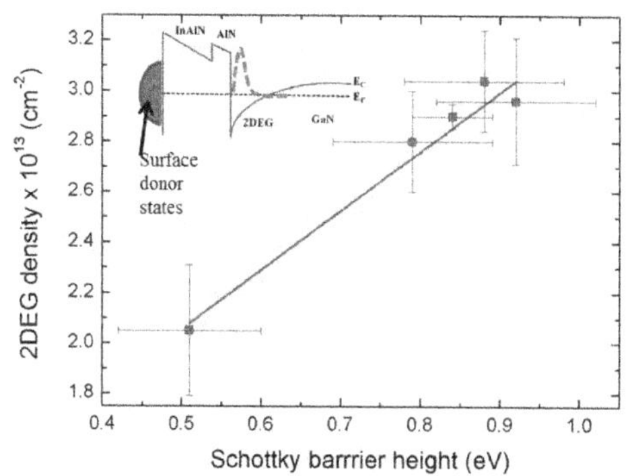

Figure.4 2DEG density as evaluated by Hall effect plotted as a function of the Schottky barrier height obtained by IV measurements of Schottky diode on AlInN/AlN/GaN HS.

Figure 4 shows that the 2DEG density, as measured by Hall effect in density in pseudomorphic $Al_{1-x}In_xN/AlN/GaN$ HS, scales linearly with the Schottky barrier height. From the linear fit of the 2DEG density vs the barrier height, the value of the surface /interface donor density has been evaluated. The value of $(2.7 \pm 0.2) \times 10^{13}$ cm^{-2} eV^{-1} was estimated [4]. The so-calculated value of surface donor density is in good agreement with earlier reports on AlGaN/GaN heterostructures [19, 20].

Defect- related degradation phenomena of the 2DEG in $Al_{1-x}In_xN/AlN/GaN$ Heterostructures

As already noted in the introduction, extended defects (threading dislocations, cracks..) and impurity segregation could induce degradation in the electrical and optical properties of the HS. Figure 5 shows AFM topographical micrographs of two $Al_{1-x}In_xN/AlN/GaN$ HSs with different AlN interlayer thickness. The two structures exhibit a quite different morphology, in the first case (fig 5a) a grain structure and V defects are visible, the V pit density in this sample ranges from $1x10^8$ to $5x10^8$ cm^{-2} and the root mean square roughness around is around 0.23 nm, in the second case (fig 5b) several cracks appear, the V-pit density and roughness significantly increase from 0.1 to $3x10^{10}$ cm^{-2} and 0.35 nm, respectively. V pit density and roughness values as a function of t_{AlN} are shown in fig.6a. In segregation phenomena at TDs and cracks have been demonstrated by dynamical and current AFM analyses [2, 3], and have been shown to be the cause of enhanced conduction through TDs and cracks.

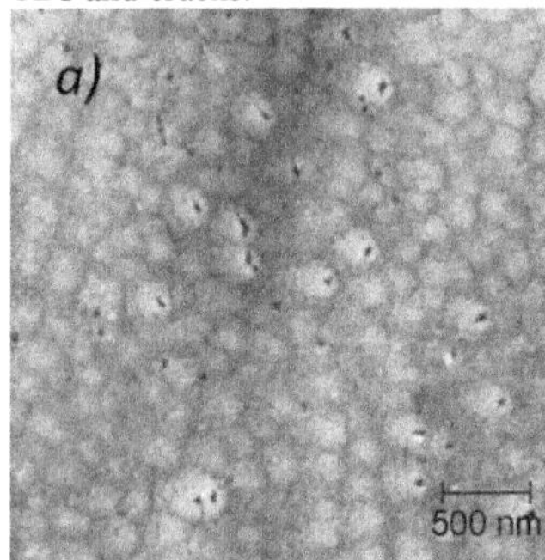

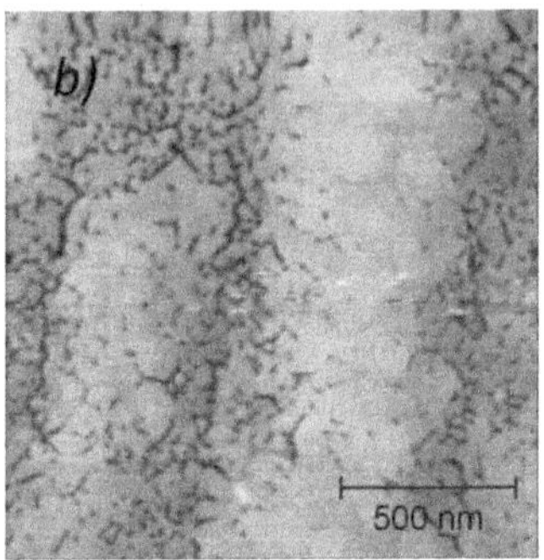

Figure.5 AFM micrographs of AlInN/AlN/GaN HS with AlN layer thickness of 1 nm (a) and of 7.5 nm (b).

In order to analyze the effect of TDs and cracks on the macroscopical properties of the 2DEG, transport measurements were performed on pseudomorphic (AlN thickness up to to 2.5 nm) and to relaxed (AlN thickness 7.5 nm) structures. Hall effect mobility values measured at 77K are plotted as a function of the AlN thickness in figure5a. It is to be noted that mobility, as well as roughness and V-pit density, show a trend vs t_{AlN}. In order to identify the dominant scattering mechanisms controlling the low temperature electron mobility of the 2DEG, we have calculated the mobility related to different scattering mechanisms. The remote surface roughness (RSR) scattering mechanism explains the low temperature mobility in AlGaN/GaN and in AlN/GaN 2DEG channel layers, respectively [21, 22], while scattering due to dislocations, alloy disorder, phonons, etc., plays an important role in limiting the 2DEG room temperature mobility [22]. We evaluated the RSR controlled mobility inserting in the model morphological parameters (the correlation length) measured by AFM and STM (Scanning Tunnelling Microscopy), the dislocation controlled mobility using the V-pit density as evaluated by AFM and finally we combined the two contribution and evaluated the total mobility [11]. The total mobility has been thus calculated using measured amounts without any fitting parameters nor a priori assumptions. In fig 6b the dislocation related, RSR and total calculated mobilities are compared with the experimentally measured values and plotted *vs* the surface roughness. The good agreement between the experimental and calculated mobility values is evident for all the roughness values except for the highest one. Indeed, the last point corresponds to samples with interlayer thickness of 7.5 nm, in this case strain relaxation induces the formation of cracks piercing the 2DEG and thus creating electrical shunts (figure 5b) [3]. The 2DEG electrical transport in such sample should be strongly affected by those cracks which are not considered in the theoretical model. We can note from Fig. 6b that the RSR scattering mechanism is the most effective one in controlling the 2DEG mobility in $Al_{1-x}In_xN/AlN/GaN$ HS, as the dislocation related mobility does not play a major role.

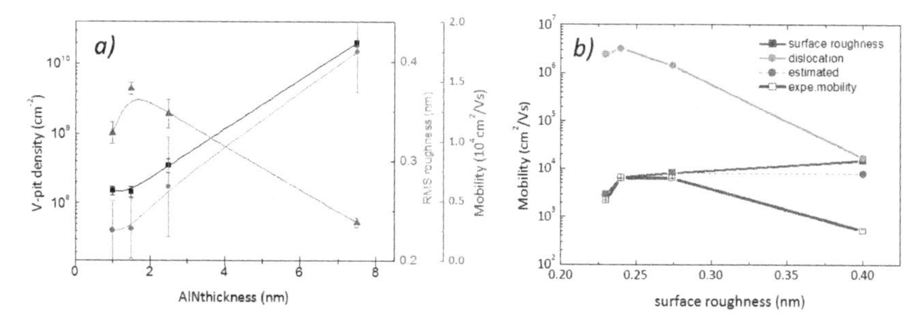

Figure.6 **(a)** V pit density (black square), Root Mean Square Roughness (red dot) and Hall mobility (blue triangle) of AlInN/AlN/GaN HS as a function of AlN layer thickness. The HS are pseudomorphic for AlN thickness below 6.5 nm, relaxed above 6.5 nm. **(b)** Measured Hall effect mobility plotted vs the RMS roughness, compared with mobilities calculated by dislocation scattering (green dot) surface roughness (blue square) and total mobilities (red dot).

In order to study the effect of strain relaxation, TDs and cracks on the electrical properties of $Al_{1-x}In_xN/AlN/GaN$ HS, current-voltage measurements were performed at room temperature on Ni-Au Schottky contacts (dots of 1 mm diameter) and Ti/Al/Ni/Au Ohmic contacts (dots of 0.6 mm diameter). The ohmic and Schottky contacts were prepared by Ti/Al/Ni/Au and Ni-Au evaporation respectively; for ohmic metallization, further annealing was performed at 850 °C for 30 s in N_2 ambient. Forward and reverse bias current-voltage measurements performed at 300 K are shown in Fig. 7a. In the pseudomorphic HS, the reverse bias leakage current is reduced by increasing the AlN interlayer thickness from 0 nm up to 2 nm; while for the relaxed HS, where t_{AlN} is 7.5 nm, the leakage current rapidly increases. The increase in the reverse bias leakage current could be related to the increase of the TD density, which is measured as V-pit density (fig.6a).

We have applied the Poole-Frenkel transport model [4] to interpret the leakage mechanisms in such AlInN/AlN/GaN HS, Poole-Frenkel emission refers to electric-field-enhanced thermal emission from a trap state into a continuum of electronic states. In the present case the trap states can be dislocation related. In fig. 7b we have plotted I/E vs $E^{1/2}$ and found for the pseudomorphic structures a linear dependence in the log scale as predicted by Poole Frenkel mechanism. These results indicate that the Poole-Frenkel mechanism is the dominant mechanism controlling the reverse leakage current in AlInN/GaN nearly lattice matched heterostructures. for the relaxed structure with 7.5 nm thick AlN layer, the current becomes nearly independent of $E^{1/2}$ suggesting that other mechanisms, such for example conduction through electrically active nanocracks, could play a major role.

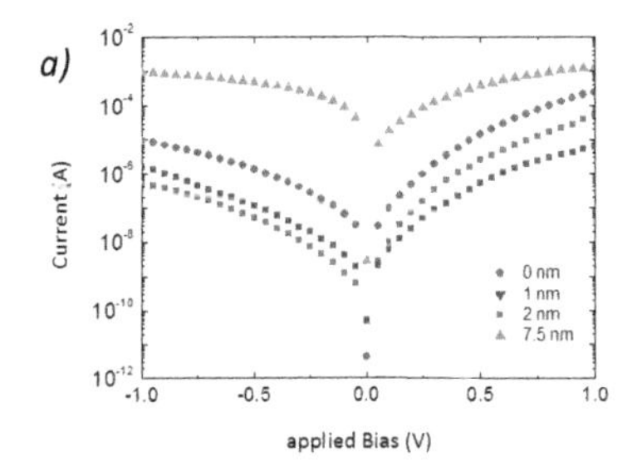

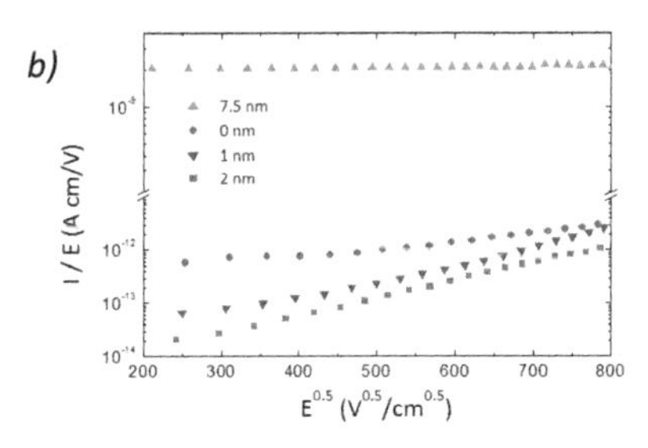

Figure 7. (a) IV curves showing the variation of reverse bias leakage current with different AlN interlayer thickness. (b) Measured reverse-bias current divided by electric field vs square root of electric field for Schottky contact on the $Al_{1-x}In_xN/AlN/GaN$ heterostructure

DISCUSSION AND CONCLUSIONS

The aim of the present contribution is to review the results obtained by the present authors on III-nitride based heterostructures, in particular on $Al_{1-x}In_xN/AlN/GaN$ HS with different AlN layer thickness. At the heterointerface between $Al_{1-x}In_xN/AlN/GaN$ a 2DEG forms. We have described the formation of the 2DEG, its origin, the analysis of its electrical and optical characteristics . Furthermore, we have investigated degradation effects of the HS and related them to defective states and morphological properties investigated at the nanoscale by AFM.

We have demonstrated that the 2DEG plays a fundamental role on the electrical and optical properties of the HS and that threading dislocations, cracks, and impurity decorated defects play a strong role on the degradation mechanisms of $Al_{1-x}In_xN/AlN/GaN$ heterostructures.

ACKNOWLEDGMENTS

H. Behmenburg, C. Giesen and M. Heuken of Aixtron SE, Herzogenrath, Germany, and P. Gamarra, and M. A. Poisson, of III-V lab, France, are gratefully acknowledged for providing the samples and for the useful discussion. T Brazzini from ISOM, ETSI Telecomunicación, Universidad Politécnica de Madrid, is gratefully acknowledged for Schottky diode preparation. D. Skuridina, P. Vogt, M. Kneissl of the Technical University of Berlin are gratefully acknowledged for useful discussion and support in STM experiments.
This work was supported by the EU under Project No. PITN-GA-2008 213238-RAINBOW.

REFERENCES

1. Th Kehagias et al. Appl. Phys. Lett. **95**, 071905 (2009).
2. A. Minj, D. Cavalcoli, A. Cavallini, Appl. Phys. Lett., **97**, 132114 (2010).
3. A. Minj, D. Cavalcoli, S. Pandey, B. Fraboni, A. Cavallini, T. Brazzini, F. Calle Script Mater, **66** 327 (2012).
4. S. Pandey, D. Cavalcoli, B. Fraboni, A. Cavallini, T. Brazzini, and F. Calle Appl. Phys. Lett.**100**, 152116 (2012).
5. F. A. Ponce, Ann. Phys. (Berlin), **1 – 2,** 523 75 (2011)
6. D. C. Look and J. R. Sizelove, Phys. Rev. Lett. **82**, 1237 (1999).
7. M Gonschorek, et al International Journal of Microwave and Wireless Technologies, , 2(1), 13–20 (2010).
8. S. F. Chichibu et al, Nature Materials **810-816** (2006)
9. T.D Moustakas Phys. Status Solidi A **210**, 1, 169–174 (2013)
10. Vilalta-Clemente A, Poisson M-A, Behmenburg H, Giesen C, Heuken M and Ruterana Phys. Status Solidi a **207** 1105 (2010)
11. S. Pandey, D. Cavalcoli, A. Minj, B. Fraboni, A. Cavallini, D. Skuridina, P. Vogt, M. Kneissl Acta Materialia **60** 3176–3180 (2012)
12. S Pandey, B Fraboni, D Cavalcoli, A Minj, A Cavallini - Applied Physics Letters, **99**, 012111 (2011)
13. D Cavalcoli, S Pandey, B Fraboni, A Cavallini Appl Phys Lett **98** 142111-142111(2011).
14. S. Pandey, D. Cavalcoli, A. Minj, B. Fraboni, A. Cavallini et al. J. Appl Phys. **112**, 123721 (2012)
15. A. Minj, D Cavalcoli and Anna Cavallini Nanotechnology **23** (2012) 115701
16. I. P. Smorchkova et al Appl. Phys. **86,** 4520 (1999)
17. M. Gonschorek et al J. of Appl Phys **103**, 093714 (2008)
18. A Rizzi et al Appl. Phys. A **87**, 505–509 (2007)
19. G. Koley, M. G. Spencer, Appl. Phys. Lett. **86**, 042107 (2005).
20. M. S. Miao, A. Janotti, C. G. Van de Walle, Phys. Rev. B **80**, 155319 (2009).
21. B. Liu et al Appl. Phys. Lett. **97**, 262111 (2010)
22. Y. Cao and D. Jena, Appl. Phys. Lett. **97**, 222116 (2010)

Mater. Res. Soc. Symp. Proc. Vol. 1617 © 2013 Materials Research Society
DOI: 10.1557/opl.2013.1157

HR XRD and Emission of $In_xGa_{1-x}As/GaAs$ quantum wells with embedded InAs quantum dots at the variation of $In_xGa_{1-x}As$ composition

Leonardo G. Vega Macotela, Ricardo Cisneros Tamayo and Georgiy Polupan

ESIME– Instituto Politécnico Nacional, México D. F. 07738, México.

ABSTRACT

The high resolution X ray diffraction (HR-XRD) diagrams have been studied in the GaAs /$In_xGa_{1-x}As$ /$In_{0.15}Ga_{0.85}As/GaAs$ quantum wells with embedded InAs quantum dots (QDs) in dependence on the composition of the capping $In_xGa_{1-x}As$ layers. The parameter x in capping $In_xGa_{1-x}As$ layers varied from the range 0.10-0.25. These technological changes have been accompanied by the variation non-monotonously of InAs QD emission. Numerical simulation of HR-XRD results has shown that the level of elastic strains and the composition of quantum layers vary none monotonously in studied QD structures. Simultaneously it was revealed that the process of Ga/In inter diffusion at the $In_xGa_{1-x}As/InAs$ QD interface are characterized by the dependence non monotonous versus parameter x in capping $In_xGa_{1-x}As$ layers. The physical reasons of the mentioned optical and structural effects in studied structures have been discussed.

INTRODUCTION

In the past twenty years, zero-dimensional quantum dot (QDs) systems with three-dimensional quantum confinement have attracted considerable attention owing to the fundamental and application reasons [1-3]. The most studded subject is the self-assembled (SA) semiconductor QDs formed by the Stranski–Krastanov (SK) growth mode. The InAs/GaAs QDs have been used as an effective active medium for the optoelectronic devices such as lasers and light emitting diodes, infrared photo-detectors and solar cells [4,5].

It was shown earlier that the InAs QD density can be enlarged significantly if the dots growth on the surface of $In_xGa_{1-x}As$ buffer layer within $In_xGa_{1-x}As/GaAs$ QWs [6]. In these structures photoluminescence (PL) has been enhanced due to the better crystal quality of layers surrounding QDs [7-9] and owing the more effective exciton capture into QWs and QDs [10-14]. It was revealed as well that the emission intensity and PL peak positions vary versus InGaAs layer composition none monotonously [7]. In this paper we try to understand the physical reasons of the emission variation non-monotonously in InAs QD structures with the different In composition in capping $In_xGa_{1-x}As$ layers.

EXPERIMENTAL CONDITIONS

The experimental set of QD structures was created using the molecular beam epitaxial growth on the (100) oriented 2''diameter semi-insulating GaAs substrates. Each structure included a 300 nm GaAs buffer layer and a 70 nm GaAs upper capping layer. Between GaAs layers where located three self-organized InAs QD arrays (formed by depositing 2.4 ML of InAs at 490 ^{o}C) embedded into an external asymmetric $In_xGa_{1-x}As/In_{0.15}Ga_{0.85}As/GaAs$ QWs separated by the GaAs spacer layers of 30 nm. The InAs QDs in all investigated structures were grown on a buffer layer of 1 nm with the composition $In_{0.15}Ga_{0.85}$ As under the InAs wetting layer (WL).

The capping $In_xGa_{1-x}As$ layers of 8.0 nm are characterized by the different In compositions with the parameter x: 0.10 (#1), 0.15 (#2), 0.20 (#3) and 0.25 (#4).

The HR-X-ray diffraction (XRD) experiments were done using the XRD equipment Model XPERT MRD with the Pixel detector, three axis goniometry and parallel collimator with the resolution of 0.0001 degree. The X ray beam was from the Cu source with $K_{\alpha 1}$ line (λ=1.5406 Å). Simulations were done by the variation of the sizes and compositions of the capping, wetting and barrier layers. The X'Pert Epitaxy software based on the dynamic diffraction theory for X-ray diffraction in layered structures has been used [15,16]. PL spectra were measured at 80 and 300K using the excitation by a 532 nm line of a solid state laser model V-5 COHERENT Verdi at an excitation power density 500 W/cm^2 [10,14]. PL spectra were dispersed by a SPEX 500M spectrometer with a Ge detector.

EXPERIMENTAL RESULTS AND DISCUSSION

It was shown earlier that the studied QD structures are characterized by the changes non monotonously of the GS PL intensity and peak positions versus In composition in the capping $In_xGa_{1-x}As$ layer (Fig.1) [17-19].

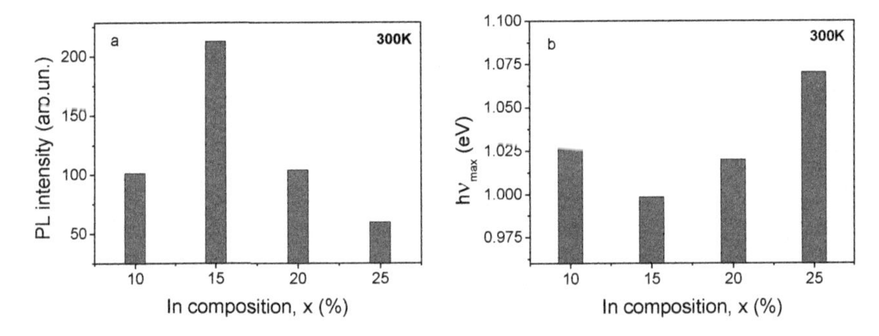

Figure 1. The variation of PL intensity (a) and peak position (b) versus $In_xGa_{1-x}As$ capping layer compositions

The "red" shift of GS PL bands is observed when x increases from 0.10 to 0.15. PL spectral red shift is accompanied by the enhancement of the PL band intensity (Fig.1a,b). The GS PL band shifts into high energy spectral range ("blue" shift) when x is changed from 0.15 to 0.20 and to 0.25 (Fig.1b). Simultaneously, the full width at half maximum (FWHM) of PL bands increases and the QD PL intensity decreases (Fig.1a). Thus the variation non monotonously of the GS integrated PL intensity and PL peak positions have been detected in studied QD structures.

Figure 2 presents the typical HR-XRD results obtained in the structures #2 and #3. The different shoulders that have seen in the main HR XRD peaks related to the interference between the diffraction from the different QW layers. As we can see there is the shape variation of HR-XRD for different samples. To explain this effect we supposed that a shape variation related to:

a) the Ga/In atom inter diffusion between InAs QDs and InGaAs capping/buffer layers and/or b) the different levels of elastic strains in studied structures.

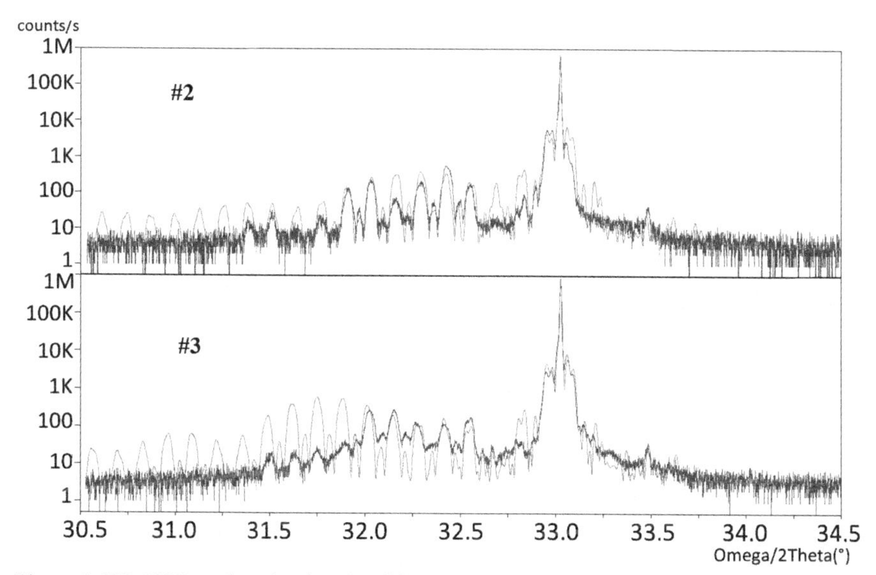

Figure 2. HR-XRD peaks related to the diffraction of $K_{\alpha 1}$ lines of the X-ray Cu source from the (400) crystal planes of GaAs and InGaAs layers in the QD structures #2 and #3. The red line presents the result of numerical simulation.

The simulations were done by the variation of the sizes and compositions of capping, wetting and barrier layers. The X'Pert Epitaxy software based on the dynamic diffraction theory for X-ray diffraction in layered structures has been used [15,16].

The comparison of fitting parameters (Table 1) has revealed that in studied QD structures the fitting parameters for the structure #2 are very close to their original technological compositions. From the other hand the fitting parameters for the structures #1,#3, #4 are a little bit different from the growth technological parameters due to the process of Ga/In inter diffusion. Note that the process of the Ga/In inter diffusion in studied structures passes non monotonously versus QD growth temperatures.

In order to understand the relationship between the HR-XRD and the variation of elastic strains, the parameters of the perpendicular mismatches (M*) were calculated for the studied structures. The perpendicular layer mismatch (as a part per million) is given by formula (1) [15,16]:

$$M^* = \left[\frac{(\omega_L - \omega_S)}{\frac{\sin(2\varnothing_S)}{2} + \tan(\theta_s)\cos^2(\varnothing_s)}\right]\left[\frac{\pi}{180}\right]10^6 \tag{1}$$

where (φ_L-φ_S) is the difference between the diffraction peaks for the two nearest layers that are considered at the numerical calculation, θ_S is the GaAs substrate Bragg diffraction angle and φ_S

is the tilt of the substrate reflection plane in the case of asymmetric reflection. M* is multiplied then by (1-ν)/(1+ν) to give the fully strained mismatch (where ν is the Poisson ratio for the layer). Results of those calculations for different layers are shown in figure 3.

Table 1. The thickness of layers chosen at the fitting

Sample	Layer	Composition	Thickness (nm)
#1	Buffer layer	$In_{0.15}Ga_{0.85}As$	1.200
	Wetting-QD layer	$In_{0.73}Ga_{0.27}As$	0.740
	Capping layer	$In_{0.098}Ga_{0.902}As$	7.500
#2	Buffer layer	$In_{0.17}Ga_{0.83}As$	0.950
	Wetting-QD layer	$In_{0.84}Ga_{0.16}As$	0.690
	Capping layer	$In_{0.13}Ga_{0.87}As$	7.200
#3	Buffer layer	$In_{0.165}Ga_{0.835}As$	1.300
	Wetting-QD layer	$In_{0.8}Ga_{0.2}As$	0.760
	Capping layer	$In_{0.228}Ga_{0.772}As$	8.100
#4	Buffer layer	$In_{0.15}Ga_{0.85}As$	1.000
	Wetting-QD layer	$In_{0.77}Ga_{0.23}As$	0.740
	Capping layer	$In_{0.245}Ga_{0.755}As$	7.300

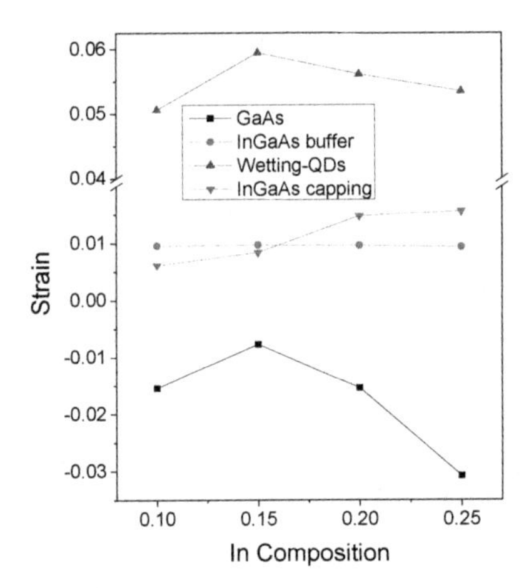

Figure 3. Calculated compression (negative) and stretch (positive) strains for the GaAs (1) and InGaAs (2) buffers, InAs wetting (3) and InGaAs capping (4) layers for the samples: #1, #2, #3 and #4

Thus the simulation of HR-XRD results (Fig.2) and the elastic strain calculation (Fig.3) permit to show that the process of Ga/In inter-diffusion is non-monotonous and strain stimulated. The higher level of compression strains in GaAs buffer layers enhances the higher Ga/In intermixing between QDs and QW layers that influents on the PL intensity and PL peak positions in the studied QD structures (Fig.1) as well.

CONCLUSIONS

The HR-XRD has been studied in strained GaAs /$In_xGa_{1-x}As/In_{0.15}Ga_{0.85}As$/GaAs QW structures with embedded InAs QDs. The parameter x in capping InxGa1-xAs layers varied from the range 0.10-0.25. Numerical simulation of HR-XRD results has shown that the level of elastic strains and the composition of quantum layers vary none monotonously in studied QD structures. Simultaneously it was revealed that the process of Ga/In inter diffusion at the $In_xGa_{1-x}As$/InAs QD interface are characterized by the dependence non monotonous versus parameter x in capping $In_xGa_{1-x}As$ layers. Thus it is shown that the different level of Ga/In inter-diffusion between the QDs and QW layers is owing to the difference in the level of compression strains in capping/buffer layers.

ACKNOWLEDGMENTS

The work was supported by CONACYT Mexico (project 130387) and by SIP-IPN, Mexico. The authors would like to thank Dr. A. Stintz from Center of High Technology Materials at University of New Mexico, Albuquerque, USA for growing the studied QD structures and Dr. Jose Alberto Andraca Adame from the Center of Nanoscience Micro and Nanotechnologies at National Polytechnic Institute of Mexico for the HR-XRD measurements.

REFERENCES

1. M. Takahasi, T.Kaizu, Journal of Crystal Growth **311**, 1761–1763 (2009)
2. D. Bimberg, M. Grundman, N. N. Ledentsov, Quantum Dot Heterostructures, Ed. Wiley & Sons (2001) 328.
3. Amtout, S. Raghavan, P. Rotella, G. von Winckel, A. Stinz and S. Krishna, J. Appl. Phys. **96**, 3782 (2004)
4. T. V. Torchinskaya, Opto-electronics Review, **6 (2)**, 121-130 (1998).
5. D. Haft, R.J. Warburton, K. Karrai, S. Huant, G. Medeiros-Ribeiro, J. Garsia, W. Schoenfeld and P.M. Petroff, Appl. Phys. Lett. **78**, 2946 (2001)
6. Stintz, G.T. Liu, L. Gray, R. Spillers, S.M. Delgado and K.J. Malloy, J. Vac. Sci. Technol. **B 18**(3), 1496 (2000)
7. T.V. Torchynska, J. Appl. Phys. **104** (7), 074315 (2008)
8. T.V. Torchynska, J.L. Casas Espinola, L. Borkovska, S. Ostapenko, M. Dybiec, O. Polupan, N. Korsunska, A. Stintz, P.G. Eliseev and K.J. Malloy, J. Appl. Phys. **101** (2), art.no. 024323 (2007)
9. H. BenNaceur , I.Moussa, O.Tottereau, A.Rebey, B.El Jani, Physica **E 41**, 1779–1783 (2009)
10. T.V. Torchynska, S. Ostapenko, M. Dybiec, Phys. Rev. **B 72,** 195341 (2005)

11. M. Dybiec, S. Ostapenko, T. V. Torchynska and E. Velasquez Losada, Appl. Phys. Lett. **84** (25), 5165-5167 (2004)
12. G.W. Shu, J.S. Wang, J.L. Shen, R.S. Hsiao, J.F. Chen, T.Y. Lin, C.H. Wu, Y.H. Huang, T.N. Yang, Mater. Scien. and Engin. B, **166**, 46–49 (2010)
13. M. Dybiec, L. Borkovska, S. Ostapenko, T.V. Torchynska, J.L. Casas Espinola, A. Stinz, K.J. Malloy, Appl. Surf. Scien. **252** (15) 5542-5545 (2006).
14. T.V. Torchynska, A. Diaz Cano, M. Dybic, S. Ostapenko, M. Mynbaeva, Physica B, Condensed Matter, **376-377** (1), 367-369 (2006).
15. H. Li, T. Mei, W. D.H. Zhang, S.F. Yoon and H.Yuan. J. Appl. Phys. **98**, 054905 (2005).
16. P. Mukhopadhyay, P Das, S Pathak, E. Y. Chang, D. Biswas. CS MANTECH Conference, Palm Springs, California, USA (2011)
17. G. Polupan , L.G. Vega-Macotela, F. Sanchez Silva, J. Luminescence, **132**, 1270–1273 (2012).
18. T.V.Torchynska, J.Palacios Gomez, G.P.Polupan, F.G.Becerril Espinoza, A.Garcia Borquez, N.E.Korsunskaya, L.Yu. Khomenkova, Appl. Surf. Science , v.**167**, 197-204 (2000).
19. T.V. Torchynska, A. Stintz, J. Appl. Phys. **108**, 2, 024316 (2010).

Mater. Res. Soc. Symp. Proc. Vol. 1617 © 2013 Materials Research Society
DOI: 10.1557/opl.2013.1158

Microwave Reflection From The Microwave Photo-Excited High Mobility GaAs/AlGaAs Two-Dimensional Electron System

Tianyu Ye[1], R. G. Mani[1] and W. Wegscheider[2]

[1]Department of Physics and Astronomy, Georgia State University, Atlanta, Georgia 30303, USA.

[2]Laboratorium fur Festkorperphysik, ETH-Zurich, 8093 Zurich, Switzerland.

ABSTRACT

We examine the microwave reflection from the high mobility GaAs/AlGaAs two-dimensional electron system (2DES). Strong correlations have been observed between the microwave induced magnetoresistance oscillations and the microwave reflection oscillations in a concurrent measurement of the microwave illuminated magnetoresistance and the microwave reflection from the 2DES. The correlations were followed as a function of the microwave frequency and the microwave power dependent. Different existing theories are considered to explain the results.

INTRODUCTION

Zero-resistance states have been a topic of interest in the two dimensional electron system (2DES) for the past several decades. It is well known that the 2DES exhibits quantum Hall effect when it is subjected to low temperature and strong perpendicular magnetic fields. Studies of QHE show that a 2DES with well-separated Landau levels periodically exhibits zero resistance states. More recently, a special type of zero resistance state – the microwave induced zero resistance state (MIZRS) - was discovered [1]. Here, a high mobility 2DES subjected to low temperature, a perpendicular magnetic field, and microwave illumination shows a magnetoresistance that periodically exhibit zero resistance states in the vicinity of $B^{-1} = [4/(4j+1)]^{-1}(2\pi f m^*/e)^{-1}$, where B is the magnetic field, f is the microwave frequency, e is the electron charge, and m^* is the electron effective mass. In a lower mobility sample and with smaller microwave power, MIZRS turn into the minima of microwave induced magnetoresistance oscillations (MIMOs) [1, 2]. In principle, such strong microwave response could have potential application in gigahertz and terahertz electronics. Theoretically, MIZRS and MIMOs could be understood by: impurity assisted inter- and intra- Landau levels transition made by microwave illumination (displacement model) [3, 4], microwave induced oscillating electron distribution (inelastic model) [5] and microwave driven oscillating electron orbit (electron orbit model) [6].

Since their discovery, MIZRS and MIMOs have attracted a lot of research interest. Some research focus on the direct measurement of electric signal in 2DES itself while changing external parameters [7-15], such as microwave frequency, power or polarization, even samples' quality. Others focus on measuring microwave wave absorption, reflection and transmission [16-19] from 2DES and their correlation with MIMOs. Although such measurements are indirect measurement from 2DES, they could clearly reveal physical contributions of MIMOs and MIZRS, and identify the relative contributions of suggested theoretical mechanisms.

Here, we examine the microwave reflection from high mobility 2DES and correlate the microwave reflection oscillation with MIMO. Then we compare our results with different theoretical models.

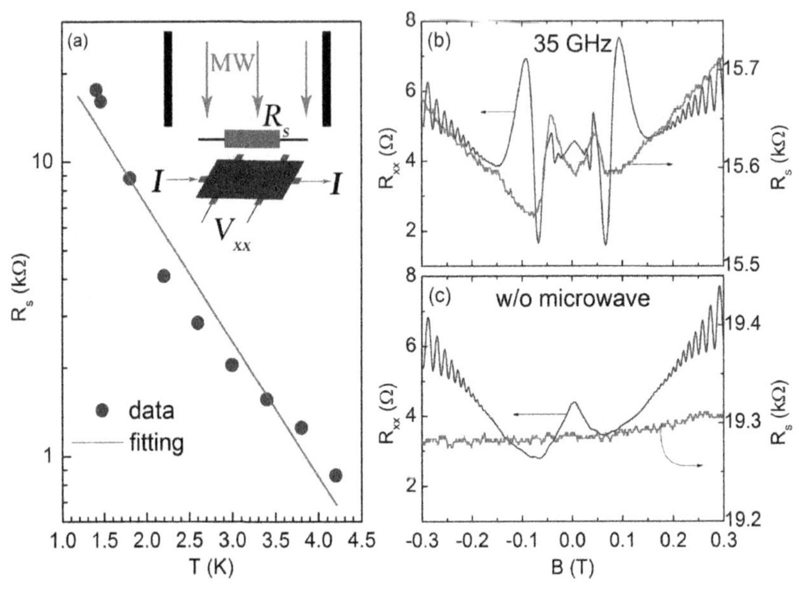

Figure 1. (a) Carbon resistance R_s versus temperature. Symbols correspond to the experimentally measured data points and the red line is a linear fit of data. Inset is a schematic diagram of sample configuration: 2DES Hall bar sample (blue) is attached at the end of microwave waveguide; carbon resistor (R_s) is above and close to the sample. (b) Under 35 GHz microwave illumination, R_{xx} (plot on the left panel) and R_s (plot on the right panel) versus magnetic field. (c) Without microwave illumination, R_{xx} and R_s versus magnetic field.

EXPERIMENT

Hall bars with gold-germanium alloyed contacts were fabricated on high mobility GaAs/AlGaAs heterojunctions by optical lithography. The specimens were mounted at the end of a long cylindrical waveguide, with the device normal oriented along the waveguide axis. A carbon resistor was placed right above but closes enough to sample (see illustration inset of Fig. 1 (a)), to remotely sense microwave reflection. The waveguide sample holder was then inserted into a variable temperature insert, inside the bore of a superconducting solenoid. A base temperature of approximately 1.5 K was realized by pumping on the liquid helium within the variable temperature insert. The specimens were briefly illuminated by a red LED light at low temperature to realize the high mobility condition. Finally, a low frequency four terminal lock-in technique was adopted to measure the signal from the specimen and the reflection detector.

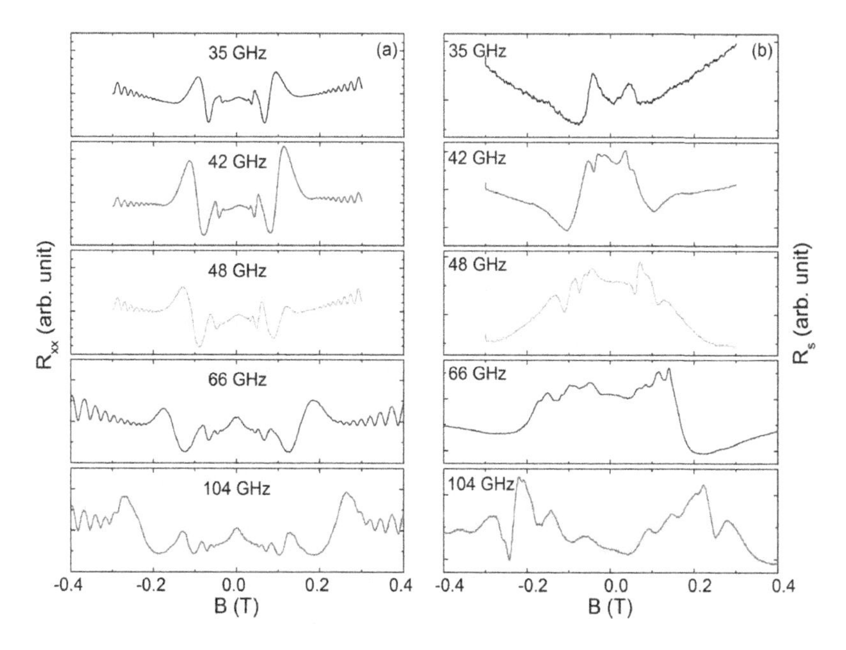

Figure 2. Microwave frequency evolution of magnetoresistance column (a) R_{xx} and column (b) R_s.

RESULTS

Figure 1 (a) exhibits the carbon resistor's negative temperature coefficient character. Fig 1(a) shows the resistance of the carbon sensor, R_s, plotted on a log scale vs. the temperature, T, in the range $1.5 < T < 4.2$ K, where our experiments were carried out. The figure shows that that R_s increases exponentially with decreasing temperature. Thus, suppose, at $T = 1.5$ K, a small change in the reflected power incrementally heats R_s. Due to the exponential dependence of R_s on T, the small change in the reflected power will lead to a relatively large measurable change in R_s, which will then become an indicator of the reflection from the 2DES.

We begin by exhibiting two sets of measurements on a high mobility specimen. Figure 1 (b) and (c) show concurrently measured longitudinal resistance R_{xx} and sensor resistance R_s as the magnetic field is swept between -0.3 T and 0.3 T. With 35 GHz microwave illumination (Figure 1 (b)) R_{xx} reveals huge MIMOs in the regime of $-0.15 < B < 0.15$ T, and R_s reveals oscillations in the same regime. Fig. 1(c) shows that when we switch-off the microwave power (Figure 1 (c)), both the MIMOs and R_s oscillations vanish. To further characterize the experimental setup, a low mobility sample was also measured, for it is well known that low mobility sample do not exhibit MIMOs. With the exact same experimental environment, neither

R_{xx} nor R_s could exhibit any oscillations in the low mobility specimen. Therefore, it is clear that the oscillations in R_s go together with MIMOs, as the R_s signal helps to remotely sense the reflection from 2DES sample. Note that at the R_s oscillations maxima, the carbon sensor reports a lower local temperature, which means that less microwave power is incident on the sensor. On the other hand, at R_s oscillations minima, local carbon sensor temperature is higher and, therefore, microwave is maximally reflected.

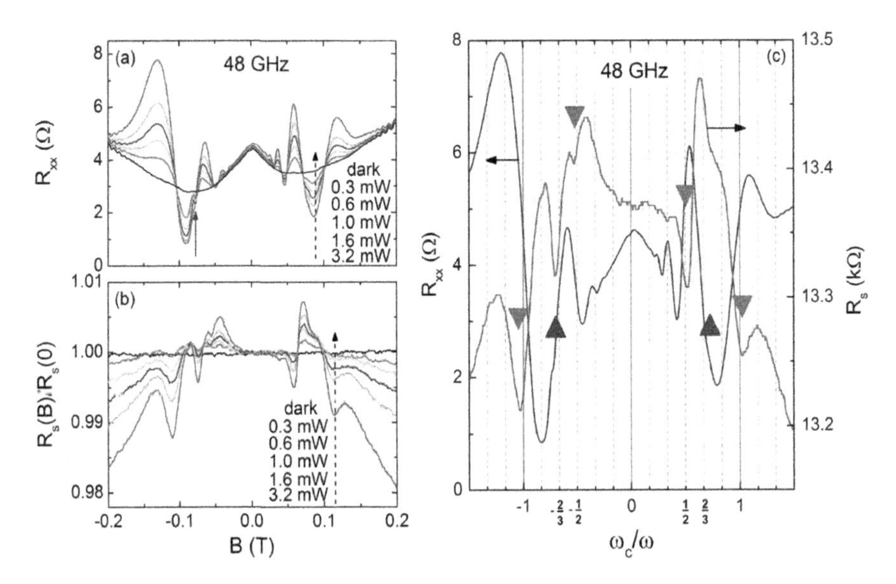

Figure 3. Microwave power dependence of (a) MIMOs and (b) R_s oscillations under 48 GHz microwave illumination. Dashed arrows in (a) and (b) indicate along the arrow direction for the microwave power change from 3.2 mW to 0 mW. The solid arrow on the left bottom of (a) indicates a two photon process. (c) R_{xx} (left panel) and R_s (right panel) plotted as functions of ω_c/ω, microwave power is 3.2 mW. Downward triangle pointers mark $\omega = j\omega_c$ and upward triangle pointers mark $2\omega = j\omega_c$.

Figure 2 shows frequency evolution of MIMOs and R_s oscillations. In the left column of Fig. 2, as microwave frequency increases from 35 GHz to 104 GHz, the MIMO active regime expands to higher magnetic field and more MIMOs are appear in R_{xx}. Similarly, in the right column, as microwave frequency increases, R_s oscillation's active regime also expand to higher magnetic field and more oscillations appear on the R_s curve. However, the B-regime of R_s oscillations always falls inside the B-regime of MIMOs for each frequency. Outside the MIMOs regime, the R_{xx} curves exhibit Shubnikov-de-Hass oscillations. However, in this range of B, R_s increase or decrease monotonically with magnetic field, without oscillations. This feature indicates that the carbon sensor is insensitive to Shubnikov-de-Hass oscillations.

Figure 3 (a) and (b) show the microwave power dependence of R_{xx} and R_s. Here, at 48 GHz, the microwave power, P, spanned $0 < P < 3.2$ mW. As the microwave power decreases, both the MIMOs in R_{xx}, and the R_s oscillations show reduced amplitudes. However, the location of the extrema on the abscissa remains unchanged. In figure 3 (a), an upward solid arrow points out a small bump on R_s oscillations, which corresponds to a two photon process.

DISCUSSION

Ref. [16], which adopts a displacement model to calculate microwave absorbed by electrons, indicates maximum absorptions at cyclotron resonance and its higher harmonics. Besides that, it also finds two photon assisted absorption peaks. Therefore, we plot, in figure 3 (c) our magnetoresistance of R_{xx} and R_s versus ω_c/ω instead of magnetic field. In this way we could clearly see microwave reflection at cyclotron resonance and its higher harmonics. Here ω_c is electron cyclotron resonance frequency. For simplicity, we adopt $m^*=0.067m_e$ as the electron effective mass. Here, ω is microwave angular frequency. In the figure, all R_s minima appear within $-1\leq\omega_c/\omega\leq1$ at around $\omega=\pm j\omega_c$, $j=1,2$ and $2\omega=j\omega_c$, $j=3$. Thus, it is clear from the figure, that microwave reflection attains its maximum value at cyclotron resonance and its second order. In the displacement model, $\omega=\pm j\omega_c$ describes a microwave excited election transition between integer times of Landau levels and $2\omega=j\omega_c$ describes two-photon-excited electrons between integer times of Landau levels. In figure 3 (a), the MIMO's do reveal a two photon process peak, which is marked by the solid arrow. This is similar to the result in ref. [16]. Figure 3 (b) also shows that as the microwave power increases, the minimum becomes deeper and microwave reflection get stronger.

Ref. [17] employed the inelastic model to calculate microwave absorption and reflection from the high mobility 2DES. The calculation showed maximum absorption at around cyclotron resonance and its higher harmonics. However, so far as the microwave reflection oscillations are concerned, this theory predicted that they are not experimentally observable due to the negligible reciprocal momentum relaxation time τ_{tr}^{-1}.

The electron orbit model also examined the microwave absorption in 2DES. The results [18] show maximum microwave absorption at only around cyclotron resonance. This theory also found that the microwave absorption is a temperature dependent quantity. Its amplitude changes as a parabolic function of temperature.

CONCLUSIONS

In this work, we employed a carbon resister to simultaneously examine the microwave reflection property of 2DES and compare the reflection with MIMOs in the high mobility 2DES. Our results confirm the existence of a detectable microwave reflection, and it strong correlation with MIMOs, through microwave frequency and power dependent measurements. Maxima in microwave reflection were found to occur at around cyclotron resonance and its higher harmonics. As well, two photon assisted inter Landau level transitions was found both on MIMOs and reflection oscillations. These results match the simulation results of ref. [16] made using the displacement model.

ACKNOWLEDGMENTS

Basic research at Georgia State University (GSU) is supported by the DOE-BES, MSE Division under DE-SC0001762. Additional support for microwave work is provided by the ARO under W911NF-07-01-0158.

REFERENCES

1. R. G. Mani, J. H. Smet, K. von Klitzing, V. Narayanamurti, W. B. Johnson, and V. Umansky, Nature **420**, 646 (2002).
2. M. A. Zudov, R. R. Du, L. N. Pfeiffer, and K. W. West, Phys. Rev. Lett. **90**, 046807 (2003).
3. A. C. Durst, S. Sachdev, N. Read, and S. M. Girvin, Phys. Rev. Lett. **91**, 086803 (2003).
4. X. L. Lei and S. Y. Liu, Phys. Rev. Lett. **91**, 226805 (2003).
5. I. A. Dmitriev, M. G. Vavilov, I. L. Aleiner, A. D. Mirlin, and D. G. Polyakov, Phys. Rev. B **71**, 115316 (2005).
6. J. Γnarrea and G. Platero, Phys. Rev. Lett. **94**, 016806 (2005); Phys. Rev. B **76**, 073311 (2007).
7. R. G. Mani, V. Narayanamurti, K. von Klitzing, J. H. Smet, W. B. Johnson, and V. Umansky, Phys. Rev. B **70**, 155310 (2004); Phys. Rev. B **69**, 161306 (2004).
8. R. G. Mani, J. H. Smet, K. von Klitzing, V. Narayanamurti, W. B. Johnson, and V. Umansky, Phys. Rev. Lett. **92**, 146801 (2004); Phys. Rev. B **69**, 193304 (2004).
9. R. G. Mani, Physica E **22**, 1 (2004); Physica E. **25**, 189 (2004).
10. R. G. Mani, Phys. Rev. B **72**, 075327 (2005); Appl. Phys. Lett. **91**, 132103 (2007); Appl. Phys. Lett. **92**, 102107 (2008); Physica E **40**, 1178 (2008)
11. R. G. Mani, W. B. Johnson, V. Umansky, V. Narayanamurti, and K. Ploog, Phys. Rev. B **79**, 205320 (2009).
12. R. G. Mani, C. Gerl, S. Schmult, W. Wegscheider, and V. Umansky, Phys. Rev. B **81**, 125320 (2010).
13. A. N. Ramanayaka, R. G. Mani, and W. Wegscheider, Phys. Rev. B **83**, 165303 (2011).
14. R. G. Mani, A. N. Ramanayaka, W. Wegscheider, Phys. Rev. B **84**, 085308 (2011); A. N. Ramanayaka, R. G. Mani, J. Γnarrea, and W. Wegscheider, Phys. Rev. B **85**, 205315 (2012).
15. R. G. Mani, J. Hankinson, C. Berger, and W. A. de Heer, Nat. Commun. **3**, 996 (2012).
16. X. L. Lei and S. Y. Liu, Phys. Rev. B **72**, 075345 (2005).
17. O. M. Fedorych, M. Potemski, S. A. Studenikin, J. A. Gupta, Z. R. Wasilewski, and I. A. Dmitriev, Phys. Rev. B **81**, 201302 (2010).
18. Jesus Inarrea and Gloria Platero, Nanotechnology **21**, 315401 (2010)
19. T. Ye, R.G Mani and W. Wegscheider, Appl. Phys. Lett. **102**, 242113 (2013)

Mater. Res. Soc. Symp. Proc. Vol. 1617 © 2013 Materials Research Society
DOI: 10.1557/opl.2013.1159

Linear Polarization Rotation Study of the Microwave-Induced Magnetoresistance Oscillations in the GaAs/AlGaAs System

A. N. Ramanayaka[1], Tianyu Ye[1], H-C. Liu[1], R. G. Mani[1], W. Wegscheider[2]
[1]Department of Physics and Astronomy, Georgia State University, Atlanta, GA 30303 U.S.A.
[2]Laboratorium für Festkörperphysik, ETH Zürich, 8093 Zürich, Switzerland.

ABSTRACT

Microwave-induced zero-resistance states appear when the associated B^{-1}-periodic magnetoresistance oscillations grow in amplitude and become comparable to the dark resistance of the two-dimensional electron system (2DES). Existing theories have made differing predictions regarding the influence of the microwave polarization in this phenomenon. We have investigated the effect of rotating, in-situ, the polarization of linearly polarized microwaves relative to long-axis of Hall bars. The results indicate that the amplitude of the magnetoresistance oscillations is remarkably responsive to the relative orientation between the linearly polarized microwave electric field and the current-axis in the specimen. At low microwave power, P, experiments indicate a strong sinusoidal variation in the diagonal resistance R_{xx} vs. θ at the oscillatory extrema of the microwave-induced magnetoresistance oscillations. Interestingly, the phase shift θ_0 for maximal oscillatory R_{xx} response under photoexcitation is a strong function of the magnetic field, the extremum in question, and the magnetic field orientation.

INTRODUCTION

In the recent past, low-B transport studies under microwave irradiation in the 2DES uncovered the possibility of eliminating backscattering by photo-excitation, without concurrent Hall quantization.[1, 2] The experimental realization of such radiation-induced zero-resistance states and associated B−1-periodic radiation-induced magneto-resistance oscillations expanded the experimental [1-14] and theoretical [15-28] investigations of light-matter coupling in low-dimensional electronic systems. Microwave-induced zero-resistance states appear when the associated B^{-1}-periodic magnetoresistance oscillations grow in amplitude and become comparable to the dark resistance of the 2DES. Such oscillations are now understood via the displacement model [15, 16, 18, 25] the non-parabolicity model [17] the inelastic model [19] and the radiation driven electron orbit model [20, 22].

A distinguishing feature between these theories is the role of the microwave-polarization. Here, the displacement model indicates that the oscillation amplitude is influenced by whether the microwave electric field, E_ω where $\omega=2\pi f$, is parallel or perpendicular to the dc-electric field, E_{dc} [16]. In contrast, the inelastic model suggests polarization insensitivity of the radiation-induced magneto-resistance oscillations [19]. The radiation-driven electron orbit model indicates a polarization immunity that depends upon the damping factor, γ, exceeding the microwave frequency, f [21, 22]. Finally, the non-parabolicity model suggests distinct polarization sensitivity for linearly polarized microwaves [17] and the absence of microwave-induced magnetoresistance oscillations for circularly polarized radiation. The polarization aspect of the microwave-induced magnetoresistance oscillations has been explored by experiment in ref.[3] and [8].

Here, we examine the effect of rotating the polarization of linearly polarized microwaves on the microwave-induced magnetoresistance oscillations in the GaAs/AlGaAs 2D electron system. Surprisingly, at low microwave power, P, experiments indicate a strong sinusoidal response as $R_{xx}(\theta) = A \pm C\cos^2(\theta - \theta_0)$ vs. the polarization rotation angle, θ, with the '+' and '−' cases describing the maxima and minima, respectively. At higher P, the principal resistance minimum exhibits additional extrema vs. θ. Notably, the phase shift θ_0 varies with *f*, B, and sgn(B).

EXPERIMENT AND RESULTS

Polarization-dependence studies utilized the novel setup illustrated in Fig. 1(a). Here, a rotatable MW-antenna introduces microwaves into a circular waveguide. The circular symmetry then allows the rotation of the polarization with respect to the stationary sample. The samples (A and B) consisted of 400 μm-wide Hall bars, see Fig. 1(b), which were characterized by n (4.2 K) = 2.2×10^{11} cm^{-2} and $\mu \approx 8\times10^{6}$ cm^2/V s. The length (L) to width (W) ratio for the R_{xx} measurements was L/W = 1. At the angle θ = 0°, the long axis of the Hall bars is parallel to the polarization axis of the MW-antenna. Thus, θ, see Fig. 1(b), represents the polarization rotation angle.

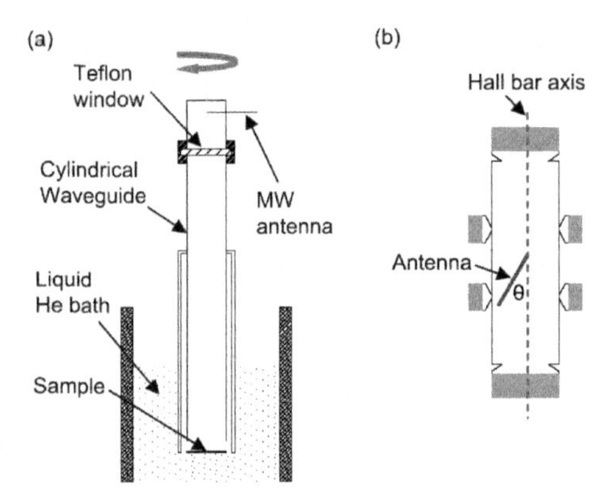

Figure 1. (Color online) (a) Schematic of the experimental setup. MW antenna is designed to rotate independently about the axis of a cylindrical waveguide. (b) A Hall bar specimen, labeled sample in (a), is oriented so that the Hall bar long axis is parallel to the MW antenna for θ = 0°.

Figure 2 exhibits the R_{xx} vs. B at *f* = 35.5 GHz, with the Hall bar device (sample A) in place at the bottom of the waveguide sample holder. Figures 2(a) and 2(b) show the results obtained at a source power P = 1mW, while Figs. 2(c) and 2(d) show the same obtained at P = 0.5 mW. Here, R_{xx}^{L} and R_{xx}^{R} represent the measurement on the left (L) and right (R) sides of the device [Fig. 1(b)]. Each panel of Fig. 2 includes three traces: a dark trace (in black) obtained in the absence of microwave photoexcitation. A $\theta = 0°$ trace in red ($\theta = 90°$ trace in green), where the MW-antenna is parallel (perpendicular) to the long axis of the Hall bar. A comparison of Figs. 2(a) and 2(c) [or Figs. 2(b) and 2(d)] shows that the 0° (red) or 90° (green) traces exhibit larger amplitude, microwave-induced magnetoresistance oscillations at P = 1 mW than at P = 0.5 mW. This feature corresponds to the usual observation that the oscillation-amplitude increases with P at modest photoexcitation.[12,13] The remarkable feature is observed when one compares the 0° (red) and 90° (green) traces within any single panel of Fig. 2. Such a comparison indicates that the amplitude of the radiation-induced magnetoresistance oscillations is reduced at the $\theta =$ 90° MW-antenna orientation. Although the magnetoresistance oscillations are reduced in amplitude, typically, they are not completely extinguished at $\theta = 90°$. Finally, the period and the phase of the radiation-induced magnetoresistance oscillations remain unchanged by MW-antenna rotation.

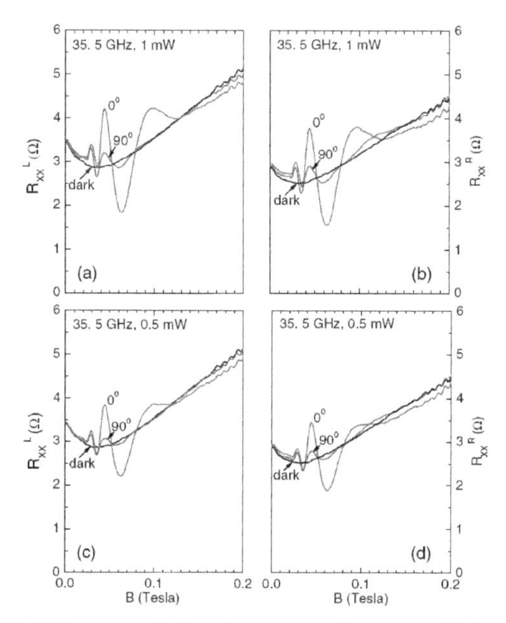

Figure 2. (Color online) Microwave-induced magneto resistance oscillations in R_{xx} at 1.5 K are shown at $f = 35.5$ GHz for P = 1 mW in panels (a) and (b), and for P = 0.5 mW in panels (c) and (d). The R_{xx} measured on the left (right) side of the Hall bar is shown as R_{xx}^{L} (R_{xx}^{R}). Each panel shows a dark trace (black), a trace (red) obtained at $\theta = 0°$, and a trace (green) obtained at $\theta = 90°$.

Figure 3(a) shows the dark and photoexcited diagonal resistance R_{xx} vs. B for a second device with similar characteristics (sample B). Here, the photoexcited measurement was carried out with microwave frequency f = 40 GHz and microwave power P = 0.32 mW and $\theta = 20°$. In Fig. 3(a), the labels $P^{-}1$, $V^{-}1$, and $P^{-}2$ identify the oscillatory extrema on the negative side of the magnetic field axis and they are examined in Figs. 3(b), 3(d), and 3(f), respectively. Similarly the labels $P^{+}1$, $V^{+}1$, and $P^{+}2$ identify the oscillatory extrema on the positive side of the magnetic field and these are examined in Figs. 3(c), 3(e), and 3(g), respectively. Figures 3(b), 3(f), and 3(c), 3(g) show that the photoexcited R_{xx} (i.e., w/MW) traces lie above the dark (i.e., w/o MW) R_{xx} traces at the resistance maxima for all θ on either side of the magnetic field. Furthermore, the photoexcited R_{xx} at oscillations maxima $P^{-}1$, $P^{+}1$, $P^{-}2$, and $P^{+}2$ fits the function $R_{xx}(\theta) = A + C \cos^2(\theta - \theta_0)$, showing a sinusoidal dependence with polarization angle θ. Figures 3(d) and 3(e) show that, at the resistance minima $V^{-}1$ and $V^{+}1$, the w/MW R_{xx} trace lies below the dark R_{xx} for all θ as the resistance minima follow $R_{xx}(\theta) = A - C \cos^2(\theta - \theta_0)$.

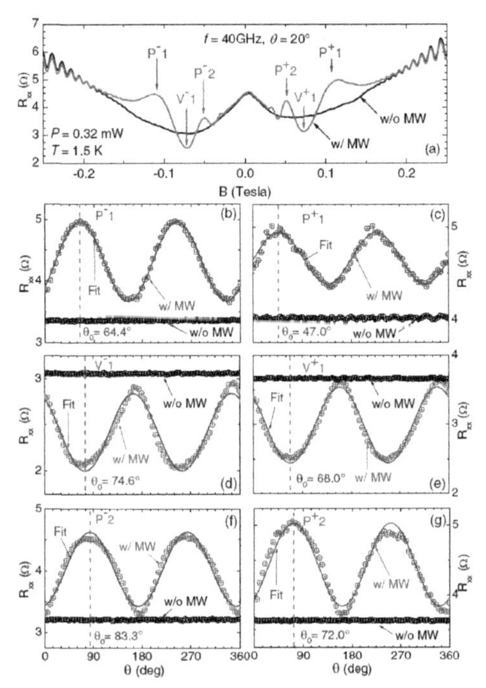

Figure 3. (Color online) This figure compares the angular response for positive and negative magnetic fields. (a) Photo-excited R_{xx} is shown at f = 40 GHz with $\theta = 0°$ over the B-range $-0.25 \leq B \leq 0.25$T. (b), (d), and (f) show the θ dependence of R_{xx} of the principal maxima $P^{-}1$, $P^{-}2$, and the minimum $V^{-}1$, respectively. (c), (e), and (g) show the θ dependence of R_{xx} of the principal maxima $P^{+}1$, $P^{+}2$, and the minimum $V^{+}1$, respectively. In these figures, the w/o MW traces indicate the sample response in the dark, while w/ MW traces indicate the response under photo-excitation. The phase shift, θ_0, is indicated by a vertical dashed line in (b) - (g).

However, the fit-extracted θ_0 differ substantially from zero and well beyond experimental uncertainty [14]. Indeed, a close inspection suggests that θ_0 depends upon the magnetic field B and its orientation sgn(B). For example, we find that $\theta_0 = 64.4°$ for P^-1 and $\theta_0 = 47°$ for P^+1. Such a large difference in θ_0 due to magnetic field reversal is unexpected. Here, note that, since the MW-antenna is far from the magnet and well isolated from the magnetic field, the magnetic field is not expected to influence the polarization of the microwaves at launch. Furthermore, the stainless-steel microwave waveguide is not known to provide a microwave frequency, magnetic field, and magnetic-field-orientation-dependent rotation to the microwave polarization. Thus, the θ_0 shift depending on B and sgn(B) looks to be a sample effect.

DISCUSSION

The main features in the exhibited results are therefore: (i) At low P, $R_{xx}(\theta) = A \pm C\cos^2(\theta - \theta_0)$ vs. the linear polarization rotation angle, θ, with the '+' and '−' cases describing the maxima and minima, respectively, see Fig. 3. (ii) The phase shift in the $R_{xx}(\theta)$ response, i.e., θ_0, varies with B, and the sign of B, see Fig. 3. Point (i) demonstrates a strong sensitivity in the radiation-induced magnetoresistance oscillations to the sense of linear microwave polarization, in qualitative agreement with the radiation driven electron orbit model when $\gamma < \omega = 2\pi f$ [21, 22]. Such results are also consistent with the displacement [15, 16] and the non-parabolicity models [17]. Point (ii) could be understood, for example, in the displacement model. Here, polarization sensitivity [16] is due to the inter-Landau level contribution to the photo-current. In these experiments, the orientation of E_ω is set by the antenna within the experimental uncertainty. The orientation of E_{DC} is variable and set by the B-dependent Hall angle, $\theta_H = \tan^{-1}(\sigma_{xy}/\sigma_{xx})$, with respect to the Hall bar long-axis. If a particular orientation between E_ω and E_{DC} is preferred, say, e.g. $E_\omega \perp E_{DC}$ or $E_\omega // E_{DC}$, for realizing large microwave induced magnetoresistance oscillations, and the Hall angle changes with B, then a non-zero θ_0 and a variation in θ_0 with B might be expected. However, the observed variations in θ_0 seem much greater than expectations since $\theta_H \approx 90°$ in this regime. The change in θ_0 upon B-reversal is also unexpected, and this feature identifies a reason for the asymmetry in the amplitude of R_{xx} under B-reversal often observed in such experiments: the typical R_{xx} vs. B measurement sweep occurs at a fixed θ. If peak response occurs at different θ_0 for the two field directions, then the oscillatory R_{xx} amplitudes should differ for positive and negative B. The observed θ_0 – variations seem to suggest an effective microwave polarization rotation in the self-response of the photoexcited Hall bar electron device. Since $\theta_0 \approx \pi/4$, see Fig. 3, $B \approx 0.1T$, and the thickness of the 2DES lies in the range of tens of nanometers, such a scenario would suggest giant effective polarization rotation in this high mobility 2DES.

CONCLUSIONS

In conclusion, experiments indicate a strong sinusoidal variation in the diagonal resistance R_{xx} vs. θ, the polarization rotation angle, at the oscillatory extrema of the microwave radiation-induced magneto-resistance oscillations. Surprisingly, the phase shift θ_0 for maximal oscillatory R_{xx} response under microwave excitation is a strong function of the microwave frequency, *f*, the extremum in question, and the magnetic field orientation.

ACKNOWLEDGMENTS

Work at Georgia State University is supported by the US Department of Energy, Office of Basic Energy Sciences, Material Sciences and Engineering Division, under Grant No. DE-SC0001762 and by D. Woolard and the ARO under Grant No. W911NF-07-01-015.

REFERENCES

1. R. G. Mani, J. H. Smet, K. von Klitzing, V. Narayanamurti, W. B. Johnson, and V. Umansky, Nature (London) 420, 646 (2002); Phys. Rev. Lett. 92, 146801 (2004); Phys. Rev. B 69, 193304 (2004); ibid. 69, 161306 (2004); ibid. 70, 155310 (2004).
2. M. A. Zudov, R. R. Du, L. N. Pfeiffer, and K. W. West, Phys. Rev. Lett. 90, 046807 (2003).
3. R. G. Mani, Physica E (Amsterdam) 22, 1 (2004); ibid. 25, 189 (2004); Appl. Phys. Lett. 85, 4962 (2004).
4. S. A. Studenikin, M. Potemski, P. T. Coleridge, A. S. Sachrajda, and Z. R. Wasilewski, Sol. St. Comm. 129, 341 (2004); S. Studenikin et al., Phys. Rev. B 71, 245313 (2005).
5. A. E. Kovalev, S. A. Zvyagin, C. R. Bowers, J. L. Reno, J. A. Simmons, Sol. St. Comm. 130, 379 (2004).
6. R. L. Willett, L. N. Pfeiffer and K. W. West, Phys. Rev. Lett. 93, 026804 (2004).
7. B. Simovic, C. Ellenberger, K. Ensslin, and W. Wegscheider, Phys. Rev. B 71, 233303 (2005).
8. J. H. Smet, B. Gorshunov, C. Jiang, L. Pfeiffer, K. West, V. Umansky, M. Dressel, R. Meisels, F. Kuchar, and K. von Klitzing, Phys. Rev. Lett. 95, 116804 (2005).
9. K. Stone, C. L. Yang, Z. Q. Yuan, R. R. Du, L. N. Pfeiffer and K. W. West, Phys. Rev. B 76, 153306 (2007).
10. D. Konstantinov and K. Kono, Phys. Rev. Lett 103, 266808 (2009).
11. Y. H. Dai, R. R. Du, L. N. Pfeiffer, and K. W. West, Phys. Rev. Lett. 105, 246802 (2010).
12. R. G. Mani, C. Gerl, S. Schmult, W. Wegscheider, and V. Umansky, Phys. Rev. B. 81, 125320 (2010); R. G. Mani, W. B. Johnson, V. Umansky, V. Narayanamurti, and K. Ploog, Phys. Rev. B 79, 205320 (2009).
13. A. N. Ramanayaka, R. G. Mani, and W. Wegscheider, Phys. Rev. B 83, 165303 (2011).
14. R. G. Mani, A. N. Ramanayaka, and W. Wegscheider, Phys. Rev. B 84, 085308 (2011); A. N. Ramanayaka, R. G. Mani, J. Inarrea, and W. Wegscheider, Phys. Rev. B. 85, 205315 (2012).
15. A. C. Durst, S. Sachdev, N. Read, and S. M. Girvin, Phys. Rev. Lett. 91, 086803 (2003).
16. V. Ryzhii and R. Suris, J. Phys.: Cond. Matt. 15, 6855 (2003).
17. A. A. Koulakov and M. E. Raikh, Phys. Rev. B 68, 115324 (2003).
18. X. L. Lei and S. Y. Liu, Phys. Rev. Lett. 91, 226805 (2003).
19. I. A. Dmitriev, M. G. Vavilov, I. L. Aleiner, A. D. Mirlin, and D. G. Polyakov, Phys. Rev. B 71, 115316 (2005).
20. J. Inarrea and G. Platero, Phys. Rev. Lett. 94, 016806 (2005).
21. J. Inarrea and G. Platero, Phys. Rev. B 76, 073311 (2007).
22. J. Inarrea and G. Platero, J. Phys. Conf. Ser. 210, 012042 (2010).
23. A. D. Chepelianskii, A. S. Pikovsky, and D. L. Shepelyansky, Eur. Phys. J. B 60, 225 (2007).
24. I. G. Finkler and B. I. Halperin, Phys. Rev. B 79, 085315 (2009).
25. I. A. Dmitriev, M. Khodas, A. D. Mirlin, D. G. Polyakov, and M. G. Vavilov, Phys. Rev. B 80, 165327 (2009).
26. D. Hagenmuller, S. de Librato, and C. Ciuti, Phys. Rev. B 81, 235303 (2010).
27. N. H. Lindner, G. Refael, and V. Galitski, Nat. Phys. 7, 490 (2011).
28. Z. Gu, H. A. Fertig, D. P. Arovas, and A. Auerbach, arXiv:1106.0302v1.

Mater. Res. Soc. Symp. Proc. Vol. 1617 © 2013 Materials Research Society
DOI: 10.1557/opl.2013.1160

Optical and electrical study of cap layer effect in QHE devices with double-2DEG

L. Zamora-Peredo[1*], I. Cortes-Mestizo[1], L. García-Gonzáez[1], J. Hernández-Torres[1], T. Hernandez-Quiroz[1], M. Peres-Caro[2], M. Ramirez-López[2], I. Martinez-Veliz[2], Y. L. Casallas-Moreno[2], S. Gallardo-Hernández[2], A. Conde-Gallardo[2], M. López-López[2].

[1] Centro de Investigación en Micro y Nanotecnología, Universidad Veracruzana, Calzada Adolfo Ruiz Cortines # 455, Fracc. Costa Verde, C.P. 94292, Boca del Río, Veracruz, México.

[2]Departamento de Física, Centro de Investigaciones y de Estudios Avanzados - IPN, México D. F., México

ABSTRACT

In this work we report on the characteristics of GaAs/AlGaAs heterostructures with a symmetric double two-dimensional electron gas (D-2DEG). Optical characterization was made by room temperature photoreflectance (PR) spectroscopy as well as electrical properties were determinated using the quantum Hall effect measurements at 2K. In order to study the surface effects on the conduction band profile, three samples with different GaAs cap layer thickness (25, 60 and 80 nm) were grown by the molecular beam epitaxy. Photoreflectance spectra at room temperature show the wide-period Franz-Keldysh oscillations between 1.42 and 1.70 eV originated by the surface electric field. The analysis of these oscillations shows that the surface electric field varies from 503 to 120 kV/cm whereas the thickness of the cap layer increases that was produced by the reduction of the depletion zone near the surface. Using QHE measurements we found that electron density increases if the surface electric field decreases.

INTRODUCTION

In the field of the electrical quantum metrology, the quantum Hall effect (QHE) in most devices is based on a two-dimensional electron gas (2DEG) realized in GaAs/AlGaAs heterostructures. The metrological calibrations are carried out by filling with factor i=2, thus providing a quantized resistance value of $R_K/2=h/2e^2 \sim 12.9$ kΩ, where R_K is the von Klitzing constant, e is the electron's charge, and h is the Planck constant [1, 2]. For practical applications, however, it is desired to have a set of resistance values besides $R_K/2$. QHE circuits consisting of devices connected in parallel or in series have been proposed for tailoring of resistance to different values [3, 4]. For parallel QHE circuits, double 2DEGs structures with two parallel transport channels stacked in the as-grown heterostructure have substantial advantages. The number of Hall bars is halved and it is possible to use the higher values of the working current. Different type heterostructures have been proposed in order to design a double-2DEG system. Some symmetric heterostructures [5] with the electron density of 5.4×10^{-11} cm^{-2} was created and others asymmetric structures [6, 7] were created with the electron density varying between 4 and 9×10^{-11} cm^{-2}.

In this work, we studied a set of heterostructures by PR and QHE measurements in order to determinate the effects of the surface electric field over the conduction band behavior.

EXPERIMENT

Three heterostructures were grown on semi-insulating GaAs (100) substrate by molecular beam epitaxy. The samples have a 1 μm-thick GaAs buffer layer (BL), a first spacer layer of 7 nm un doped $Al_xGa_{1-x}As$, followed by a 80-nm-thick Si-doped AlGaAs barrier, next a second spacer layer of 7 nm un doped $Al_xGa_{1-x}As$, finally the structure was capped with S= 25, 60 and 80 nm of undoped GaAs for the samples M379, M380 and M381, respectively. See figure 1. Si and Al concentrations were $1.4x10^{18}$ atoms/cm^3 and 32%, respectively.

Mobility and electron concentration were determined by quantum Hall effect measurements at 2K using a quantum design PPMS dynacool system. The room temperature photoreflectance (PR) measurements were carried out by employing the experimental setup similar to those that is described elsewhere [8]. A sciencetech monochromator of 0.5 m focal distance was used, with a 543 nm line of solid-state laser as the modulation source and with a maxima output power of 80 mW, chopped at a frequency of 200 Hz.

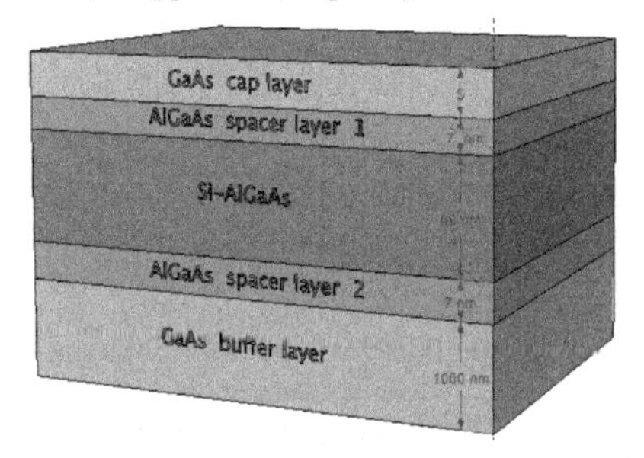

Figure 1. Sketch of AlGaAs/GaAs heterostructures studied in this work.

RESULTS AND DISCUSSION

Figure 2 shows PR spectra at room temperature. Is possible to see a wide-period oscillation between 1.37 and 1.55 eV for all samples, which is originated by surface electric field. Also we can see a small oscillation at 1.42 eV associated to the GaAs energy bandgap. There is a clear contradiction of the oscillation period as the thickness of cap layer increase associated to the diminution of the surface electric field produced by the reduction of the depletion zone near to the surface.

In order to determine the internal electric fields magnitude associated to the FKO we employed the asymptotic Franz-Keldysh modulation theory. In this model the energies E_j of the FKO extremes can be fitted to [9]

$$j\pi = \frac{\pi}{2} + \left(\frac{4}{3}\right)[(E_j - E_g)/\hbar\Theta]^{3/2} \quad (1)$$

where j is the index and E_j is the photon energy of the jth extreme, E_g is the energy gap of GaAs. $\hbar\Theta$ is the characteristic electro-optic energy and it can be calculated by

$$\hbar\Theta = (e^2F^2\hbar^2/2\mu)^{\frac{1}{3}} \quad (2)$$

where e is the electron charge, F is the electric field strength and μ the interband reduced mass involved in the transition.

Equation (1) can be rearranged as

$$E_j = \hbar\Theta X_j + E_g \tag{3}$$

where $X_j = \left[\frac{3\pi}{4}(j - \frac{1}{2})\right]^{2/3}$

As we can see, the equation (3) corresponds to a linear function with a slope $\hbar\Theta$ and the intersection E_g, which can be determined by a linear fitting the experimental dates.

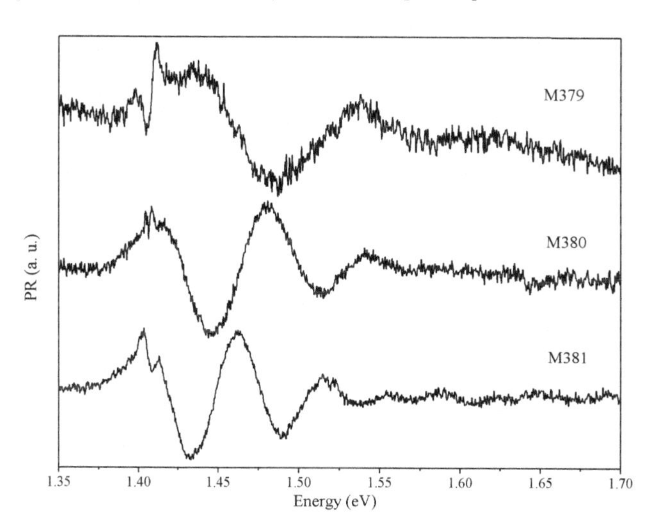

Figure 2. PR spectra of AlGaAs/GaAs heterostructures at room temperature.

Finally, the electric field magnitude can be calculated as

$$F = \sqrt{\frac{2\mu(\hbar\Theta)^3}{e^2\hbar^2}} \tag{4}$$

By using the Gauss's law and considering that on the sample surface there is a charged sheet with the electron density N_S, we can calculate it by use the electric field strength F_S obtained by PR through the following equation [10]:

$$N_S = \varepsilon\varepsilon_0 F_S \tag{5}$$

where ε and ε_0 are the relative and vacuum permittivity.

As we can see from the Table 1, the surface electric field decreases from 503 to 120 kV/cm whereas the thickness of the capping layer varies from 25 to 80 nm and the charge density at the surface decrease from 36.40 x10 $^{-11}$ to 8.68x10 $^{-11}$ cm^{-2}. This surface charge density modifies the conduction band profile and the electron population in the 2DEG at AlGaAs/GaAs heterojuntions, see figure 3.

In order to determine the electronic properties, we experimentally studied the low temperature electron magneto transport in all considered samples. The integer quantum Hall effect (IQHE) was observed in all devices. Figure 4 shows the behavior of Hall resistivity ρ_{xx} for M379 and M380 in the range from 0 to 9 T. It is clear seen a Hall plateaus at 5.5 and 6.5 T, respectively, corresponding to the filling factor i=4. Figure 5 shows the results of measurements of the magnetoresistance ρ_{xx} of two samples, M380 and M381, in this case we can see that the Hall plateaus i=4 for M381 is localized near 7 T. The behavior of the steadily varying component of the longitudinal magnetoresistance $\rho_{xx}(B)$ corresponds to the case of 2D systems. In all ~~of the~~ samples, we observe SdH oscillations. Electron density and mobility were extracted from the ρ_{xy} and $\rho_{xx}=V_{xx}W/IL$ measured values at ~~a~~ constant temperature of 2 K.

Sample	F_S (kV/cm)	N_S (10^{11} cm^{-2})	N_{2DEG} (x10^{11} cm^{-2})	μ (x10^3cm^2/V.s)
M379	503	36.40	5.30	101
M380	138	9.97	6.02	107
M381	120	8.68	6.57	112

Table 1. Surface electric field F_S and charge density N_S obtained from PR measurements. Concentration N_{2DEG} and electron mobility μ obtained from QHE measurements.

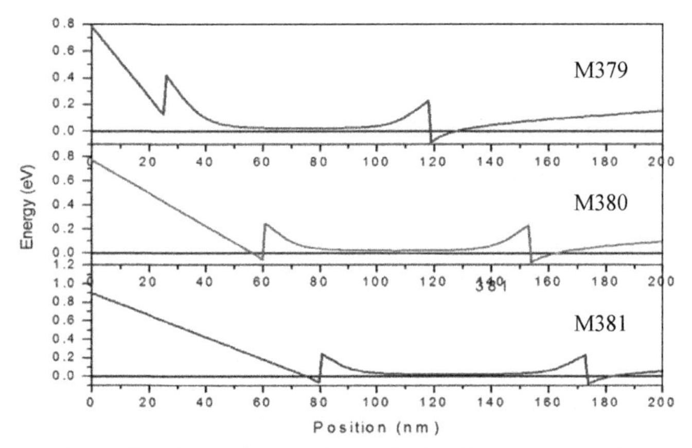

Figure 3. Potential profile of the conduction band calculated by nextnano software considering that the surface electric field is found by PR.

Table 1 shows the observed values of the concentration N_{2DEG} and electron mobility μ. As we can see, N_{2DEG} increase from 5.30 to 6.57 x $10^{-11}cm^{-2}$ as the cap layer thickness increase that is originated owing to flatting the conduction band profile. Moreover, the decrease in surface electric field allows reaching greater mobility.

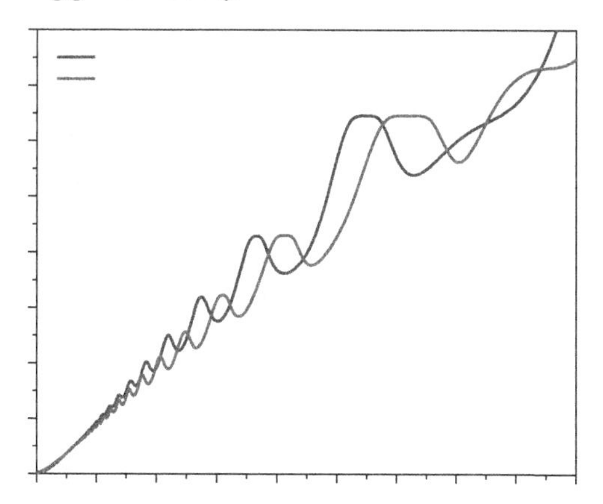

Figure 4. Hall resistances ρ_{xy} versus an externally applied magnetic field showing IQHE

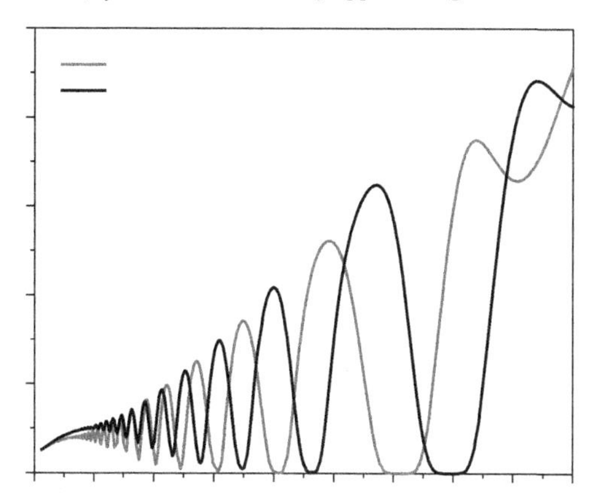

Figure 5. Magnetoresistance ρ_{xx} versus an externally applied magnetic field showing IQHE

CONCLUSIONS

In this study, the effect of the surface electric field and the optical and electrical properties of QHE devices are analyzed. The effect of the variation of the GaAs cap layer thickness on the energy band profile is numerically simulated. With the use of PR measurements we found the strength of the surface electric fields between 520 and 120 kV/cm^{-2} when the thickness of the cap layer is varied from to 25 to 80 nm, respectively. By QHE measurements we have found the electron density between 5.3 and 6.57 $x10^{-11}$ cm^{-2} that decreases at the surface electric field decreases. This dependence can be attributed to change the variations of the potential profiles of the conduction band.

ACKNOWLEDGMENTS

We acknowledge the partial financial supports of SEP-CONACyT, Mexico, through the contract 106268.

REFERENCES

1. Von Klitzing, G. Dorda, and M. Pepper, Phys. Rev. Lett. **45** (1980) 494.
2. Delahaye and B. Jeckelmann, Metrologia **40** (2003) 217.
3. W. Poirier, A. Bounouh,K. Hayashi, H. Fhirna, F. Piquemal, G. Genevès and J. P. André, J. Appl. Phys. **92** (2002) 2844.
4. W. Poirier, A. Bounouh,F. Piquemal, and J. P. André, Metrologia **41** (2004) 285
5. Bounouh, W. Poirier, F. Piquemal, G. Genevès, and J. P. André, IEEE Trans. Instrum. Meas. **52** (2003) 555.
6. K. Pierz, G. Hein, E. Pesel, B. Schumacher, H. W. Schumacher, and U. Siegner, Appl. Phys. Lett. **92** (2008) 133509.
7. K Pierz, G Hein, B Schumacher, E Pesel and H WSchumacher, Semicond. Sci. Technol. **25** (2010) 035014.
8. J. Misiewicz, P. Sitarek, G. Sek, Opto-electronics review **8(1)**, 1-24 (2000).
9. H. Shen and F. H. Pollak, Phys. Rev. B **42,** (1990) 7097.
10. B. Jayant Baliga, Fundamentals of Power Semiconductor Devices, Springer, USA, 2008, pp. 306.

Mater. Res. Soc. Symp. Proc. Vol. 1617 © 2013 Materials Research Society
DOI: 10.1557/opl.2013.1161

Exciton Gas versus Electron Hole Liquid in the Double Quantum Wells

Vladimir S. Babichenko[1] and Ilya Ya. Polishchuk[1,2,3]
[1]RNC Kurchatov Institute, Kurchatov Sq.1, 123182, Moscow, Russia
[2] Max Planck Institute for the Physics of Complex Systems, Nöthnitzer Str. 38, D-01187 Dresden, Germany
[3]Moscow Institute of Physics and Technology, Institutskii per 9, Dolgoprudnii, Moscow Reg, 141700, Russia

ABSTRACT

The many-body correlation effects in the spatially separated electron and hole layers in the coupled quantum wells (CQW) are investigated. A special case of the many-component electron-hole system is considered, $\nu \gg 1$ being the number of the components. Keeping the main diagrams in the parameter $1/\nu$ allows us to justify the selection of the RPA diagrams. The ground state of the system is found to be the electron-hole liquid with the energy smaller than the dense exciton gas phase. The possible connection is discussed between the results obtained and the experiments in which the inhomogeneous state in the CQW is found.

INTRODUCTION

Investigation of spatially separated electrons and holes in double quantum wells was induced by the fact that such systems support the formation of bound electron-hole states (excitons) with a long lifetime [1] what makes possible observation of the Bose-Einstein condensation of the excitons. However, the phase diagram of such systems can be quite complicated (see, e.g., [2]). In this work we investigate how the Coulomb correlations influence the origin of the ground state of the system under consideration.

The model of CQW assumes that electron move in one 2D layer and the holes move in the other 2D layer, these layers being separated by the distance l. At sufficiently low density and temperature, the electron-hole system may be considered as a degenerated exciton gas with interaction. However, with an increase in the exciton density n, when the distance between the particles $n^{-1/2}$ becomes smaller or of the order of magnitude of the isolated exciton in-layer radius R_{ex}, the system transforms into degenerated strongly correlated spatially separated electron-hole plasma. It is assumed that there exist $\nu \gg 1$ kinds of the electrons (holes). This approach was proposed in [3,4] to investigate the electron-hole plasma in many-valley semiconductors. Then, the diagrammatic expansion in $\nu \gg 1$ was used to select the main diagrams.

In this work, it is shown that the many-body Coulomb correlations result in the negative minimum in the ground state energy of the electron-hole system in the CQW. This minimum takes place at the certain density n_{eq}, and $n_{eq}{}^{-1/2} < R_{ex}$. It occurs that this minimum is below the ground state energy of the exciton gas with the same density. It is shown that, if initial density $n > n_{eq}$, the system stays homogeneous; otherwise it decays into droplets of liquid phase. The possibility of the existence of the electron-hole droplets in conventional semiconductors was predicted in [5] and later confirmed in numerous theoretical and experimental works [6]. Finally, we discuss the possible relation of the results obtained with a number of the experiments available [7-15].

MODEL

To describe the spatially separated electron-hole system in the CQW, it is assumed that the electrons are confined within one infinitely narrow 2D layer, while the holes are confined within the other 2D layer. For the sake of simplicity, it is supposed that the effective mass of both the electrons and holes are the same. Below the system of units is used with the effective electron (hole) charge $e = 1$, the effective electron (hole) mass $m = 1$, and the Planck constant ħ=1. Then, the effective Bohr radius is $a_B = 1$ and the energy is measured in the Hartree units. Let n be concentration of electrons or holes. Then, the Fermi momentum $p_F = 2\pi^2)(n/v)^{1/2}$, and the Fermi energy $\varepsilon_F = 2\pi n/v$ are the same for both the electrons and the holes. We confine ourselves to the case of strongly degenerated 2D plasma, that is, the temperature $T \ll \varepsilon_F$.

The Hamiltonian of the system is $H = H_0 + V, H_0$ being the kinetic energy and V being the Coulomb interaction. In the second quantization one has

$$H_0 = \sum_{\alpha\sigma k} \frac{k^2}{2} a^+_{\alpha\sigma}(\boldsymbol{k}) a_{\alpha\sigma}(\boldsymbol{k});$$

$$V = \frac{1}{2S} \sum_{\alpha\alpha'\sigma\sigma' \boldsymbol{k}\boldsymbol{k}'\boldsymbol{q}} V_{\alpha\alpha'}(\boldsymbol{q})\, a^+_{\alpha\sigma}(\boldsymbol{k}) a^+_{\alpha'\sigma'}(\boldsymbol{k}') a_{\alpha'\sigma'}(\boldsymbol{k}' - \boldsymbol{q})\, a_{\alpha\sigma}(\boldsymbol{k} + \boldsymbol{q}). \tag{1}$$

Here $\alpha = e(h)$ stands for the electrons (holes); $\sigma = 1,2, \ldots v$ denotes the kind of the electron or the hole component; $a^+_{\alpha\sigma}(\boldsymbol{k})$ and $a_{\alpha\sigma}(\boldsymbol{k})$ are the creation and annihilation operators, respectively, $\boldsymbol{k}$ is the in-plane 2D momentum. The Coulomb interaction in the momentum representation reads

$$V_{\alpha\alpha'}(k) = \begin{cases} \frac{2\pi}{k}, \alpha = \alpha', \\ -\frac{2\pi}{k} e^{-kl}, \alpha \neq \alpha' \end{cases}. \tag{2}$$

Let $G^0_{\alpha\sigma}(\omega, \boldsymbol{p}) = (-i\omega + \mu - \varepsilon_{\boldsymbol{p}})^{-1}$ be the electron (hole) free Matsubara Green function and $G_{\alpha\sigma}(\omega, \boldsymbol{p}) = (-i\omega + \mu - \Sigma_{\alpha\sigma}(\omega, \boldsymbol{p}) - \varepsilon_{\boldsymbol{p}})^{-1}$ be the electron (hole) total Green function. Here $\boldsymbol{\omega}$ is the Matsubara frequency, $\varepsilon_p = p^2/2$; $\Sigma_{\alpha\sigma}(\omega, \boldsymbol{p})$ is the self-energy part, and μ is the chemical potential. Let us consider the system of the self-consistent equations whose diagram representation is shown in Figure 1. Among all the diagrams of a given order in the interaction, only the diagrams are retained which contain the maximal number of the fermion loops what corresponds to Figure 1. Formally, this results in keeping the RPA-like diagram only.

The first diagrammatic equation in Fig. 1 reads

$$\Sigma_{\alpha\sigma}(\varepsilon, \boldsymbol{p}) = \Sigma_{\alpha\sigma}{}^{(1)}(\varepsilon, \boldsymbol{p}) + \Sigma_{\alpha\sigma}{}^{(2)}(\varepsilon, \boldsymbol{p}), \tag{3}$$

where

$$\Sigma_{\alpha\sigma}{}^{(1)}(\varepsilon, \boldsymbol{p}) = \frac{T}{S} \sum_{\alpha'\sigma' \boldsymbol{k}\omega} U_{\alpha\alpha'}(0) G_{\alpha'\sigma'}(\boldsymbol{k}, \omega), \tag{4}$$

$$\Sigma_{\alpha\sigma}{}^{(2)}(\varepsilon, \boldsymbol{p}) = -\frac{T}{S} \sum_{\alpha'\sigma' \boldsymbol{k}\omega} U_{\alpha\alpha}(k, \omega) G_{\alpha'\sigma'}(\boldsymbol{p} + \boldsymbol{k}, \omega). \tag{5}$$

Here T is the temperature and S is the area of the layers.

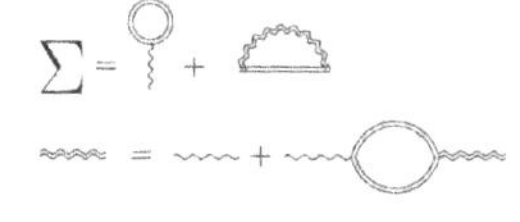

Figure 1. Self-consistent set of the diagrammatic equations for the self-energy part of the Green function. The double solid line represents the total Green function. The wavy line represents the inter-action V. The double wavy line corresponds to the renormalized interaction U.

The renormalized interaction $U_{\alpha\alpha}(k,\omega)$ entering Eq. (5) obeys the second diagrammatic equation in Figure.1 which reads

$$U_{\alpha\alpha}(\boldsymbol{p},\omega) = V_{\alpha\alpha}(\boldsymbol{p}) + V(\boldsymbol{p})\Pi_{\alpha\sigma}(\boldsymbol{p},\omega)U_{\alpha\alpha}(\boldsymbol{p},\omega)\,, \tag{6}$$

$\Pi_{\alpha\sigma}(\boldsymbol{p},\omega)$ being the polarization operator.

Our goal is to find the equation of the state at T=0. Let us first calculate the chemical potential μ of the electrons (holes) as a function of the density n using the self-consistent system of Eqs. (3) – (6). Let us remind the exact relation

$$\mu - \Sigma_{\alpha\sigma}\ \ (0, p_F) = {p_F}^2/2. \tag{7}$$

The contribution of the direct Coulomb interaction to the self-energy part is calculated exactly

$${\Sigma_{\alpha\sigma}}^{(1)}(\varepsilon,\boldsymbol{p}) = 2\pi ln. \tag{8}$$

Then, using the convergence method, one can show that ${\Sigma_{\alpha\sigma}}^{(2)}(\varepsilon,\boldsymbol{p})$ mainly comes from the momenta $k - k_0 \gg p_F$ and from the frequencies $\omega - \omega_0 \gg \varepsilon_F$. With this in mind, one finds

$${\Sigma_{\alpha\sigma}}^{(2)}(\varepsilon,\boldsymbol{p}) = -\frac{1}{(2\pi)^3}\int d\omega \int_{k\gg k_F} dk \frac{(\frac{2\pi}{k})^2 n(\frac{k^2}{\omega^2+(k^2/2)^2})^2}{1+\frac{2\pi}{k}\frac{nk^2}{\omega^2+(k^2/2)^2}} = -cn^{1/3}. \tag{9}$$

Here $c = 2.528$ if $l \gg 1$, and $c = 2^{1/3}2.528$ if $l \ll 1$.
Thus,

$$\mu(n) \approx 2\pi n/v + 2\pi nl - cn^{1/3} \tag{10}$$

THE EQUATION OF STATE

Using Eq. (10) one finds the energy per an electron (hole)

$$\varepsilon(n) = \frac{\pi n}{v} + \pi nl - \frac{3}{4}cn^{1/3}. \tag{11}$$

Then the pressure of the strongly degenerated electron-hole plasma reads

$$p = 2[\mu(n) - \varepsilon(n)]n = \frac{2\pi n^2}{v} + 2\pi n^2 l - \frac{c}{2}n^{4/3}. \tag{12}$$

Let us first turn to the case $l \gg 1$. The most energetically favourable is the state with the density n for which the energy $\varepsilon(n)$ possesses the minimal value. This value is given by the condition $\frac{d\varepsilon(n)}{dn} = 0$ (what is equivalent to the condition $p = 0$). This condition determines the density for the energy favorable phase

$$n_{eq} = (c/4\pi l)^{3/2}. \tag{13}$$

The energy per particle for this density is given by the expression

$$\varepsilon_{eq} = \varepsilon_{eq}(n_{eq}) = -\frac{c^{3/2}}{4(\pi l)^{1/2}}. \tag{14}$$

On the other hand, the binding energy for the dense exciton gas state per particle exceeds the binding energy of an isolated exciton $\varepsilon_0 \approx -1/l$. Comparing this estimate with (14) we conclude that, for $l \gg 1$, the electron-hole liquid with density (13) is energetically more favourable than the exciton state.

If $l \ll 1$, one obtains that

$$n_{eq} \simeq -(\frac{\nu}{1+\nu l})^{3/2}, \tag{15}$$

and

$$\varepsilon_{eq} \simeq -(\frac{\nu}{1+\nu l})^{1/2} . \tag{16}$$

On the other hand, if $l \ll 1$, the exciton binding energy $\varepsilon_{ex} \sim -1$. Comparing this value with Eq.(16) one concludes that the electron-hole state is energetically more preferable, as well.

For the intermediate case $l \sim 1$, both the electron-hole phase energy and the exciton phase energy are of the same order of magnitude $\varepsilon_{eq} \sim \varepsilon_{ex} \sim -1$. However, in this case the distance between the excitons is of the order of the in-plane exciton radius, i.e. $n_{ex}{}^{-1/2} \sim R_{ex} \sim 1$. In this case the concept of the exciton becomes senseless.

DICUSSION

Thus, in this work, a multicomponent $\nu \gg 1$ two-dimensional electron-hole plasma with spatially separated charges is considered. It is shown that, irrespective of the layer spacing l the electron-hole liquid possesses a lower energy per particle than the exciton gas. If $l \ll 1$, according to Eq. (11), the contribution of the kinetic energy to the chemical potential can be neglected. Thus, the energy of the electron-hole system is a sum of two terms. The first one is $\pi n^2 l$. It is associated both with the intralayer *e-e* and *h-h* direct Coulomb interaction and the interlayer *e-h* interaction. The *e-e* and the *h-h* contributions are positive and diverging. The *e-h* contribution is negative and also is diverging. These three terms yield the finite value $\pi n^2 l$ owing to the electric neutrality. The second contribution is the negative correlation energy $cn^{4/3}/4$. It is noteworthy that at $l \gg 1$ this term is associated with the *e-e* and *h-h* interactions. The correlation energy is negative, although the particles repel within each layer. A similar situation takes place in a three dimensional electron gas on a homogeneous positive background [6,16].

We nor turn to a possible relation between the present results and a number of experiments in which fragmented luminescent regions were observed in double quantum wells [7-15]. The presence of such regions is sometimes regarded as a manifestation of the Bose-Einstein condensation of excitons. In [17, 18], the luminescent fragments are attributed to the exciton liquid, in which the excitons are considered as Bose particles with effective attraction at short distances. In real experiments, the fragmentation is observed at $n_{ex}R_{ex}^2 \sim 1$. In this case, the emergence of exciton correlations near the Fermi energy is interpreted in some works as the presence of excitons which are of the bosonic nature. However, in our opinion, the luminescent fragments are unrelated to the excitons, being rather two-dimensional (2D) electron-hole droplets in which the strong electron-hole correlations of excitonic kind can occur at the Fermi surface [19].

In the experiments, the electron-hole plasma in a double quantum well is created by an external optical source. If the average charge density in the created 2D plasma is lower than n_{eq}, the homogeneous state of the system is unstable and the neutral 2D electron-hole droplets appear. Positive and negative charges of the droplets are concentrated in different layers. The charge density of each sign is n_{eq}. and the charges regions are situated just one above another.

To estimate the characteristic radius R of the droplet, we take into account that the lifetime τ of the spatially separated electrons and holes is finite. Thus, the decrease in the total charge of a certain sign in the droplet (under the conservation of neutrality) is $\pi R^2 n_{eq}/\tau$. At the same time, there exist a flux j of excitons outside the droplets, which is continuously created by the external source. This flux entering the 2D droplet increases its charge in each layer by $2\pi Rj$. In the state of dynamic equilibrium, the loss and the income of charges balance each other, yielding the relation $R \approx 2j\tau/n_{eq}$.. Obviously, the concept of the electron-hole droplets becomes senseless if its radius is $R < a_B = 1$. Owing to this reason, there is a minimal threshold value for the flux

$$j > j_c = \frac{n_{eq}a_B}{2\tau}, \tag{17}$$

which leads to the formation of the droplets. As follows from Eq. (17), the electron-hole droplets are formed if the flux j exceeds the critical value j_c. Thus, if the optical pumping is weak, $j < j_c$ a droplet does not have time to be formed. This agrees with the above-mentioned experimental findings of the minimal threshold pumping above which the fragmentation is observed. If the pumping is sufficiently strong, the electron-hole plasma with the uniform density $n > n_{eq}$ is formed. In this case, droplets do not appear and the stable state of the system is a homogeneous electron-hole liquid. Obviously, in this case, the pressure is positive and the implementation of this state requires the presence of the walls.

CONCLUSION

In this work, it is shown that the correlation effects are caused primarily by scattering in the screening channel. It is noteworthy that Lozovik and Berman [20] studied another kind of electron-hole liquid. In their work, the minimum of the energy as a function of the charge density n in a double quantum well is found by the variation approach and is associated with the exciton correlation in the electron-hole scattering channel. In our paper, taking into account of these correlations would correspond to taking into consideration the diagram that are small in the parameter $\frac{1}{\nu}$. Strictly speaking, our results are valid only in the case $\nu \gg 1$.At the same time, the results of [20] in fact correspond to the case $\nu = 2$ Therefore, a formal comparison of our results with those of [20] is incorrect. In our opinion, in [20] the role of the correlation effects considered in this paper is underestimated. However, the consistent inclusion of the correlation effects in the electron-hole scattering channel on the background of the screening channel at $\nu \gg 1$ can be performed only numerically, which lies beyond the scope of the present work. At the same time exactly the correlations in the electron-hole scattering channel are responsible for the appearance of anomalous averages of the excitonic type, which lead to the emergence of a dielectric band gap at the Fermi surface

ACKNOWLEDGMENTS

The study is supported by the Russian Fund for Basic Research (Grants 13-02-00472a) and by the Ministry of education and science of Russian Federation, project 8364.

REFERENCES

1. Yu. E. Lozovik and V.I. Yudson, Zh. Exp.Theor. Fiz. 71, 738 (1976) [Sov. Phys. JETP 44, 389 (1976)].
2. Yogesh N. Joglekar, Alexander V. Balatsky, and S. Das Sarma, 74, 233302 (2006).
3. E.A. Andrushin, V.S. Babichenko, L.V. Keldysh, et all. JETPh Lett, 24, 210 (1976).
4. L.V. Keldysh, Electron-Hole liquid in Semiconductors, in Morden Problems of Condense Matter Science, v. 6, C.D. Jeffries and L.V. Keldysh (North Holland, Amsterdam, 1987).
5. L. V. Keldysh, "Excitones in Semiconductors", (Nauka, Moscow, 1971)..
6. T. M. Rice, "The electron-hole liquid in semiconductors", Solid State Physics V 32, ed.; H. Ehrenreich, F. Zeitz, D. Turnbull, Academic Press, INC. 1977.
7. L.V. Butov, A.C. Gossard, and D.S. Chemla, Nature (London) **418**, 751 (2002).
8. D. Snoke, S. Denev, Y. Liu, L. Pheiffer, and D.S. Chemla, Nature (London) 418, 754 (2002)..
9. Butov, Solid State Commun. 127, 89 (2003).
10. D. Snoke, Y. Liu, S. Denev, L. Pheiffer, and K. West. Solid State Commun. 127, 187 (2003).
11. L.V. Butov, L.S. Levitov, A.V. Mintsev, B. D. Simons, A.C. Gossard, and D.S. Chemla, Phys. Rev. Lett. 92, 117404 (2004)
12. A.V. Larionov, V.B. Timofeev, P.A. Ni, S.V. Dubonos, I. Hvam, and K. Soerensen, Pis'ma Zh. Ekp. Theor. Fiz. 75, 233 (2002) [JETP Lett. 75, 570 (2002)].
13. A.A. Dremin, A.V. Larionov, and V. B. Timofeev, Fiz. Tverd. Tela (St. Peterburg) 46, 168 (2004) [Solid. State. Phys. 46, 170 (2004)].
14. V.B. Timofeev, Usp. Phys. Nauk 175, 315, (2005) [Phys. Usp. 48, 295 (2005)].
15. L.S. Levitov, B.D. Simons, and L.V. Butov, Phys. Rev. Lett, 94, 176404 (2005).
16. . D. Pines, P. Nozieres, Quantum theory of liquids, New York (1973)
17. A.A. Chernuk and V.I. Sugakov, Phys. Rev. B 74, 085303 (2006).
18. V.I. Sugakov, Phys. Rev. B 76, 115303 (2007).
19. L.V. Keldysh, and Yu.V. Kopaev, Sov. Phys. Solid State 6, 2219 (1965).
20. Yu. E. Lozovik and O.L. Berman, JETP Lett. 6**4**, 573(1996) V.I. Yudson, Zh. Exp.Theor. Fiz. 71, 738 (1976) [Sov. Phys. JETP 44, 389 (1976)].

Mater. Res. Soc. Symp. Proc. Vol. 1617 © 2013 Materials Research Society
DOI: 10.1557/opl.2013.1162

Emission and X ray diffraction in AlGaAs/ InGaAs Quantum wells with embedded InAs Quantum dots

R. Cisneros Tamayo[1], I.J. Gerrero Moreno[1], A. Vivas Hernandez[1], J.L. Casas Espinola[2] and L. Shcherbyna[3]

[1]ESIME– Instituto Politécnico Nacional, México D. F. 07738, México
[2]ESFM– Instituto Politécnico Nacional, México D. F. 07738, México
[3]V. Lashkaryov Institute of Semiconductor Physics at NASU, Kiev, Ukraine

ABSTRACT

The photoluminescence (PL), its temperature dependence and X-ray diffraction (XRD) have been studied in MBE grown GaAs/AlGaAs/InGaAs/AlGaAs /GaAs quantum wells (QWs) with InAs quantum dots embedded in the center of InGaAs layer in the freshly prepared states and after the thermal treatments during 2 hours at 640 or 710 °C. The structures contained two buffer ($Al_{0.3}Ga_{0.7}As/In_{0.15}Ga0_{.85}As$) and two capping ($In_{0.15}Ga0_{.85}As$ / $Al_{0.3}Ga_{0.7}As$) layers. The temperature dependences of PL peak positions have been analyzed in the temperature range 10-500K with the aim to investigate the QD composition and its variation at thermal annealing. The experimental parameters of the temperature variation of PL peak position in the InAs QDs have been compared with the known one for the bulk InAs crystals and the QD composition variation due to Ga/Al/In inter diffusion at thermal treatments has been detected. XRD have been studied with the aim to estimate the capping/buffer layer compositions in the different QW layers in freshly prepared state and after the thermal annealing. The obtained emission and XRD data and their dependences on the thermal treatment have been analyzed and discussed.

INTRODUCTION

Self-assembled InGaAs/GaAs quantum dots (QDs) are especially attractive for the application in lasers, photodiodes and memory devices [1-4]. The realization of efficient light-emitting devices operated at room temperature requires understanding the QD PL temperature dependences and the study of the reasons of PL variation at thermal annealing. The PL intensity decay in InAs QDs, as a rule, attributed to thermal escape of excitons from the QDs into a wetting layer (WL) or into the GaAs barrier [5-11], or to a thermally activated capture of excitons by the nonradiative defects in the GaAs barrier or at the GaAs/InAs interface [5, 8, 9]. It was shown experimentally that the main reason for the PL thermal decay in QD structures related to the thermal escape of the excitons, or correlated electron-hole pairs, from QDs [10 - 15]. In QD structures introducing the additional AlxGayAs layers into InGaAs/GaAs quantum wells (QWs) will, as it is expected, to increase the height of potential barrier for the exciton thermal escape from QDs into the barrier and can permit the application of these QD structures at higher temperatures. Improved understanding of the operation and the design peculiarities of InAs QDs embedded into InGaAs/AlGaAs QWs could be obtained from the study of the variation of PL spectra and XRD pattern at thermal annealing.

EXPERIMENTAL DETAILS

The solid-source molecular beam epitaxy (MBE) in V80H reactor was used to grow the waveguide structures consisting of the layer of InAs self-organized QDs inserted into 9 nm $In_{0.15}Ga_{0.85}As/Al_{0.30}Ga_{0.70}As/GaAs$ QWs. The thickness of the buffer $In_{0.15}Ga_{0.85}As$ layer was 1nm, which was grown on the 300 nm $Al_{0.30}Ga_{0.70}As$ buffer layer, the 200 nm GaAs buffer layer and the 2 inch (100) GaAs SI substrate. Then an equivalent coverage of 2.4 monolayers of InAs QDs were confined by the first capping (8 nm) $In_{0.15}Ga_{0.85}As$ layer, by the second 100 nm $Al_{0.30}Ga_{0.70}As$ capping layer, and by the 10 nm AlAs and 2 nm GaAs layers. Investigated QD structures are grown under As-stabilized conditions at the temperature 510°C, during the deposition of the InAs active region and InGaAs wells, and at 590-610 °C for the rest of layers. The in-plane density of QDs, estimated from previous AFM study, was 4 $\times 10^{10}$cm^{-2}.

The samples were mounted in a closed-cycle He cryostat where the temperature is varied in the range of 10 - 500 K. PL spectra are measured under the excitation of the 488 nm line of a cw Ar+-laser at an excitation power density of 500 W/cm^2 in the temperature range 10-500K. The setup used for PL study was presented earlier in [10,11]. The freshly prepared states are labeled by the letter A (#1A). Thermal annealing was carried out for some part of the structure #1 at 640 °C during 2 hours (state labeled by B, #1B) and for other part of the structure #1 at 710 °C during 2 hours as well (state labeled by C, #1C).

The X-ray diffraction (XRD) experiments were done using the XRD equipment Model XPERT MRD with the Pixel detector, three axis goniometry and parallel collimator with the resolution of 0.0001 degree. The X ray beam was from the Cu source with $K_{\alpha 1}$ line (λ=1.5406 Å).

RESULTS AND DISCUSSION

Typical PL spectra of the freshly prepared structure #1A measured at different temperatures in the range 10-500K are shown in Fig.1a.

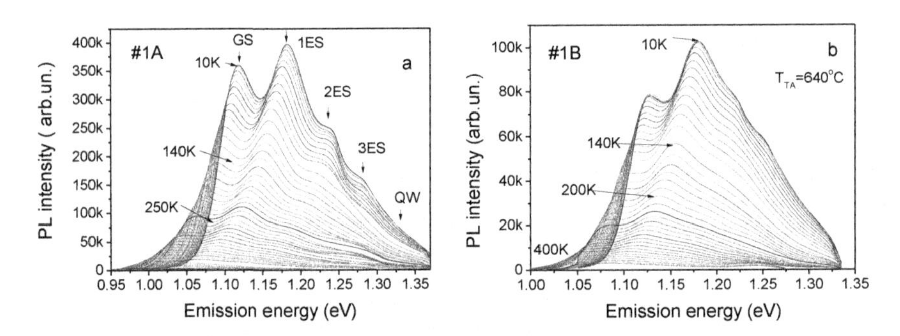

Figure 1. PL spectra of the QD structure #1 measured at different temperature in a freshly prepared state (a) and after the thermal annealing at 640 °C during 2 hours (b).

Five overlapping PL bands appear due to the recombination of excitons localized at a ground state (GS), at the first (1ES), second (2ES) third (3ES) excited states in InAs QDs and in QWs

(Fig.1a). The peak position of GS at 10K is 1.118 eV and the 1ES is 1.182 eV (#1). As one can see the full width at half maximum (FWHM) of GS and 1ES states and their overlapping in the structure #1A is not higher and these PL bands are definite very well. The thermal annealing stimulates the shift of GS PL peaks into the high energy range and the increase of a half width of PL bands and their overlapping (Fig.1b and Fig.2).

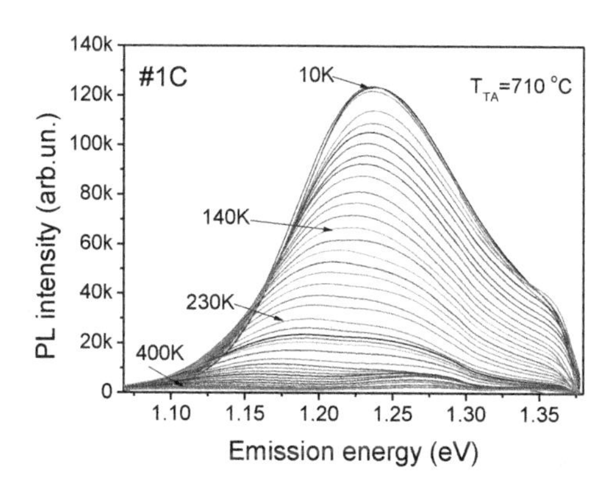

Figure 2. PL spectra measured at different temperatures for the QD structure #1C after the thermal annealing at 710 °C during 2 hours.

To analysis the process of Ga/Al/In inter diffusion the PL spectra of InAs QDs have been studied at different temperatures in a freshly prepared state (A) and after the thermal annealing (states B, C, Fig.1 and 2). Then the dependences of GS PL peak positions versus temperatures in all studied states have been obtained (Fig. 3a). It can be seen from Fig. 3a that the PL peaks shift to lower energies with increasing temperature due to the shrinkage of the band gap in QDs. Generally, due to the temperature induced lattice dilatation and electron–lattice interaction, the band gap energy follows the well-known Varshni formula [16]:

$$E(T) = E_o - \frac{aT^2}{T+b} \quad , \qquad (1)$$

where E_0 is the band gap at the absolute temperature T = 0 K; "a" and "b" are the Varshni thermal coefficients. The lines in Fig.3a present the fitting results for the structures #1A, #1B and #1C. The comparison of fitting parameters "a" and 'b' with the variation of the energy band gap versus temperature in the bulk InAs crystal (Table 1) has revealed that in both QD structures (#1A and #1B) the fitting parameter "a" and 'b' are close to their values in the bulk InAs crystal (Table 1).

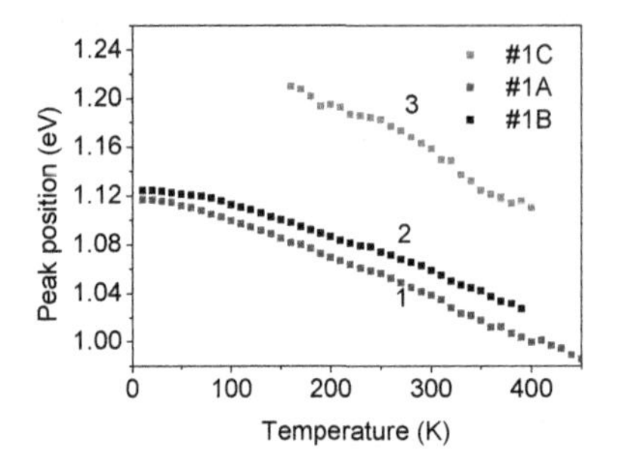

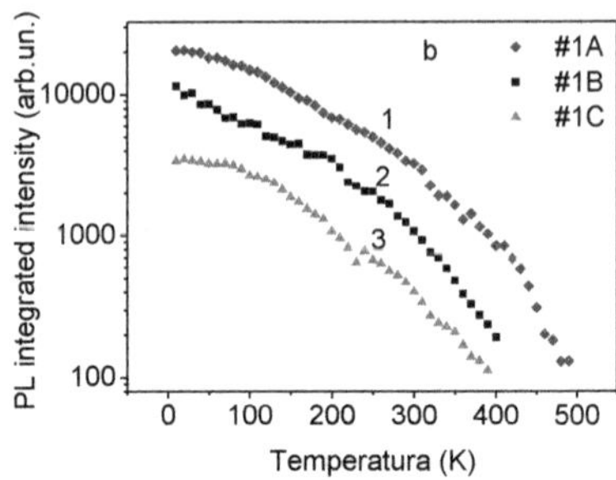

Figure 3. The variation of PL peak positions (a) and PL integrated intensity (b) versus temperature in the structures #1A, #1B and #1C. The lines in figure 3a represent the Varshni fitting results.

Table 1. Varshni fitting parameters

samples	#1A	#1B	#1C	InAs [17]	GaAs[17]
E_0 [eV]	1.117	1.128	1.220	0.400	1.519
α [meV/K]	0.345	0.334	0.350	0.276	0.540
β [K]	96	97	104	93	204

The transformation of PL spectra of InAs QDs in #1B and #1C structures after the annealing (Fig.1b and Fig.2) has shown that the GS peak positions shift into a higher energy range owing to changing, apparently, the QD sizes and/or the QD material composition [18-20]. Fitting parameters obtained for #1B and #1C structures after the annealing are shown in Table 1 as well. The comparison of QD fitting parameters with those for the bulk InAs and GaAs crystals (Table 1) has shown that a QD material composition in the structure #1C varies essentially in comparison with the structure #1B in the process of annealing due to the more efficient Ga/In inter mixing at higher temperature.

The variation of integrated PL intensity versus temperature in QD structures #1A, #1B and #1C has been presented in figure 3b. As it is clear the integrated PL intensity decays faster versus temperature in the structures #1B and #1C than in #1A.Thus we can conclude that the process of Ga/Al/In inter diffusion is accompanied by the creation on nonradiative recombination centers which stimulate the faster PL decay with temperature in the structures #1B and #1C.

To confirmed the variation of material composition owing the Ga/Al/In inter diffusion in QD structures the X-ray diffraction technique has been used. The diffraction of $K_{\alpha 1}$ and $K_{\alpha 2}$ lines of a Cu source from the (400) crystal planes in the cubic GaAs (AlGaAs) QW layers and the GaAs substrate has been detected (Fig.4). Simultaneously, a set of low intensity XRD peaks related to the diffraction from (400) crystal planes in the InGaAs QW layers has been revealed as well (Table 2). It was surprise that XRD testifies concerning two (#1A) types of the InGaAs QW layers such as: $In_{0.06}Ga_{0.94}As$ capping layers and $In_{0.62}Ga_{0.38}As$ buffer/wetting layer, intermixed completely in #1A.

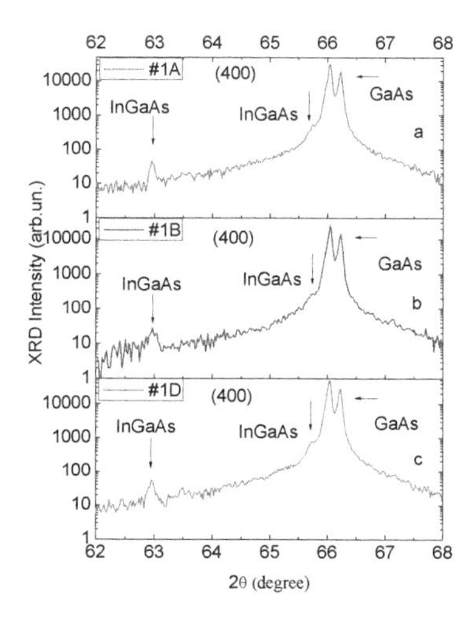

Figure 4. XRD pattern obtained in the structures #1A (a), #1B (b) and #1C (c)

The $In_{0.06}Ga_{0.94}As$ capping layer can be created owing to the Ga/Al/In intermixture between $In_{0.15}Ga_{0.85}As$ and AlGaAs capping layers at QD growth temperatures. At the same time the $In_{0.62}Ga_{0.38}As$ buffer/wetting layers are created at the Ga/In inter diffusion between the $In_{0.15}Ga_{0.85}As$ buffer and InAs wetting layers at QD growth temperatures. Note that in the structure #1A the Ga/In inter diffusion leads to complete intermixing with formation only one buffer/wetting layer.

Table 2. XRD data of InGaAs layer compositions

Samples	#1A, 2θ, degree	InGaAs Composition	#1B, T_{an}=640°C 2θ, degree	InGaAs Composition	#1C, T_{an}=710°C 2θ, degree	InGaAs Composition
	65.7477	$In_{0.06}\,Ga_{0.94}\,As$	65.7477	$In_{0.06}\,Ga_{0.94}\,As$	65.7477	$In_{0.06}\,Ga_{0.94}\,As$
	62.9477	$In_{0.62}\,Ga_{0.38}\,As$	62.9677	$In_{0.61}\,Ga_{0.39}\,As$	62.9677	$In_{0.61}\,Ga_{0.39}As$

After the thermal annealing the capping layer composition $In_{0.06}Ga_{0.94}As$ did not change, but the buffer layer composition a little bit change (Table 2) due to Ga/Al/In interdiffusion. As it is clear from Table 2, the In concentration in the buffer/wetting layers decreases after the thermal annealings in both QD structures #1B and #1C.

CONCLUSION

The photoluminescence (PL), its temperature dependence and high resolution X-ray diffraction (HR-XRD) have been studied in MBE grown GaAs/AlGaAs/InGaAs/AlGaAs /GaAs quantum wells (QWs) with InAs quantum dots embedded in the center of InGaAs layer in the freshly prepared states and after the thermal treatments during 2 hours at 640 or 710 °C. It is shown that a QD material composition change at thermal annealing owing to the Ga/Al/In inter diffusion at higher temperatures a little bit for the 640 °C and essentially for 710 °C . Thus the studied QD structures can be used at high temperatures up to 640 °C.

ACKNOWLEDGMENT

The authors would like to thank the CONACYT (project 130387) and SIP-IPN, Mexico, for the financial support, as well as the Dr. Andreas Stinz from the Center of High Technology Materials at University New Mexico, USA, for growing the studied QD structures and Dr. Jose Alberto Andraca Adame from the Center of Nanoscience Micro and Nanotechnologies at National Polytechnic Institute of Mexico for the XRD measurements.

REFERENCES

1. D. Bimberg, M. Grundman, N. N. Ledentsov, Quantum Dot Heterostructures, Ed. Wiley & Sons (2001) 328.
2. V. M. Ustinov, N. A. Maleev, A. E. Shukov, A. R. Kovsh, A. Yu .Egorov, A. V. Lunev, B. V. Volovik, I. L. Krestnikov, Yu. G. Musikhin, N. A. Bert, P. S. Kopev, Zh .I. Alferov, N. N. Ledentsov, D. Bimberg, Appl.Phys.Lett. **74,** 2815 (1999).
3. G. T. Liu, A. Stintz, H. Li, K. J. Malloy and L. F. Lester, Electron Lett, **35,** 1163 (1999).
4. A. Stintz, G. T. Liu, L. Gray, R. Spillers, S. M. Delgado, K. J. Malloy, J. Vac. Sci. Technol. B. **18(**3), 1496 (2000).
5. T. V. Torchynska, J. L. Casas Espínola, E. Velazquez Losada, P. G. Eliseev, A. Stintz, K. J. Malloy, R. Peña Sierra, Surface Science **532,** 848 (2003).
6. C. M. A. Kapteyn, M. Lion, R. Heitz, and D. Bimberg, P. N. Brunkov, B. V. Volovik, S. G. Konnikov, A. R. Kovsh, and V. M. Ustinov, Appl. Phys. Lett. **76,** 1573 (2000)
7. M. Dybiec, S. Ostapenko, T. V. Torchynska, E.Velasquez Losada, Appl. Phy. Lett. **84,** 5165-5167 (2004).
8. T.V. Torchynska, Superlattice and Microstructure, **45,** 349-355 (2009).
9. L. Seravalli, P. Frigeri, M. Minelli, P. Allegri, V. Avanzini, S. Franchi, Appl. Phys. Lett. **87,** 063101 (2005).
10. T. Torchynska, J. Appl. Phys., **104,** 074315, n.7 (2008).
11. T.V. Torchynska, A. Stintz, J. Appl. Phys. **108,** 2, 024316 (2010).
12. T. V. Torchynska, J. L. Casas Espinola, L. V. Borkovska, S. Ostapenko, M. Dybic, O. Polupan, N. O. Korsunska, A. Stintz, P. G. Eliseev, K. J. Malloy, J. Appl. Phys**. 101**, 024323 (2007).
13. S. Sanguinetti, M. Henini, M. Grassi Alessi, M. Capizzi, P. Frigeri, S. Franchi, Phys. Rev. B **60,** 8276 (1999).
14. E. C. Le Ru, J. Fack and R. Murray, Rhys. Rev. B. **67,** 245318 (2003).
15. G. Bacher, C. Hartmann, H. Schweizer, T. Held, G. Mahler, H. Nickel, Phys. Rev. B. **47,** 9545 (1993).
16. Y. P. Varshni, Physica **34,** 149 (1967).
17. A. Landolt-Boernstein, Numerical Data and Functional Reationship, in: Science and Techology, Springer, Berin, v. 22 (1987) p. 118.
18. T.V. Torchynska, A. Vivas Hernandez, G. Polupan, E. Velazquez Lozada, Material Science and Engineering B. **176**, 331 (2011).
19. T.V. Torchinskaya, Opto-Electronics Review, **6(2),** 121 (1998).
20. T. Torchynska, G. Polupan, F. Conde Zelocuatecatl, E. Scherbina, Modern Physics Letter, **15,** 593 (2001).

Si based materials

Mater. Res. Soc. Symp. Proc. Vol. 1617 © 2013 Materials Research Society
DOI: 10.1557/opl.2013.1163

EPR and emission study of silicon suboxide nanopillars

V. Bratus'[1], I. Indutnyi[1], P. Shepeliavyi[1], T. Torchynska[2]
[1]V. Lashkaryov Institute of Semiconductor Physics, NAS of Ukraine, Kyiv 03680, Ukraine
[2]ESFM-Instituto Politécnico Nacional, México D.F. 07738, Mexico

ABSTRACT

The results of correlated electron paramagnetic resonance (EPR) and photoluminescence (PL) study of obliquely deposited porous SiO_x films after step-by-step 15 min annealing within 105 min in vacuum at 950^0C are presented. The low intensity symmetrical and featureless EPR line with a g-value g=2.0044 and a linewidth of 0.77 mT has been detected in as-sputtered films and attributed to dangling bonds (*DB*) of silicon atoms in amorphous SiO_x domains with x=0.8. Successive annealing results in decreasing this line and the appearance of an intense EPR line with g=2.0025, linewidth of 0.11 mT and a hyperfine doublet with 1.6 mT splitting. According to the parameters this spectrum has been attributed to the *EX* center, a hole delocalized over four non-bridging oxygen atoms grouped around a Si vacancy in SiO_2. The impact of chemical treatment before annealing and duration of anneals on the defect system, and a correlation of the PL intensity with decreasing of the *DB* EPR signal are discussed.

INTRODUCTION

Owing to intense emission at room temperature, the structures consisting of Si nano-crystallites (*nc*-Si) embedded in silicon oxide SiO_2 or suboxide SiO_x ($1< x <2$) show considerable promise for optoelectronic and photonic applications [1-4]. To the fabrication of these structures several techniques like the plasma-enhanced deposition, ion implantation, laser ablation, magnetron sputtering, evaporation in vacuum *etc.* are used. All these methods allow to fabricate SiO_x films with required x. Subsequent thermally induced decomposition of a suboxide and the formation of Si nanoparticles in SiO_x matrix are governed by a relation $y\text{SiO}_x \rightarrow x\text{SiO}_y + (y - x)\text{Si}$, where $y > x$, and SiO_y will be evidently consisted of SiO_2 and SiO_x. The temperature of annealing determines the structure of inclusions: annealing below 1000^0C favors a coalescence of Si atoms into amorphous clusters, at higher temperature silicon nanocrystals are formed [1-5].

Thermally induced formation of *nc*-Si results in considerable dispersion of nanocrystallite sizes, that in turn decreases the intensity of photoluminescence (PL) and increases the half-width of PL band. Recently it has been shown that thermal evaporation of silicon monoxide on a substrate obliquely oriented to direction of evaporated substance stream and subsequent annealing in vacuum lead to the formation of porous SiO_x films with columnar structure and *nc*-Si inclusions (see [6, 7] and references there). Depending on the angle of evaporation and other technological parameters the diameter of deposited columns varies from 10 to 100 nm. Limited volume of SiO_x columns results in smaller dimensions of *nc*-Si than ones in normally deposited dense films with the same x. Relatively high porosity of obliquely deposited films makes it possible to control their light-emitting properties. The dependences of PL band position on pre-annealing chemical treatment in ammonia or acetone saturated vapor [8], on selective etching of *nc*-Si/SiO_x structures in HF solution [7] and on post-annealing chemical treatment in HF and H_2O_2 vapors [9] have been recently observed.

In above-mentioned publications [6-9] the obliquely deposited films were studied mainly with optical methods. The temperature of anneals in vacuum was restricted to 950-975^{0}C to avoid the process of film sublimation and the duration of thermal annealing was limited to 15 minutes. Electron paramagnetic resonance (EPR) is an indispensable tool for the defect identification in crystalline and amorphous materials; it is well suited for a quantitative study of volume and surface defects. Studying with EPR the normally deposited dense SiO_x films it has been shown that paramagnetic defects and their evolution may reflect the structural transformations of films [2]. Annealing duration was found an important parameter for properties of *nc*-Si/SiO_x structures in dense films [2, 10]. This work is aimed to elucidating the basic types of paramagnetic defects and their evolution during thermal annealing of obliquely deposited films. The results of correlated EPR and PL study of porous SiO_x films after step-by-step 15 min anneals within 105 min in vacuum at 950°C are presented.

EXPERIMENT

Thin (about 960 nm) SiO_x films were deposited by thermal evaporation of silicon monoxide SiO (Cerac Inc.) in vacuum $(1 \div 2) \times 10^{-3}$ Pa on polished (100) Si substrates arranged at an angle β=75° between the normal to the substrate surface and the direction of an evaporator. Step-by-step 15 min anneals within 105 min were carried out in a vacuum chamber at 950°C and a residual pressure of 1×10^{-3} Pa. Three types of samples labeled Ias, Iam and Iac were annealed simultaneously. Samples of the Ias group represented as-deposited untreated films, while samples of the Iam and Iac groups were kept during 120 hours in ammonia and acetone saturated vapor before the first anneal, respectively.

The EPR measurements were carried out at 300 K using an X-band (ν = 9.32 – 9.54 GHz) spectrometer with a lock-in signal detection at the magnetic field modulation frequency of 100 kHz. Signal averaging facilities were used to measure low intensities and linewidths without overmodulation and saturation effects. Amount of paramagnetic defects was determined relative to a MgO:Mn^{2+} standard sample with a known number of spins through double numerical integration of the respective derivative absorption signal. The g-values were determined with a precision of $\Delta g = \pm 0.0001$ via a microwave frequency counter and a calibration of the magnetic field by a proton nuclear magnetic resonance probe.

The photoluminescence spectra of SiO_x samples were measured at room temperature using a MDR-23 grating spectrometer with a cooled FEU-62 photomultiplier and a lock-in registration system. The discrete line 488 nm of an Ar ion laser and the line 532 nm of a light-emitting diode were used for excitation.

A scanning electron microscope (SEM) Zeiss EVO 50XVP was used to observe the cross-section of deposited films.

RESULTS and DISCUSSION

A typical cross-sectional view of SiO_x film obliquely deposited at β=75° on silicon wafer is shown in figure 1. As can be seen in the figure, the investigated SiO_x films have a porous inclined pillar-like structure with the pillar diameters of 10-100 nm. The porosity of films depends on the angle at the deposition and equals to 53% for β=75°. High-temperature annealing of these films does not change the porosity and pillar-like structure of samples.

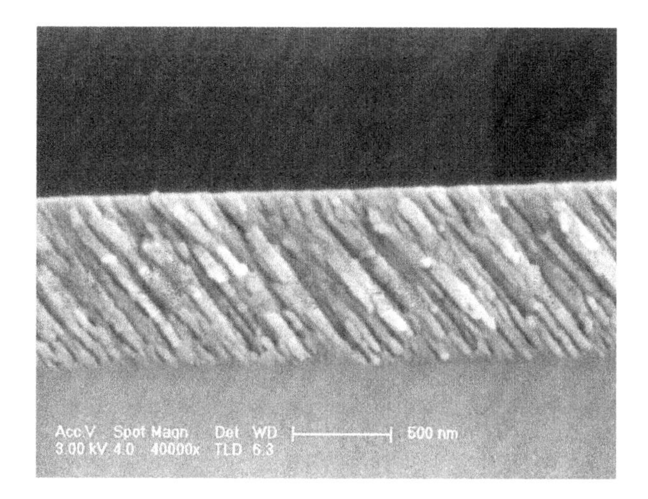

Figure 1. The SEM micrograph of SiO_x film cross-section for a sample deposited at $\beta=75^{\circ}$.

EPR spectra were found to be identical for unannealed samples of all three groups, for the Ias group of SiO_x films they are shown in figure 2a. Before annealing the samples are characterized by a broad structureless EPR line with g-value of 2.0044±0.0002 and width ΔB_{pp}=0.77 mT.

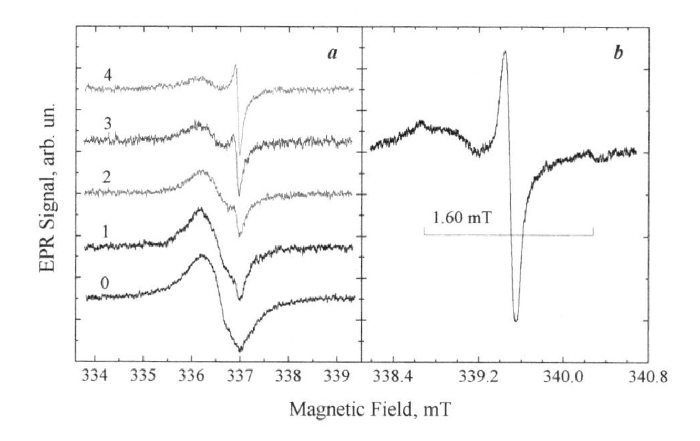

Figure 2. ***a*** – Room temperature EPR spectra for the Ias group of SiO_x films before (0) and after annealing at 950^{0}C for 15 (1), 30 (2), 75 (3) and 105 (4) min, microwave frequency ν = 9.442 GHz; intensities of spectra are normalized to unit sample area. ***b*** – Hyperfine doublet of the *EX* center detected on an Iac sample after anneals for 105 min, ν = 9.515 GHz, T = 300K.

Its slightly asymmetric line shape indicates that the spectrum may be a superposition of several components, the parameters are close to those of the EPR line observed in amorphous SiO_x films with $x \approx 0.8$ [11]. This line can be attributed to dangling bonds (*DB*) of Si atoms in structural tetrahedrons Si–Si_3O and, probably, in amorphous Si precipitates [2]. On the assumption of a uniform distribution of paramagnetic centers across the film thickness, their volume density is found to be 1.5×10^{19} cm^{-3}. Notice that due to optical measurements the main body of as-deposited porous SiO_x films corresponds to $x \approx 1.5$ [6, 7].

Annealing of the films studied gives rise to a narrow EPR line with g=2.0025 ± 0.0001 and ΔB_{pp}=0.10 mT, its intensity increases considerably with prolongation of annealing time while the *DB* line drops (figure 2*a*). Measurements on a sample with the greatest intensity of the narrow line provided a possibility to reveal its hyperfine doublet with 1.60 mT splitting (figure 2*b*). The observed parameters of EPR line are characteristic of so-called the *EX* center – a specific defect in SiO_2 pictured as a hole delocalized over 4 nonbridging oxygen atoms grouped around a Si vacancy [12]. It is pertinent to note that the hyperfine structure of the *EX* center in SiO_2 has been previously observed mainly at 4.2K [12, 13], for nanometer-sized silica particles it has been revealed at 100K [14]. To the best of our knowledge, hyperfine doublet of the *EX* center has been distinctly recorded at room temperature for the first time.

Chemical treatment in acetone and ammonia vapor before thermal annealing of porous SiO_x films furnishes different results. The density of paramagnetic defects in acetone-treated samples is comparable to that of untreated ones while for ammonia-treated samples it is three times less (figure 3). We believe that hydrogen atoms arising by ammonia dissociation can passivate a considerable amount of the *DB* defects.

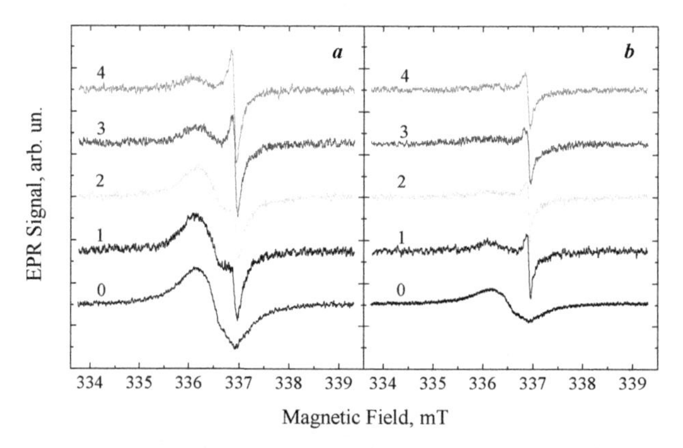

Figure 3. EPR spectra for the Iac (***a***) and Iam (***b***) groups of SiO_x films before (0) and after annealing at 950^0C for 15 (1), 30 (2), 75 (3) and 105 (4) min, ν = 9.442 GHz, T = 300K, intensities of spectra are normalized to unit sample area.

Transformations with annealing of EPR spectra for the Iac group is similar to the Ias one with the exception of higher growing of the *EX* line. Contrary to them, relatively intense *EX* line appears after the first anneal for 15 min of samples of the Iam group remaining practically unchanged after the following annealing steps (figure 3b). Comparing the total number of paramagnetic defects before and after annealing for 105 min, about fourfold decrease of them was found for all groups.

Simultaneously with decreasing of defect amounts a rise of PL intensity has been observed for all groups of samples, for the Ias group it is presented in figure 4a.

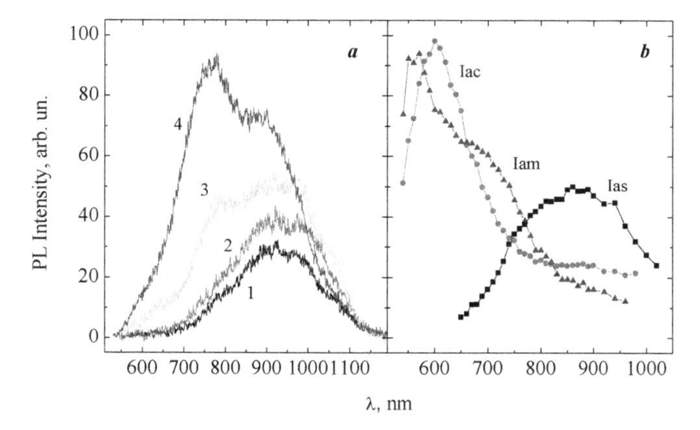

Figure 4. Room temperature PL spectra: ***a*** – for a sample of the Ias group after annealing at 950^0C for 15 (1), 30 (2), 75 (3) and 105 (4) min, $\lambda_{exc.}$ = 532 nm; ***b*** – for samples of different groups of SiO_x films annealed at 950^0C for 15 min, $\lambda_{exc.}$ = 488 nm.

Overall, the gain of integral PL intensity correlates with reduction in the number of *DB* centers. Defects of the dangling bond type are usually considered as non-radiative recombination centers. In any case, optically-detected magnetic resonance technique must be used to elucidate their role since a variety of non-paramagnetic defects can be annealed simultaneously with paramagnetic ones. Qualitatively, the increase of PL intensity can be explained by the passivation of non-radiative recombination centres at *nc*-Si/SiO_x interface [15].

Excitation with 488 nm provides a way for observation of new short wavelengths PL bands for chemically treated samples (figure 4b). It is assumed that nitrogen in ammonia and carbon in acetone promotes passivation of *nc*-Si surface by nitrogen or carbon and prevents formation of Si=O bonds during annealing in vacuum and further exposure on air. Moreover, the presence of acetyl and methyl influences on oxidation (or oxicarbonization) of silicon and formation of *nc*-Si with smaller sizes takes place [16].

Comparing the EPR results of present study with those obtained on normally deposited dense SiO_x films [2] we can find several main distinctions. First, for as-deposited dense films (where $x \approx 1.3$) the volume defect density was estimated as 4×10^{20} cm^{-3}, that is an order of magnitude higher than for porous films. Evidently, this is a direct consequence of higher value of x for the latter films. Second, thermal annealing of dense SiO_x films at 900^0C gave rise to the emergence of EPR line with g = 2.0055, typical for dangling bonds in amorphous Si precipitates. The lack of this signal in EPR spectra of porous films implies smaller concentration of Si nanocrystallites there. Third, the *EX* center has never been observed in dense SiO_x films, it is due to porosity of obliquely deposited films that promotes oxidation of pillars surface.

CONCLUSIONS

An EPR study of obliquely deposited SiO_x films reveals the presence of the *DB* and *EX* paramagnetic centers there. The emergence of these defects and their behavior with thermal annealing furnish important information about the structural peculiarities of the films. The impact of chemical treatment before annealing and the duration of anneal on the defect system, and a correlation of the PL intensity with decreasing of *DB* EPR signal have been observed.

REFERENCES

1. M. Ya. Valakh, V. A. Yukhimchuk, V. Ya. Bratus', A. A. Konchits, P. L. F. Hemment and T. Komoda, J. Appl. Phys. **85**, 168 (1999).
2. V. Ya. Bratus', V. A. Yukhimchuk, L. I. Berezhinsky, M. Ya. Valakh, I. P. Vorona, I. Z. Indutnyi, T. T. Petrenko, P. E. Shepeliavyi, I. B. Yanchuk, Fiz. Tekh. Poluprovodn., **35**, 854 (2001) [Semiconductors **35**, 821 (2001)].
3. H. Rinnert, M. Vergant, A. Burneau, J. Appl. Phys. **89**, 237 (2001).
4. D. Nesheva, C. Raptis, A. Perakis, I. Bivena, Z. Aneva, Z. Levi, S. Alexandrova, H. Hofmeister, J. Appl. Phys. **92**, 4678 (2002).
5. S. Boninelli, F. Iacona, G. Franzo, C. Bongiorno, C. Spinella, F. Priolo, J. Phys.: Condens. Matter **19**, 225003 (2007).
6. I.Z. Indutnyi, I.Yu. Maidanchuk, V.I. Min'ko, P.E. Shepeliavyi, V.A. Dan'ko, J. Optoelectron. Adv. Mater. **7**, 1231 (2005).
7. I.Z. Indutnyi, E.V. Michailovska, P.E. Shepeliavyi, V.A. Dan'ko, Fiz. Tekh. Poluprovodn., **44**, 218 (2010) [Semiconductors **44**, 206 (2010)].
8. I.Z. Indutnyi, I.Yu. Maidanchuk, V.I. Min'ko, P.E. Shepeliavyi, V.A. Dan'ko, Fiz. Tekh. Poluprovodn., **41**, 1265 (2007) [Semiconductors **41**, 1255 (2007)].
9. V.A. Dan'ko, V.Ya. Bratus', I.Z. Indutnyi, I.P. Lisovskyi, S.O. Zlobin, K.V. Michailovska, P.E. Shepeliavyi, Semicond. Phys. Quant. Electronics & Optoelectronics **13**, 413 (2010).
10. B.G. Fernandez, M. Lopez, C. Garcia, A. Perez-Rodriguez, J. R. Morante, C. Bonafos, M. Carrada, A. Claverie. J. Appl. Phys. **91**, 798 (2002).
11. E. Holzenkampfer, F.-W. Richter, J. Stuke, U. Voget-Grote, J. Non-Cryst. Solids, **32**, 327 (1979).
12. A. Stesmans, F. Scheerlinck, Phys. Rev. B **50**, 5204 (1994); J. Appl. Phys. **75**, 1047 (1994).
13. M. Jivanescu, A. Romanyuk, A. Stesmans, J. Appl. Phys. **107**, 114307 (2010).
14. A. Stesmans, K. Clemer, V.V. Afanas'ev, Phys. Rev. B **72**, 155335 (2005).
15. M. Lopes, B. Garrido C. Garsia P. Pellegrino, A. Pérez-Rodríguez, J. R. Morante, C. Bonafos, M. Carrada, A. Claverie, Appl. Phys. Lett. **80**, 1637 (2002).
16. I.Z. Indutnyy, V.S. Lysenko, I.Yu. Maidanchuk, V.I. Min'ko, A.N. Nazarov, A.S. Tkachenko, P.E. Shepeliavyi, V.A. Dan'ko, Semicond. Phys. Quant. Electronics & Optoelectronics **9**, 9 (2006).

Mater. Res. Soc. Symp. Proc. Vol. 1617 © 2013 Materials Research Society
DOI: 10.1557/opl.2013.1164

New Paramagnetic Center and High Conductivity in a-$Si_{1-x}Ru_x$:H Thin Films

Jian He, Wei Li*, Rui Xu, An-ran Guo,Yin Wang and Ya-dong Jiang
State Key Laboratory of Electronic Thin Films and Integrated Devices, University of Electronic Science and Technology of China, Chengdu 610054, China
E-mail: wli@uestc.edu.cn

ABSTRACT

In this work, the metallic element Ru is introduced into a-Si:H. The structural and electrical properties of the films doped with Ru have been investigated. Raman spectra reveal that the addition of Ru disarranges further the intrinsically disordered amorphous network and generates more coordinated defects. Meanwhile, a new paramagnetic signal, associated with the holes localized in valence band tail, has been observed. Moreover, the conductivity increases by about nine orders of magnitude with the increase of doping concentration, and the temperature coefficient of resistance (TCR) results show that this material may have a potential application in the infrared detectors.

INTRODUCTION

Over the last few decades, the doping of crystalline semiconductor with metallic elements, such as Gd [1], Tm and Yb [2], has been of considerable interest due to its potential application in stimulated emission and infrared detection. The key scientific concerns, in particular, focus on the electronic configuration, the local coordination of impurities in the lattice and their interaction with the host. Because of the low solubility in lattice, it becomes more difficult for metal atoms to get into the covalent semiconductor like silicon and germanium. In contrast, this problem can be partially overcome in the amorphous counterpart because the disorder of the amorphous network allows it to accommodate more metal atoms with different size. Hydrogenated amorphous silicon (a-Si:H), a technologically important material, is widely used in many practical applications, such as the production of solar cells [3], infrared detectors in night vision systems [4], and thin film transistors in flat panel display devices [5]. In addition, a-Si:H is also an ideal theory model in amorphous semiconductor physics. Consequently, the metal doping of a-Si:H has been an interesting area and there are many investigations on the a-Si:H doped by metallic elements, especially rare-earth elements. It is reported that the electrical properties of the doped a-Si:H with La, Nd, Er, Lu or Pr exhibit a large degree of variation [6]. Kumeda et. al [7, 8] found that in the electronic spin resonance (ESR) measurement the resonance signal of Mn, Fe and Ni impurities in a-Si:H can be observed and the paramagnetic neutral dangling bond (D^0) density decreases as the impurities concentration increases. However, the effects of the incorporation of metal impurities on the microstructure and properties of disordered semiconductors are not completely known and understood. In recent years, metallic element Ru has attracted increasing attention due to its several applications: (i) the barrier layer for Cu metallization in electronic devices [9]; (ii) the capping material in Mo/Si multilayer optical systems [10]; (iii) the suitable capacitor electrode material for the device based on high-*k* dielectrics to hamper the reaction between the semiconductor and the insulator [11], and (iv) the photo-sensitizing Ru complexes for solid state dye solar cells [12]. Nevertheless, the Ru doping of a-Si:H is rarely reported. With the above ideas in mind, we introduce metallic element Ru into a-Si:H thin films by rf co-sputtering technique for the first time, and study the coordination of Ru impurities in the amorphous network as well as its effect on the electrical properties of a-Si:H.

EXPERIMENT

The Ru-doped a-Si:H thin films were deposited by co-sputtering using a high-purity (99.999%) c-Si target and several Ru slices in an Ar atmosphere. The chamber was pumped to a base pressure of 3×10^{-4} Pa and the flow rate of Ar was 20 sccm (standard cubic centimeter per minute). The substrate temperature was kept at 300 °C and the process gas pressure was 0.52 Pa. The flow rate of H_2 was 3 sccm. The nominal doping concentration was evaluated from the coverage area of Ru slices. The laser used in Raman measurement for excitation was 532 nm and the power was set below 2mW while the beam was defocused with a diameter of 2μm to avoid the laser-induced crystallization. The ESR measurement was on a BRUKER ESP4105 spectrometer. The electrical properties of samples were analyzed using Keithley4200 semiconductor characterization system. The optical band gap E_g was obtained from SE850 spectroscopic ellipsometer.

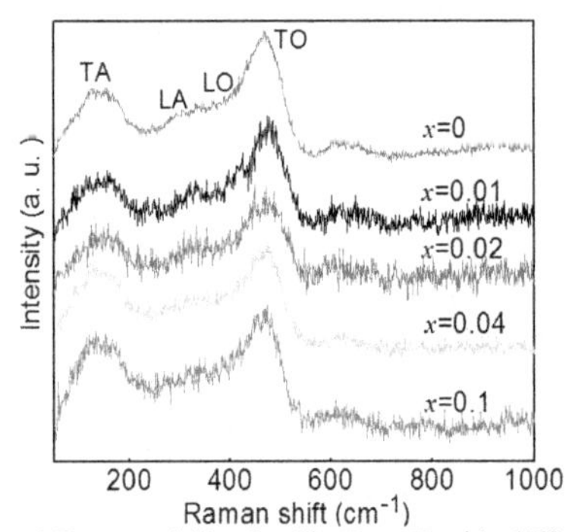

Figure 1. Raman shift of a-$Si_{1-x}Ru_x$:H with different x.

DISCUSSION

Raman scattering has been used extensively to examine the amorphous network due to its intensity sensitive to the structural disorder in solids. The Raman spectrum of a-Si:H thin film consists of several vibrational modes, as shown in the of Fig. 1.The TA band at about 150 cm^{-1} is proportional to the density of dihedral angle fluctuations in materials, reflecting the medium-range order (MRO) of amorphous network [13].The TO phonon band at about ω_{TO}=480 cm^{-1} is sensitive to the short-range order (SRO) [14].The presences of the LA band at 300 cm^{-1} and the LO band at 410 cm^{-1} originate the coordination defects in thin films [15]. The deconvolution shows that root-mean-square bond-angle variation $\Delta\theta$, determined from the linear relation $\Gamma_{TO}=15+6\Delta\theta$, where Γ_{TO} is the peak width of TO band [14], increases from 6.71° to 8.35° as the dopant concentration x increases from 0 to 0.1. The integrated intensity ratios I_{TA}/I_{TO} and

I_{LA+LO}/I_{TO} increase from 0.70 and 0.88 to 3.19 and 6.65, respectively, with the increase of x. The above data reveal that the incorporated Ru atoms disarrange further the intrinsically disordered amorphous network on the medium and short ranges, and result in more defects. It can be attributed to the different size and electronegativity between the Si and Ru atoms.

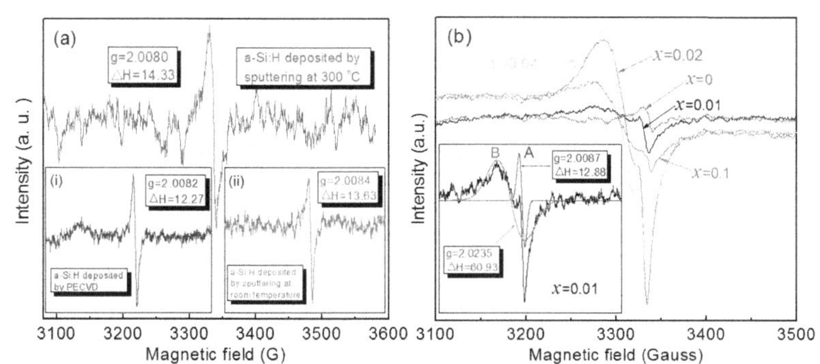

Figure 2. (a) ESR lines of undoped a-Si:H deposited under various conditions; (b) ESR lines of a-$Si_{1-x}Ru_x$:H with different x, the inset is the resonance line of x=0.01.

ESR is an important tool to investigate the local environment of paramagnetic impurities in amorphous materials. Before analyzing the ESR signals in a-$Si_{1-x}Ru_x$:H, it is necessary to examine first the ESR paramagnetic centers in undoped a-Si:H deposited under the identical condition without Ru, as shown in Fig 2 (a). Measurement gives the g value of 2.0080, which is inconsistent with the commonly accepted g=2.0055 identified as a D^0 signal in undoped a-Si:H. As we known, a-Si:H has complex microstructure. In order to exclude the possibility that a particular microstructure induces the deviation of g value, and confirm this stability of g=2.0080, we have prepared undoped a-Si:H thin films using sputtering at room temperature and plasma enhanced chemical vapor deposition (PECVD) at 250 °C, respectively, and performed ESR measurement on them. Results show that the g values are still near 2.0080, as shown in the inset of Fig 2 (a). Consequently, this g=2.0080 does stem from the D^0 in undoped a-Si:H. Fig 2(b) gives the ESR spectra of a-$Si_{1-x}Ru_x$:H with different x. It can be seen that the resonance lines of a-Si:H doped with Ru are strikingly different from that of the undoped a-Si:H, and the resonance signal varies dramatically with the doping concentration. Take the sample with x=0.01 for example, the signal can be deconvoluted into two independent components, as shown in the inset of Fig 2 (b). According to the obtained g value and linewidth △H as well as lineshape, the D^0 in a-Si:H is responsible for the component at higher field, labeled as A. The normalized intensity of A increases from 3.75×10^5 to 1.02×10^6, and then decreases to 8.58×10^4 with the doping concentration increasing from 0 to 0.02 then to 0.1. This nonmonotonic trend will be discussed later. The other component of the ESR signal with g being near 2.0230, labeled as B, is a new paramagnetic signal and associated with the holes localized in valence band tail, which will be further analyzed below.

Fig 3includes the DC dark conductivity and temperature coefficient of resistance (TCR) of a-$Si_{1-x}Ru_x$:H thin films. As x increases from 0 to 0.1, the conductivity increases by about nine orders of magnitude, which exceeds the conductivity of the conventional p-type and n-type a-

Si:H. It is worth point out that we do not present the data of undoped a-Si:H at higher temperature due to the instability induced by its high resistance during measurement. The conductivity of undoped a-Si:H given here is only used as reference for order magnitude. It is clear from Fig 3 (a) that the predominant conductivity mechanism in a-$Si_{1-x}Ru_x$:H is still an activated-like conductivity. For the sake of further analysis, we fit the obtained experimental data using the relation

$$\sigma(T) = \sigma_0 \exp\left(-\frac{E_a}{kT}\right) \qquad (5)$$

where the prefactor σ_0 is the conductivity extrapolated to $1/T=0$, k is Boltzman constant, E_a is the activation energy that represents the threshold energy of conductance. As shown in Fig 3 (a), the E_a of a-$Si_{1-x}Ru_x$:H is significantly less than that of intrinsic a-Si:H, which in turn leads to its higher conductivity. A key application of Si-based amorphous material is infrared detector, and TCR is an important characterization of material sensitivity to heat. It can be seen from Fig 3 (b) that TCR of a-$Si_{1-x}Ru_x$:H decreases with the increase of x. However, the samples with x=0.02 and x=0.04, their conductivity being above 10^{-4} S/cm, remain above 2%/K even at 370 K. The related research shows that the requirement for the TCR of sensitive material used in infrared detector is above 2%/K [16, 17]. This exciting result indicates that if we do not consider other factors, a-$Si_{1-x}Ru_x$:H may have the potential application in the infrared detector.

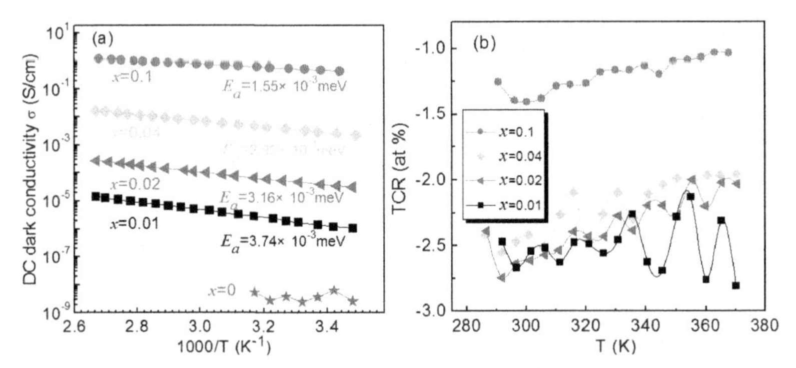

Figure 3. Temperature-dependent DC dark conductivity (a) and TCR (b) of a-$Si_{1-x}Ru_x$:H.

As mentioned above, the addition of Ru into a-Si:H further disarranges the amorphous network and generates more defects. It appears that the normalized intensity of component A in ESR signal should increase with the doping concentration. However, our results display a nonmonotonic trend. The light-induced and equilibrium ESR experiments in intrinsic a-Si:H and P- as well as B-doped a-Si:H confirm the existence of charged dangling bonds [18, 19]. Also, the increase of conductivity in our case reveals that the Ru-induced impurities energy level in band gap is electrically active and hence away from the energy region of D^0.On the other hand, the g value of the new paramagnetic centers in our case is close to that of the holes in valence band tail of B-doped a-Si:H (2.012), away from that of the electrons in conduction band tail of P-doped a-

Si:H (2.0044). Besides, the activation energy E_a and the optical band gap E_g (see Fig 4) of a-$Si_{1-x}Ru_x$:H decrease sharply with the increase of x. Based on these considerations, we suggest the following possible mechanism to explain the observed features. It is well known that there are seven 4*d* electrons and one 5*s* electron in the outer orbits of Ru atom. The energy of 4*d* orbit is higher than that of 5*s* orbit. So the Ru incorporated into a-Si:H may form the acceptor-like states by sd_3 hybridization produced by the one 5s electron and three 4*d* electrons due to the local crystal field. The remaining four 4*d* electrons locate at the non-bonding states. Subsequently, the acceptor-like states should be occupied by the electrons from the valence band tail and the D^0, yielding the holes localized in valence band tail and the positively charged dangling bonds that cannot be detected by ESR. Compared to the generated acceptor-like states, the number of dangling bonds by virtue of the addition of Ru is greater at low doping concentration. As the doping concentration increases, the impurities level should split into impurities band overlapping with the valence band tail, and thus the charged dangling bonds are predominant. This is self-consistent with the increase of conductivity and the obtained nonmonotonic behavior of component A in ESR signals. The above model is illustrated in Fig 5.

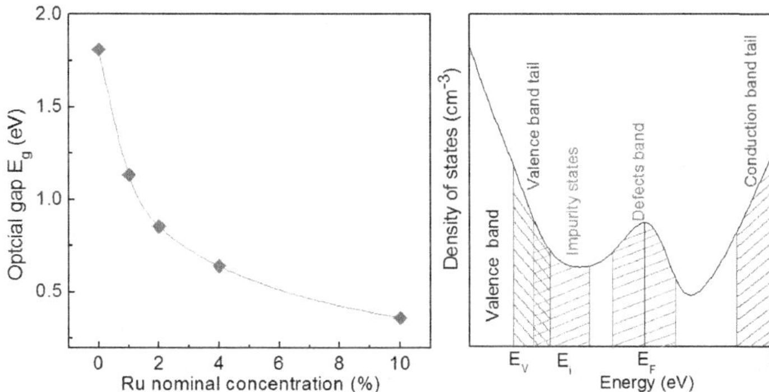

Figure 4. Dependence of optical gap of a-$Si_{1-x}Ru_x$:H on the Ru doping concentration.

Figure 5. Sketch of density of states in a-$Si_{1-x}Ru_x$:H, where E_v, E_c, E_F and E_i are valence band mobility edge, conduction band mobility edge, Fermi level and impurities level, respectively.

CONCLUSIONS

In summary, we have introduced Ru element into a-Si:H by rf co-sputtering technique for the first time and examined the structural and electrical properties of a-$Si_{1-x}Ru_x$:H. Experimental results suggest that the basic amorphous network become more disordered with the addition of Ru, and more coordinated defects appear. Meanwhile, the new paramagnetic centers with g value of near 2.023 are observed, which is ascribed to the holes localized in valence band tail. The conductivity of a-$Si_{1-x}Ru_x$:H increases dramatically with the doping concentration, while the E_a and E_g decreases. The calculated TCR indicates that the a-$Si_{1-x}Ru_x$:H with x=0.02 and x=0.04 may have a potential application in the infrared detector. Finally, according to these obtained data, we have proposed a possible mechanism for the doping of Ru in a-Si:H, which can describe the observed experimental behavior well.

ACKNOWLEDGMENTS

This work was supported by the Fundamental Research Funds for the Central Universities (Grant No. ZYGX2012YB024), People's Republic of China.

REFERENCES

1. J. Mandelkorn, L. Schwartz, J. Broder, H. Kautz and R. Ulman: J. Appl. Phys. **35,** 2258 (1964).
2. N. Sclar: Infrared Phys. **17,** 71 (1977).
3. E. Klimovsky, A. Sturiale and F. A. Rubinelli: Thin Solid Films **515,** 4826 (2007).
4. D. Knipp, R. A. Street, H. Stiebig, M. Krause, J. P. Lu, S. Ready and J. Ho: IEEE Trans. Electron. Dev. **53,** 1551 (2006).
5. A. J. Flewitt, S. Lin, W. I. Milne, R. B. Wehrspohn and M. J. Powell: J. Non-Cryst. Solids **352,** 1700 (2006).
6. J. H. Castilho, I. Chambouleyron, F. C. Marques, C. Rettori and F. Alvarez: Phys. Rev. B **43,** 8946 (1991).
7. M. Kumeda, Y. Jinno, I. Watanabe and T. Shimizu: Solid State Commun. **23,** 833 (1977).
8. T. Shimizu, M. Kumeda, I. Watanabe and Y. Jinno: J. Non-Cryst. Solids **35,** 36 (1980).
9. M. Damayanti, T. Sritharan, S. G. Mhaisalkar and Z. H. Gan: Appl. Phys. Lett. **88,** 044101 (2006).
10. T. E. Madey, N. S. Faradzhev, B. V. Yakshinskiy and N. V. Edwards: Appl. Surf. Sci. **253,** 1691 (2006).
11. Y. Matsui, Y. Nakamura, Y. Shimamoto and M. Hiratani: Thin Solid Films **437,** 51 (2003).
12. Y. Saito, T. Azechi, T. Kitamura, Y. Hasegawa, Y. Wada and S. Yanagida: Coordin. Chem. Rev. **248,** 1469 (2004).
13. W. S. Wei, G. Y.Xu, J.L. Wang and T. M. Wang: Vacuum **81,** 656 (2007).
14. M. Marinov and N. Zotov: Phys. Rev. B **55,** 2938 (1997).
15. N. Zotov, M. Marinov, N. Mousseau and G. Barkema: J. Phys. : Condens. Matter. **11,** 9647 (1999).
16. D. Murphy, M. Ray, R. Wyles, J. Asbrock, N. Lum, A. Kennedy, J. Wyles, C. Hewitt, G. Graham, T. Horikiri, J. Anderson, D. Bradley, R. Chin and T. Kostrzewa: Pro. of SPIE **4454,** 147 (2001)
17. S. K. Ajmera, A. J. Syllaios, G. S. Tyber, M. F. Taylor and R. E. Hollingsworth: Proc. of SPIE **7660,** 766012 (2010).
18. M. Stuzmann, D. K. Biegelsen and R. A. Street: Phys. Rev. B **35,** 5566 (1987).
19. S. Hasegawa, T. Kasajima and T. Shimizu: Philos. Mag. B **43,** 149 (1981).

Mater. Res. Soc. Symp. Proc. Vol. 1617 © 2013 Materials Research Society
DOI: 10.1557/opl.2013.1165

Effect of magnetic field on the formation of macroporous silicon: structural and optical properties

E. E. Antunez[1], J. O. Estevez[2], J. Campos[3], M. A. Basurto[1], V. Agarwal[1*]
[1]Centro de Investigación en Ingeniería y Ciencias Aplicadas, UAEM, Av. Universidad 1001, Col. Chamilpa, Cuernavaca, Morelos, CP 62210, México.
[2]Instituto de Física, B. Universidad Autónoma de Puebla, A.P. J-48, Puebla 72570, México.
[3]Instituto de Energías Renovables, UNAM, Priv. Xochicalco S/N, Temixco, Morelos 62580, México.

ABSTRACT

The conventional method to fabricate porous silicon with n-type substrates requires light assisted generation of holes used in the electrochemical reaction. Recently, two different methods have been proposed to fabricate some similar structures: Hall effect [1] and lateral electrical field [2]. Hall effect assisted etching involves the application of a perpendicular electric and magnetic field to achieve the concentration of holes at the HF/silicon interface to assist the electrochemical reaction, while the other involves the application of a lateral electrical field across the silicon wafer. In this work, the electrochemical etching of high resistivity n-type silicon wafers under the combined effect of magnetic and lateral electrical field to produce photoluminescent macroporous structures under dark conditions, is reported. A lateral gradient in pore sizes as well as in light emission is observed. Optical and structural properties were studied for their possible applications as a biosensor.

INTRODUCTION

Electrochemical anodisation is the most commonly used technique for the fabrication of porous silicon (PSi). It is well known that in order to carry out an anodic oxidation at the HF/silicon interface and consequently PSi formation, the presence of holes is required to assist the electrochemical reaction. For p-type Si, the majority charge carriers are holes, and etching process is not limited by their availability, so p-type porous silicon (p-PSi) layers are easily produced. On the other hand, light assisted anodisation (where illumination of the wafer helps in the generation of the required holes) is a conventional technique for obtaining PSi from n-type silicon wafers [3-4]. However, light assisted etching is depth limited. Recently, Lin et al.[1] proposed a new method for fabricating photoluminescent structures in n-type Si under dark conditions (without light assistance), using Hall effect which involves the application of perpendicular electric and magnetic fields, to drive holes to the HF/silicon interface. Li et al. [2], on the other hand, reported macropore formation by applying a lateral electrical field, which results in a current flow across the substrate, producing photoluminescent n-PSi in the dark.

However, it is well known that PSi has a large range of morphologies. The macroporous silicon (MPSi) morphology is strongly dependent on the doping type level of the Si substrate and the electrochemical etching parameters [5]. One of the main applications of PSi based optical devices is biosensing, and PSi as a biomaterial has been reported in the literature [6-9]. In particular, results have shown that graded PSi films can promote cell adhesion, viability and act as sensor when a particular object is infiltrated inside its pores [7]. Nevertheless, one of the limitations of the PSi structures aforementioned is that only objects smaller than the pore size can be detected which suggest that pores of higher dimensions are needed for biosensing. Conventionally, macropore formation in low doped n-type Si are usually formed using light assisted etching.

In this research, we report the fabrication of luminescent MPSi structures in n-type Si under dark conditions. We propose the use of the combined effect of magnetic and lateral electrical field during the etching process as an alternative for fabricating optically active MPSi structures, with a lateral gradient, for possible biosensing applications.

EXPERIMENTAL DETAILS

The MPSi samples were electrochemically fabricated using low doped n-type <100> silicon substrate with a resistivity of 8–12 Ω cm. A mixture of 48 wt. % aqueous HF and 99.7 vol. % ethanol in volumetric ratio of 1:4 was used as an electrolyte to perform the etching at room temperature over a period of time of 10 minutes for all the samples.

Two different experimental setups (i.e. with and without magnetic field) were used to fabricate the MPSi samples under dark conditions. Figure 1 schematically describes these two configurations. In setup A a lateral potential field (V_x) is applied across the silicon wafer in x-direction (I_x). The applied potential as reported in [2], is believed to draw the electrons away from the substrate's surfaces, thereby increasing the holes concentration at the HF/silicon interface allowing etching to proceed. Whereas, in the setup B the same lateral potential is applied and in addition a magnetic field is placed perpendicular to the V_x in the y-direction. A Lorentz force will be experienced by the charge carriers flowing in the x-direction and consequently the electrons will be swept down by this force (F_z). By applying a sufficiently large bias-voltage and a magnetic field, the upper layers of the n-type silicon wafer are inverted from an n-type to a p-type semiconductor [1] (see inset in setup B). This combined effect should supply enough holes in the upper layers to participate in the chemical reaction during the etching process. Thus, the outcome will be a sample with a surface displaying a lateral gradient in pore sizes across the electrical field direction.

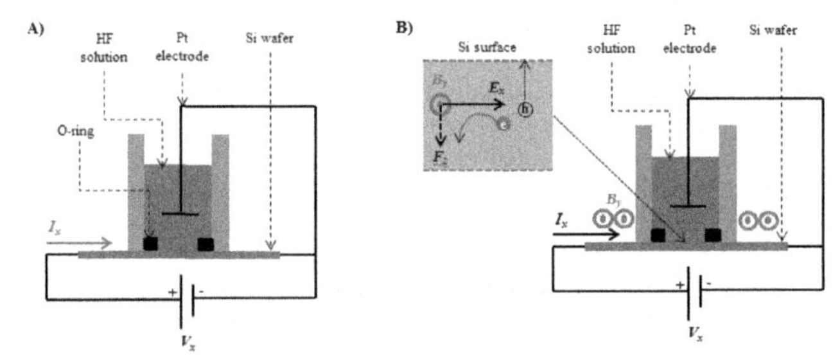

Figure 1. Schematic diagrams of the two different experimental setups used to etch the n-type Si. The difference between setup A and setup B is the magnetic field placed perpendicular to the applied electrical field in the former.

The morphologies of the etched pores were observed with a field-emission scanning electron microscope. Top-view micrographs show the (100) planes, and the micrographs with cross-section, after cleaving the samples, show the pores in the (110) cleavage planes. An excitation wavelength of 250 nm was used for photoluminescence (PL) measurements.

RESULTS AND DISCUSSION

Top-view of MPSi sample formed in the presence of a lateral potential of 50 V fabricated with setup A are shown in figure 2. The distribution of the pores at positive and negative terminal (figures 2(a) and (b), respectively) exposes a higher density of pores at the positive terminal. It is attributed to the high lateral potential which acts in such a way that the majority charge carriers (e^-) are swept down from the sample surface, giving rise to the accumulation of holes required for the etching reaction. As this effect is expected to be stronger at the positive terminal of the sample, a lateral increase in MPSi formation at the positive end of the specimen is observed. A magnified view shown in the inset of figure 2(b) reveals a quasi-quadratic pore morphology with a pore diameter of approximately 1.5 μm on the entire sample.

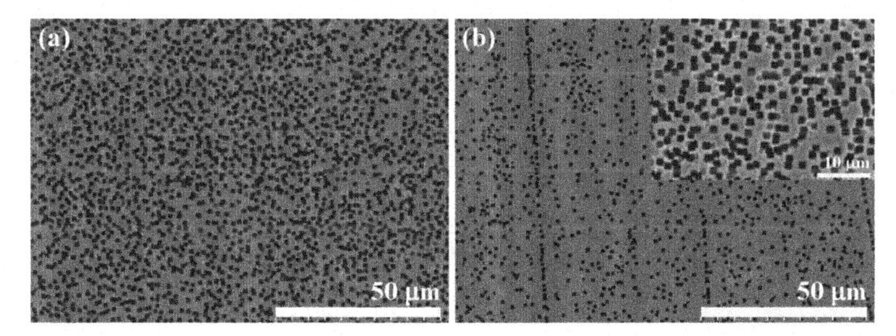

Figure 2. SEM micrographs (top view) of n-type Si wafer etched by applying a lateral potential of 50 V. Image taken close to the (a) positive and (b) negative terminal of the sample. Magnified view revealing the quasi-quadratic morphology of the pores is shown in the inset.

Cross-sectional view of the MPSi sample, prepared using setup A (figure 3), reveals the continuity of the pore propagation from the top to the bottom of the porous layer as well as the microstructure present on the pore walls. Similar to the surface image, higher pore density (figure 3(a)) is observed at the positive terminal. Distance between the pores is found to increase towards the negative contact of the sample (figure 3(b)). Gradient in terms of MPSi film thickness is observed i.e. the layer is thicker near the positive terminal (~54 μm) compared to the thickness of the layer towards the negative contact (~48 μm). Moreover, needle-like trenches are observed at the tip of the pores. Apart from that, a regular wave-shape pattern is formed on the walls of the pores which seems to increase the roughness of the pore walls as shown in a magnified cross-section view of the pores in the inset of figure 3(b).

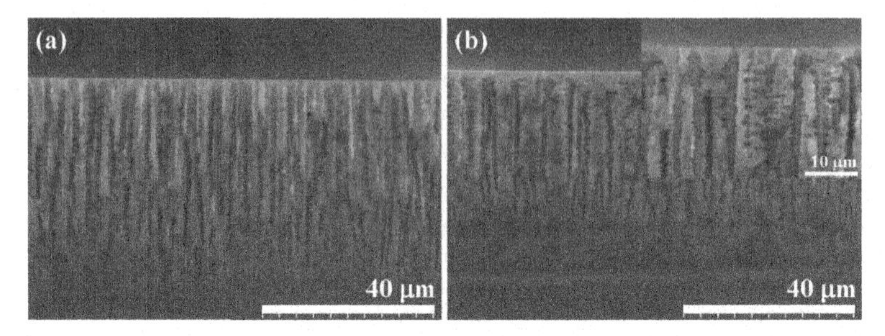

Figure 3. Cross-section SEM micrographs of n-type Si biased at 50 V using setup A in dark conditions. (a) Positive and (b) negative terminal across the electrical field direction of the etched sample. Magnified view of cross-section of the pores is shown in the inset.

Figure 4 shows top-view of MPSi formed by etching the sample in the presence of a lateral potential of 50 V and a perpendicularly applied magnetic field of 30 mT using setup B. It can be observed that pore dimension and morphology varies across the electrical field direction, being smaller and circular-shaped at the negative terminal, and larger and quasi-quadratic shaped pores towards the positive terminal of the MPSi sample (figure 4(a) and (b), respectively). Pore diameter is found to be approximately in the range of 0.6 -1 μm from the negative to the positive terminal. Moreover, some pore nucleation sites are also observed on the surface of the sample.

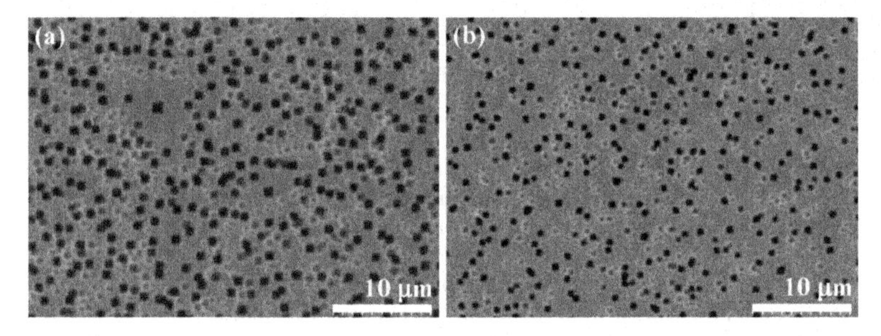

Figure 4. Top-view SEM micrographs of n-type Si sample etched by applying a lateral potential of 50 V and a magnetic field density of 30 mT. Density, size and morphology of the pores close to the: (a) positive and (b) negative terminal of the sample.

Figure 5 shows the SEM micrographs corresponding to the cross-section of n-type Si sample etched using setup B. The sample is biased at 50 V and placed in a 30 mT magnetic field, perpendicular to the electrical field direction. From the sequence of images shown in figure 5, smoother pores are obtained near the positive terminal of the sample (figure 5(a)), the pore walls does not show a marked trend to be branched compared to sample formed using setup A. Across the electrical field direction we observed an increase of the roughness in the pore walls and a more marked tendency to needle-like pore tips as well as a decrease of the density of pores.

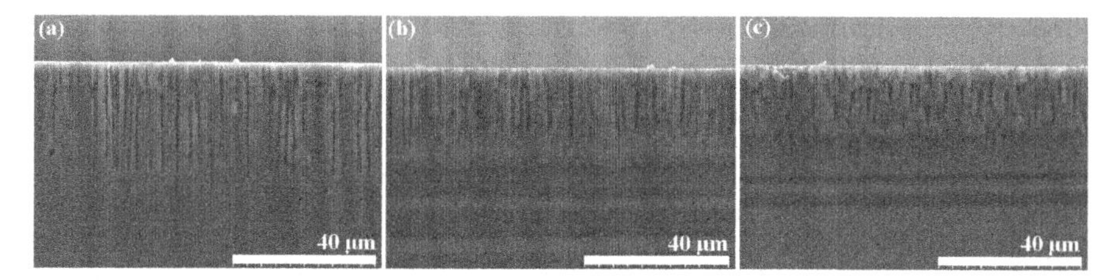

Figure 5. SEM micrographs revealing the cross sectional view of n-type MPSi formed using setup B (i.e. electric potential of 50V and a magnetic field of 30 mT). Morphology studied across the applied electric field direction. (a) Positive, (b) center and (c) negative regions of the etched MPSi sample.

Photoluminescence (PL) spectra at different locations of samples formed with the two setups are presented in figure 6. N-type MPSi sample fabricated with setup A shows the same PL emission peak at approximately 580 nm for the three different locations accompanied by a decrease in the intensity response, whilst for sample formed using setup B shows a redshift in the PL. Moreover, the PL response is higher in sample fabricated with setup B. A redshift can be attributed to the difference in the microstructure inside the MPSi structure.

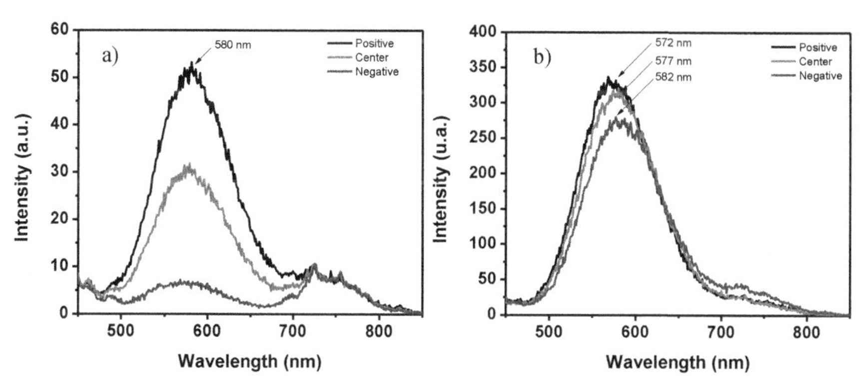

Figure 6. PL spectra collected at different locations on n-type MPSi sample etched using (a)setup A and (b) setup B.

CONCLUSION

Under the effect of high lateral potential across the n-type silicon wafer, the formation of a quasi-quadratic macropore has been demonstrated. With the application of only lateral electric field (setup A), approximately 1.5 μm wide pores are formed on the entire sample with the higher pore density towards the positive terminal of the fabricated sample. On the other hand, using setup B and varying the magnitude of the lateral potential and magnetic field opens the possibility to tune the pore sizes (ranging from approximately 0.6 to 1 μm) and structural morphology (from circular to square pores). Apart from the formation of a well defined pore tip with no tapering end, perpendicular magnetic field enhanced the presence of holes at the positive end of the sample, allowing the formation of perfectly straight macropores at this location with an average pore dimension of 1 μm. However, PL emission response is found to be higher for samples fabricated under the magnetic field assisted etching than for the specimens formed using just lateral potential.

REFERENCES

1. J. C. Lin, P. W. Lee and W. C. Tsai, *J. Appl. Phys. Lett.* **89**, 121119 (2006).
2. S. Q. Li, T. L. Sudesh, L. Wijesinghe and D. J. Blackwood, *J. Adv. Mater.* **20**, 3165-3168 (2008).
3. A. G. Cullis, L. T. Canham and P. D. J. Calcott, *J. Appl. Phys.* **82**, 909 (1997).
4. N. Koshida and H. Koyama, *Jpn. J. Appl. Phys. Part 2* **30**, 1221 (1991).
5. H. Ouyang, M. Christophersen, and P. M. Fauchet, *Phys. Stat. Sol. (A)* **202**, 8, 1396–1401 (2005).
6. B. E. Collins, K. P. S. Dancil, G. Abbi and M. J. Sailor, A*dv. Funct. Mater.* **12**, 187 (2002).
7. Y. L. Khung, G. Barritt and N. H. Voelcker, *Exp. Cell Res.* **314**, 789 (2008).
8. L. R. Clements, P.-Y. Wang, F. Harding, W.-B. Tsai, H. Thissen and N. H. Voelcker, *Phys. Status Solidi A* **208**, 1440 (2011).
9. P. Y. Wang, L. R. Clements, H. Thissen, A. Jane , W. B. Tsai and N. H. Voelcker, *Adv. Funct. Mater.* **22**, 3414–3423 (2012).

Mater. Res. Soc. Symp. Proc. Vol. 1617 © 2013 Materials Research Society
DOI: 10.1557/opl.2013.1166

Memory effect in nanostructured Si-rich hafnia films

L. Khomenkova[1], X. Portier[2], F. Gourbilleau[2], A.Slaoui[3]

[1]V. Lashkaryov Institute of Semiconductor Physics, 45 Pr. Nauky, Kyiv 03028, Ukraine
[2]CIMAP, CEA/CNRS/ENSICAEN/UCBN, 6 Blvd. Maréchal Juin, 14050 Caen cedex 4, France
[3]ICube, 23 rue du Loess, BP 20 CR, 67037 Strasbourg Cedex 2, France

ABSTRACT

Microstructral and charge-trap properties of single Hf-silicate dielectric films are presented versus annealing treatment. The as-grown films were found to be homogeneous and amorphous. It is shown that annealing treatment results in the formation of alternated Hf-rich and Si-rich layers. The mechanism responsible for this phenomenon is found to be surface directed spinodal decomposition. The increase of annealing temperature up to 1000-1100°C resulted in the crystallization of Hf-rich phase. The stability of its tetragonal phase caused an enhancement of film permittivity was observed. The evolution of charge trapping properties of the films results in the memory effect which nature was discussed.

INTRODUCTION

Non-volatile memory devices (NVM) with discrete charge-trapping layers have drawn much attention due to their wide-spread applications and compatibility with standard CMOS processing [1,2]. Among NVM structures, silicon oxide layers with embedded semiconductor or metallic nanoclusters (NCs) have been widely investigated [2-7]. These structures demonstrated higher scalability, larger memory capacitance and good retention properties, operating under relatively lower voltage than conventional floating gate memories.

High-k dielectrics are considered mostly to be used instead of conventional silicon oxide gate in NVM devices [2,8-10]. However, their charge-trap role is not often addressed. Meanwhile, the high-k layers with embedded discrete dielectric NCs offer promising properties as a charge-storage node in NVM devices. For example, an incorporation of RuO discrete NCs in amorphous ZrHfO layers led to a significant memory effect in comparison with NCs-free ZrHfO layers [11]. Furthermore, more interesting issue is the elaboration and investigation of NVM structures fully based on high-k dielectrics of the same nature. Recently, charge storage properties of NVM prototype fully-based on Zr-silicate with modulated composition were demonstrated in [12]. It was shown that the formation of ZrO_2 NCs in $Zr_{0.8}Si_{0.2}O_4$ layer upon annealing at 800°C allowed the memory effect to be achieved [12]. It is worth to note that Zr-based dielectrics were rather considered then Hf-based materials. However, the isovalent similarity of Hf and Zr ions as well as of their oxides and silicates opens the ways to wider application of Hf-based dielectrics for NVM devices. In this paper we demonstrate NVM properties of metal-insulator-semiconductor (MIS) structures based on single Hf-silicate layers with modulated composition.

EXPERIMENT

Radio frequency (RF) magnetron sputtering technique was used for layers fabrication, whereas an annealing treatment allowed their microstructure and electrical properties to be tuned. The sputtering of 4-inches HfO_2 target (99.9%) topped by Si chips in pure argon plasma was used for the fabrication of thin Hf-silicate ultrathin films with $Hf_{0.2}Si_{0.2}O_{0.6}$ composition. More details can be found in [6,8]. The anneal treatments in a conventional furnace in nitrogen flow at temperatures T_A=800-1100°C during 15 min were applied for tuning film microstructure, optical and electrical properties.

Several techniques were used to analyze the properties of the layers. Attenuated total reflection Fourier transmission infrared (ATR-FTIR) spectra were measured in the range 600-4000 cm^{-1} by means of a 60° Ge Smart Ark accessory inserted in a Nicolet Nexus spectrometer. X-ray diffraction analysis was performed using a Phillips XPERT HPD Pro device with Cu $K_{\alpha 1}$ radiation (λ=0.15406 nm) at a fixed grazing angle incidence of 0.5°. The microstructure of the films was observed by transmission electron microscopy (TEM). Cross-sectional specimens were prepared by the standard procedure involving grinding, dimpling and Ar^+ ion beam thinning until electron transparency. The samples were examined by conventional (CTEM) and high resolution electron microscopy (HRTEM) using a FEG 2010 JEOL instrument, operated at 200 kV. Image processing was done with the commercial Digital micrograph GATAN software. The top and back-side Al contacts with a diameter of 150 μm were deposited by thermal evaporation to fabricated simple MOS capacitors. Capacitance-voltage (C-V) properties were examined at different frequencies using a HP 4192A LF Impedance Analyzer.

RESULTS AND DISCUSSION

ATR-FTIR and XRD study

Among nondestructive methods to study thin films, ATR technique holds an important place, since it allows very thin layers to be analysed. These experiments have been performed on as-deposited and annealed films (Fig. 1). The ATR spectra of as-deposited films demonstrate featureless broad band peaked at 1050 cm^{-1} with a shoulder at about 900 cm^{-1} with a tail till 600 cm^{-1} (Fig. 1). This featureless band is the evidence of the formation of Hf-silicate amorphous matrix. An annealing at T_A=800°C leads to the increase of the intensity as well as to the slight narrowing and to the shift of the broad ATR peak from 1050 cm^{-1} to 1070 cm^{-1}. On the contrary, the lower wavenumber part of the ATR spectrum is featureless (Fig. 1a).

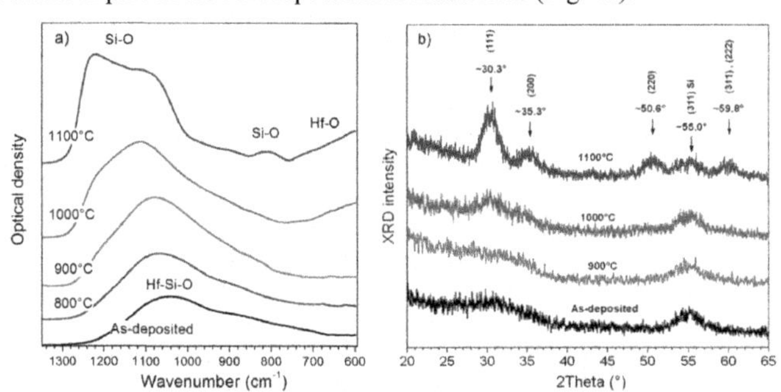

Fig.1. ATR-FTIR spectra (a) and XRD patterns (b) of as-deposited and annealed films.

When T_A=900-1000°C, the continuous increase of the peak intensity, the shift of peak position to 1120 cm^{-1} and the appearance of a shoulder at about 1220 cm^{-1} take place. A treatment at T_A=1100°C causes the narrowing of this vibration band, a redistribution of the intensities of its components and an appearance of the band at about 820 cm^{-1}. All these three bands can be ascribed to the SiO_2 phase [9]. The intensity increase and the peak shift to higher wavenumbers are due to increasing contribution of the SiO_2 phase in the film structure as well as an realignment of Si-O bonds followed the SiO_2 phase formation. The latter is due an interfacial SiO_2 layer formation confirmed by the higher intensity of LO_3 phonon (~1240cm^{-1}) exceeding that of TO_3 phonon (~1080cm^{-1}) (Fig.1a). At the same time, a contribution of TO_3 phonon and TO_4-LO_4 doublet into ATR spectrum allowed assuming the presence of SiO_2 phase inside the film volume due to the phase separation of the HfSiO film on SiO_2 and HfO_2 phases.

The phase separation at T_A=1000°C is confirmed by an increase of the ATR intensity in the range of 780-600 cm^{-1}, corresponding to Hf-O vibrations. As we mentioned above, a broad and featureless band is usually ascribed to amorphous HfO_2 [11,12]. However, it is hardly believable that such a high annealing temperature (1100°C) will lead to an amorphous HfO_2 phase. To get more information about the structure of HfO_2 phase, we performed a XRD study for these samples. The as-deposited films show a broad peak in the range of 2Θ=25-35° with a maximum intensity at about 2Θ=32° (Fig.1b) that is an evidence of the amorphous nature of the films investigated.

This annealing at T_A≤900°C does not lead to the crystallization of the films. At higher T_A, a formation of the HfO_2 phase occurs that is confirmed by the appearance of the peaks at 2Θ=30.3°, 2Θ=35.3° and 2Θ=50.6° (Fig.1b). These peaks can be ascribed to the (111), (200) and (220) crystalline planes of the tetragonal HfO_2 phase. The peak at 2Θ=59.8° is an overlapping of the reflections from (311) and (222) planes of the same phase.

It is worth to note that the absence of any transformation of ATR band in the range 780-600 cm^{-1} upon annealing at T_A=900°C testifies the thermal stability of the HfSiO layers. The most probable reason is the high flexibility of Si-O bonds conserving the amorphous nature of fused silica. However, even if HfSiO can be stable at such annealing temperatures, the transformation of Si-O vibration band can be due to a better silicon oxide structure and/or due to the formation of a SiO_2 interfacial layer. Such a transformation occurs at lower temperatures (less than 900°C) than the phase separation in HfSiO film (1000-1100°C). To clarify this issue, we performed TEM observations of cross-sections of the samples.

TEM investigation

Figures 2 show TEM images of as-deposited and annealed films. All the samples have smooth surfaces with the roughness below than 1 nm. As-deposited layers are amorphous and homogeneous with an interfacial SiO_x layer thinner than 1 nm (Fig.2a). An annealing at higher T_A values leads to the rise of interfacial SiO_2 layer thickness (from 2.3 nm (T_A=900°C) up to 5.0 nm (T_A=1100°C)) as well as to the expansion of the silicate film itself (from 14.5 nm (T_A=900°C) up to 16.1 nm (T_A=1100°C)) (Fig.2,b-d). Although the film annealed at T_A=900°C stays amorphous, the appearance of alternated bright and dark lines illustrates the efficient diffusion processes in the film volume and the formation of HfO_2-rich (darker regions) and SiO_2-rich (brighter regions) multilayer structure (Fig.2b). The line intensity profile clearly shows the formation of this "multilayer" structure (Fig.2b, right-side insert corresponding to selected rectangle area).

Further T_A increase results in the same HfO_2-rich/SiO_2-rich multilayer structure with higher contrast (Figs.1c,d, right-side inserts). Lattice fringes are visible in the darker lines demonstrating the presence of tetragonal HfO_2 nanocrystals in the films (Figs.1(c-d)).

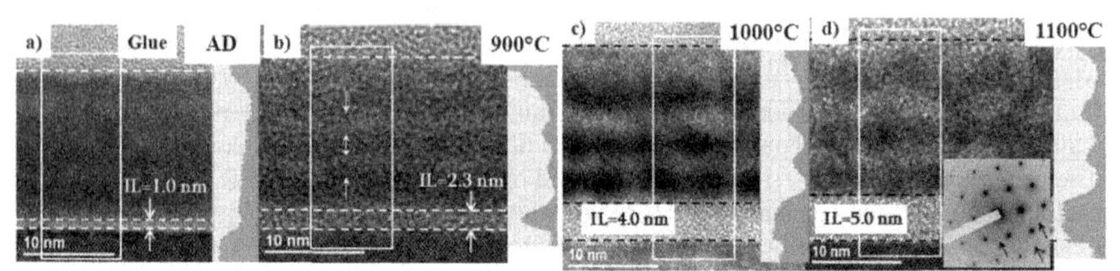

Figure 2. Cross-sectional TEM images of HfSiO layer taken for as-deposited (a), annealed at 900°C (b), 1000°C (c) and 1100°C (d) in nitrogen flow during 10 min. IL means interfacial layer. The right-side inserts in (a-d) show the intensity profiles integrated over the width of rectangles drawn in the bright-field images.

The selected area electron diffraction pattern, taken for the sample annealed at T_A=1100°C, confirms this statement (Fig.1d, inserts). It shows weak rings (marked by the arrows) corresponding to the following *d*-spacings: 2.95 Å, 1.80 Å and 1.53 Å. These values are compatible with lattice planes of the HfO_2 tetragonal phase (JPCD 08-0342): $d_{(111)}$=2.95 Å, $d_{(220)}$=1.82 Å and $d_{(311)}$=1.59 Å. Since any other HfO_2 phase is not revealed, the tetragonal phase is stable. The formation and stability of HfO_2 tetragonal phase in comparison with monoclinic one occurred due to silicon incorporation in the films [8,9,13]. Besides, an enhancement of dielectric constant (k=25) in comparison with that of monoclinic HfO_2 phase (k=16) was also predicted [13]. Thus, the films with stable tetragonal HfO_2 phase can be considered as high-k materials with enhanced dielectric constant that is promising from electrical point of view.

Evolution of electrical properties with annealing treatment

To study the relationship between structure evolutions and the electrical properties of the films, C-V curves were recorded for MIS capacitors fabricated from the same as-deposited and annealed layers. The typical high-frequency (1 MHz) C-V curves demonstrate well-defined accumulation, inversion and depletion regions (Fig. 2). The as-deposited samples show C-V curves, stretch-out along the gate voltage axis due to the presence of interface traps (not shown here). The significant negative shift of flat-band voltage (V_{fb}) down to -4 V indicates the existence of high amount of positive charge in the films. Its introduction can be caused by interface traps and, more probably, by some defects present in the film volume (for instance, oxygen vacancies). An annealing treatment leads to the decrease of the stretch-out effect due to a recovering of slow traps at the film/substrate interface, and, as a consequence, a sharp transition from the depletion to the accumulation is observed (Figs. 2). An appearance of a C-V loop was observed for all annealed films whereas most pronounced memory effect was observed for samples annealed at T_A=1000°C. It increases from ΔV_{fb}=0.3V (V_G=±3V) to ΔV_{fb}=6.6V (V_G=±12V) (Fig. 2a). No significant effect of the frequency on the C-V curves is observed for all annealed samples as it is shown for the film annealed at 1000°C (Fig. 2b). Since interface traps are usually recovered upon high temperature annealing, one can assume that the hysteresis memory window is rather due to the silicate film itself than the traps located at the film/p-Si interface. Meanwhile, the C-V data linked with the microstructure analysis give the evidence of a charge trapping mechanism by defects located inside HfO_2-rich and SiO_2-rich phases, as well as at their interfaces.

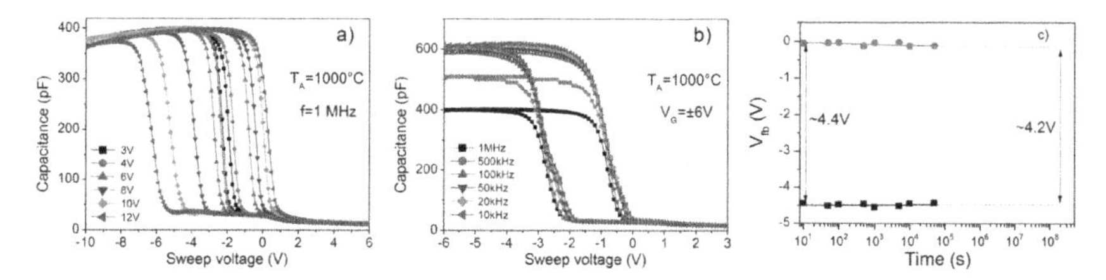

Figure 3. C-V curves measured for the film annealed at 1000°C versus applied sweep voltage (a) and frequency (b). The evolution of ΔVfb versus storage time (c) is measured for f=1MHz and V_G=±6V (c) and interpolated for 10 years threshold.

Moreover, in spite of the increase of the SiO_2 interfacial layer thickness from 2.3 nm (T_A=900°C) up to 4.0 nm (T_A=1000°C), the sharp depletion-accumulation transition confirms the low number of interfacial traps located in the film/substrate interfacial layer (if any). Although the presence of some defects inside this SiO_2 interfacial layer cannot be ruled out, one can assume that the main contribution to the C-V loop originates from the HfO_2-rich phase and the formation of HfO_2 crystalline grains. This is confirmed by the C-V loop obtained for the sample annealed at 1000°C (Fig.2c) and showed that the memory effect is due to trapping and de-trapping of electrons. Usually, the counter-clockwise nature of C-V hysteresis curves indicates the negative charge trapping in the MIS capacitors and it is assigned to charge storage through substrate injection mechanism. When a positive bias voltage is applied, electrons are being injected from the inversion layer of the Si substrate into the gate dielectric matrix. When a negative voltage is applied, electrons are ejected back into the Si substrate (equivalent to hole injection from the deep accumulation layer of the substrate), resulting in a shift of the C-V curve towards negative voltages.

The variation of the memory window with annealing treatment can be explained taking into account the conduction band (CB) offsets of dielectrics with silicon channel. Thus, it is ~1.8 eV (for $HfSiO_4$), ~1.4 eV (pure HfO_2) and ~3.5 eV (pure SiO_2). The decrease of the CB offset due to the formation of HfO_2-rich phase has to stimulate the appearance of memory effect at lower V_G values for annealed films. On the other hand, the formation of SiO_2-rich phase as well as SiO_x interfacial layer would require higher sweep voltages to achieve the memory capacitance. In this regard, higher memory effect is expected to be observed at lower V_G values for the samples annealed at lower T_A values, whereas their counterparts annealed at higher temperatures will demonstrate memory effect at higher V_G values. The carrier trap density versus annealing temperature was estimated from the accumulation regime of C-V curves recorded at V_G=±8V (Fig.2) and found to be in the range (2.5-8.0)$\times10^{12}$ cm^{-2}, the highest value being for T_A=950°C. The k values were also obtained from the analysis of the C-V data. An enhancement of k from 20 (as-deposited film) to 35-40 (film annealed at 1100°C) was observed. Thus, from electrical point of view, an application of the films annealed at high temperatures is preferable due to the formation of tetragonal HfO_2 phase and enhancement of dielectric constant. Furthermore, such films can be also used as a gate dielectric for NVM structures with Si-NCs requiring a processing at 1100°C for Si-NCs formation [5,6]. In the meantime, the films annealed at 950-1000°C can be used as charge storage node in NMV devices due to the formation of discrete HfO_2 NCs leading to a large memory effect. It is worth to note this effect was found to be stable and demonstrated only about 10% narrowing of memory window for interpolated 10 years storage time.

CONCLUSION S

The analysis of the film's properties revealed that the anneal treatments of as-grown homogeneous films lead to the formation of alternated Hf-rich and Si-rich layers due to a spinodal decomposition process. In addition, the crystallization of Hf-rich phase and the formation of stable HfO_2 tetragonal grains occurred at T_A=1000-1100°C. The phase separation results in the enhancement of the permittivity of these films up to 30 due to the formation of tetragonal HfO_2 phase. The analysis of capacitance-voltage curves showed the correlation between film decomposition and their charge storage properties. No significant frequency dependence was observed testifying the low contribution of slow traps. The investigation of retention properties showed the stability of obtained memory effect of the MIS structures with storage time. This work opens perspectives to use single Si-rich-HfO_2 dielectric in the different nodes of NVM devices.

ACKNOWLEDGMENTS

This work was partly supported by the French National Research Agency (ANR) via NOMAD Project n°.ANR-07-NANO-022-02 and, for one of the authors (L.K.) by the Basse Normandie Region through the CPER project - Nanoscience axe (2007-2013).

REFERENCES

1. Wilk G.D., Wallace R.M., Anthony J.M., J. Appl. Phys. **89,** 5243 (2001).
2. He G., Zhu L.Q., Sun Z.Q., Wan Q., Zhang L.D., Progress in Materials Science **56** , 475 (2011).
3. Tiwari S., Rana F., Hanafi H., Hartstein A., Crabbe E.F., Chan K., Appl. Phys. Lett. **68,** 1377 (1996).
4. Lee C.H., Meeter J., Narayanan V., Kan E.C., J. Electron. Mater. **34**, 1 (2005).
5. Perego M., Seguini G., Wiemer C., Fanciulli M., Coulon P.-E., Bonafos C., Nanotechnology **21**, 055606 (2010).
6. Khomenkova L., Sahu B.S., Slaoui A., Gourbilleau F., Nanoscale Research Letters **6**, 172 (2011).
7. Lu T.Z., Alexe M., Scholz R., Appl. Phys. Lett. **87**, 202110 (2005).
8. Khomenkova L., Dufour C., Coulon P.-E., Bonafos C., Gourbilleau F., Nanotechnology **21**, 095704 (2010).
9. Khomenkova L., Portier X., Cardin J., Gourbilleau F., Nanotechnology **21,** 285707 (2010).
10. Lui J., Wu X., Lennard W.N., Landheer D., Dharma-Wardana M.W.C., J. Appl. Phys.**107,** 123510 (2010).
11. Lin C.-H., Keo Y., J. Appl. Phys. **110**, 024101 (2011).
12. LV Sh.-C., Ge Zh.-Y., Zhou Y., Xu B., Gao L.-G., Yin J., Xia Y.-D., Liu Zh.-G., Chin. Phys. Lett. **27**, 068502 (2010).
13. Fischer D., Kersch A., Appl. Phys. Lett. **92**, 012908 (2008).

Mater. Res. Soc. Symp. Proc. Vol. 1617 © 2013 Materials Research Society
DOI: 10.1557/opl.2013.1167

Interrelation between Light Emitting and Structural Properties of Si Nanoclusters Embedded in SiO_2 and Al_2O_3 Hosts

L. Khomenkova[1], O. Kolomys[1], V. Strelchuk[1], A. Kuchuk[1], V. Kladko[1], M. Baran[1], J. Jedrzejewski[3], I.Balberg[3], P. Marie[2], F. Gourbilleau[2], N. Korsunska[1]

[1]V. Lashkaryov Institute of Semiconductor Physics, 45 Pr. Nauky, Kyiv 03028, Ukraine
[2]CIMAP/ENSICAEN, 6 Blvd. Maréchal Juin, 14050 Caen cedex 4, France
[3]Racah Institute of Physics, Hebrew University, 91904 Jerusalem, Israel

ABSTRACT

The present work deals with the comparative investigation of Si-ncs embedded in SiO_2 and Al_2O_3 dielectrics grown by RF magnetron sputtering on fused quarts substrate. The effect of post-deposition processing on the evolution of microstructure of the films and their optic and luminescent properties was investigated. It was observed that photoluminescence (PL) spectra of $Si_x(SiO_2)_{1-x}$ films showed one PL band, which peak position shifts from 860 nm to 700 nm when the x decreases from 0.7 to 0.3. It is due to exciton recombination in Si-ncs. For $Si_x(Al_2O_3)_{1-x}$ films, several PL bands peaked at about 570-600 nm and 700-750 nm and near-infrared tail or band peaked at about 800 nm were found. Two first PL bands were ascribed to different oxygen-deficient defects of oxide host, whereas near-infrared PL component is due to exciton recombination in Si-ncs. The comparison of both types of the samples showed that the main radiative recombination channel in $Si_x(SiO_2)_{1-x}$ films is exciton recombination in Si-ncs, while in $Si_x(Al_2O_3)_{1-x}$ films the recombination via defects prevails due to higher amount of interface defects in the $Si_x(Al_2O_3)_{1-x}$ caused by stresses.

INTRODUCTION

The realization of low-cost integrated optoelectronic devices fully based on well-developed Si-based CMOS technology (i.e. all-in-one Si chip) is important task. In this regard, silicon nanocrystallites (Si-ncs) attract considerable interest due to significant transformation of their optical and electrical properties caused by quantum-confinement effect [1,2]. Being embedded in dielectric hosts, Si-ncs offer potential applications in optoelectronic devices that were demonstrated during last decades for Si-ncs-SiO_2 systems [3-6].

However, the downscaling of microelectronic devices requires the elaboration of novel materials to overcome bottleneck of silicon oxide as a gate material. In this regard, other dielectrics such as ZrO_2, HfO_2 and Al_2O_3 are considered as promising gate dielectrics [7]. It was also demonstrated that Si-ncs embedded in such high-k host offer a wider application for non-volatile memories due to the higher performance of the corresponding devices [8,9].

Among different dielectrics, Al_2O_3 is not well addressed as photonic material. At that, it has relatively higher refractive index (1.73 at 1.95 eV) in comparison with that of SiO_2 (1.46 at 1.95 eV) at similar band gap energies offering better light confinement, making compact device structures possible. It was shown an application of alumina-based waveguides fabricated by sol-gel techniques for optical communication. Few reports on Si-ncs-Al_2O_3 materials fabricated by ion implantation or electron beam evaporation are also available [10-12]. At the same time, magnetron sputtering was not often considered for fabrication of Al_2O_3 materials with embedded

Si-ncs [13-15] in spite of the relative simplicity of this approach and its wide application for the fabrication of Si-ncs-SiO_2 films [5,9].

The present paper demonstrates the application of magnetron sputtering for the fabrication of Si-rich-Al_2O_3 and Si-rich-SiO_2 films with different Si content. The effect of post-deposition processing on the evolution of microstructure of the films and their optic and luminescent properties was observed.

EXPERIMENT

The $Si_x(Al_2O_3)_{1-x}$ and $Si_x(SiO_2)_{1-x}$ films with $0.15 \le x \le 0.7$ were deposited by radio frequency magnetron co-sputtering of two spaced-apart targets (pure Si and pure oxide (Al_2O_3 or SiO_2)) in pure argon plasma on a long silicon oxide substrate at room temperature. More details can be found elsewhere [15,16]. The as-deposited original films were annealed at 1150°C during 30 min in nitrogen flow to form the Si-ncs in oxide hosts.

To investigate the microstructure and luminescent properties of the films, a Horiba Jobin-Yvon T-64000 Raman spectrometer equipped with confocal microscope and automated piezo-driven XYZ stage was used. The Raman scattering and photoluminescence spectra were detected in 100-900-cm^{-1} and in 500-900-nm spectral ranges, respectively. A 488.0-nm line of Ar-Kr ion laser was used as the excitation source. The laser power on the sample surface was always kept below 5 mW to obtain the best signal-to-noise ratio, preventing a laser heating of the investigated sample. The spectral resolution of the spectrometer was less than 0.15 cm^{-1}. To study the chemical composition of the films (x), their refractive index and thickness, the spectroscopic ellipsometry measurement was performed by means of a Jobin-Yvon ellipsometer (UVISEL), where the incident light was scanned in the range 1.5-4.5 eV under an incident angle of 66.3°. [7,13]. The electron paramagnetic resonance (EPR) spectra were measured by means Varian-12 spectrometer to obtain the information about the defect structure of the samples. The investigations were performed at 300 K.

RESULTS

Raman scattering study of as-deposited $Si_x(Al_2O_3)_{1-x}$ and $Si_x(SiO_2)_{1-x}$ films with the $x \ge 0.35$ revealed the presence of amorphous silicon (a-Si) phase (Fig.1a). Besides, the shift of peak position of the transverse optic (TO) band to $\omega_{TO\text{-}a\text{-}Si}$=460$cm^{-1}$ was observed for $Si_x(Al_2O_3)_{1-x}$ films contrary to that detected for $Si_x(SiO_2)_{1-x}$ counterparts ($\omega_{TO\text{-}a\text{-}Si}$=480$cm^{-1}$). This latter corresponds to the TO phonon peak position of relaxed amorphous silicon. The low-frequency shift observed for $Si_x(Al_2O_3)_{1-x}$ samples can be ascribed to tensile stresses between the film and fused quarts substrate due to mismatching between lattice parameters of fused quarts and the film. It is obvious that this effect is negligible for the $Si_x(SiO_2)_{1-x}$ films.

Annealing treatment at T_A=1150°C results in the increase of TO phonon band intensity and its narrowing that is the evidence of Si-ncs formation in both types of the samples (Fig.1b). When the x decreases, the shift of the $\omega_{TO\text{-}nc\text{-}Si}$ to the lower wavenumbers occurs for $Si_x(SiO_2)_{1-x}$ (Fig.1b, inset) that can be ascribed to the decrease of Si-ncs sizes.

In all $Si_x(Al_2O_3)_{1-x}$ samples, the $\omega_{TO\text{-}nc\text{-}Si}$ is shifted to lower wavenumbers (517.3-518.7 cm^{-1}) in comparison with the peak position of TO phonon band of bulk Si ($\omega_{TO\text{-}bulk\text{-}Si}$=521 cm^{-1}). But contrary to $Si_x(SiO_2)_{1-x}$ films, for $Si_x(Al_2O_3)_{1-x}$ samples with x=0.55-0.7 only slight shift of the ω_{TO} towards the higher wavenumbers is detected with x decrease (Fig.1b, inset). It is worth to note that along with Si crystalline phase, the amorphous Si phase was also detected in annealed samples. However, for the samples with the same x values its contribution is lower for the $Si_x(Al_2O_3)_{1-x}$ samples than for $Si_x(SiO_2)_{1-x}$ counterparts.

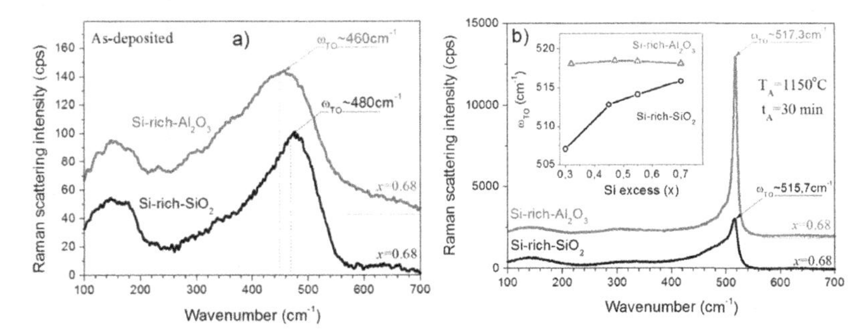

Fig.1. Raman scattering spectra as-deposited (a) and annealed (b) $Si_x(SiO_2)_{1-x}$ and $Si_x(Al_2O_3)_{1-x}$ films with x=0.68. The inset in Fig.1b shows variation of TO phonon peak position versus x for both types of samples.

The presence of amorphous Si phase in as-deposited samples was also revealed by EPR measurements. They showed for both types' samples with $x \geq 0.35$ a significant contribution of the signal with g=2.0055. This latter corresponds to the silicon dangling bonds. Its intensity reflects the total number of these centers and decreases with x decrease. Annealing treatment results in the transformation of EPR spectra of both types of samples.

For $Si_x(SiO_2)_{1-x}$ films with $x>0.35$, an asymmetric signal with g=2.0057, which intensity increase with x, was found. Besides, for the layers with $x>0.55$ slight dependence of EPR spectra on the orientation of magnetic field was detected. Since annealed $Si_x(SiO_2)_{1-x}$ films contain both Si-ncs and amorphous Si phase, this signal is obviously a superposition of two signals, i.e. P_b-like centers that are the feature of Si/SiO_2 interface formation, and Si dangling bonds. Slight anisotropy of EPR signal and the increase of the g-factor value are additional arguments for the formation of P_b-like centers. Annealed $Si_x(Al_2O_3)_{1-x}$ samples showed also the signal with g_1=2.0057 attributed to the superposition of Si dangling bonds and P_b-like centers that can be the feature of both Si/SiO_2 and Si/Al_2O_3 interfaces [16].

Any emission was not observed for as-deposited $Si_x(SiO_2)_{1-x}$ films, whereas weak PL emission in orange spectral range was detected from $Si_x(Al_2O_3)_{1-x}$ films with $x<0.5$. Similar PL emission was also observed in pure Al_2O_3 film and can be assigned to F_2^{2+} centers in Al_2O_3 [17]. Annealing of $Si_x(SiO_2)_{1-x}$ films results in the appearance of one broad PL band in red-near-infrared spectral range (Fig.2a). Its peak position shifts from 1.4 eV to 1.8 eV when the x decreases from 0.45 to 0.3 and does not change for $x>0.5$ (Fig.2a, inset). Annealed $Si_x(Al_2O_3)_{1-x}$ films demonstrate the PL spectrum in wider spectral range (Fig.2b). These spectra contain two broad PL bands with maxima at 2.06-2.18 eV and 1.65-1.77 eV accompanied by near-infrared tail or weak band (1.55-1.60 eV). These bands can be well-separated (for x=0.45-0.5) or strongly overlapped. The first band consists of two components with maxima positions at ~2.06 eV and ~2.18 eV. The latter one is clearly seen in the sample with x=0.3 and is similar to PL emission from F_2^{2+} centers in Al_2O_3 [18]. Furthermore, this PL band presents in other spectra also, testifying that the Si-ncs are incorporated into Al_2O_3 matrix. At the same time, both components are strongly overlapped in the samples with x>0.3 (Fig.2).

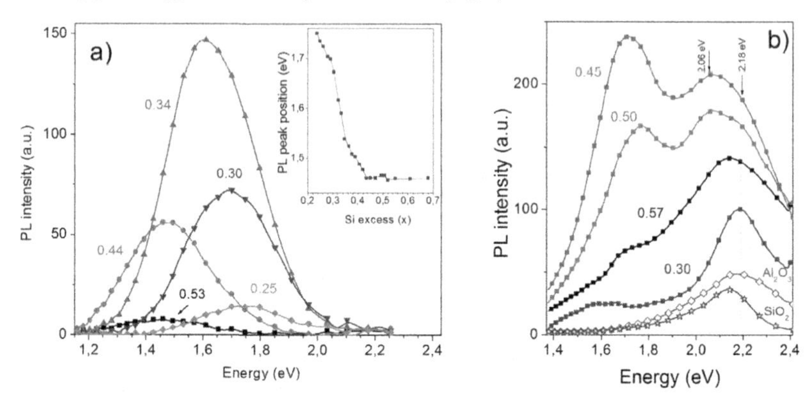

Figure 2. Room-temperature PL spectra of $Si_x(SiO_2)_{1-x}$ (a) and $Si_x(Al_2O_3)_{1-x}$ (b) films. The values of the x are mentioned in the figures. Excitation wavelength is 488 nm.

DISCUSSION

The investigation of structural properties of as-deposited $Si_x(SiO_2)_{1-x}$ and $Si_x(Al_2O_3)_{1-x}$ films showed that one of their specific features is the presence of amorphous Si phase for the samples with x>0.35. Besides, the $Si_x(Al_2O_3)_{1-x}$ films are stressed contrary to $Si_x(SiO_2)_{1-x}$ ones. These stresses are tensile and caused by the mismatching in the lattice constants of fused quarts substrate and the film. Raman scattering data shows also that after annealing treatment the Si-ncs in $Si_x(Al_2O_3)_{1-x}$ samples are stressed. In fact, the peak position of TO phonon band of the Si-ncs for the samples with x>0.5 is shifted to the lower frequency side ($\omega_{TO\text{-}nc\text{-}Si}$=517-518 cm^{-1}) in comparison with that for bulk Si ($\omega_{TO\text{-}bulk\text{-}Si}$=521 cm^{-1}). At the same time, the mean size of Si-ncs, estimated from XRD data, is about 14 nm [15]. It is obvious that the contribution of phonon quantum confinement effect is negligible in this case. This means that the Si-ncs in the $Si_x(Al_2O_3)_{1-x}$ samples are under tensile stress contrary to the Si-ncs in the $Si_x(SiO_2)_{1-x}$ films. This is in agreement with Raman scattering data obtained for as-deposited samples.

As it was mentioned the peak position of Raman band of the Si-ncs for $Si_x(Al_2O_3)_{1-x}$ in the samples with x>0.55 shifts slightly to high frequency side with the x decrease that cannot be caused by the change of crystallite sizes because the decrease of Si content should result in the decrease of Si crystallites and lead to opposite shift of Raman line. The observed shift is obviously caused by the decrease of amorphous Si phase content that is in agreement with the decrease of intensity of TA phonon of amorphous Si ($\omega_{TA\text{-}a\text{-}Si}$=150 cm^{-1}). Thus, the sizes of Si-ncs in $Si_x(Al_2O_3)_{1-x}$ films cannot be estimated from Raman data.

Another situation occurs in the $Si_x(SiO_2)_{1-x}$ films. With the x decrease the shift of the $\omega_{TO\text{-}nc\text{-}Si}$ to the lower wavenumbers occurs (Fig.1b, inset). Besides, the increase of full-width at half maximum of this phonon band is observed (not shown). The sizes of Si-ncs embedded in SiO_2 host can be estimated from the fitting of Raman scattering spectra. Based on such analysis the increase of Si-ncs mean size from ~2.7 nm to 6.0 nm was found for $Si_x(SiO_2)_{1-x}$ samples when the x increases from 0.3 to 0.5, whereas for $x>0.5$, Si-ncs size does not change practically. These results are in a good agreement with XRD data obtained earlier.

Present results showed also that mean size of Si-ncs in Al_2O_3 exceeds that for Si-ncs in SiO_2 for the films with same x values. One of the reasons of this phenomenon can be faster diffusion of Si in alumina than that in silica in the case when Si-ncs formation is determined by Si diffusion towards Si-nuclei and their Ostwald ripening. Another reason can be lower temperature required for phase separation in $Si_x(Al_2O_3)_{1-x}$ than that for $Si_x(SiO_2)_{1-x}$. In spite of the difference in Si-ncs sizes these films have one trait in common. For the samples with $x>0.5$ the mean Si-ncs sizes do not change with x. This can be connected with presence of amorphous Si inclusions in as-deposited films. In this case their crystallization can contribute to appearance of Si-nc besides the process of phase separation. For $x>0.5$ this contribution can be crucial. If these inclusions are big enough (that can be expected for high Si excess) the crystallite sizes will be determined by the temperature and duration of annealing only as for amorphous Si films crystallization. Indeed, rapid thermal annealing of $Si_x(Al_2O_3)_{1-x}$ samples results in the formation of smaller Si-ncs, but their mean size was found to be independent on x for $x>0.5$ [13,14].

Raman scattering spectra of annealed films showed also the higher relative contribution of amorphous Si phase in $Si_x(SiO_2)_{1-x}$ than in $Si_x(Al_2O_3)_{1-x}$. This can be due to lower temperature for the crystallization of amorphous Si clusters in Al_2O_3 host compared with that in SiO_2 and is in the agreement with the data of Ref. [10].

Significant difference in PL properties of $Si_x(SiO_2)_{1-x}$ and $Si_x(Al_2O_3)_{1-x}$ films was observed. For $Si_x(SiO_2)_{1-x}$ films, evolution of PL peak position versus x correlates with the variation of Si-ncs mean size. This allows ascribing it to exciton recombination in Si-ncs. Thus, in these films exciton recombination in Si-ncs is dominant radiative channel.

At the same time several radiative channels are observed in $Si_x(Al_2O_3)_{1-x}$ films. The investigation of temperature behavior of PL spectra showed that the peak positions and the intensities of PL components peaked at 2.06-2.15 eV and 1.65-1.77eV do not change with cooling [13]. This allowed ascribing them to radiative recombination of carriers through the defects of matrix and/or Si-ncs/host interface states. They are F-like centers in Al_2O_3 emitted at ~2.06 eV [19] and ~2.18 eV [18]. It is worth to note that PL components at ~1.65-1.77 eV and ~2.06 eV were observed only when Si-ncs are present in the film. This can be explained by their location near Si-ncs or at Si-ncs/host interface.

The contribution of near-infrared tail or band peaked at about 1.55 eV increases with cooling [13,14] that is typical feature of Si-ncs exciton PL band. However, its PL intensity is much lower than the emission of oxide-related defects contrary to that observed in $Si_x(SiO_2)_{1-x}$ films. This can be due to high number of non-radiative defects at Si-ncs/Al_2O_3 interface which, in particular, can appear due to mechanical stress in $Si_x(Al_2O_3)_{1-x}$ films.

CONCLUSIONS

RF magnetron sputtering approach was used for deposition of $Si_x(SiO_2)_{1-x}$ and $Si_x(Al_2O_3)_{1-x}$. The investigation of structural and light emitting properties of these films allowed getting the information about the Si-ncs formation and the nature of the emitting centers in the films with different Si content. Comparative analysis of PL spectra of both types' samples showed that the main contribution to PL spectra of $Si_x(SiO_2)_{1-x}$ films is given by exciton recombination in the Si-ncs whereas PL emission of $Si_x(Al_2O_3)_{1-x}$ films is caused mainly by carrier recombination either via defects in matrix or via electron states at the Si-ncs/matrix interface.

ACKNOWLEDGMENTS

This work was supported by the National Academy of Sciences of Ukraine, National Center of Scientific Research (CNRS) of France and Ministry of Art and Science of Israel.

REFERENCES

1. L. T. Canham, *Appl. Phys. Lett.* **57**, 1046 (1990).
2. V. Lehman, U. Gosele, *Appl. Phys. Lett.* **58**, 856 (1991).
3. X. Y. Chen, Y. F. Lu, L. J. Tang, Y. H. Wu, B. J. Cho, X. J. Xu, J. R. Dong, W. D. Song, *J. Appl. Phys.* **97**, 014913 (2005).
4. L. Khomenkova, N. Korsunska, V. Yukhimchuk, B. Jumaev, T. Torchynska, A. Vivas Hernandez, A. Many, Y. Goldstein, E. Savir, J. Jedrzejewski, *J. Lumin* **102-103**, 705 (2003).
5. N. Baran, B. Bulakh, Ye. Venger, N. Korsunska, L. Khomenkova, T. Stara, Y. Goldstein, E. Savir, J. Jedrzejewski, Thin Solid Films **517**, 5468 (2009).
6. G. G. Qin, X. S. Liu, S.Y. Ma, J. Lin, G. Q. Yao, X.Y. Lin, K.X. Lin, *Phys. Rev. B*, **55**, 12876 (1997).
7. L. Khomenkova, X. Portier, J. Cardin, F. Gourbilleau, *Nanotechnology* **21,** 285707 (2010).
8. R. F. Steimle, R. Muralidhar, R. Rao, M. Sadd, C.T. Swift, J. Yater, B. Hradsky, S. Straub, H. Gasquet, L. Vishnubhotla, E. J. Prinz, T. Merchant, B. Acred, K. Chang, B. E. White Jr., Microelectronics Realibility, **47,** 585 (2007).
9. T. Baron, A. Fernandes, J.F. Damlencourt, B. De Salvo, F. Martin, F. Mazen, S. Haukka, *Appl. Phys. Lett.* **82,** 4151 (2003).
10. A. N. Mikhaylov, A. I. Belov, A. B. Kostyuk, I. Yu. Zhavoronkov, D. S. Korolev, A. V. Nezhdanov, A. V. Ershov, D. V. Guseinov, T. A. Gracheva, N. D. Malygin, E. S. Demidov, D. I. Tetelbaum: *Physics of the Solid State (St.Petersburg, Russia)* **54,** 368 (2012).
11. S. Yerci, U. Serincan, I. Dogan, S. Tokay, M. Genisel, A. Aydinli, R. Turan, *J. Appl. Phys.*, **100**, 074301 (2006).
12. S. Núñez-Sánchez, R. Serna, J. García López, A. K. Petford-Long, M. Tanase, B. Kabius, *J. Appl. Phys.* **105,** 013118 (2009).
13. N. Korsunska, T. Stara, V. Strelchuk, O. Kolomys, V. Kladko, A. Kuchuk, B. Romanyuk, O. Oberemok, J. Jedrzejewski, P. Marie, L. Khomenkova, I. Balberg, *Nanoscale Research Letters*, **8**, 273 (2013).
14. N. Korsunska, T. Stara, V. Strelchuk, O. Kolomys, V. Kladko, A. Kuchuk, L. Khomenkova, J. Jedrzejewski, I. Balberg, *Physica E* **51**, 115 (2013).
15. L. Bi, J. Y. Feng, *J. Lumin.* **121,** 95 (2006).
16. B. J. Jones, R. C. Barklie, *J. Phys. D.:Appl.Phys.*, **38**, 1178 (2005).
17. S. Yin, E. Xie, C. Zhang, Z. Wang, L. Zhou, I.Z. Ma, C.F. Yao, H. Zang, C.B. Liu, Y.B. Sheng, J. Gou, *Nucl. Instrum. Meth. B* **12-13**, 2998 (2008).
18. Y.Song, C.H.Zhang, Z.G. Wang, Y.M. Sun, J.L. Duan, Z.M. Zhao, *Nucl. Instrum. Meth. B*, **245**, 210 (2006).
19. I. Dogan, I. Yildiz, R. Turan, *Physica E*, **41,** 976 (2009).

Mater. Res. Soc. Symp. Proc. Vol. 1617 © 2013 Materials Research Society
DOI: 10.1557/opl.2013.1168

Transformation of optical properties of Si-rich Al_2O_3 films at thermal annealing

E. Vergara Hernandez[1], B. Perez Miltan[1], J. Jedrzejewski[2] and I.Balberg[2]

[1]UPIITA-Instituto Politecnico Nacional, Mexico DF, 07320, Mexico

[2]Racah Institute of Physics, Hebrew University, 91904 Jerusalem, Israel

ABSTRACT

The effect of thermal annealing on the optical properties of Al_2O_3 films with different Si content was investigated by the photoluminescence method. Si-rich Al_2O_3 films were prepared by RF magnetron co-sputtering of the silicon and alumina targets on long quarts glass substrates. Photoluminescence (PL) spectra of freshly prepared Si-rich Al_2O_3 films are characterized by three PL bands with the peak positions at 2.97-3.00, 2.25-2.29 and 1.50 eV. The thermal annealing of the films at 1150 °C during 30 min stimulates the formation of Si nanocrystals (NCs) in the film area with Si content exceeded 60%. After the thermal annealing the PL intensity of all mentioned PL bands decreases and the new PL band appears with the peak position at 1.67 eV. The new PL band is attributed to the photo currier recombination inside of Si NCs. The size of NCs estimated from the PL peak position 1.67 eV of Si NC emission is about ~-4.5-5.0 nm.

The temperature dependences of PL spectra of Si-rich Al_2O_3 films have been studied in the range of 10-300K with the aim to reveal the mechanism of recombination transitions for mentioned above PL bands 2.97-3.00, 2.25-2.29 and 1.50 eV in freshly prepared films. The thermal activation of PL intensity and permanent PL peak positions in the temperature range 10-300K permit to assign these PL bands to defect related emission in Al_2O_3 matrix.

INTRODUCTION

Light emitting Si nanocrystalls (NCs) were studied intensively during the last 20 years due to their attractive aspects for electronics and photonics related to the possibility to joint in a single chip the Si base devices for the realization of short scale optic interconnections [1-3]. Other promising application of Si NCs is in biology owing to the low toxicity of Si in comparison with II –VI semiconductor quantum dots widely used for the image production [4,5]. Si NCs in dielectric matrix are interesting as well for the charge trapped nonvolatile memory devices [6]. The different types of matrices (SiO_2, ZrO_2, HfO_2, Al_2O_3) have been studied for the application of Si NCs in nonvolatile memory devices [7,8]. In the following, the luminescence properties of Si-rich Al_2O_3 films obtained by RF magnetron spattering will be presented.

EXPERIMENTAL DETAILS

Si-rich Al_2O_3 films were deposited by radio frequency (RF) magnetron co-sputtering from two targets (Si and Al_2O_3) in argon plasma on a long glass substrate a 10-mm width and a 140-mm length as it is shown in figure 1. Five groups of samples, entitled as it is shown in figure 1 (2A, 4A, 6A, 8A, 10A), with the different contents of Si were chosen for the investigation. The freshly prepared Si-rich Al_2O_3 films were then annealed at 1150°C during 30 min in ambient air to form the Si NCs inside of the Al_2O_3 matrix. Annealed samples in the text mentioned as 2B,

4B, 6B, 8B, 10B. PL spectra were measured at the excitation by a He-Cd laser with a wavelength of 325 nm and a beam power of 80 mW at 10-300K using a PL setup on a base of spectrometer SPEX500 described in [9,10].

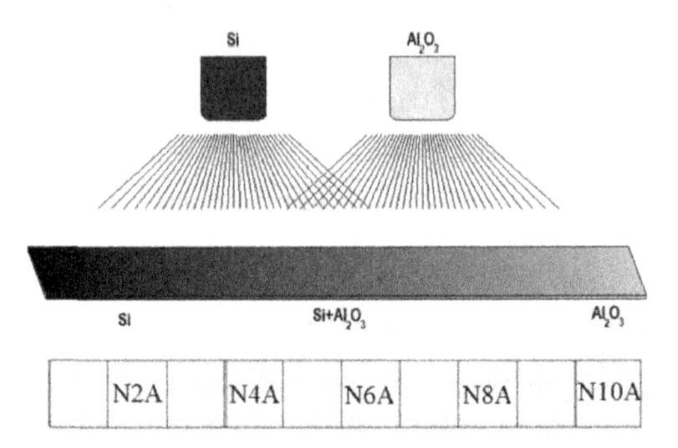

Figure 1. The scheme of RF spattering of Si-rich Al_2O_3 films

EXPERIMENTAL RESULTS AND DISCUSSION

PL spectra of all studied groups of samples both the freshly prepared (A) and thermal annealed (B) have been presented in figure 2.

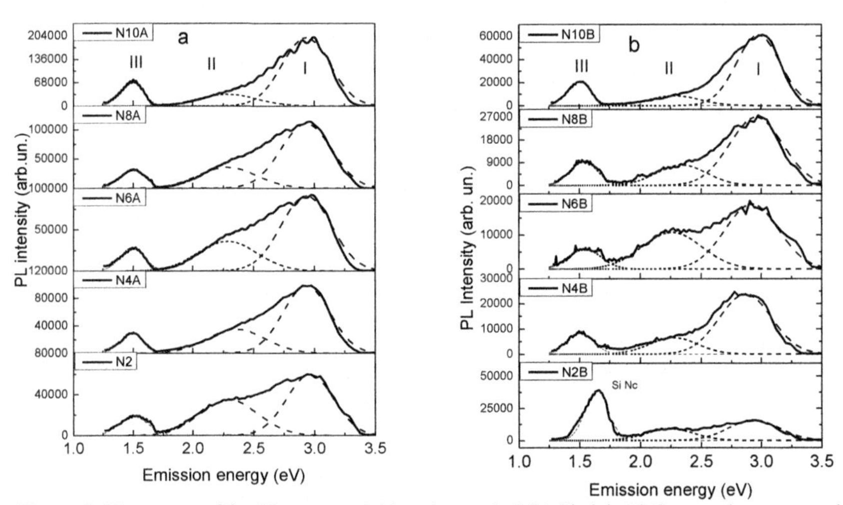

Figure 2. PL spectra of freshly prepared (a) and annealed (b) Si-rich Al_2O_3 samples measured at 300K

In the freshly prepared states three overlapped PL bands with the peak positions at 2.97-3.00 (I), 2.25-2.29 (II) and 1.50 (III) eV clearly have been seen in PL spectra. It was shown early [11,12] that the Si concentration C_{Si} in long range Si-rich Al_2O_3 or SiO_2 films varied from 22 vol.% (10A) to 60-80 vol.% (1A). Thus all three PL bands are characterized by highest PL intensities in the sample 10A with the minimum Si content. With increasing Si content in films the PL intensity of all PL bands decreases monotonically.

It is well known that thermal annealing of the Si-rich Al_2O_3 films at 1150 °C during 30 min stimulates the formation of Si nanocrystals [11,13]. After the thermal annealing the PL intensity of all mentioned PL bands decreases (Fig.2) and the new PL band appears with the peak position at 1.67 eV in the film with Si content around 60%. The temperature dependences of PL spectra of Si-rich Al_2O_3 films have been studied in the range of 10-300K with the aim to reveal the mechanism of recombination transitions for the PL bands centered at 2.97-3.00 (I), 2.25-2.29 (II) and 1.50 (III) eV in the freshly prepared film (Fig.3a).

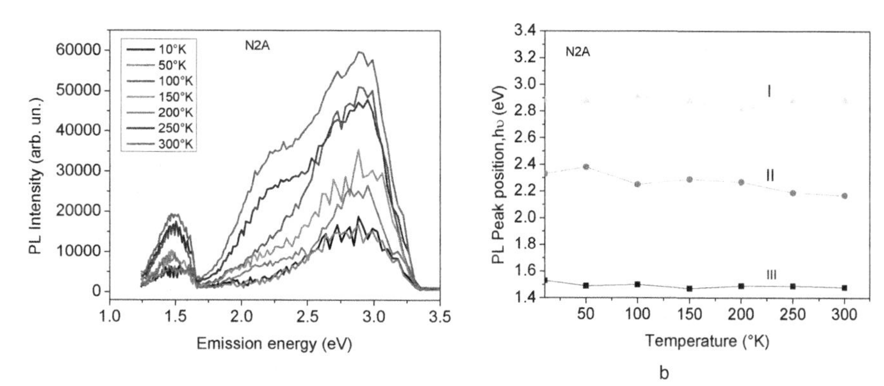

Figure 3. PL spectra (a) and the peak positions of PL bands (b) in the sample 2A measured at different temperatures

The PL intensity of all PL bands increases monotonously versus temperature due to the activation of optical transitions as it is clear from figure 3a. The variation of PL peak positions with temperature has been presented in figure 3b as well. Permanent PL peak positions in the temperature range 10-300 K permit to assign three PL bands to defect related emission in the Al_2O_3 matrix. Actually the orange emission (~2.06 - 2.18 eV) of Al_2O_3 films was investigated earlier and attributed to the F^{2+} centers in Al_2O_3 [14,15]. The blue defect related emission of Al_2O_3 films was reported as well [16, 17]. The emission at 3 eV in Al_2O_3 has been ascribed to F centers [17]. The nature of 1.50 eV PL band is not clear. Actually the synchronous decreasing the PL intensities of 1.50 and 3.00 eV bands with the Si content rise permits to assigned the 1.50 eV PL band to second order diffraction of the 3.0 eV emission.

The annealing of Si-rich Al_2O_3 films in ambient air stimulates both the oxidation process in the Al_2O_3 matrix and the occupation of Al vacancies by the Si atoms. These processes are accompanied by falling down the defect concentrations in the Al_2O_3 matrix and, as a result, by decreasing the PL intensity of all defect related PL bands. The new PL band with a peak at 1.67 eV is attributed to the photocurrier recombination inside of Si NCs. The average size of Si NCs estimated from the peak position 1.67 eV of Si NC emission is about ~-4.5-5.0 nm [13].

CONCLUSIONS

The effect of thermal annealing on photoluminescence of Al_2O_3 films with different Si content was investigated. PL bands in freshly prepared Si-rich Al_2O_3 films have been studied and assigned to defect related emission in the Al_2O_3 matrix. Thermal annealing of the Si-rich Al_2O_3 films at 1150 °C during 30 min stimulated the formation of Si nanocrystals with the size 4.5-5.0 nm. Simultaneously the new PL band with peak at 1.67 eV related to the exciton emission in Si NCs appears in the PL spectrum of samples with Si content 60%.

ACKNOWLEDGMENTS

The authors would like to thank the CONACYT (project 130387) and SIP-IPN, Mexico, for the financial support, as well as the Dr. J.L. Casas Espinola for PL measurements.

REFERENCES

1. T.V. Torchynska, "Si and Ge quantum dot structures", in book:" *Nanocrystals and quantum dots of group IV semiconductors"* Editors T. V. Torchynska and Yu. Vorobiev, American Scientific Publisher, Stevenson Ranch, CA, USA, 42-84 (2010).
2. M. Salib, M. Morse, L. Liao, *Intel Technology Journal*, **8,** n. 2, (2004).
3. G. T. Reed, "Device physics: the optical age of silicon," *Nature*, **427**, n. 6975, 595–596, (2004).
4. X. Michalet, F. F. Pinaud, L. A. Bentolila, Science, **307**, n. 5709, 538–544 (2005).
5. Z. F. Li and E. Ruckenstein, Nano Letters, **4**, n. 8, 1463–1467 (2004).
6. Zs.J. Horváth, P. Basa, "Nanocrystal memory structures" in the book "Nanocrystals and quantum dots of group IV semiconductors", Editors: T. V. Torchynska and Yu. Vorobiev, American Scientific Publisher, Stevenson Ranch, CA, USA, 225-252 (2010).
7. E. Talbot, M. Roussel, C. Genevois, P. Pareige, L. Khomenkova, X. Portier, F. Gourbilleau, J. Appl. Phys. **111**, 103519 (2012).
8. L.Bi, J.Y. Feng: J Lumin. **121**, 95 (2006).
9. T. V. Torchynska, Nanotechnology, **20,** 095401 (2009)
10. T.V.Torchynska, A. Diaz Cano, M.Dybic, S. Ostapenko, M.Mynbaeva, Physica B, **376-377,** 367-369 (2006).
11. N. Korsunska, T. Stara, V. Strelchuk, O. Kolomys, V. Kladko, A. Kuchuk, L. Khomenkova, J. Jedrzejewski, I. Balberg, Physica E, **51,** 115-119 (2013).
12. I. Balberg, Physica E, **51**, 2-9 (2013).
13. T.V. Torchynska, Physica E, **44** (1), 56-61 (2011)
14. S. Yin, E. Xie, C. Zhang, Z. Wang, L. Zhou, I.Z. Ma, C.F. Yao, H. Zang, C.B. Liu, Y.B. Sheng, J. Gou, Nucl Instr Methods B, **12-13**, 2998 (2008).
15. I. Dogan, I. Yildiz, R. Turan, Physica E , **41,** 976 (2009).
16. Y.-T.Chen, Ch.-L. Cheng and Y.-F. Chen, Nanotechnology, **19** n. 44 , 5707 (2008).
17. K.H. Lee, J.H. Crawford, Jr., Phys. Rev. B **15,** 4065 (1977).

Mater. Res. Soc. Symp. Proc. Vol. 1617 © 2013 Materials Research Society
DOI: 10.1557/opl.2013.1169

Light-Emitting and Structural Properties of Si-rich HfO_2 Thin Films Fabricated by RF Magnetron Sputtering

D. Khomenkov[1], Y.-T. An[2], X. Portier[2], C. Labbe[2], F. Gourbilleau[2] and L. Khomenkova[3]

[1]Taras Shevchenko National University of Kyiv, Faculty of Physics, 4 Pr. Hlushkov, Kyiv 03022, Ukraine
[2]CIMAP/ENSICAEN/UCBN, 6 Blvd. Maréchal Juin, 14050 Caen Cedex 4, France
[3]V. Lashkaryov Institute of Semiconductor Physics at NASU, 41 Pr. Nauky, Kyiv 03028, Ukraine

ABSTRACT

Structural, optical and luminescent properties of Si-rich HfO_2 films fabricated by RF magnetron sputtering were investigated versus annealing treatment. Pronounced phase separation process occurred at 950-1100°C and resulted in the formation of hafnia and silica phases, as well as pure silicon clusters. An intense light emission of annealed samples in visible spectral range was obtained under broad band excitation. It was ascribed to exciton recombination inside silicon clusters as well as host defects. To confirm the formation of Si clusters, the structures were co-doped with Er^{3+} ions and effective light emission at 1.54μm was obtained under non-resonant excitation due to energy transfer from Si clusters towards Er^{3+} ions. The interaction of Si clusters, host defects and Er^{3+} ions under is discussed.

INTRODUCTION

During the last years high-*k* hafnia-based materials are mainly considered as alternative gate dielectrics to silicon oxide in complementary metal-oxide semiconductor technology (CMOS) [1]. Recently, nanomemory applications of Si-doped hafnia films were reported [2,3]. Besides promised electrical properties, HfO_2 demonstrates enhanced hardness, high refractive index (almost 2.1 at 632 nm), high optical transparency in the ultraviolet-infrared spectral range, wide optical bandgap (~ 5.8 eV) and low phonon cut-off energy (~about 780 cm^{-1}) offered low probability of phonon assisted relaxation. However, despite these advantages, optical applications of HfO_2-based materials are not numerous. HfO_2 films as optical coatings were investigated in [4,5]. The intrinsic 4.2-4.4 eV luminescence of pure HfO_2 was ascribed to self-trapped exciton [6,7], whereas the 2.5-3.5 eV emission was attributed to different oxygen vacancies with trapped electrons [6]. Rare-earth doped HfO_2 materials were also studied in [8,9], but the mechanism of the excitation of rare-earth ions and their interaction with the host defects were not clarified. It is worth to note that the development of rare-earth doped materials suffers from lower absorption cross-sections of rare-earth ions for 4f-4f transitions (10^{-18}-10^{-20} cm^{-2}) required high-power excitation sources. Meanwhile 4f-5d transitions have higher cross-section (~10^{-12} cm^{-2}), but corresponding excitation levels belong to UV and vacuum UV spectral range, restricting many applications of these materials. Thus, to enhance an excitation of 4f-4f transitions a host mediated excitation via energy transfer is needed.

Among different rare-earth elements, the Er^{3+} ion is one of the most popular due to its radiative transitions in the green ($^4S_{3/2} \rightarrow {}^4I_{15/2}$) and infrared ($^4I_{13/2} \rightarrow {}^4I_{15/2}$) being extensively used as an eye-safe source in atmosphere, laser radar, medicine and surgery ($^4I_{11/2} \rightarrow {}^4I_{13/2}$) [10,11]. A lot of efforts were concentrated on Er^{3+}-doped Si-rich-SiO_2 materials [12-14]. An enhancement of

Er^{3+}absorption cross-section from 10^{-21} cm^{-2} [15] up to 10^{-16} cm^{-2} [13,14] was achieved due to effective energy transfer from Si-nanoclusters (either crystallized or amorphous) towards Er^{3+}ions. For this purpose, visible broad-band excitation was used offering the safe applications of these materials. It is worth to note that only a few studies on Er-doped SiO_2-HfO_2 materials were reported [8,16,17], whereas the mechanism of the Er^{3+} ion excitation was not addressed. In the present paper, we study the effect of annealing treatment on the formation of Si nanoclusters (Si-ncs) in Si-rich hafnia-based host (either doped or not doped with Er^{3+} ions) as well as the mechanism of the interaction of host defects, Si-ncs and rare-earth ions to achieve efficient light emission of these materials.

EXPERIMENT

Doped layers were grown on Si wafers by RF magnetron sputtering of pure HfO_2 target (99.9%) topped by calibrated chips of silicon and/or Er_2O_3. The deposition was carried out at 100°C in pure argon plasma with RF power density of 0.74 W/cm^2. The samples were annealed in a conventional furnace in nitrogen flow at T_A=800-1100 °C and t_A=10-60 min. The chemical composition of the layers was studied by means of FTIR spectroscopy. The spectra were recorded in the range of 600–4000 cm^{-1} using a Nicolet Nexus spectrometer under normal and Brewster incidence angle (65°). Raman scattering spectra were investigated with back-scattered geometry using a dispersive Raman spectrometer equipped with a CCD camera and a laser source at 532 nm. The PL and PL excitation spectra in the 200-800 nm spectral range were studied with a Jobin–Yvon Fluorolog3-22 setup equipped by a Xe lamp as excitation source and R928 photomultiplayer tube.

RESULTS and DISCUSSION

Microstructure properties of Si-rich HfO_2 materials versus annealing treatment

The evolution of film microstructures with annealing was investigated by means of FTIR and Raman scattering spectra. Both methods allowed following the phase separation process upon annealing. FTIR spectra of as-deposited samples (AD) show two broad bands (Fig.1a,b). One of them is observed in the range of 700-1200 cm^{-1} with a maximum at ~1000-1030 cm^{-1} and corresponds to Si-O-Hf stretching vibration mode. Another one, detected in the 460-700 cm^{-1} range, is due to Hf-O vibrations. An annealing at T_A=800-950 °C during t_A=60 min in nitrogen flow leads to an increase of the intensity of Si-O-Hf band and its spectral shift towards higher wavenumbers up to 1070-1080 cm^{-1} (Fig.1a). Further T_A increase up to 1000 °C causes the appearance of several peaks at about 1240 cm^{-1} and 1090 cm^{-1} due to Si-O vibrations as well as at 650 cm^{-1} caused by Hf-O vibrations. This evolution of FTIR spectra is an evidence of a phase separation process and a formation of silica and hafnia phases [2]. However, the presence of a shoulder at about 950 cm^{-1} confirms that this process is not completed yet and HfSiO phase is still present in the annealed films. A treatment at T_A=1100 °C results in the narrowing of all vibration bands. The well-defined peaks appear at 630 cm^{-1} and 770 cm^{-1} due to crystallization of hafnia phase [2]. The bands peaking at 820 cm^{-1}, 1090 cm^{-1} and 1250 cm^{-1} (Fig.1a) correspond to LO_2-TO_2, TO_3 and LO_3 vibration modes of pure silica, respectively [2].

For additional investigation of microstructure evolution, the comparison with Raman scattering spectra of as-deposited and annealed samples was performed. For this purpose, the films were deposited on silica fused substrates to eliminate the contribution of Si TO-phonon observed usually at 521 cm^{-1}. As-deposited samples demonstrated only a broad band with a maximum at about 499 cm^{-1}. After annealing at T_A=800°C this peak position shifted up to ~502 cm^{-1} due to the formation of silicon phase (Fig.2). Higher annealing temperatures results in the continuous shift of the Raman peak position up to 516 cm^{-1} (T_A=900°C) as well as increasing its intensity. This is due to increasing of Si-ncs number and their sizes. These latter were estimated at about 3.7-4.0 nm using the model described in [18].

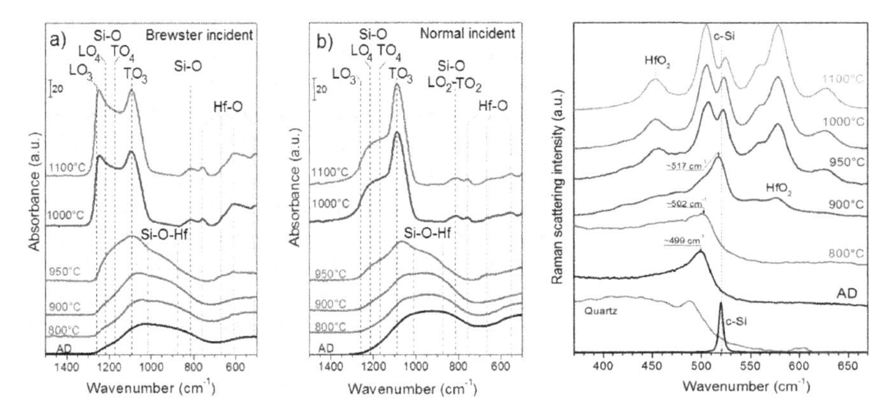

Fig.1 FTIR spectra of as-deposited (AD) and annealed Si-rich-HfO_2 films grown on Si substrate versus T_A; t_A=60min. Spectra were measured at Brewster (65°) angle (a) and normal incident (b) of the light. The T_A values are pointed in the figure; annealing time is.

Fig.2. Raman scattering spectra of Si-rich-HfO_2 films grown on fused silica substrates versus T_A; t_A=60min. Spectra for quartz and silicon substrates are given for the comparison.

Further T_A increase up to 1000-1100°C results in a significant rise of the Raman signal intensity as well as in the appearance of well-defined peaks at 455cm^{-1}, 506 cm^{-1} and 523 cm^{-1}. All these bands correspond to the crystallized HfO_2 phase. The Si-ncs can be also formed upon such treatment similar to the case of Si-ncs formation in SiO_2 host [3,19]. However, corresponding Raman signals from Si-ncs and HfO_2 phase overlap significantly, hampering to conclude about the size variation of Si-ncs (Fig.1b).

It is worth to note that the crystallization of Si-ncs in SiO_2 matrix occurs usually at 1100-1150°C and results in the bright visible PL emission [19]. It is known for Si-rich-SiO_2 materials that the peak position of PL band corresponding to the Si-ncs depends on their sizes and demonstrates the "blue" shift with the size decrease [3,19]. By analogy, the similar behavior of PL emission from Si-ncs embedded in HfO_2 can be also expected Thus, the comparison of light emitting properties of Si-rich-HfO_2 samples with that of Si-rich-SiO_2 films, grown with similar approach [3,19], can give information about the formation of Si-ncs in HfO_2-based host.

Light emission from the films

PL emission of Si-rich-HfO_2 samples, appeared only after annealing and under the visible light excitation (450-550 nm), of two broad bands with maxima at ~580 nm and ~720-780 nm are observed (not shown here). The comparison of pure and Si-rich HfO_2 films [17] showed that Si incorporation results in the appearance of orange-red emission upon annealing treatment. The T_A increase up to 900°C results in the enhancement of PL intensity as well as in the "blue" shift of total PL spectrum. Meanwhile, the t_A increase from 10 min to 60 min at T_A=900°C leads the "red" shift of PL band. Further increase of T_A for any t_A caused PL quenching and a shift of PL spectrum towards lower wavelength. Such a behavior of PL emission of Si-rich HfO_2 films was found to be similar to that observed for Si-rich-SiO_2 materials [20]. It is known that Si-ncs embedded in SiO_2 host doped with rare-earth ions are effective sensitizers of these latter. Thus, an achievement of the efficient PL emission from rare-earth elements under non-resonant excitation can be an additional argument of Si-ncs presence in HfO_2 host. For this purpose, pure and Si-rich HfO_2 films were co-doped with Er^{3+} ions and their PL properties have been studied versus annealing treatment.

Figure 3 shows a comparison of PL spectra from Er-doped and (Er,Si)-codoped HfO_2 films annealed at the same conditions. Besides usually addressed Er^{3+} emission in 1.4-1.7 μm from $^4I_{13/2} \rightarrow ^4I_{15/2}$ transitions, the spectra were recorded in the 630-700-nm range corresponding to $^4F_{9/2} \rightarrow ^4I_{15/2}$ (Fig.3) at resonant (488 nm or 532 nm) and non-resonant (476 nm) excitation.

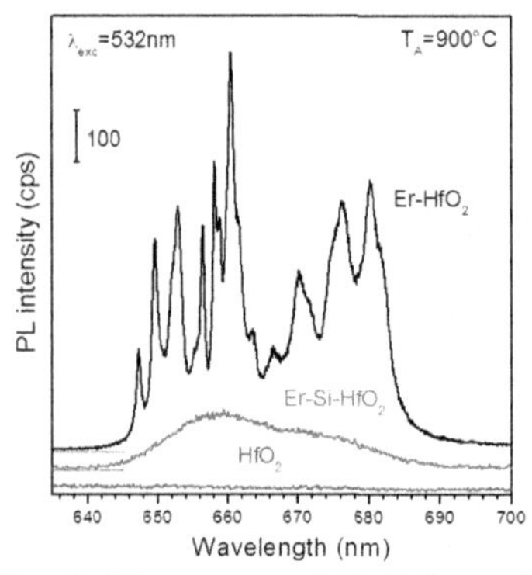

Fig. 3. PL spectra of Er-HfO_2 and (Er,Si)-HfO_2 films annealed at T_A=900°C, t_A=60 min; λ_{exc}=532 nm.

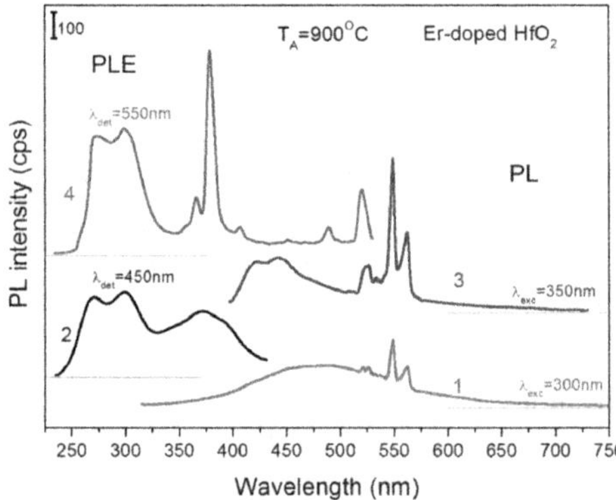

Fig. 4. PL (1,3) and PL excitation (2,4) spectra of Er-HfO_2 samples annealed T_A=900°C, t_A=60 min. λ_{exc}=300 (1), 350 nm (3); λ_{det}=450 (2), 550 nm (4).

Since pure HfO_2 does not emit under all mentioned excitation wavelength (Fig.3) this means that, if any oxygen vacancies are present in Er-doped HfO_2 films, their emission cannot be excited by these light wavelengths. Moreover, self-trapped excitons, required an ultraviolet excitation [6], cannot be also considered as a sensitizer of Er^{3+} ions, since used optical excitation is not efficient in this case. Thus, well-defined PL peaks observed for Er-doped HfO_2 films, corresponding to $^4F_{9/2} \rightarrow ^4I_{15/2}$, under 488-nm and 532-nm excitation, (Fig.3) are due to direct excitation of Er^{3+} ions. This latter was confirmed by very low PL intensity (as a noise level) observed under "non-resonant" 476-nm-illumination [17].

Figure 4 represents PL spectra of Er-doped HfO_2 films under UV excitation (curves 1,3) and corresponding PL excitation (PLE) spectra (curves 2,4). Several broad PLE bands are seen in PLE spectra of Er-doped HfO_2 materials at 280 nm (4.42 eV), 300 nm (4.13 eV) and 370 nm (3.30 eV), when detected at λ_{det}=450 nm (Fig.4, curve 2). The overlapping of all these bands can affect their peak positions and slight shift of their maxima can be observed. Two first peaks are usually attributed to negatively charged Hf-related oxygen vacancies with different coordination, i.e. $V_{O(3)}^{-2}$ (4.42-4.44 eV) and $V_{O(4)}^{-2}$ (4.2-4.23 eV) [21]. The third PLE band can be ascribed to either neutral or positively charge oxygen vacancy, i.e $V_{O(3)}^{0}$ or $V_{O(3)}^{+2}$ based on the data of Ref.[21].

The broad "blue" PL band peaked at 400-460 nm is due to carrier recombination via different oxygen vacancies (Fig.4). This emission is usually observed for HfO_2 based materials under UV light excitation, whereas its peak position depends on the excitation light wavelength specific for different types of oxygen vacancies. Besides this blue PL band, the sharp peaks in the 520-570 nm range are also observed (Fig.4, curves 1,3). They are due to $^4S_{3/2} \rightarrow {}^4I_{15/2}$ transitions in Er^{3+} ions that is confirmed by PLE spectrum recorded at 550 nm and demonstrated corresponding sharp excitation lines (Fig.4, curve 4). The PLE spectrum detected at 450 nm (curve 2) shows also the presence of broad PLE bands peaked at 280 nm and 300 nm, whereas the peak at about 380 nm is overlapped with the absorption bands corresponded to $^4I_{15/2} \rightarrow {}^4G_{11/2}, {}^4H_{11/2}$ transitions. Thus one can conclude that Er^{3+} ions can be effectively excited not only via direct excitation but also due to energy transfer from oxygen vacancies of HfO_2 matrix.

Similar host mediate excitation was observed for Er-doped Si-rich-HfO_2 samples. Figure 5 shows the comparison of PL and PLE spectra obtained for Er-free (curves 1, 5) and Er-doped films (curves2-4, 3-6). The variation of the excitation wavelength in the range 260-380 nm does not result in the spectral shift of PL peak position. Thus, this PL band in due to defects of Si-rich-HfO_2 matrix.

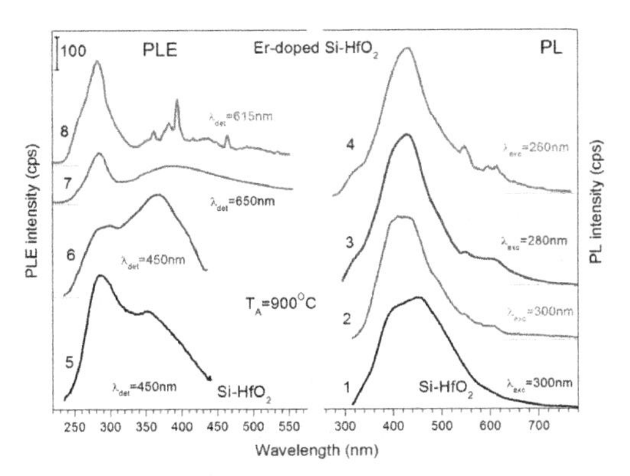

Figure 5. PL (1-4) and PL excitation (5-8) spectra of Si-rich-HfO_2 (1,5) and Er-doped-Si-rich-HfO_2 (2-4,6-8) samples. The excitation and detenciton wavelength are mentioned in the figure.

In general, PLE spectra of all samples are broad. Two main peaks are observed similar to that for Er-free Si-rich-HfO_2 samples (Fig.5). The first one is situated at about 290 nm (4.27 eV), whereas the position of the second one depend on the detection wavelength. Meanwhile it is observed at 370-400 nm (3.35-3.10 eV). Similar shift of PLE peak position was reported in [20] and explained by depth distribution of the emitting defects. It is interesting that the Er-free samples show broader PL spectrum (Fig.5, curve 1) than their Er-doped counterparts (curves, 2-4). This can be due to quenching of some radiative channels due to energy transfer towards Er^{3+} ions. This assumption is confirmed by the observation of some PL peaks in the green (~570 nm) and orange (~620 nm) spectral ranges (Fig.5, curves 2-4) when 260-280 nm-excitation is used.

The PLE spectrum, detected at 615-nm wavelength, confirms the Er-related nature of green and orange emission. It is worth to note that PLE spectrum, measured at 720 nm detection wavelength that corresponds to PL maximum of Si-NCs emission [17], was found to be similar to PLE spectrum presented by curve 6 in Fig.5. Similar data were obtained in [22] and demonstrated that oxygen-related defects can take part in the excitation of Si nanoclusters in Si-rich-SiO_2 films. Similar PL properties were demonstrated for the Er-doped Si-rich-ZrO_2 films [23,24]. In this system, the achievement of Er emission under non-resonant excitation was ascribed to energy transfer from Si-ncs. However, the presence of these latter was not demonstrated clearly. Meanwhile, it was shown that oxygen-deficient silicon centers (Si-ODCs) and oxygen interstitial defects are effective excitation paths for Er^{3+} ions [23]. It is also known that Si-(Ge) ODCs can be effective sensitizers of rare earth ions embedded in Si-rich [22] or Ge-rich [25] silicon oxide host.

Thus, one can conclude that the energetic balance of interacted "Er-ions, Si-NCs and defects" system is complicated. All three of them can emit light under "corresponding" excitation, whereas Si-Ncs and defects can be effective sensitizers for rare-earth ions. This can allow achieving broad-band emission from hafnium-silicates doped by other rare earth ions.

CONCLUSIONS

In the present study, the properties of RF magnetron sputtered hafnia-based films were investigated by means of FTIR, Raman scattering and PL techniques. It was observed that high temperature annealing governs a phase separation process and the formation of silica and hafnia phases, as well as silicon phase. The appearance of a PL emission in the visible-near-infrared spectral range occurred. The evolution of the PL peak position was found to be correlated with Si-ncs sizes. The properties of Si-doped-HfO_2 films are compared with those of their counterparts doped with Er^{3+} ions. The investigation of the effect of annealing treatment on luminescent properties revealed that the enhancement of Er^{3+} PL emission occurs due to an effective energy transfer from Si-nanoclusters and defects of hafnia based host (for example, oxygen vacancies).

ACKNOWLEDGMENTS

This work is partially supported by French National Agency (ANR) through Nanoscience, Nanotechnology Program (NOMAD project, No.ANR-07-NANO-022-02) and the Conseil Regional de Basse Normandie through CPER project - Nanoscience axe (2007-2013).

REFERENCES

1. G. He, L.Q. Zhu, Z.Q. Sun, Q. Wan, L.D. Zhang, *Progress in Materials Science* **56** (2011) 475.
2. L. Khomenkova, X. Portier, J. Cardin, F. Gourbilleau. *Nanotechnology* **21** (2010) 285707.
3. L. Khomenkova, B.S. Sahu, A. Slaoui, F. Gourbilleau, *Nanoscale Research Letters* **6**, 172 (2011).
4. J.M. Khoshman, A. Khan and M.E. Kordesch, *Surf. Coat. Technol.*, **202** , 2500 (2008).
5. O. Stenzel, S. Wilbrandt, S. Yulin, N. Kaiser, M. Held, A. Tünnermann, J. Biskupek and U. Kaiser, Opt. Mater. Express, **1**, 278 (2011).
6. M. Kirm, J. Aarik, M. Jürgens and I. Sildos, *Nucl.Instr.Meth.A*, **537**, 251 (2005).
7. K. Smits, L. Grigorjeva, D. Millers, A. Sarakovskis, J. Grabis and W. Lojkowski, *J. Lumin.*, **131**, 2058 (2011).
8. V. Kiisk, I. Sildos, S. Lange, V. Reedo, T. Ta¨tte, M. Kirm and J. Aarik, *Appl. Surf. Sci.* **247**, 412 (2005).
9. L. X. Liu, Z. W. Ma, Y. Z. Xie, Y. R. Su, H. T. Zhao, M. Zhou, J. Y. Zhou, J. Li, E. Q. Xie, *J. Appl. Phys.* **107** (2010) 024309.
10. C. Stoneman, L. Esterowitz, *Opt. Lett.* **15** (1990) 486.
11. L. Feng, J. Wang, Q. Tang, L.F. Liang, H.B. Liang, Q. Su, *J. Lumin.* **124** (2007) 187.
12. A.J. Kenyon, *Semicond. Sci. Technol.*, **20**, R65 (2005).
13. M. Wojdak, M. Klik, M. Forcales, O. B. Gusev, T. Gregorkiewicz, D. Pacifici, G. Franzò, F. Priolo, and F. Iacona, Phys. Rev. B 69, 233315 (2004).
14. S. Cueff, C. Labbé, J. Cardin, J.-L. Doualan, L. Khomenkova, K. Hijazi, O. Jambois, B. Garrido and R. Rizk, *J. Appl. Phys.*, 108 (2010), p. 064302
15. J. Miniscalco, J. Lightwave Technol. 9 (1991) 234.
16. G.C. Righini, S. Berneschi, G. Nunzi Conti, S. Pelli, E. Moser, R. Retoux, P. Féron, R.R. Gonçalves, G. Speranza, Y. Jestin, M. Ferrari, A. Chiasera, A. Chiappini, C. Armellini. *J. Non-Cryst. Sol.* **355** (2009) 1853.

17]. L. Khomenkova, Y.-T.An, C. Labbé, X.Portier, F. Gourbilleaua, ECS Trans., 45 (2012) 119.

18. H. Richter, Z.P. Wang and L. Ley, *Solid State Comm.* **39** (1981) 625.
19. E. Talbot, R. Lardé, F. Gourbilleau, C. Dufour, P. Pareige. *Eur. Phys. Lett.* **87**, (2009) 26004.
20. L. Khomenkova, N. Korsunska, V. Yukhimchuk, B. Jumayev, T. Torchynska, A. Vivas Hernandez, A. Many, Y. Goldstein, E. Savir, J. Jedrzejewski. *J. Lumin.*, **102-103**, (2003) 705.
21. H.-K. Noh, B. Ruy, E.-A. Choi, J. Bang, and K.J. Chang, Appl. Phys. Lett. **95**, 082905 (2009).
22. S. Cueff,C. Labbé, B. Dierre, F. Fabbri, T. Sekiguchi, X. Portier, R. Rizk, J. Appl. Phys. **108** (2011) 113504.
23. C. Rozo, L. F. Fonseca, *J. Phys.: Condens. Matter* **20** (2008) 315003
24. C. Rozo, D. Jaque, L.F. Fonseca, J.G. Solé, J. Lumin. **128** (2008) 1197.
25. A. Kanjilal, L. Rebohle, M. Voelskow, W. Skorupa, and M. Helm, *Appl. Phys. Lett.* **94**, 051903 (2009).

ZnO nanocrystals

Mater. Res. Soc. Symp. Proc. Vol. 1617 © 2013 Materials Research Society
DOI: 10.1557/opl.2013.1170

Size dependent optical properties in ZnO nanosheets

Brahim El Filali and Aaron I. Díaz Cano,
UPIITA-Instituto Politécnico Nacional, México D.F.07738, México.

ABSTRACT

Photoluminescence (PL), scanning electronic microscopy (SEM) and Raman scattering have been studied in crystalline ZnO nanosheets with different sizes after the thermal annealing at 400 °C for 2 hours in ambient air. ZnO nanosheets were created by the electrochemical (anodization) method using the variation of the etching durations with obtained ZnO nanosheet sizes from the range 40-360 nm. Earlier it was shown using the X ray diffraction (XRD) method that thermal annealing performed the ZnO oxidation and crystallization with the creation of the wurtzite crystal lattice. Four PL bands are revealed in PL spectra with the PL peaks at 1.60, 2.08, 2.50 and 3.10 eV. Size decreasing of ZnO nanosheets stimulates tremendous changes of ZnO optical parameters. It is shown that decreasing the ZnO nanosheet sizes is accompanied by the intensity increase of a set of Raman peaks and the surface defect related PL bands. The reasons of emission transformation and the nature of optical transitions have been discussed as well.

INTRODUCTION

There is a great interest in growing the ZnO nanosystems using the anodization technology developed early for the PSi production [1-3]. Nano crystalline ZnO – wide band gap semiconductors with the direct band gap (3.37eV) and high exciton binding energy (60meV) at room temperature, promises the numerous applications in optical and electronic devices [4] such as room temperature white light-emitting diodes [5] or ultraviolet nanolasers [6], field emission cathodes [7] or field – effect transistors [8]. However, even high-quality ZnO samples contain a variety of point defects contributing to the visible luminescence [9]. Actually it is not clear the relation between the radiative defects appeared and the size of ZnO nanosystems. In this work the dependence of optical properties of ZnO nanosheets versus their sizes has been studied.

EXPERIMENTAL DETAILS

The electrochemical anodization of Zn foils was performed in a two electrode system including the Zn electrodes exposed in the electrolyte. The pieces of a Zn foil (Aldrich 99.99% purity, 0.25 mm thickness) with a radius of 6 mm were treated in acetone and ethanol for 15 min with ultrasonic cleaning and then washed with deionized water. The electrolyte composition was a 1:10 volume mixture of HF and deionized water without surfactants. The applied voltage was 5 V and the reaction time ranges from 1 min to 10 min (Table 1). Then these samples were annealed at 400 °C during 2 hours in ambient air.

The morphology of ZnO films has been studied using the scanning electron microscopy (SEM) (Quanta 3D FEG-FEI). Raman scattering spectra were measured in Jobin-Yvon LabRAM HR 800UV micro-Raman system using an excitation by a solid state LED with a light wavelength of 785 nm [10,11]. PL spectra were measured at the excitation by a He-Cd laser with a wavelength of 325 nm and a beam power of 80 mW at 300K using a PL setup on a base of spectrometer SPEX500 described in [12,13].

EXPERIMENTAL RESULTS AND DISCUSSION

1. SEM study

The SEM image of typical ZnO nanosheets obtained at the anodization during 10 min after the thermal annealing is presented in figure 1. The size of ZnO nanosheets estimated from SEM images depends on the duration of etching process. In Table 1 the average sizes of all studied ZnO nanosheets estimated from SEM images have been presented. Early we have shown that the thermal annealing at 400 °C for 2 hours in ambient air stimulates the process of ZnO oxidation and crystallization with the formation of the wurtzite ZnO crystal structure with the hexagonal lattice parameters of a=3.2498A and c=5.2066A [14].

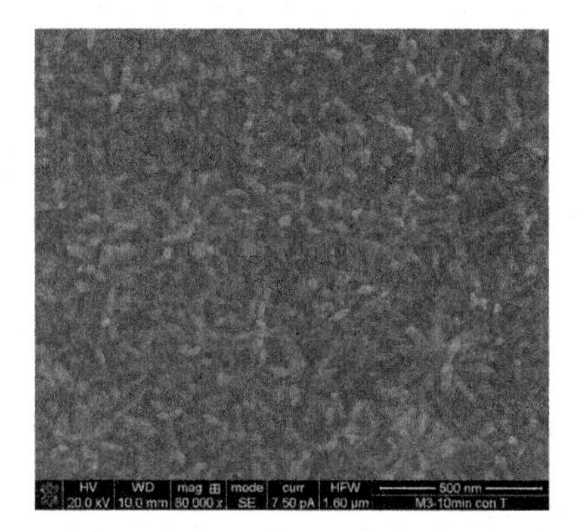

Figure 1. The SEM image of ZnO nanosheets obtained at the anodization time 10 min after thermal annealing.

Table 1. The dependence of ZnO nanosheet sizes on technological parameters

Number of samples	Etching duration, min	Etching voltage, V	Average nanosheet sizes, nm
#1	1	5	170x320
#2	3	5	120x220
#3	6	5	60x110
#4	10	5	30x50

2. Raman scattering study

Raman scattering spectra measured in the range 100-800 cm^{-1} are presented in figure 2. The ZnO oxidation and crystallization at annealing are accompanied by appearing four Raman peaks at 327, 379, 434 and 549-556 cm^{-1} (Fig.2 and Table 2). With increasing the anodization duration (decreasing the ZnO nanosheet sizes) the intensity of Raman peaks enlarges (Fig.2, curves 1-4). The last effect can be connected with increasing the width of ZnO nanosheet layers and the change of the geometry of Raman scattering as well.

The group theory predicts for the wurzite ZnO crystal structure the Raman active optic phonons in Brillouin zone center as: i) A_1 and E_1 symmetry polar phonons with the different frequencies for the transversal (TO) and longitudinal (LO) optic phonons, and ii) E_2 symmetry non-polar phonon mode with two frequencies E_2 (low) and E_2 (high) (Table 2). Note E_2 (high) mode is associated with oxygen sublattice and E_2 (low) mode is attributed to the Zn sublattice [15,16]. Additionally some Raman peaks, such as 330 and 437 cm $^{-1}$, are attributed to the second order Raman peaks arising from the zone boundary phonons $3E_{2H}$-E_{2L} and E_{2H} modes of ZnO nanocrystals [17].

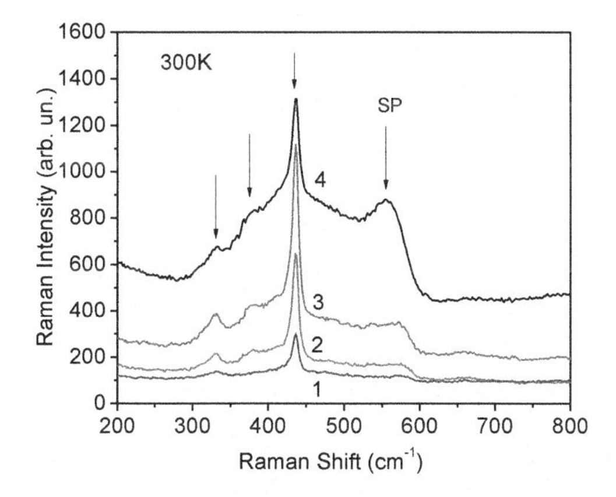

Figure 2. Raman scattering spectra of thermal annealed ZnO nanosheets obtained at the anodization times: 1 min (curve 1), 2 min (curve 1), 6 min (curve 3) and 10min (curve 4).

The Raman peaks at 379 and 434 cm^{-1} detected in studied ZnO nanocrystals after thermal annealing can be attributed to the first order Raman peaks in the wurzite crystal lattice related to the A_1(TO) and E_2(high) phonon modes (Table 2). The Raman peak at 327cm^{-1} is consistent with the second order Raman peak arising from the zone boundary phonons at $3E_{2H}$-E_{2L}. The nature of the Raman peak at 551-570 cm^{-1} is not clear. Its variable positions in different samples (Fig.4, curves 2, 3, 4), its intensity increasing and the shift with decreasing the ZnO nanosheet sizes, the location in the range 551-570 cm^{-1} between E1(TO) and E1(LO) optic phonons permit to attribute this peak to the surface phonon (SP) Raman scattering. The value of the surface phonon frequency (w_{SP}) in ZnO was calculated using the formula [18]:

$$w_{sp} = w_{TO}\sqrt{\frac{\varepsilon_o l+\varepsilon_M(l+1)}{\varepsilon_\alpha l+\varepsilon_M(l+1)}} \quad , \qquad (1)$$

where w_{TO} is the frequincy of the TO phonon, ε_o and ε_∞ are the static and high-frequency dielectric constants of the bulk ZnO and ε_M is the static dielectric constant of surrounding medium (air). The values of ε_o , ε_α, ε_M equal to 8.36 [19,25], 3.75[25] and 1[25] respectively, and the lowest (l=1) mode were choosen at the calculation. The frequency of the surface phonon 550 cm^{-1} was obtained that is very close to the experimental value (Table 2).

Table 2. Raman peaks in ZnO

Samples	E_2 (low) cm^{-1}	$3E_{2H}$-E_{2L} cm^{-1}	A_1(TO) cm^{-1}	E_1(TO) cm^{-1}	E_2(high) cm^{-1}	SP cm^{-1}	A_1(LO) cm^{-1}	E_1(LO) cm^{-1}
Bulk ZnO[18]	102		379	410	439	550	574	591
ZnO nanosheet		332.5	375.8		435.8	551-570		

3. ZnO emission study

PL spectra of ZnO nanosheets of different sizes are shown in figure 3. These PL spectra are complex and can be represented as a set of elementary PL bands with the peaks at 1.58, 2.05, 2.50 and 3.08-3.15 eV (Fig.4). The last PL band at 3.08-3.15 eV is related to the near-band-edge (NBE) luminescence in ZnO. The two PL bands at 2.05 and 2.50 eV are attributed, as a rule, to the defects in ZnO nanosheets and the PL band at 1.58 eV was assigned earlier to the second order diffraction peaks of 3.15 eV PL bands, respectively [20].

The near-band-edge emission at 3.15 eV at 300K is attributed to the optical transition between the shallow donor and valence band [21]. However, the position of near-band-edge emission at room temperature can vary significantly due to the variation of relative contributions of free exciton emission and phonon replicas, which will be different for different growth conditions [22].

With decreasing the ZnO nanosheet sizes the intensity of defect related PL band at 2.50 eV increases mainly and this PL band dominates in the spectrum of the small size nanosheets (Fig.3, curve 4). The defect related green PL band in the spectral range 2.40-2.50 eV in ZnO is assigned ordinary to the oxygen vacancies [23], Cu impurities [24] or surface defects [25,26], similar to those proposed earlier for the porous Si [27]. Decreasing the ZnO nanosheet size is accompanied by the rise of the surface to volume ration that permits to assign 2.50 eV to the surface defects.

CONCLUSIONS

The Raman and PL spectra have been studied in the thermal annealed ZnO nanocrystals of different sizes. It is shown that decreasing the ZnO nanosheet sizes is acompaned by the intensity increase of a set of Raman peaks and the surface defect related PL band. The reasons of emission transformation and the nature of optical transitions have been discussed as well.

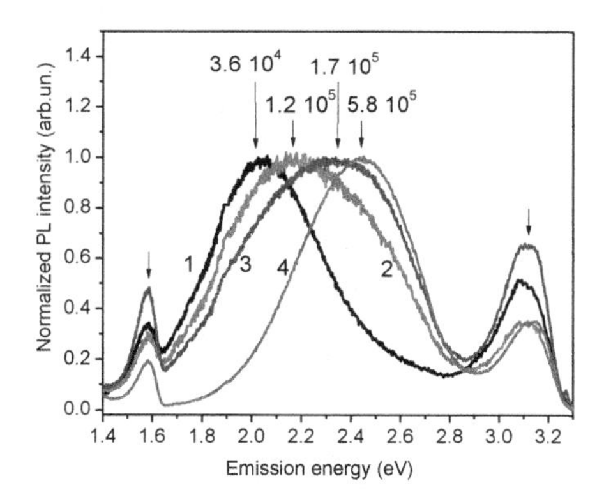

Figure 3. Normalized PL spectra of thermal annealed ZnO nanosheets obtained at the anodization durations of 1min (1), 3min (2), 6min (3) and 10min (4). The coefficients used at the normalization for dividing the experimental PL intensity have shown for every curve.

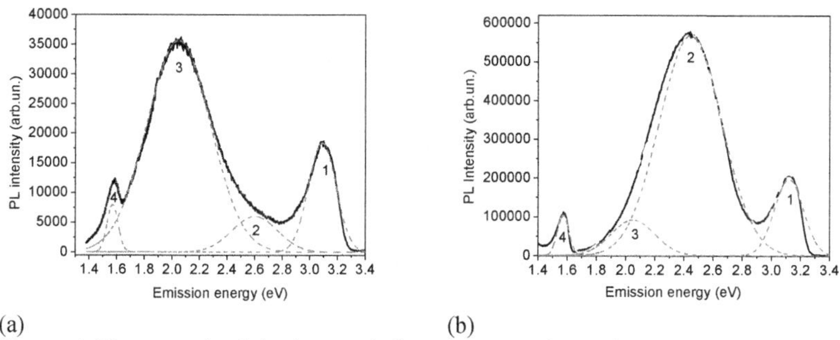

(a) (b)

Figure 4. The example of the deconvolution of the experimental PL spectrum on elementary PL bands 1 min (Figure 4.a) and 10min (Figure 4.b).

ACKNOWLEDGMENTS

The authors would like to thank the CONACYT (project 130387) and SIP-IPN, Mexico, for the financial support, as well as the Dr. T.V. Torchynska for the fruitful discussion and the Dr. J.L. Casas Espinola and CNMN-IPN for the PL and SEM measurements, respectively.

REFERENCES

1. T.V. Torchynska, "Nanocrystals and quantum dots. Some physical aspects" in the book "Nanocrystals and quantum dots of group IV semiconductors", Editors: T. V. Torchynska and Yu. Vorobiev, American Scientific Publisher, 1- 42 (2010).
2. P.G. Li, W.H. Tanga, X. Wang, J. of Alloys and Compounds, **479**, 634-637 (2009).
3.T.V.Torchynska, J.Palacios Gomez, G.P.Polupan, F.G.Becerril Espinoza, A.Garcia Borquez, N.E.Korsunskaya, L.Yu. Khomenkova, Appl. Surf. Science, **167,** 197-204 (2000).
4. S.J. Pearton, D.P. Norton, K. Ip, Y.W. Heo, T. Steiner, Prog. Mater. Sci., **50**, 293 (2005).
5. N.H. Alvi, S.M. Usman Ali, S. Hussain, O. Nur and M. Willander, Scripta Materialia **64,** 697-700 (2011).
6. M.H. Huang, S. Mao, H. Feick, Science, **292**, 1897 (2001).
7. Y.B. Li, Y. Bando, D. Golberg, Appl. Phys. Lett. **84**, 3603 (2004).
8. W.I. Park, J.S. Kim, G.C.Yi, M.H. Bae, H.J.Lee, Appl. Phys. Lett. **85**, 5052 (2004).
9. M. A. Reshchikov, H. Morkoc, B. Nemeth, J. Nause, J. Xie, B. Hertog, A. Osinsky, Physica B. Condensed Matter **401–402**, 358-361 (2007).
10. T. V. Torchynska, J. Douda, S. S. Ostapenko, S. Jimenez-Sandoval, C. Phelan, A. Zajac, T. Zhukov, T. Sellers, J. of Non-Crystal. Solid. **354,** 2885 (2008).
11. T. V. Torchynska, A. I. Diaz Cano, M. Dybic, S. Ostapenko, M. Morales Rodrigez, S. Jimenes Sandoval, Y. Vorobiev, C. Phelan, A. Zajac, T. Zhukov, T. Sellers, phys. stat. sol. (c), **4**, 241 (2007).
12. M. Dybic, S. Ostapenko, T.V. Torchynska, E. Velazquez Lozada, Appl. Phys. Lett. **84** (25), 5165-5167 (2004).
13. T. V. Torchynska, A.I. Diaz Cano, M. Dybic, S. Ostapenko, M. Mynbaeva, Physica B, Condensed Matter, **376-377**, 367-369 (2006)
14. A.I. Diaz Cano, B. El Filali, T.V. Torchynska, J.L. Casas Espinola, J. of Phys. and Chem. of Solids, **74**, 431–435, (2013).
15. N. Ashkenov, B. N. Mbenkum, C. Bundesmann, V. Riede, M. Lorenz, D. Spemann, E. M. Kaidashev, A. Kasic, M. Schubert, M. Grundmann, G. Wagner, H. Neumann, V. Darakchieva, H. Arwin, and B. Monemar, J. Appl.Phys. **93**, 126 (2003).
16. J.F. Scott, Phys. Rev. B, **2,** 1209 (1970).
17. S. Ghoopum, N. Hongsith, P. Mangkorntong, Physica E, **39,** 53-56 (2007)
18. G. Polupan, T.V. Torchynska, Thin Solid Films **518**, S208–S211 (2010).
19. B. Gil, Al. V. Kavokin Appl. Phys. Let**. 81**, n. 4, (2002)
20. A.I. Diaz Cano, B. El Filali, T.V. Torchynska, J. L.Casas Espinola, Physica E, **51**, 24–28 (2013).
21. A. B. Djuris, A.M.C. Ng, X.Y. Chen. Progress in Quantum Electronics **34**. 191-259 (2010).
22. T. Voss, C. Bekeny, L. Wischmeier, H. Gafsi, S. Borner, W. Schade, A.C. Mofor,
A. Bakin, A. Waag, Appl. Phys. Lett. **89**, 182107 (2006).
23. M.K. Patra, K. Manzoor, M. Manoth, S.P. Vadera, N. Kumar, J. Lumin. **128** (2) 267–272 (2008).
24. N.Y. Garces, L. Wang, L. Bai, N.C. Giles, L.E. Halliburton, G. Cantwell, Appl. Phys. Lett. **81** (4) 622–624 (2002).
25. A.B. Djurišic, W.C.H. Choy, V.A.L. Roy, Y.H. Leung, C.Y. Kwong. K.W. Cheah, T.K. Gundu Rao, W.K. Chan, H.F. Lui, C. Surya, Adv. Funct. Mater. **14,** 856-864 (2004).
26. O. Madelung, "Semiconductors-Basic Data", Springer (1996) 336p.
27. T.V.Torchinskaya, N.E.Korsunskaya, B.Dzumaev, B.M.Bulakh, O.D.Smiyan, A.L.Kapitanchuk, S.O.Antonov, Semiconductors, **30,** 792-796 (1996).

Mater. Res. Soc. Symp. Proc. Vol. 1617 © 2013 Materials Research Society
DOI: 10.1557/opl.2013.1171

Photoluminescence Study of ZnO Nanosheets with embedded Cu Nanocrystals.

Aaron I. Diaz Cano and Brahim El Filali,
[1]UPIITA – Instituto Politécnico Nacional, México D. F. 07738, México.

ABSTRACT

In this work a simple method to produce the ZnO nanosheets (NSs) with inclusions of Cu nanocrystals by means of electrochemical etching without the necessity of any surfactant has been presented. The Raman spectroscopy demonstrates that the amorphous samples of ZnO-Cu present appreciable changes in its vibrational behavior after the thermal treatment at 400°C in ambient atmosphere. The study of Photoluminescence (PL) shows monotonous increasing the bands centered in 3.07, 2.41, 2.03 and 1.57 eV versus etching time in freshly prepared samples. The intensity variation of the PL bands, the changes in vibrational behavior, as well as the impact of the copper content and preparation conditions allow identifying emission inside the visible spectral range related to the surface defects that is interesting for the future possible application this ZnO system in room temperature "white" light-emitting diodes.

INTRODUCTION

ZnO nanostructures have been one of the most popular materials in recent years [1,2]. ZnO nanostructures have attracted considerable attention due to their promising applications in electronic, photoelectronic and sensing devices, mainly due to the potential for engineering the properties not obtained in the bulk materials. For this reason, the electrochemical technique has greater potential to control the dimensionality through the variations of electrolyte composition, the process duration and applied potential at room temperature. Thus it allows the precise control of reaction conditions.

Recently, researchers have found that some doping elements can influent on the luminescence properties of ZnO nanostructures [3–5]. Copper is a prominent luminescence activator for the II–VI compounds [8]. The diffusion of Cu into ZnO can cause the formation of defects (Cu_{Zn}, Cu_i). The rich defect chemistry of porous ZnO-Cu NSs permits to expect a great variety of photoluminescence (PL) bands in the visible spectral range [6,7]. However, the majority of these defects remain unidentified and optical transitions responsible for the PL bands need to be verified. Actually it is not clear the relations between the defects appeared and the structural properties of ZnO-Cu nanosystems. In this work ZnO-Cu nanosheets were prepared by the anodization method and their optical properties have been studied before and after thermal annealing.

EXPERIMENT

Electrochemical anodization of Zn foils was performed in a two-electrode system including the Cu electrodes exposed to the electrolyte. The distance between them was 9 mm. The pieces of a Zn foil (Aldrich 99.99% purity, 0.25 mm thickness) with a radius of 6 mm were treated in acetone and ethanol for 15 min with ultrasonic cleaning and then washed with deionized water. All electrolytes were prepared from chemical reagents (Aldrich) and deionized water. The electrolyte composition was a 1:10 volume mixture of HF and deionized water without any surfactant. The reaction durations were 1, 3, 6, and 10 min. (Table 1) and the applied voltage was 5 V. Obtained nanosheets (NSs) were washed in deionized water to eliminate remanent compounds and dried in air at room temperature. Then the part of the samples was annealed at 400 °C during 2 h in ambient air.

Table 1. Technological parameters of studied ZnO-Cu nanosheets.

Sample	*Time, (min.)*	*Voltage, (V)*	*Annealing*
1	1	5	+
2	3	5	+
3	6	5	+
4	10	5	+

Raman scattering spectra were measured in a Jobin-Yvon LabRAMHR800 UVmicro-Raman system using excitation by a solid state LED with a light wavelength of 633 nm [8-10]. PL spectra, excited by a He–Cd laser with a wavelength of 325 nm and a beam power of 80 mW, were measured at 300 K using a PL setup based on a spectrometer SPEX500 described in [10,11].

RESULTS AND DISCUSSION

Raman scattering spectra measured in the range 80-650 cm^{-1} are presented in figure 1. Raman spectra of fresh prepared ZnO-Cu nanosheets show peaks centered in 96, 228, 277 and 423 cm^{-1}, (Fig.1a). The high band in the range 220-230 cm^{-1} is related, apparently, to defect-induced modes [12]. No Raman peaks of CuO or Cu_2O Appeared in the spectrum, indicating no secondary phase in our samples.

Thermal annealing at 400 °C for 2 hours in ambient atmosphere was done for stimulates the crystallization process of copper-doped samples. This crystallization is accompanied by the appearance in Raman scattering spectra of five Raman peaks at 96, 331, 379, 435 and 576-586 cm^{-1} (Fig.1b). With increasing the anodization duration the intensity of Raman peaks enlarges, apparently, due to increasing the crystalline ZnO-Cu NC volume (Fig.1b). The group theory predicted for the wurtzite crystal structure the Raman active optic phonons in Brillouin zone center: i) A_1 and E_1 symmetry polar phonons with different frequencies for the transverse (TO) and longitudinal (LO) optic phonons, and ii) E_2 symmetry non-polar phonon mode with two frequencies E_2 (low) and E_2 (high) (Table 2). Note E_2 (high) mode is associated with oxygen sublattice and E_2 (low) is attributed to the Zn sublattice [13, 14]. Additionally some Raman peaks, such as 330 and 437 cm^{-1}, are attributed to the second order Raman peaks arising from the zone boundary phonons $3E_{2H}$-E_{2L} and E_2H modes of ZnO crystal [15].

Raman peaks at 379 and 435 cm^{-1} detected in ZnO nanosheets after the thermal annealing which accompanied by the ZnO crystallization can be attributed to first order Raman peaks in wurtzite crystall lattice related to the A_1(TO) and E_2(high) phonon modes. The Raman peak at 331 cm^{-1} is consistent with the second order Raman peak arising from the zone boundary phonons at $3E_{2H}$-E_{2L}. The nature of the Raman peak at 576-586 cm^{-1} is not clear. Its variable position in different samples (Fig.1b, curves 2,3 and 4) in the range 576-586 cm^{-1} and its location between TO and LO optic phonons permit to attribute this peak to the surface phonon (SP) Raman scattering.

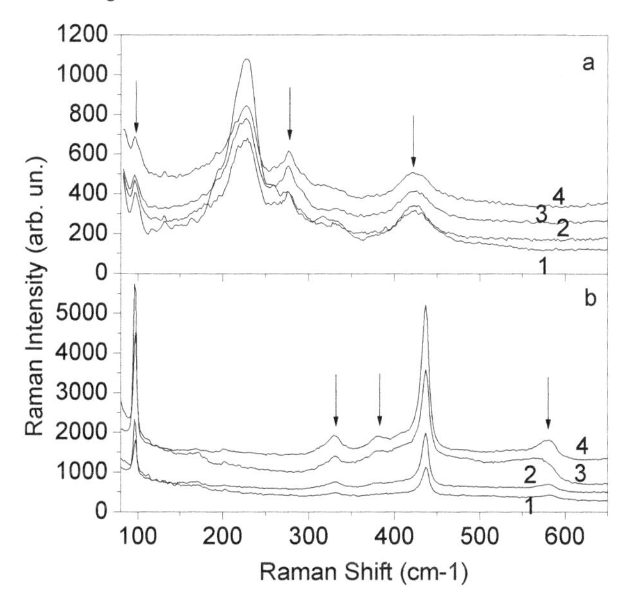

Fig. 1. Raman signal for ZnO-Cu nanosheets a) 1-1 min., 2-3 min., 3-6 min., 4-10 min. without thermal annealing and b) 1-1 min., 2-3 min., 3-6 min., 4-10 min. with thermal annealing.

Table 2. Raman peaks in ZnO-Cu

Samples	E_2(low) cm^{-1}	$3E_{2H}$-E_{2L} cm^{-1}	A_1(TO) cm^{-1}	E_1(TO) cm^{-1}	E_2(high) cm^{-1}	SP cm-1	A_1(LO) cm^{-1}	E_1(LO) cm^{-1}
Bulk ZnO[14]	102		379	410	439		574	591
ZnO-Cu nanosheet		331	380		435	576		

PL spectra of freshly prepared copper-doped samples are presented in figure 2 for samples with different anodization durations used. As it is clear the PL spectra are complex and can be represented as superposition of, at least, 4th elementary PL bands with the peaks at 1.43, 2.20, 2.58 and 2.86 eV (Fig.2, curves 5, 6, 7 and 8). Mentioned elementary visible PL bands are connected with the defect-related emission in ZnO-Cu nanosheets of the amorphous phase. With increasing the anodization duration up to 10 min the PL band at 2.86 eV decreases and the PL band at 2.40- 2.58 eV monotonically increases. Last PL band can be associated with the Cu impurity [16-18] or surface defects [19]. The main reason, why the band centered at 2.58 eV increases with increasing the time of etching, apparently, is connected with a bigger quantity of defects associated to the Cu owing to using the Cu electrode. So it can be supposed that the emission center, responsible for the 2.40- 2.58 eV PL band in studied ZnO-Cu nanosheets, is related to Cu impurities and/or surface defects.

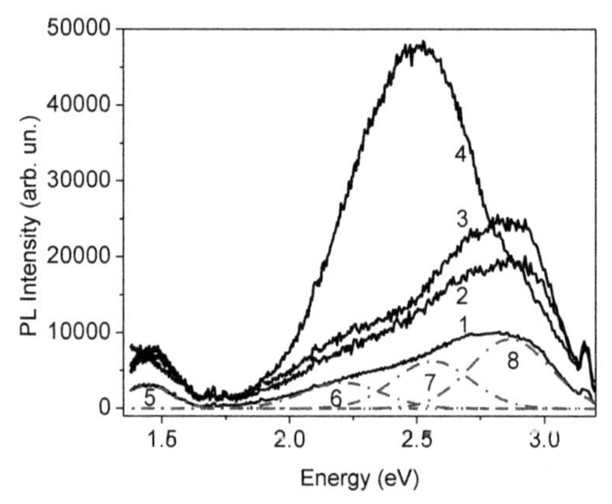

Fig. 2. PL spectra of fresh prepared ZnO-Cu nanosheets obtained at the anodization durations 1min (1), 3 min (2), 6 min (3) and 10 min (4). Dashed curves present the deconvolution of the experimental PL spectrum (1) on elementary PL bands (5, 6, 7, 8).

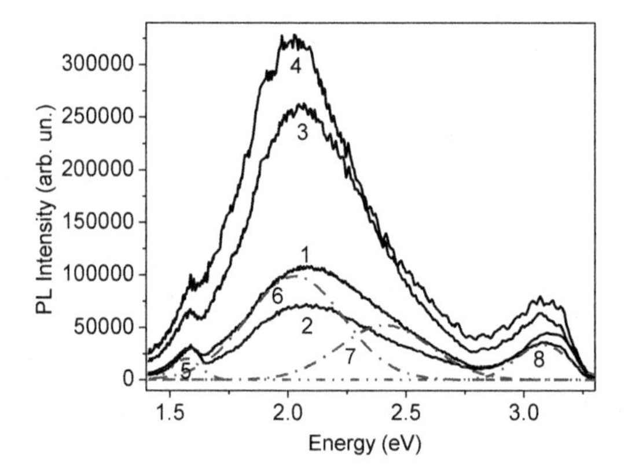

Fig. 3. PL spectra of thermal annealed ZnO-Cu nanosheets obtained at the anodization durations 1min (1), 3 min (2), 6 min (3) and 10 min (4). Dashed curves present the deconvolution of the experimental PL spectrum (1) on elementary PL bands (5, 6, 7, 8).

Thermal annealing at 400 °C for 2 hours in ambient atmosphere was performed, which is accompanied by the essential transformation of PL spectra (Fig. 3). PL spectra of annealed ZnO-Cu nanosheets are complex as well (Fig.3, curves 1,2,3 and 4) and can be presented as a set of elementary PL bands (Fig.3, curves 5,6,7,8) with the peaks at 1.54, 2.03, 2.41 and 3.07 eV. As it is clear the defect-related PL band with the peak at 2.86 eV disappeared completely at the process of ZnO-Cu crystallization. In comparison with the fresh samples, the PL bands (1.54, 2.03, 2.41 eV) suffer a shift wit constant energy by ±0.17 eV.

The PL band with the peak at 2.86 eV is attributed to the Cu related defects [20], to Zn interstitials [21] or to the donor-acceptor pairs including the shallow donor and oxygen vacancy [22]. In our samples the 2.86 eV PL band completely disappeared after the oxidation at thermal annealing in ambient air. Thus we can conclude that 2.86 eV PL band detected in fresh prepared ZnO-Cu nanosheets related to the native defects and the concentration of these defects decreases at the oxidation and crystallization that is the reason of PL band disappearing.

The near-band-edge emission at 3.07 eV at 300K is attributed to the optical transition between the shallow donor and valence band [23]. However, the peak position at room temperature can vary significantly due to overlapping with free exciton emission and phonon replicas, and will be different for different growth conditions [24].

CONCLUSIONS

Raman scattering and emission of thermal annealed crystalline ZnO-Cu nanosheets, as well as their dependences on electrochemical etching parameters and annealing, have been studied. From Raman studies show that no secondary phase in Cu-doped ZnO nanostructures. For some PL bands (2.86 and 2.58 eV) the nature of radiative defects has been clarified. It is shown that the creation of crystalline ZnO-Cu nanosheets is accompanied by the enlargement of the PL intensity and surface defect related visible PL bands. The anodization method has been recognized as one of the simplest, economic, non-vacuum and low temperature effective method to prepare ZnO-Cu nanosheets and this method permits in different technological regimens wide range of PL emission compared with another methods.

ACKNOWLEDGMENTS

The authors would like to thank CONACYT (Project 130387) and SIP-IPN, Mexico, for the financial support, as well as the Dr. T.V. Torchynska for the fruitful discussion and the Dr. J.L. Casas Espinola and CNMN-IPN for the PL and Raman measurements, respectively.

REFERENCES

1. A. I. Diaz Cano, B. El Filali, T. V Torchynska and J. L. Casas Espinola, *J. of Phys. and Chem. of Sol.*, **74**,431 (2013).

2. A. I. Diaz Cano, B. El Filali, T. V Torchynska and J. L. Casas Espinola, *Physica E*, **51**, 24 (2013).
3. Z.F. Liu, F.K. Shan, J.Y. Sohn, S.C. Kim, G.Y. Kim, Y.X. Li, Y.S. Yu, J. Electroceram. **13** 183 (2004).
4. D.J. Qiu, H.Z. Wu, A.M. Feng, Y.F. Lao, N.B. Chen, T.N. Xu, Appl. Surf. Sci. **222** 263 (2004).
5. F.K. Shan, B.I. Kim, G.X. Liu, Z.F. Liu, J.Y. Sohn, W.J. Lee, B.C. Shin,Y.S. Yu, J. Appl. Phys. **95** 4772 (2004).
6. P. Dahany, V. Fleurovy, P. Thurianz, R. Heitzz, A. Hoffmannz, I. Broserz, J. Phys.: Condens. Matter **10** 2007 (1998).
7. D. Gruber, F. Kraus, J. Muller, Sensors and Actuators B **92** 81 (2003).
8. M.A. Reshchikov, H. Morkoc, B. Nemeth, J. Nause, J. Xie, B. Hertog, A. Osinsky, Physica B. Condensed Matter **401–402** 358(2007).
9. M. Dybic, S. Ostapenko, T.V. Torchynska, E.Velazquez Lozada, App. Phys. Lett. **84** (25) 5165 (2004).
10. T.V. Torchynska, A.I. Diaz Cano, M. Dybic, S. Ostapenko, M. Mynbaeva, Physica B: Condensed Matter **376–377** 367 (2006).
11. M. Dybic, S. Ostapenko, T.V. Torchynska, E. Velazquez Lozada, Appl. Phys. Lett. **84** (25), 5165-5167 (2004).
12. F.J. Manjon, B. Mari, J. Serrano, A.H. Romero, J. Appl. Phys. **97** 053516 (2005).
13. N. Ashkenov, B. N. Mbenkum, C. Bundesmann, V. Riede, M. Lorenz, D. Spemann, E. M. Kaidashev, A. Kasic, M. Schubert, M. Grundmann, G. Wagner, H. Neumann, V Darakchieva, H. Arwin, and B. Monemar, J.Appl.Phys. **93,** 126 (2003).
14. J.F. scott, Phys. Rev. B, **2**, 1209 (1970).
15. S. Ghoopum, N. Hongsith, P. Mangkorntong, Physica E, **39,** 53-56 (2007)
16. C. West, D.J. Robbins, P.J. Dean, W. Hays, Physica B, **116,** 492 (1983).
17. R. Dingle, Phys. Rev. Lett. **23,** 579 (1969).
18. N.Y. Garces, L. Wang, L. Bai, N.C. Giles, E. Halliburton, G. Cantwell, App. Phys. Lett. **81**, 622 (2002).
19. A.B. Djurišic, W.C.H. Choy, V.A.L. Roy, Y.H. Leung, C.Y. Kwong. K.W. Cheah, T.K. Gundu Rao, W.K. Chan, H.F. Lui, C. Surya, Adv. Funct. Mater. **14,** 856 (2004).
20. M.A. Reshchikova, H. Morkoc, B. Nemeth, J. Nause, J. Xie, B. Hertog, A. Osinsky, Physica B, Condensed Matter, **401–402**, 358 (2007).
21. M.K. Patra, K. Manzoor, M. Manoth, S.P. Vadera, N. Kumar, J. Lumin. **128** (2) 267–272 (2008).
22. D.H. Zhang, Z.Y. Xue, Q.P. Wang, J. Phys. D: Appl. Phys. 35 (21) 2837–2840 (2002).
23. A. B. Djuris, A.M.C. Ng, X.Y. Chen, Progress in Quantum Electronics **34**. 191 (2010).
24. T. Voss, C. Bekeny, L. Wischmeier, H. Gafsi, S. Borner, W. Schade, A.C. Mofor, A. Bakin and A. Waag, Appl. Phys. Lett. **89**, 182107 (2006).

Mater. Res. Soc. Symp. Proc. Vol. 1617 © 2013 Materials Research Society
DOI: 10.1557/opl.2013.1172

Microstructural evolution during mechanical treatment of ZnO and black NiO powder mixture

M. Kakazey,[1,*] M. Vlasova[1], Y. Vorobiev[2], I. Leon[3], M. Cabecera Gonzalez[4], E.A. Chávez Urbiola[2]
[1]CIICAp-Universidad Autonoma del Estado de Morelos, Cuernavaca, Mexico
[2]Unidad Querétaro del CINVESTAV-IPN, Querétaro, Mexico
[3]CIQ-Universidad Autonoma del Estado de Morelos, Cuernavaca, Mexico
[4]FCQI-Universidad Autonoma del Estado de Morelos, Cuernavaca, Mexico

ABSTRACT

Kinetics of the microstructural evolution in ZnO and NiO black powder mixture during prolonged mechanical processing (MP) was investigated by Scanning electron microscopy (SEM), Laser Particle Sizer (LPS), X-ray diffraction, electron paramagnetic resonance (EPR), infrared absorption (FTIR) and UV-Visible diffuse reflection methods.

INTRODUCTION

ZnO-NiO system has attracted particular attention as the basis for creating ultraviolet light-emitting diodes [1, 2], transparent p-NiO/n-ZnO semiconductors [3, 4], seeking opportunities to create the room temperature ferromagnetism [5] etc. This study provides information about the production and behavior of native defect states in ZnO-NiO system with increasing mechanical treatment time. The structural defects in initial components of samples as well as the defects which are formed in zones of destruction are subjected to: a local hyper-rapid ultrahigh-temperature spike (at the moment of destruction), significantly prolonged collective high-temperature influences (at the moment of ball loading), and accumulative thermal effects.

EXPERIMENTAL RESULTS

The basic starting materials were the commercially available ZnO powder (Reasol, Milan, Italy) and the nickel oxide black powder (99.26 % purity; Fluka Chemie J.). The ZnO + 1% wt. NiO mixture was prepared. MP of samples was carried out in a Planetary Ball-Mill (type PM 400/2, Retsch Inc.) using the rotation speed of 400 rev×min^{-1}. The durations of mechanical processing, t_{MP}, were: 1, 3, 9, 30, 90 and 390 min.

The Scanning electron microscopy (SEM), model LEO 1450 VP; the Laser Particle Sizer (LPS), Analysette 22 COMPACT, FRITSCH; the X-ray diffractometer, Siemens D-500 with CuKa radiation; the Electron Paramagnetic Resonance equipment (EPR), SE/X 2547 – Radiopan; an infrared (IR) spectrophotometer (Bruker Vector 22, FTIR) and the UV-Vis diffuse reflectance spectroscope (DRS), with Pelkin–Elmer Lamda 2 spectrometer, were used in this work.

SEM results. The initial samples consisted of ZnO-particles with a spherical-like shape. The average ZnO particle size, D_{SEM}^{ZnO} is estimated of ~ 250 nm. The NiO large granules are characterized by a spherical-like shape as well with the size from 3 to 10 mm. Thus, the initial NiO (black) powder is a granular powder. A consecutive reduction of D_{SEM}^{ZnO} with increasing t_{MP} has been detected.

LPS results. The initial sample consisted of the particles with sizes: $D_{LPS}^{ZnO} \sim 3\times10^2 - 5\text{-}8\times10^3$ nm. For t_{MT} equal to 30 min the quantity of fine fraction particles with $D_{LPS}^{ZnO} \leq 300$ nm increased from 1 to 3%) and the maximum of size distribution curve has shifted to smaller, D_{LPS}^{ZnO}, values: from 1500 nm to 850 nm. Actually the size distributions of ZnO particles are the plots of the size distributions of strength ZnO particle agglomerates which cannot be destroyed by ultrasonic processing used in the equipment LPS Analysette 22 COMPACT (Fig. 1a).

X-ray diffraction results. The NiO black particles are characterized by the NaCl-type crystal structure with the average crystallite size of about $D_X^{NiO} \sim 40$ nm. XRD patterns of the ZnO powders as a function t_{MP} shows decreases the amplitude of the XRD peaks and increases its width. The change of the ZnO crystallite size [6], D_X^{ZnO}, depending on the duration of MP is shown in Fig. 1b. The variation of the average ZnO particle size during MP can be presented using the following equation: $D_X^{ZnO} \sim 1/\, t_{MP}^{\ 0.45}$.

EPR results. The EPR signals were not detected in the initial ZnO sample. The EPR spectrum of the original NiO sample can be well interpreted as a superposition of two signals (A and B). The signal A is characterized by g ≈ 2.12 and a line width of around 940 G. The stoichiometric NiO material is antiferromagnetic with the Néel temperature equal to 523 K [7] and does not show the EPR signals at room temperature. The non-stoichiometric NiO material can present paramagnetic properties at room temperature, due to the presence of Ni^{3+} ions in the samples [8]. Some thermal changes of the EPR signal (g_1, g_2, ΔB) were assigned earlier to the different ratios of ions with a valence of Ni^{2+} and Ni^{3+} [9]. The signal B has g ~ 2.03 and a line width of around 430 G. Rapid decreasing the B signal intensity (see Fig.1c) takes place with increasing t_{MP}, and after the MP duration of $t_{MP} > 30$ min the EPR spectrum is determined by the A signal only. This signal may be due to the some ferromagnetic by-product, localized on the surface of NiO crystallites, which was derived from a granulation process. The formation and evolution of electron-hole paramagnetic centers in ZnO during MP were detected in EPR spectra earlier [10, 11]. The signal **I** with $g_\perp = 2.0190$, $g_\parallel < g_\perp$ was attributed to the $V_{Zn}^- : Zn_i^o$ center [12], the signal **II** with $g_\perp = 2.0130$, $g_\parallel = 2.0140$ was assigned to the V_{Zn}^- center [13], and the signal **III** with $g_1 = 2.0075$, $g_2 = 2.0060$, $g_3 = 2.0015$ was assigned to the $(V_{Zn}^-)_2^-$ center [12]. These centers were formed in the destruction zone of ZnO particles. Fig. 1d shows the change of the EPR signal intensity with increasing t_{MP}. The EPR signals **I**, **II** and **III** appeared and disappeared with increasing t_{MP} in the range of 1 to 90 min. Using the variation of EPR signals **I**, **II** and **III** (see Fig. 1d) as a temperature probe, can allow monitoring the changes of sample temperature, T_{MP}, depending on the duration of MP (Fig. 1e). The average temperature of a sample achieves the maximum value (equilibrium temperature, T_{eq}) around 800 K [14]. Thus, the ensemble of defects (the active centers) formed in a final material will be substantially defined by the temperature conditions developed at MP.

IR absorption results. There are two strong absorption peaks in the initial ZnO sample: the ZnO stretching mode at ~ 450 cm^{-1} and a peak at 3500-3300 cm^{-1} that can be attributed to hydroxyl groups (adsorption of H_2O and the formation of $Zn(OH)_2$). The analysis of the form of ZnO IR absorption spectra [15, 16] testifies on the existence of a set of particles with $L_\perp/L_\parallel = 1 \div$

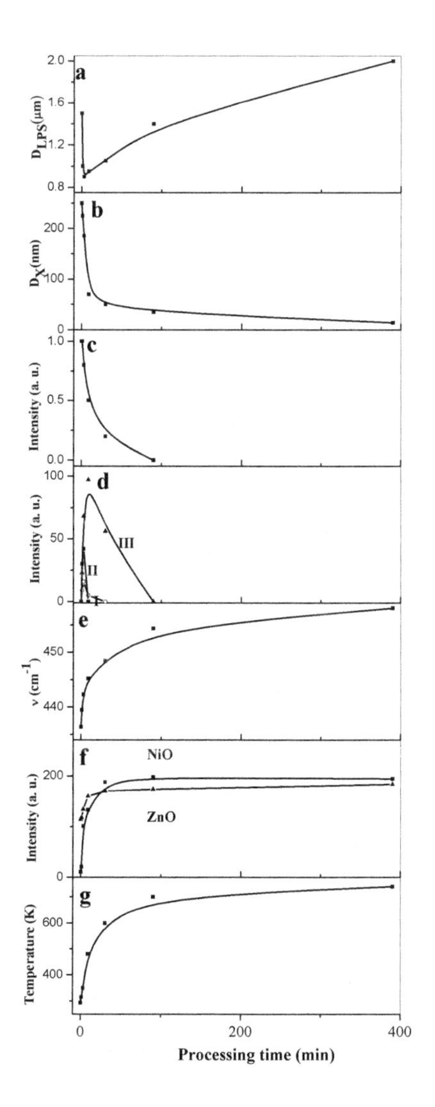

Fig. 1. Summary of consecutive changes that taken place in a mixture ZnO + 1 % wt. NiO black during longtime MP. a) Average particle sizes, D_{LPS}^{ZnO}, defined by a Laser Particle Sizer; b) Crystallites average size, D_X^{ZnO}; c) Intensity of the EPR signal B in NiO black; d) Intensity of the EPR spectra of electron-hole centers **I, II** and **III** in ZnO; e) Shift of IR ZnO-peak absorption; f) Intensity of the 380 nm and 410 nm UV-Vis DR peaks; g) Average sample temperature, T_{MP}.

1.4 in initial samples, where $L_{\parallel}$ and $L_{\perp}$ are the form factors ($L_{\parallel}$ is connected with the crystallographic axis c and $L_{\parallel} + 2L_{\perp} = 1$) [15]. In the Fig. 1e it is shown that with increasing t_{MP}

≥ 30 min the IR peak shifted towards the higher wave numbers. The broadening of the ZnO-IR peak at $t_{MP} \geq 30$ min indicates on the agglomeration of nanoparticles.

UV-VIS DRS results. The actual optical characteristics are the absorption coefficient α and the scattering coefficient σ, and their sum determines the total attenuation of light in the material. These characteristics could be ascribed to individual particles, agglomerates or granules; therefore the size evolution might be important. The standard way of optical characterization of granular material is the Diffuse Reflectance Spectroscopy (DRS) measuring the diffuse reflection spectra (that were actually measured in our experiments) and presenting the results in units of Kubelka-Monk function [17, 18] defining the ratio of absorption to scattering coefficients:

$$F(R) = (1 - R)^2/2R = \alpha/\sigma.$$

Here R is the diffuse reflection coefficient. Fig. 2 presents the corresponding data taken after different MP processing times. We take that the scattering coefficient σ has spectral dependence of the type λ^{-4}, the absorption coefficient a has rapid spectral changes near semiconductor's fundamental absorption edge thus showing corresponding photon's energy in the F(R) spectra. Fig. 2 shows that in the untreated sample we see only two peaks: narrow at 380 nm (3.26 eV) and wide with a maximum at 410 nm (3.03 eV). The first one represents the transition from near conduction band edge to valence band of ZnO, although there are different points of view [19, 20] on the nature of this peak. The peak at 410 nm (with wide long wavelength tail at 425 – 800 nm) is connected with the NiO sample component, it is absent in pure ZnO samples. The value of the fundamental energy-band gap of stoichiometric NiO samples is in the range of 3.4–3.8 eV [21]. For nonstoichiometric NiO the fundamental energy band gap varied (data of DRS) from 2.818 eV to 4.133 eV by changing the preparation $NiCO_3$ firing temperature from 800 °C to 1200 °C [8]. The change $I_{DR}^{ZnO}(t_{MP})$ (see Fig. 1f) we refer to the destruction of large aggregates of particles of ZnO present in the original samples. The observed change $I_{DR}^{NiO}(t_{MP})$ we refer to intense grinding of NiO grains. Grinding pure ZnO powders initiates the emergence of scattering in the long-wavelength region of the spectrum due to the formation of vacancy defects (V_{Zn}, V_O, Zn_{in}) in ground ZnO particles [22, 23].However, the increase in the intensity of the scattering bands with increasing t_{MP} is much slower than the increase in the intensity of the scattering of non-stoichiometric NiO. Thus we ascribe this tail in the ZnO + 1%NiO samples mainly to the scattering by the defect states in non-stoichiometric NiO, but some participation of ZnO defects is also possible.

DISCUSSION

Initial samples ZnO + 1% NiO black are a mixture of single and agglomerated particles of ZnO ($D_X^{ZnO} \sim 250$ nm and $D_{LPS}^{ZnO} \sim 3$ μm) and large granules ($D_{SEM}^{NiO} \sim 3$-10 μm) spherical-like shape powders of high defects nonstoichiometric NiO black ($D_X^{NiO} \sim 40$ nm). During MP the ZnO particles grinding ($D_X^{ZnO} \sim 16$ nm at $t_{MP} = 390$ min) and the deagglomeration ($t_{MP} \leq 30$ min) and "secondary" agglomeration ($t_{MP} > 30$ min) processes taken places. Data of UV-VIS diffuse reflectance spectroscopy (Fig. 1g and Fig. 2) show that the significant increasing of NiO surface takes place at $t_{MT} \rightarrow 30$ min. That is, the NiO particle size changes from the original size of

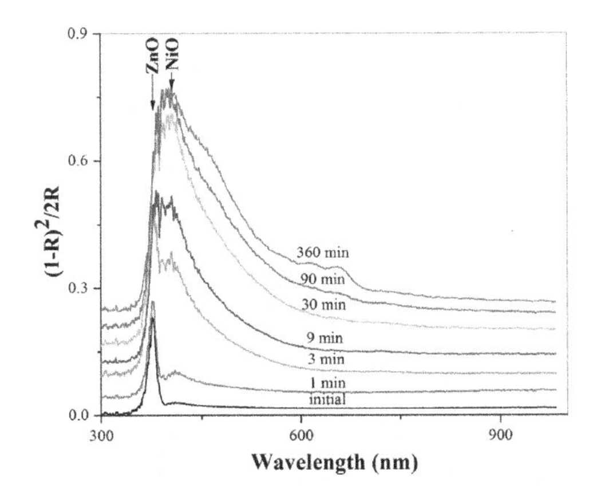

Fig. 2. UV-Vis diffuse reflection spectra of ZnO + 1%NiO samples depending on duration of mechanical treatment. Spectra are shifted vertically by $(R - 1)^2/2R = 0.04$ from one another.

granules ~ 3-10 μm down to $D_{NiO} << \lambda_{UV\text{-}Viz}$, or $D_{NiO} < 100$ nm, or $D_{NiO} \sim D_X^{NiO} \sim 40$ nm. The different point defects ($V_{Zn}^{-} : Zn_i^{o}$ (**I**), V_{Zn}^{-} (**II**), and $(V_{Zn}^{-})_2^{-}$ (**III**) centers) are forming in the deformation-destruction zones of ZnO particles. These defects have different physical and chemical properties, and have different activation energies of annealing, E_{act}. In turn, the formation of defects is accompanied by the development of various mechano-thermal processes, which increase the sample temperature, T_{MP}, for longer duration of MP, t_{MP}. The combination of all local pulse-mechano-thermal processes results in an increase of the average temperature, T_{MP}, of the sample at an initial stage of MP. In this case the consecutive annealing of defects befalls. The annealing of defects, which have a minimal value of E_{ac}, take place at first (**II**, I and **III**). The equilibrium temperature of sample, T_{eq}, determines the type of defects accumulated in the sample with increasing t_{MP} (see Fig. 1g). The preconditions are created for the occurrence of the solid-state interactions between the components of the treated mixture. The absence of EPR $ZnO:Ni^{3+}$ signals shows that induced defect states and the temperature in the MP ZnO + 1% NiO black sample are insufficient for the diffusion of nickel atoms in the zinc oxide lattice.

CONCLUSIONS

Microstructural evolution in ZnO and NiO black powders mixture during prolonged mechanical processing (MP) was investigated by SEM, LPS, X-ray, EPR, FTIR and UV-Vis diffuse reflection methods. During MP, the ZnO particles grinding and the deagglomeration (starting period of MP) and "secondary agglomeration" (after prolonged time of MP) processes taken places. Becides, an intensive grinding of NiO granules (at $t_{MT} > 30$ min $D_{NiO} < \lambda_{UV\text{-}Viz}$) and formation of mixed NiO and ZnO nanoparticles take place. Some point defects (namely, $V_{Zn}^{-} : Zn_i^{o}$ (**I**), V_{Zn}^{-} (**II**), and $(V_{Zn}^{-})_2^{-}$ (**III**) centers) having different physical and chemical properties are forming in deformation-destruction zones of ZnO particles. Formation of defects is accompanied by development of various mechanothermal processes, which increase the sample temperature, T_{MP}. The increase of t_{MP} (and T_{MP}) activates reaction processes: promotes consecutive annealing of «low-temperature» defects with small values of E_{act} (**II**, **I** and **III**).

Thermal effects on the defect structure of the surface of the particles (dehydration, interaction etc.) take place. During MP the defect structure modification in NiO black was noted, but the formation of a solid solution of Ni in ZnO was not observed.

REFERENCES

[1] Y. Y. Xi, Y. F. Hsu, A. B. Djurišić, A. M. C. Ng, W. K. Chan, H. L. Tam, K. W. Cheah, Appl. Phys. Lett. **92,** 113505 (2008).

[2] H. Long, G. Fang, H. Huang, X. Mo, W. Xia, B. Dong, X. Meng, X. Zhao, Appl. Phys. Lett. **95**, 013509 (2009).

[3] H. Ohta, M. Hirano, K. Nakahara, H. Maruta, T. Tanabe, M. Kamiya, T. Kamiya, H. Hosono, Appl. Phys. Lett. **83,** 1029 (2003).

[4] M. Cavas, R. K. Gupta, A. A. Al-Ghamdi, O.A. Al-Hartomy, F. El-Tantawy, F. Yakuphanoglu, J. Sol-Gel. Sci. Technol. 64, 219–223 (2012).

[5] P. Srivastava, S. Ghosh, B. Joshi, P. Satyarthi, P. Kumar, D. Kanjilal, D. Buerger, S. Zhou, H. Schmidt, A. Rogalev, F. Wilhelm, J. Appl. Phys. **111,** 013715 (2012).

[6] H.P. Klug, L.E. Alexander, X-ray Diffraction Procedures for Polycrystalline and Amorphous Materials, 2nd edition, J. Wiley & Sons, New York, 1974.

[7] W.L. Roth, Phys. Rev. B 110, 1333 – 1341 (1958).

[8] S. A. Gad, M. Boshta, A. M. Aboel-Soud, M. Z. Mostafa, Fizika A 18, 173–184 (2009).

[9] M. Rubinstein, R.H. Kodama, S.A. Makhlouf, JMMM, 234, 289–293 (2001).

[10] M.G. Kakazey, T.V. Sreckovic, M.M. Ristic, J. Mater. Sci. **32,** 4619 – 4622 (1997).

[11] M. G. Kakazey, M. Vlasova, M. Dominguez-Patiño, G. Dominguez-Patiño, G. Gonzalez-Rodriguez, B. Salazar-Hernandez, J. Appl. Phys. **92,** 5566 (2002).

[12] B. Schallenberger, A. Hausmann, Z.Physik **B 23,** 177 - 181 (1976).

[13] D. Galland, A. Herve, Sol. St. Comm. **14,** 953 – 956 (1974).

[14] M. Kakazey, M. Vlasova, M. Dominguez-Patiño, I. Leon, M. Ristic, J. Mater. Sci. **42** 7116-7122 (2007).

[15] S. Hayashi, N. Nakamori, H. Kanamori, J. Phys. Soc. Jpn. **46,** 176 – 183 (1979).

[16] M.G. Kakazey, V.A. Melnikova, T. Sreckovic, T.V. Tomila, M.M. Ristic, J. Mater. Sci. **34** 1691 – 1697 (1999).

[17] C.R. Bamford, Color Generation and Control in Glasses, Elsevier, Amsterdam (1977).

[18] P. Kubelka, F. Monk, Z. Tech. Phys. **12,** 593- 608 (1931).

[19] V.S. Yalishev, Y.S. Kim, X.L. Deng, B.H. Park, Sh.U. Yuldashev, J. Appl. Phys. **112** 013528 (2012).

[20] S. T. Tan, B. J. Chen, X. W. Sun, W.J. Fan, J. Appl. Phys. **98,** 013505 (2005).

[21] S. Nandy, U.N. Maiti, C.K. Ghosh, K.K. Chattopadhyay, J. Phys.: Condens. Matter. **21** 115804 (2009).

[22] A. B. Djurišić, W. C. H. Choy, V. A. L. Roy, Y. H. Leung, C. Y. Kwong, K. W. Cheah, T. K. Gundu Rao, W. K. Chan, H. Fei Lui, C. Surya, Adv. Funct. Mater. **14,** 856–864 (2004).

[23] M. Scepanovic, M. Grujic-Brojcin, K. Vojisavljevic, T. Sreckovic, J. Appl. Phys. **109** 034313 (2011).

Mater. Res. Soc. Symp. Proc. Vol. 1617 © 2013 Materials Research Society
DOI: 10.1557/opl.2013.1173

Photoluminescence Variation With Temperature in ZnO:Ag Nanorods obtained by Ultrasonic Spray Pyrolysis

E. Velázquez Lozada[1*], S. Mera Luna[2], and L. Castañeda[3]
[1]ESIME – Instituto Politécnico Nacional, México D.F. 07738, México.
[2]ESIQE – Instituto Politécnico Nacional, México D.F. 07738, México.
[3]ESIME – Ticomán – Instituto Politécnico Nacional, México D.F. 07340, México.

ABSTRACT

The photoluminescence, its temperature dependences, as well as structural characteristics obtained by the method of Scanning electronic microscopy (SEM) have been studied in ZnO:Ag nanorods prepared by the ultrasonic spray pyrolysis (USP). PL spectra of ZnO:Ag NRs in the temperature range from 10 K to 300 K are investigated. Three types of PL bands have been revealed: i) the near-band-edge (NBE) emission, ii) defect related emission and iii) IR emission. It is shown that IR emission corresponds to the second-order diffraction of near-band-edge (NBE) emission bands. The study of NBE PL temperature dependences reveals that the acceptor bound exciton (ABE) and its second-order diffraction peak disappeared at the temperature higher than 200 K. The attenuation of the ABE peak intensity is ascribed to the thermal dissociation of ABE with appearing a free exciton (FE). The PL bands, related to the LO phonon replica of FE and its second-order diffraction, dominate in the PL spectra at room temperature that testify on the high quality of ZnO:Ag films prepared by the USP technology.

INTRODUCTION

Zinc oxide (ZnO) nanocrystals (NCs) with wide band gap energy (3.37 eV) have attracted the great attention due to exceptional exciton properties (high exciton binding energy equal to 60 meV at 300K) and a number of deep levels that emit in the whole visible range and, hence, can provide intrinsic “white” light emission. ZnO NCs are promising candidates for the different optoelectronic devices such as light emitting diodes and lasers [2-7]. The control of the ZnO defect structure in ZnO nanostructures is a necessary step in order to improve the device quality. Since the structural imperfection and defects generally deteriorate the exciton related recombination process, it is necessary to grow the high quality films for the efficient light-emitting applications. The ultrasonic spray pyrolysis (USP) method is a simple, inexpensive, non-vacuum and a low temperature technique for the film synthesis [8]. It will be interesting to study emission of the USP produced ZnO nanostructures doped with Ag versus temperature in order to identify the best regimes for obtaining bright emitting NCs and the nature of optical transitions.

EXPERIMENTAL DETAILS

ZnO:Ag thin solid films were prepared by the USP technique on the surface of soda-lime glass substrate for the substrate temperatures 400^0 C and the different deposition times of 3, 5 and 10 min. Using this technique the nanoparticle's size can be easily controlled by changing a concentration of starting solution and the atomization parameters. The deposition system includes a piezoelectric transducer operating at variable frequencies up to 1.2 MHz and the ultrasonic power of 120 W. ZnO:Ag thin solid films were deposited from a 0.4 M solution of zinc (II) acetate [$Zn(O_2CCH_3)_2$] (Alfa), dissolved in a mix of deionized water, acetic acid [CH_3CO_2H] (Baker), and methanol [CH_3OH] (Baker) (100:100:800 volume proportion). Separately, a 0.2 M solution of silver nitrate [$Ag(NO_3)$] (Baker) dissolved in a mix of deionized water and acetic acid [CH_3CO_2H] (Baker) (1:1 volume proportion) was prepared, in order to be used as doping source. A constant [Ag]/[Zn] ratio of 2 at. % was applied at the ZnO Ag film preparation. The morphology of ZnO:Ag films has been studied using the scanning electron microscopy (SEM) Dual Beam, FEI brand, model Quanta 3D FEG with field emission gun. PL spectra were measured in the temperature range 10-300K at the excitation by a He-Cd laser with a wavelength of 325 nm and a beam power of 20 mW at 300K using a PL setup on a base of spectrometer SPEX500 described in [9-11].

EXPERIMENTAL RESULTS AND DISCUSSION

SEM images of the typical ZnO:Ag nanorods (NRs) obtained at the deposition times 3 and 10 min are presented in figure 1. It is clear that the ZnO nanorods have the hexagonal cross section and the rod orientation along the c-axis. The cross section size of ZnO nanorods increases with the duration of UPS process from 50-70 nm (for the duration of 3min), 100-150 nm (at 5min) and 150-200 nm (at 10 min). PL spectra of ZnO:Ag NRs are shown in figure 2 for all studied samples.

Figure 1. SEM images of the samples prepared at the substrate temperatures 400^0 C and the durations of 3 (a) and 10 (b) min.

It is clear that the PL spectra are complex and can be represented as a superposition of elementary PL bands with the peaks in the spectral ranges: 2.90-3.25 eV (I, II), 2.00-2.50 eV (III, IV) and 1.45-1.61 eV (V, VI). The deconvolution procedure has been applied to the PL spectra with the aim to separate the elementary PL bands (Fig.3). The analysis of figures 3 permits to distinguish six elementary PL bands with the peaks at: 3.25 (1), 2.92 (2), 2.70 (3), 2.10 (4), 1.62 (5) and 1.46 (6) eV at 10K (Fig.3).

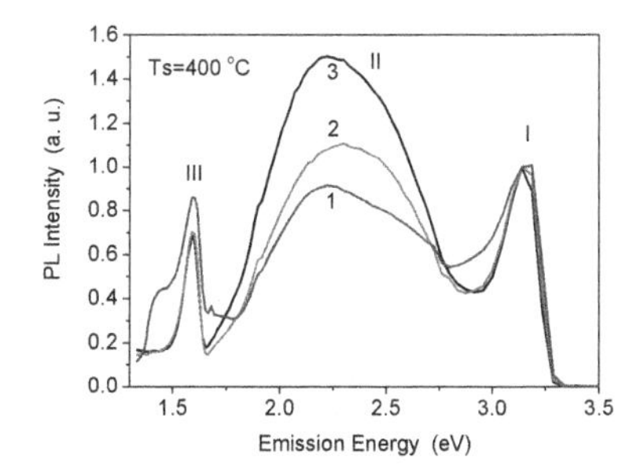

Figure 2. Normalized PL spectra of samples prepared on the substrate with T = 400° C at the deposition times: 1-3 min, 2-5 min, 3-10 min.

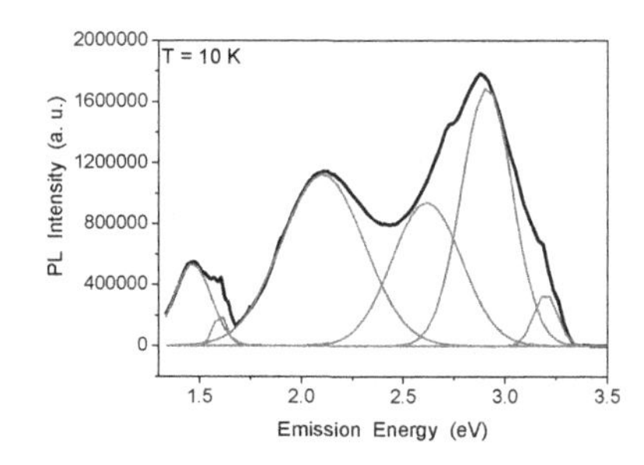

Figure 3. PL spectrum deconvolution on six elementary PL bands for the film obtained at 400^0 C at the duration 3 min.

It well known that the UV-visible PL bands (I, II) in ZnO are near-band-edge (NBE) or free exciton emissions [12]. The PL band in the spectral range 2.00-2.70 eV related to the defect emission [12] and the nature of IR PL bands has to be clarified. With increasing the USP duration the intensity of defect (III, IV) related PL bands rose mainly in comparison with the intensity of NBE PL bands (Fig.2).

PL spectra of ZnO:Ag NRs measured at different temperatures in the range of 10-300K are shown in Fig.4. The same rates of the variation of PL peak intensities versus temperature have been revealed for the PL bands 1 and 5, as well as for the emission bands 2 and 6 (Fig.4). This fact, as well as the PL peak positions, permits to assign the IR PL bands 5 and 6 to the second order diffraction of NBE PL bands 1 and 2, respectively [13].

A great variety of luminescence bands in the UV and visible spectral ranges have been detected in ZnO crystals earlier [12]. The near-band-edge (NBE) emission at 3.0-3.37 eV is attributed to the free (FE) or bound (BE) excitons, their LO phonon replicas, such as FE-1LO or FE-2LO, to optical transition between the free to bound states, such as the shallow donor and valence band, or to donor-acceptor pairs [14]. However, the position of the near-band-edge emission at room temperature can vary significantly due to the variation of relative contributions of free exciton emission and phonon replicas [15].

The defect related PL band in the spectral range 2.40-2.70 eV in ZnO is assigned ordinary to oxygen vacancies [16], Cu impurities [17] or surface defects [18]. The orange PL band with the peak at 2.02-2.10 eV was attributed earlier to oxygen interstitial atoms (2.02 eV) [19] or to the hydroxyl group (2.10 eV) [20, 21]. Taking into account that the PL intensity of this PL band increases with raising the USP duration (Fig.2) the assumption that the corresponding defects are related to oxygen interstitial atoms (or to the hydroxyl group) looks very reliable.

Finally, we need to discuss the nature of the near-band-edge PL bands at 2.92 and 3.25 eV. Studied ZnO films were doped by Ag and, therefore, have the acceptor type defects, Ag_{Zn}, which were formed when the Ag atoms substitute Zn atoms in the ZnO crystal lattice. The intensity of 2.92 eV PL band decreases essentially with temperature and this PL band and its IR second-order diffraction peak (1.46eV) disappeared completely at 200K (Fig.4).

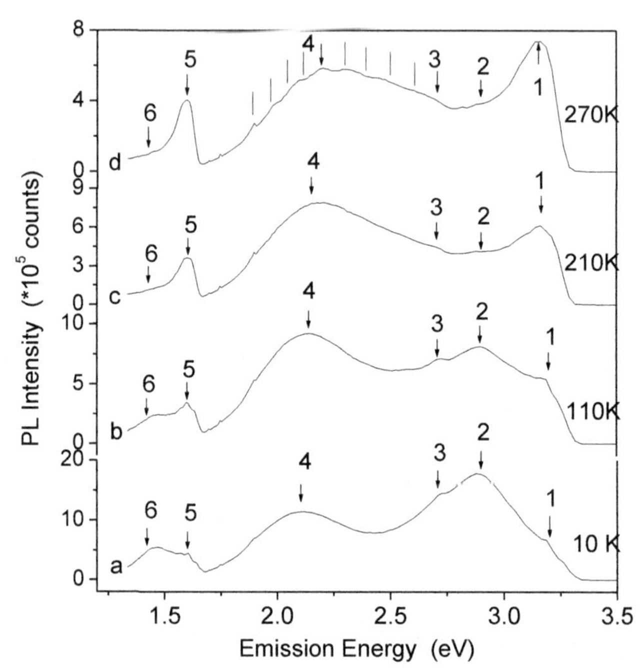

Figure 4. PL spectra at different temperatures for the sample prepared at the duration 5 min.

Thus it is possible to assume that this PL band (2.92 eV) owes to emission of the acceptor bound excitons (ABE), involving the acceptor Ag_{Zn}, or its complexes such as donor-acceptor pairs, in the ZnO:Ag nanorods. The thermal decay of 2.92 eV PL intensity is assigned to the thermal dissociation of ABE to free exciton. In this case the 3.16 eV PL band, which dominates in the room temperature PL spectrum (Fig.4) can be attributed to the 3LO phonon replica of FE emission.

CONCLUSIONS

ZnO:Ag nanorods with hexagonal structures have been successfully synthesized by the USP method. The PL spectra of the ZnO:Ag NRs over the temperature range from 10 K to 300 K have been investigated. The temperature dependence of PL spectra shows that IR emission corresponds to the second-order diffraction of NBE emission band. The study of NBE PL temperature dependences reveals that ABE and its second-order diffraction peak disappeared at the temperature higher than 200 K. The attenuation of the ABE peak intensity is ascribed to the thermal dissociation of ABE with appearing a free exciton. PL bands, related to the LO phonon replica of free exciton and its second-order diffraction peak, dominate in the PL spectrum at room temperature that testify on the high quality of the ZnO:Ag films prepared by USP.

ACKNOWLEDGMENTS

The authors would like to thank the CONACYT (project 130387) and SIP-IPN, Mexico, for the financial support, as well as the CNMN-IPN for the SEM measurement.

REFERENCES

1. S.J. Pearton, D.P. Norton, K. Ip, Y.W. Heo, T. Steiner, Prog. Mater. Sci. **50**, 293 (2005).
2. M.H. Koch, P.Y. Timbrell, R.N. Lamb, Semicond. Sci. Technol. **10,** 1523 (1995).
3. K. Vanheusden, C.H. Seager, W.L. Wareen, D.R. Tallant, J. Caruso, M.J. Hampden-Smith, T.T. Kodas, J. Lumin.**75** , 11 (1997).
4. R. Scheer, T. Walter, H.W. Schock, M.L. Fearheiley, H.J. Lewerenz, Appl. Phys. Lett. **63,** 3294 (1993).
5. Y. Chen, D.M. Baghall, H. Koh, K. Park, K. Hiraga, Z. Zhu, T. Yao, J. Appl. Phys. **84,** 3912 (1998).
6. Y.B. Li, Y. Bando, D. Golberg, Appl. Phys. Lett. **84** , 3603 (2004).
7. J. Ding, T.J. McAvoy, R.E. Cavicchi, S. Semancik, Sens. Actuat. B **77** , 597 (2001).
8. W. Tang, D.C. Cameron, Thin Solid Films **238,** 83 (1994).
9. M. Dybic, S. Ostapenko, T.V. Torchynska, E. Velazquez Lozada, Appl. Phys. Lett. **84** (25), 5165 (2004)
10. T. V. Torchynska, A.I. Diaz Cano, M. Dybic, S. Ostapenko, M. Mynbaeva, Physica B, Condensed Matter, **376-377**, 367 (2006)
11. T.V. Torchynska, J. Palacios Gomez, G.P. Polupan, F.G. Becerril Espinoza, A. Garcia Borquez, N.E. Korsunskaya, L.Yu. Khomenkova, Appl. Surf. Science, **167**, 197-204 (2000).
12. A. B. Djuris, A.M.C. Ng, X.Y. Chen. Progress in Quantum Electronics **34**. 191-259 (2010).
13. J. Lv, Ch. Liu, W. Gong, Zh. Zi, X. Chen, K. Huang,T. Wang, G. He, Sh. Shi, X. Song, Zh. Sun, Optical Materials, **34,** 1917 (2012).
14. T. Mahalingam, K.M. Lee, K.H. Park, S. Lee, Y. Ahn, J.Y. Park, K.H. Koh, Nanotechnology **18** , 035606 (2007).
15. T. Voss, C. Bekeny, L. Wischmeier, H. Gafsi, S. Borner, W. Schade, A.C. Mofor, A. Bakin and A. Waag, Appl. Phys. Lett. **89**, 182107 (2006).
16. M.K. Patra, K. Manzoor, M. Manoth, S.P. Vadera, N. Kumar, J. Lumin. **128** (2) 267–272 (2008).
17. N.Y. Garces, L. Wang, L. Bai, N.C. Giles, L.E. Halliburton, G. Cantwell, Appl. Phys. Lett. **81** (4) 622–624 (2002).
18. A.B. Djurišic, W.C.H. Choy, V.A.L. Roy, Y.H. Leung, C.Y. Kwong. K.W. Cheah, T.K. Gundu Rao, W.K. Chan, H.F. Lui, C. Surya, Adv. Funct. Mater. **14,** 856-864 (2004).
19. X. Liu, X. Wu, H. Cao, R.P.H. Chang, J. Appl. Phys. **95** (6) 3141–3147 (2004).
20. J. Qiu, X. Li, W. He, S.-J. Park, H.-K. Kim, Y.-H. Hwang, J.-H. Lee, Y.-D. Kim, Nanotechnology **20** 155603 (2009).
21. T.V. Torchinskaya, N.E. Korsunskaya, B. Dzumaev, B.M. Bulakh, O.D. Smiyan, A.L. Kapitanchuk, S.O. Antonov, Semiconductors, **30**, 792-796 (1996).

Mater. Res. Soc. Symp. Proc. Vol. 1617 © 2013 Materials Research Society
DOI: 10.1557/opl.2013.1174

A new mild synthesis and optical properties of colloidal ZnO nanocrystals in dimethylformamide/ethanol solutions

Yaroslav V. Panasyuk, Oleskandra E. Rayevska, Oleksandr L. Stroyuk, and Stepan Ya. Kuchmiy

L.V. Pysarzhevsky Institute of Physical Chemistry of National Academy of Sciences of Ukraine, 31 Nauky av., 03028, Kyiv, Ukraine

ABSTRACT

A green and mild synthesis of colloidal zinc oxide nanocrystals in ethanol/dimethylformamide mixtures was introduced which allows to produce stable crystalline ZnO particles and tailor their average size in the range of 2.8–4.5 nm by varying temperature and duration of post-synthesis ageing. An increase in dimethylformamide fraction in the mixture results in acceleration of ZnO nanocrystals ripening. Colloidal ZnO nanocrystals emit broadband photoluminescence in the range of 2–3 eV with the quantum yields of up to 12 %.

INTRODUCTION

Interest in colloidal zinc oxide nanocrystals (NCs) is stimulated by a broad variety of their application in advanced technologies such as light-emitting devices, luminescent biomarkers, solar cells, etc. [1]. Introduction of ZnO NCs into these technologies requires simple and green synthetic ways to ZnO NCs with controlled average size and photoluminescence (PL).

Today among the mild syntheses of ZnO NCs probably the most popular remains an approach introduced in 1980s based on reaction of zinc salts with inorganic bases in absolute aliphatic alcohols [2–7]. Despite the popularity the approach has a number of limitations impeding considerably progress in application of ZnO NCs. In particular, strict requirements to the alcohol dryness are dictated by a low aggregation stability of colloidal ZnO NCs in the presence of even small amounts of water and by the fact that in wet alcohols only $Zn(OH)_2$ forms which then only slowly dehydrates and converts into ZnO. Another limitation is the necessity of prolonged post-synthesis thermal treatment of zinc oxide colloids at around 60 °C which imparts ZnO NCs with high aggregation stability and crystallinity but makes impossible controlled and smooth variation of the ZnO NC size. Finally, the quantum yields of the broadband PL in the visible part of the spectrum, which is typical for colloidal ZnO NCs produced by this method, are typically not higher than 4–5%.

The present paper shows that using mixtures of ethanol with a second moderately polar solvent – dimethylformamide (DMFA) as dispersive media for preparation of colloidal ZnO NCs allows to simplify considerably the synthetic procedure allowing at the same time to tailor the average size and the absorption band threshold of ZnO NCs, as well as to increase the PL quantum yields of up to 10–12%.

EXPERIMENTAL DETAILS

Colloidal ZnO NCs were synthesized via reaction between $Zn(CH_3COO)_2$ and NaOH in DMFA/ethanol mixtures containing a varied amount of DMFA. Typically, two types of colloids were prepared containing either 3 v% or 40 v% of DMFA. In the first case, 0.3 mL of 1.0 M zinc acetate solution in DMFA were added to 5.7 mL absolute ethanol and then this solution was mixed at vigorous stirring with 4 mL of 0.1 M NaOH solution in dry ethanol. The resulting colloid was then allowed to age at room or an elevated temperature in the dark. In the second case, 0.3 mL of 1.0 M zinc acetate solution in DMFA was mixed with 5.7 mL of DMFA. This solution was mixed at vigorous stirring with 4.0 mL of 0.1 M NaOH solution in absolute ethanol.

Absorption spectra were registered using a Specord 220 spectrophotometer. PL spectra were registered on a Perkin-Elmer LS55 luminescence spectrometer, typically using 320-nm light for excitation of the samples placed in standard 1.0 mm quartz cuvettes. PL yields (PL QY) were determined using solid anthracene as a PL reference standard (QY = 100%) maintaining identical conditions for spectra registration of both tested ZnO NC and a reference. Transmission electron microscopy (TEM) experiments were carried out on a Selmi S-100 microscopy with an accelerating voltage of 100 kV. The solvodynamic size of ZnO NC was determined by dynamic light scattering technique on a Malvern Zetasizer Nano setup.

RESULTS AND DISCUSSION

Interaction between zinc acetate and NaOH in ethanol/DMFA mixtures at room temperature (RT) followed by ageing of the as-prepared solutions at RT for 120 min yields zinc oxide colloids with a typical for ZnO NCs fundamental absorption band with an edge at 3.64–3.65 eV and a distinct exciton absorption maximum at 3.94–3.95 eV (Fig. 1a, curve 1).

The PL spectrum of such colloidal solution exhibits a sole broad band (spectral width of ~0.7 eV) in the range of 2–3 eV with a pronounced maximum at ~2.5 eV (Fig. 1a, curve 2) separated from the absorption threshold by more than 1 eV. By the spectral parameters the PL band of ZnO NCs can be assigned to the radiative electron-hole recombination with the participation of charge carriers captured by the NC surface traps [1, 3–6].

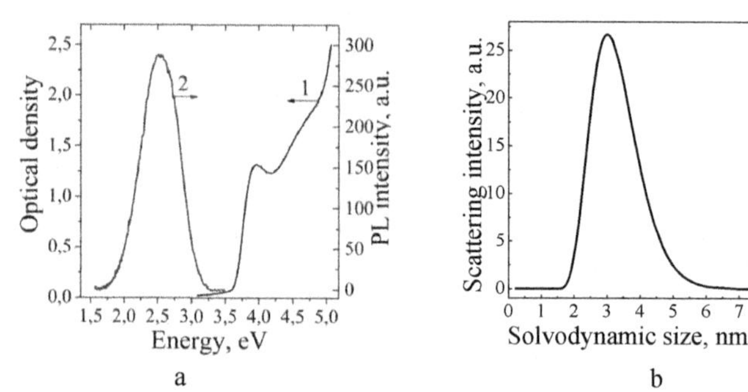

Figure 1. (a) Absorbance (curve 1) and PL (2) spectra of colloidal ZnO NCs in a DMFA/ethanol mixture (3 v.% DMFA). (b) Distribution of colloidal ZnO NCs by the solvodynamic size.

The average size *d* of ZnO NCs in colloidal solutions were determined from the position of the fundamental absorption band edge (corresponding to the band gap E_g of ZnO NCs) using the reported empiric calibration curves similarly to [7]. The estimations showed that the ZnO NCs exhibiting the absorption spectrum shown by the curve 1 in Fig. 1a have an average size of 3.5 nm. This value fits well with the results of determination of the average solvodynamic size d_{SD} of ZnO NCs made directly in the colloidal solutions by the dynamic light scattering technique which indicate d_{SD} being smaller than 5 nm with a distribution summiting at ~3 nm (Fig. 1b). According to the bright-field TEM measurements (Fig. 2a) the colloidal ZnO solution corresponding to Fig. 1a contains particles with the average size of 3.5–4.5 nm. The electron diffraction pattern of ZnO NCs (Fig. 2b, inset) reveals a set or reflexes characteristic for the hexagonal zinc oxide (ICDD card #036-1451). A high crystallinity degree of the sample is proved by the dark-field TEM (Fig. 2b) indicating the average crystal size in the sample be 3.5–4.5 nm similarly to the results of optical measurements and light-field TEM.

Recently it has been found by some of the present authors [7] that upper concentration limit in the conventional synthesis of ZnO NCs in ethanol is around 0.02 M. In case of the present DMFA/ethanol system, due to a high solubility of zinc acetate in DMFA and NaOH – in ethanol the maximal concentration of stable colloid is 0.06–0.07 M that is at least 3 times higher that in the conventional synthesis. The average size and PL intensity of colloidal ZnO NCs were found to be almost independent on the ZnO concentration in the studied range.

The ZnO NCs form in DMFA/ethanol mixtures only when a molar ratio of Zn(II) to OH^- in solution is higher than 2. As can be seen in Fig. 3a, in the presence of NaOH in the amount higher than necessary to bind zinc(II) according to the reaction $Zn^{2+} + 2OH^- \rightarrow Zn(OH)_2 \rightarrow ZnO + H_2O$, no spectral signs of ZnO NCs formation can be observed (curves 1, 2).

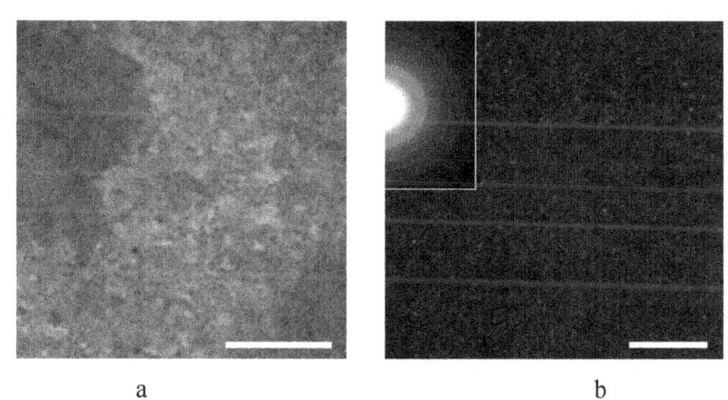

a b

Figure 2. Bright-field (a) and dark-field (b) TEM images of colloidal ZnO NCs. Inset in (b) Electron diffraction pattern of ZnO NCs. The scale is 100 nm in both cases.

At the stoichiometric and a higher ratio of Zn(II) to OH^- (Fig. 3a, curves 3, 4) the characteristic absorption band appears in the spectrum indicating formation of ZnO NCs. Such NCs probably have a core/shell structure with a ZnO core and a $Zn(OH)_2$ shell. Gradual transformation of zinc hydroxide shell to ZnO, simultaneously with the Oswald ripening, result in coarsening of ZnO NCs at the colloid ageing at RT (compare curves 4 and 5, Fig. 3a).

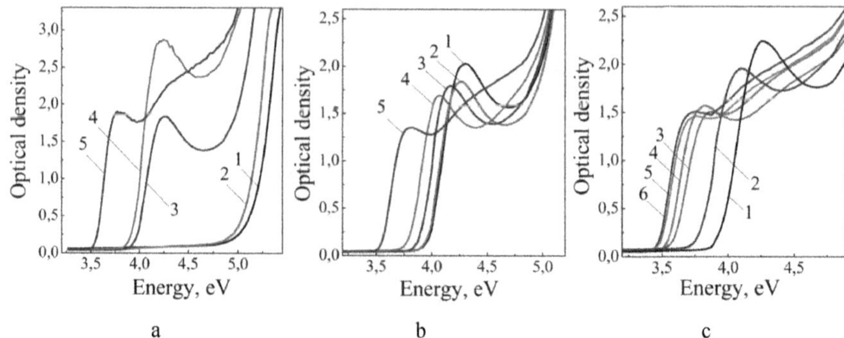

Figure 3. (a) Absorption spectra of ZnO colloids prepared in DMFA/ethanol mixtures (3 v.% DMFA) at [NaOH] = 4×10^{-2} M and different concentration of zinc acetate – 1×10^{-2} M (curve 1), 2×10^{-2} M (2), 3×10^{-2} M (3) and 4×10^{-2} M (4). Curve 5 is the solution 4 after 4 days of ageing at room T. (b, c) Absorption spectra of ZnO colloids prepared in DMFA/ethanol mixtures (3 v.% DMFA) at [NaOH] = 4×10^{-2} M and [$Zn(CH_3COO)_2$] = 3×10^{-2} M and (b) at different temperature – 5 °C (curve 1), 18 °C (2), 40 °C (3), 60 °C (4), curve 5 – solution (4) after 4 days ageing at room T; (c) at room T immediately after synthesis (curve 1), after 120 min ageing at room T (2) or after ageing at 60 °C for 15 min (3), 30 min (4), 60 min (5), and 120 min (6).

A study of the influence of synthesis and post-synthesis treatment conditions on the optical properties of ZnO NCs showed that the most potent factors of affecting the average NC size are (i) varying of temperature of synthesis and (ii) varying of duration and/or T of the post-synthesis ageing of colloidal solutions. Fig. 3b shows that elevation of temperature at which the reaction between zinc acetate and sodium hydroxide proceeds from 5 to 60 ° results in a shift of the fundamental absorption band edge of resulting ZnO NCs from ~4 to 3.8 eV (see Table I). The shift corresponds to the average NC size increase from 2.8 to 3.2 nm. The RT post-synthesis ageing of the solution synthesized at 60 °C for 4 day is accompanied by a more pronounced absorption threshold shift, to 3.56 eV indicating the ZnO NC size growth to 4 nm (Table I).

In a series of colloids synthesized at the same T the average NC size depends on the duration and T of the post-synthesis ageing. As can be seen from Fig. 3c and Table I, the ageing of the as-synthesized ZnO colloid at RT for 120 min results in only slight E_g variation – from 3.94 to 3.81 eV corresponding to the average NC size increase from 2.9 to 3.1 nm. As opposite, at 60 °C the ripening of the colloidal NC is considerably accelerated and an increment of the average NC size becomes much more pronounced. Fig. 3c and Table I demonstrate that by incrementally increasing the ageing duration at this T we can precisely control the average ZnO NC size in the range of 2.9 – 4.5 nm.

Another important parameter affecting both the average size of ZnO NCs and the rate of their ageing at elevated temperature is the volume fraction of DMFA in the solvents mixture. As can be seen in Fig. 4a, an increase in DMFA fraction from 3 to 40 v.% results, with other synthesis conditions being equal, only in a small shift of the ZnO absorption edge, from 3.98 to 4.09 eV indicating a decrease of the average NC size from 2.8 to 2.7 nm. However, further ageing of the colloid with 40 v.% DMFA at 60 °C proceeds by an order of magnitude faster than in a similar system containing only 3 v.% DMFA.

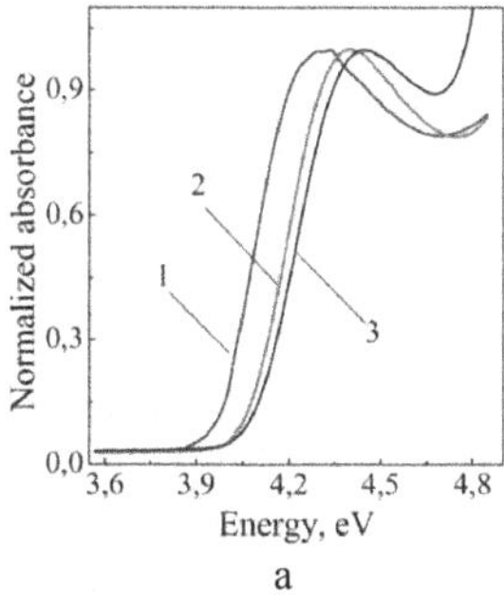

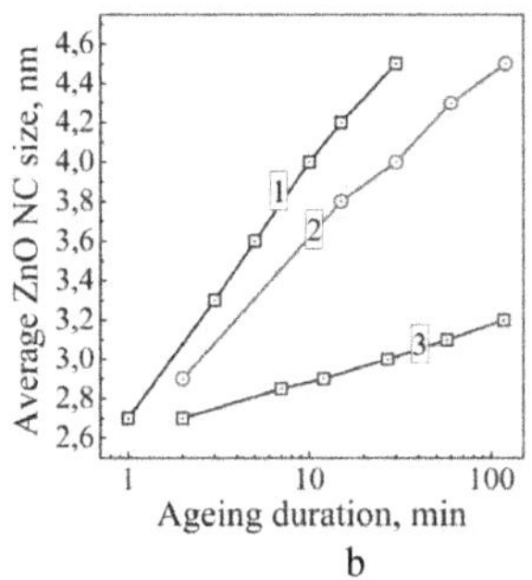

a b

Figure 4. (a) Normalized absorption spectra of ZnO colloids synthesized in DMFA/ethanol mixtures at 18 °C and a different volume fraction of DFMA – 3 % (curve 1), 67 % (curve 2), 40 % (curve 3). The spectra were normalized to the optical density in the ZnO band maximum. (b) Average size of ZnO NCs as a function of post-synthesis ageing at 60 (curves 1,2) and 18 °C (curve 3). The DMFA volume fraction is 40 % (curves 1, 3) and 3 % (curve 2). [ZnO] = 0.03 M.

Fig. 4b shows that the largest ZnO NCs among the synthesized in the present work, with the average size of 4.5 nm form in the first case after 12 min ageing of the zinc oxide colloid at the elevated T (curve 1) while in the second case, at 3 v.% DMFA content, similar process requires 120 min of post-synthesis thermal treatment at 60 °C (curve 2).

Table I. The band gap E_g, average size d, PL band maximum energy E_{PL} and relative PL quantum yield (PL QY) for colloidal ZnO NCs in DMFA

Sample	E_g (eV)	d (nm)	E_{PL} (eV)	PL QY (%)
1.1	3.98	2.8	2.65	9.7
1.2	3.96	2.8	2.55	12.0
1.3	3.92	2.9	2.55	10.0
1.4	3.80	3.2	2.50	8.0
1.5	3.56	4.0	2.40	4.5
2.1	3.94	2.9	2.60	7.2
2.2	3.81	3.1	2.50	6.7
2.3	3.59	3.8	2.40	7.6
2.4	3.55	4.0	2.40	7.5
2.5	3.49	4.3	2.35	7.2
2.6	3.47	4.5	2.30	5.0

Note: the absorption spectra for the samples 1.1 – 1.5 are given by the curves (1) – (5) in Fig. 3b, for the samples 2.1 – 2.6 – by the curves (1) – (6) in Fig. 2c. Accuracy of E_g, E_{PL}, and PL QY determination is 0.01 eV, 0.05 eV, and 0.1%, respectively.

The increase in DMFA content in the dispersive medium accelerates the RT ageing as well. In these conditions, as mentioned above, a prolonged (120 nm) ageing of the colloidal solutions with 3 v.% DMFA, according to the optical data, results in an increment of the average ZnO NC size of 0.2 nm. Similar procedure done for the colloidal solution containing 40 v.% DMFA allows to increase d by 0.5 nm (Fig. 4b, curve 3). It should be noted that at storing the both types of ZnO solutions at –5-0 °C no noticeable changes of the average NC size can be observed after tens of days of observations.

Table I shows that an increase in the average ZnO NC size is accompanied by a shift of the PL band maximum to lower energies, the absolute shift values being roughly twice smaller than the corresponding shifts of E_g. Such tendency is typical for colloidal ZnO NCs [3,5,6] and originates from the fact that at least one of the charge carriers participating in the radiative recombination is captured by a surface trap. The traps have a localized nature and the energy of the trap states lying in the band gap is affected only slightly by changes of the NC size and spatial exciton confinement. A general tendency of lowering of the PL quantum yield with increasing the ZnO NC size can be deduced from the data of Table I. The highest PL QY, 12 %, is observed just after the synthesis of ZnO NCs at RT (sample 1.2 in Table I). The NCs subjected to prolonged ripening demonstrate lower QY, around 4.5–5.0 % (samples 2.5 and 2.6). Decreasing of the PL QY after the post-synthesis treatments in the present systems may originate from the presence of a layer of $Zn(OH)_2$ on the surface of as-synthesized ZnO NCs, as discussed earlier. Such a shell can act as the surface capping agent blocking the radiationless recombination sites increasing the PL QY. Similar trend of the increase of ZnO NC PL efficiency has recently been reported in case of NC surface passivation with a layer of $Cd(OH)_2$ [8]. Ageing of the primary ZnO@$Zn(OH)_2$ nanoparticles is accompanied by gradual dehydration of the zinc hydroxide layer and its transformation into zinc oxide and NC coarsening. At that the passivating layer vanishes leaving the NC surface available for the adsorption of various species, which can then act as the radiationless recombination sites and quench PL.

CONCLUSIONS

A mild and green synthesis of colloidal ZnO nanocrystals in dimethylformamide/ethanol mixtures was reported. The method allows to produce stable crystalline ZnO nanocrystals with concentration of zinc oxide up to 0.06–0.07 M and to tailor the average nanocrystal size in the range of 2.8–4.5 nm by varying temperature and duration of the post-synthesis ageing of colloidal solutions. After increasing the DMFA fraction in the solutions from 3 to 40 v.% the rate of ZnO nanocrystals ripening rate grows by an order of magnitude. Colloidal ZnO nanocrystals exhibit broadband photoluminescence in the range of 2–3 eV with a maximum at around 2.5 eV and the quantum yields of up to 12 %.

REFERENCES

1. Semiconductor nanocrystal quantum dots: synthesis, assembly, spectroscopy and applications, ed. by A. Rogach, Springer-Verlag GmbH, Vienna, 2008.
2. M. Haase, H. Weller, A. Henglein, *J. Phys. Chem.* **92**, 482-487 (1988).
3. D.W. Bahnemann, C. Kormann, M.R. Hoffmann, *J. Phys. Chem.* **91**, 3789-3789 (1987).
4. L. Spanhel, M.A. Anderson, *J. Amer. Chem. Soc.* **112**, 2278-2284 (1990).
5. A. van Dijken, E.A. Meulenkamp, D. Vanmaekelbergh, A. Meijerink, *J. Lumin.* **87-89**, 454-456 (2000).
6. P.V. Kamat, B. Patrick, *J. Phys. Chem.* **96**, 6829-6834 (1992).
7. O.L. Stroyuk, V.M. Dzhagan, V.V. Shvalagin, S.Ya. Kuchmiy, *J. Phys. Chem. C* **114**, 220225 (2010).
8. J.-F. Zhou, J. Ao, Y.-Y. Xia, H.-M. Xiong, *J. Colloid Interface Sci.* **393**, 80-86 (2013).

II-VI materials

Mater. Res. Soc. Symp. Proc. Vol. 1617 © 2013 Materials Research Society
DOI: 10.1557/opl.2013.1175

Efficacy of Tobramycin Conjugated to Superparamagnetic Iron Oxide Nanoparticles in Treating Cystic Fibrosis Infections

Marek Osiński[1], Yekaterina I. Brandt[1], Leisha M. Armijo[1], Michael Kopciuch[1], Nathan. J. Withers[1], Nathaniel C. Cook[1], Natalie L. Adolphi[2], Gennady A. Smolyakov[1], and Hugh D. C. Smyth[3]
[1]Center for High Technology Materials, University of New Mexico,
1313 Goddard SE, Albuquerque, NM 87106-4343, U.S.A.
Tel. +1 (505) 272-7812; Fax +1 (505) 272-7801; E-mail: osinski@chtm.unm.edu
[2]Department of Biochemistry and Molecular Biology, School of Medicine,
University of New Mexico, Albuquerque, NM 87131, U.S.A.
[3]College of Pharmacy, University of Texas at Austin, Austin, TX 78712, U.S.A.

ABSTRACT

Cystic fibrosis (CF) is an inherited childhood-onset life-shortening disease. It is characterized by increased respiratory production, leading to airway obstruction, chronic lung infection and inflammatory reactions. The most common bacteria causing persisting infections in people with CF is *Pseudomonas aeruginosa*. Superparamagnetic Fe_3O_4 iron oxide nanoparticles (NPs) conjugated to the antibiotic (tobramycin), guided by a gradient of the magnetic field or subjected to an oscillating magnetic field, show promise in improving the drug delivery across the mucus and *P. aeruginosa* biofilm to the bacteria. The question remains whether tobramycin needs to be released from the NPs after the penetration of the mucus barrier in order to act upon the pathogenic bacteria. We used a zero-length 1-ethyl-3-[3-dimethylaminopropyl] carbodiimide hydrochloride (EDC) crosslinking agent to couple tobramycin, via its amine groups, to the carboxyl groups on Fe_3O_4 NPs capped with citric acid. The therapeutic efficiency of Fe_3O_4 NPs attached to the drug versus that of the free drug was investigated in *P. aeruginosa* culture.

INTRODUCTION

Cystic fibrosis (CF) is a life-shortening debilitating inherited disease, occurring in 1 out of every 3,900 children born in the United States [1]. It currently affects 30,000 people in the Untied States, and 70,000 worldwide [2]. In the 1950s, few children with CF lived long enough to attend elementary school. Today, advances in research and medical treatments have significantly enhanced and extended life expectancy of CF patients. Nonetheless, even with intensive treatment, current median life expectancy is only 36.8 years.

The disease is mediated by an autosomal recessive genetic defect, specifically by a mutation in the gene encoding the cystic fibrosis transmembrane conductance regulator (CFTR) protein [3]. CFTR is a membrane-bound glycoprotein, consisting of 1480 amino acids with a molecular mass of 170,000 Da [4]. It is present in the cells of secretory epithelia and exocrine glands, where it functions as a chloride ion channel. The CFTR-mediated influx of Cl ions in the lung can regulate absorption of ions in excess of water, creating hypotonic water outside the cells. In epithelial cells of pancreas and intestine, CFTR regulates water secretion via Cl ion efflux. It also disrupts the functioning of the epithelial sodium channel (ENaC), resulting in an increased influx of extracellular sodium into the cell. Most importantly, the epithelial dysfunction mediated by defective CFTR affects lungs and the airways, as well as pancreas, digestive tract, genitourinary tract, and sweat glands [5]. More than 1,000 different mutations of the CFTR gene have been identified. In CF lungs, abnormal salty "viscous" secretions resulting from increased sodium and chloride levels inside the cells, together with reduced mucociliary clearance, cause mucus adhesion to the airway surface, plaque formation, and entrapment of bacteria. These conditions promote bacterial infections in patients from an early age, and later on set up the stage for chronic infections, inflammation, and bronchiectasis (abnormal widening of lung airways caused by destruction of the muscle and elastic tissue). Respiratory complications, including obstruction of the airways, inflammatory reactions, and chronic lung infections eventually lead to progressive lung damage, respiratory failure, and death in most CF patients [6], [7]. Therefore, the presence and the severity of complications in the lungs determine the quality of life and life expectancy of the CF patients [8].

Normal airway mucus covers the epithelial surfaces and provides an important innate immune function by detoxifying noxious molecules and by trapping and removing pathogens and particulates from the airways via mucociliary clearance mechanism. It is primarily composed of mucin glycoproteins, which are large, heavily glycosylated proteins [9]. The abnormal mucus in CF is deficient in intact mucin, and instead contains extracellular polymeric DNA, filamentous actin, lipids, and proteoglycans [10]. The CFTR dysfunction leads to low water and low salt content of airway surface fluid, resulting in an increased mucin concentration [11]. Depletion of the periciliary liquid layer leads to a decrease in both mucociliary and cough transportability [12]. Mucus dehydration results in an increased mucin concentration and, in association with decreased mucin pH, decreased reduced glutathione and increased myeloperoxidase, which in turn leads to the formation of additional inter-chain mucin bonds, and consequently to increased viscoelasticity, poor mucus clearance, and vulnerability to persistent infection, as it supports bacterial colonization [13].

Pseudomonas aeruginosa, a Gram-negative bacterium, has been identified as a primary pathogen responsible for the bacterial colonization of the lungs in CF. It is an opportunistic human pathogen, ubiquitous in the environment, but infecting only the individuals with a compromised immune system. Once contracted, the *P. aeruginosa* lung infection cannot be eliminated even by a long-term antibiotic therapy [6], [7], and it persists in a chronic form. Patients with CF who acquire *P aeruginosa* have 2.6 times higher risk of death [14]. Thriving of *P. aeruginosa* in the altered mucus of the CF airways is mediated by phenotypic transition from nonmucoid planktonic growth (free single bacteria) to a mucoid biofilm growth mode [15]-[17], which is innately resistant to phagocytosis [18], [19] and antimicrobial treatment [20]-[22]. Alginate biofilm represents structured communities (microcolonies) of bacteria, embedded in self-produced extracellular polymeric substance (EPS) matrix and frequently surrounded by inflammatory cells. A major part of the *P. aeruginosa* biofilm matrix in CF lungs is the polysaccharide alginate. In early stages of infection, *P. aeruginosa* resembles the environmental strain, it grows fast, is non-mucoid, and relatively susceptible to antibiotics [23]. During the later, chronic stages of biofilm growth, the bacteria adapt to survive in the airway environment via increased frequency of mutations, slowing down their growth and acquiring antibiotic resistance [24]. Moreover, the polymorphonuclear leukocytes, activated by the inflammatory response, induce the production of alginate.

Antibiotics administered through inhalation aerosols show great promise in treatment of *P. aeruginosa* infection [25], [26]. Not only do they ensure drug delivery directly to the airway lumen, eliminating systemic exposure and toxicity, but also they are significantly cheaper than IV therapy at home or clinic. However, presence of multiple biological and physical diffusion barriers, such as CF mucus and bacterial biofilm, significantly inhibits the efficacy of drug delivery. Our novel solution to this problem is to use the superparamagnetic iron oxide nanoparticles (SPIONs) for the magnetic-field-guided or hyperthermia-induced delivery through the mucus. The Fe_3O_4 (magnetite) nanoparticles (NPs) are conjugated to the model drug for treatment of *P. aeruginosa* infections (tobramycin, derived from *Streptomyces tenebrarius*) and can serve as nano-pullies upon application of a nonuniform external magnetic field. The Fe_3O_4 NPs have no toxicity and have been approved by the U.S. Food and Drug Administration (FDA) for biomedical use as contrast agent for magnetic resonance imaging (MRI) [27] and as an iron supplement for treatment of iron deficiency in patients with renal failure [28]. In this paper, we evaluate the efficacy of tobramycin acting on *P. aeruginosa* colonies, while attached to the Fe_3O_4 NPs.

EXPERIMENT

Colloidal synthesis and characterization of Fe_3O_4 nanoparticles

Hydrophilic SPIONs were synthesized using a modified aqueous alkaline hydrolysis procedure of Sahoo *et al.* [29] by coprecipitation of ferrous/ferric ions mixture with ammonium hydroxide NH_4OH as a base. It has been shown previously [30] that an increase in the molar percentage of ferrous to ferric ions in the reaction mixture can promote the increase of the overall particle size from 9 nm to ~37 nm, whereas ferrous salt used alone results in the production of the largest size particles of 37 nm. Since our earlier experiments on the use of SPIONs for hyperthermia indicated that the optimal size of the NPs was about 20 nm in diameter [31], we chose the molar ratio of ferrous to ferric ions to be 9:1.

Briefly, Milli-Q water was deoxygenated using freeze-pump-thaw degassing method [32]. 0.36 g (1.8 mmol) of ferrous salt $FeCl_2{\bullet}4H_2O$ and 0.06 g (0.22 mmol) of ferric salt $FeCl_3{\bullet}6H_2O$ were added to the 250 mL three-neck flask at room temperature. The flask was connected to the Schlenk line, evacuated, and backfilled with dry nitrogen three times. 40 mL of deoxygenated Milli-Q water were added to the flask using syringe, the temperature was increased to 80 °C and the mixture was stirred at high speed (~400 rpm) until dissolved. Using syringe, 5 mL of NH_4OH were rapidly introduced to the iron salt solution under vigorous stirring, resulting in the solution turning black immediately and the formation of a precipitate. The mixture was allowed to react for 30 min under dry nitrogen flow. After that, 2 mL of 0.5 g/mL of citric acid solution in deoxygenated Milli-Q water was introduced using syringe, the solution was heated to 95 °C and stirred for additional 90 min. The Fe_3O_4 NPs capped with citric acid were collected by applying a magnet to the bottom of the flask, washing three times with Milli-Q water, and pipetting out the obtained supernatant. Finally, dried NPs were weighed, Milli-Q water was added to obtain their desired concentration, and the NPs were re-dispersed in water by sonication.

Our selected method of synthesis produced water-soluble and stable colloidal dispersions of the Fe_3O_4 NPs in water, rendering them directly compatible with biological applications. The main advantages of the citrate anion as a stabilizing ligand for surface functionalization are that not only does it make the NP surface hydrophilic, but it also possesses an active carboxylate group, amenable to the subsequent conjugation reactions, and it is non-toxic.

To obtain TEM images of the Fe_3O_4 NPs, we used high-resolution transmission electron microscope (HR-TEM) JEOL-2010F, operated at an acceleration voltage of 200 kV, equipped with an Oxford Instruments 200 energy-dispersive spectroscopy (EDS) analytical system fitted with an Inca X-Site Ultra Thin Window EDS detector. The samples were prepared by dropping the colloidal solution of the NPs onto a 200-mesh carbon-coated copper grid using a pipette, and allowing the solvent to evaporate.

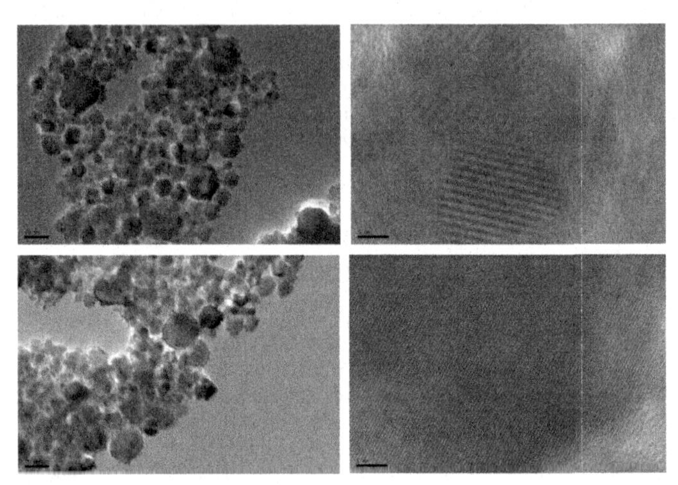

Figure 1. TEM (left panels, scale bar 50 nm) and HR-TEM (right panels, scale bar 5 nm) images of Fe_3O_4 NPs.

As shown in Fig. 1 (left panels), the average size of the Fe_3O_4 NPs was ~20 nm. HR-TEM images (Fig. 1, right panels) show fringing of the NPs, proving their high crystallinity. The characteristic lines shown on EDS spectrum (Fig. 2) confirm the presence of iron and oxygen in the elemental composition of the NPs. The carbon and copper peaks originate from the carbon-coated copper grid.

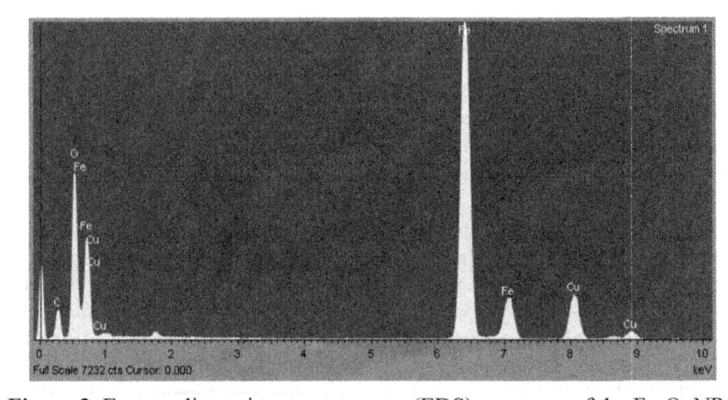

Figure 2. Energy-dispersive spectroscopy (EDS) spectrum of the Fe_3O_4 NPs.

Conjugation of tobramycin to Fe_3O_4 nanoparticles

Coupling of the model drug for treating *P. aeruginosa* infections, tobramycin, to the magnetite NPs was performed via aqueous conjugation reaction assisted by 1-ethyl-3-[3-dimethylaminopropyl] carbodiimide hydrochloride (EDC). EDC is a zero-length cross-linker, meaning that it acts by bringing the two molecules of interest together, but does not become a part of the final chemical structure. It is a frequently used reagent in a variety of applications, including preparation of biomolecular probes, labeling the 5' end of the nucleic acids, crosslinking of peptides and proteins, and immobilizing macromolecules, among others. In the first step of the reaction, the carboxylated particles are activated by addition of the EDC and formation of an intermediate, O-acylisourea reactive esters. This ester then reacts with an amine group, forming an amide (or hydrazide) derivative, but it is unstable and will hydrolyze and regenerate the carboxyl group if it does not encounter an amine. Addition of sulfo-NHS (5 mM) stabilizes the amine-reactive intermediate by converting it to an amine-reactive sulfo-NHS ester, thus increasing the efficiency of EDC-mediated coupling reactions. Specifically in our case, the amine groups on the tobramycin molecule (see Fig. 3) were connected by the amide bond to the carboxyl groups of the citric acid, the capping agent used for the Fe_3O_4 NPs. The coupling procedure was adapted from [33]. 41 mg of Fe_3O_4 NPs capped with citric acid (the product of a single synthesis process described in Section 2) were washed twice with 5 mL of coupling buffer (50 mM sodium phosphate, pH 7.2) and collected by applying a magnet to the bottom of the centrifuge tube. The NPs were then resuspended in 5 mL of coupling buffer per 100 mg of pellet weight. To provide an excess of the ligand, 50 mg of tobramycin sulphate for each 100 mg of the Fe_3O_4 NP pellet were weighed and dissolved in coupling buffer to make a 10 mg/mL tobramycin solution. While stirring, the NP solution was dropped slowly using pipettor into a beaker containing the tobramycin solution, and allowed to stir for 2 min at 450 rpm. 100 mg of EDC for each 100 mg of the pellet were weighed out, immediately added to the reaction mixture, and stirred to dissolve. The conjugation reaction was allowed to proceed for 4 hours with mixing. The NPs were then washed once with 5 mL of the coupling buffer, resuspended in the coupling buffer containing 35 mM Tris to block excess reactive sites, and washed twice again. For the experiment, the NPs were resuspended in double-distilled water (ddH_2O).

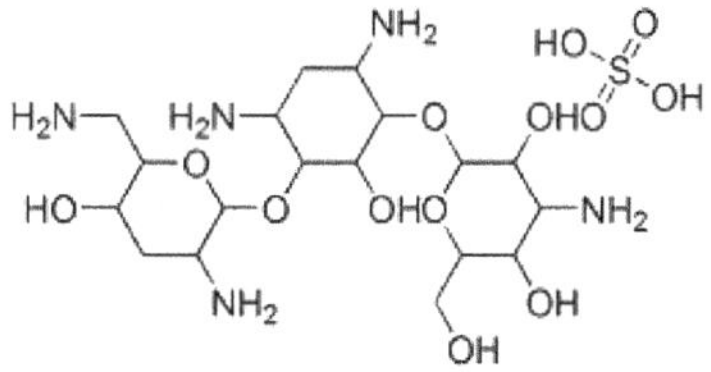

Figure 3. Chemical structure of tobramycin sulphate.

Determination of minimum inhibitory concentration of tobramycin

Cultures of *Pseudomonas aeruginosa* (ATCC 27853) were maintained as a frozen stock (in 75% glycerol) in a liquid nitrogen tank. Two days before the experiment, they were plated on the tryptic soy (TS) nutrient agar plate and grown for 24 hours at 37 °C. The next day, 10 mL of the tryptic soy broth medium (TSB) (VWR) were inoculated with one isolated colony from the plate and grown overnight on a rotary shaker at 37 °C and 150 rpm until the optical density at 600 nm (OD_{600}) reached 0.5-0.6. OD_{600} was determined using Cary 5000 UV-VIS-IR spectrophotometer against a blank cuvette containing the same volume of the TSB medium. OD_{600} measures the amount of light absorbed by a suspension of bacterial cells, which is a routine way to measure their turbidity. Higher turbidity corresponds to the higher number of bacterial cells in the solution.

For the measurement of the minimum inhibitory concentration (MIC), tobramycin sulphate was diluted to 1 mg/mL (stock solution) with sterile ddH_2O. Then, tobramycin was serially diluted into 1 mL of ddH_2O and added to the 1 mL aliquots of overnight bacterial culture to the final concentrations of tobramycin ranging from 25 to 250 μg/mL, with 25 μg/mL increments. For the control, 1 mL of sterile ddH_2O was added to the aliquot of the culture. The cultures were then grown overnight on a rotary shaker at 37 °C and 150 rpm. The next day, 50 μL aliquots of the overnight cultures were diluted 1:2 with TSB, plated on the TS nutrient agar plates, and grown for 24 hours at 37 °C. The next-day plates were checked for the presence of bacterial colonies. The MIC was narrowed down by using the dilution series with 5 μg/mL increments of tobramycin concentration, ranging between its highest concentration that still allowed the growth of *P. aeruginosa* colonies on the plate and the next lowest concentration that completely inhibited their growth.

Using this procedure and the dilution series of tobramycin, we determined the MIC of this particular strain of P. aeruginosa to be between 35 μg/mL and 50 μg/mL.

Effectiveness of tobramycin coupled to Fe_3O_4 nanoparticles against *P. aeruginosa*

To determine the effect of tobramycin-coupled Fe_3O_4 NPs on viability of *P. aeruginosa* cultures, 1 mL aliquots of liquid overnight *P. aeruginosa* cultures grown to the OD_{600} of 0.5-0.6 were diluted with 1 mL of ddH_2O alone (control), or 1 mL of ddH_2O containing either resuspended Fe_3O_4 NPs, previously subjected to conjugation reaction with tobramycin, or unconjugated Fe_3O_4 NPs. The cultures were put into a rotating shaker and grown overnight at 37 °C and 150 rpm. A well-mixed aliquot of the overnight culture was diluted 1:2 with TSB and its optical density at 600 nm (OD_{600}) was determined using Cary 5000 UV-VIS-IR spectrophotometer against a blank cuvette containing the same volume (1.5 mL) of the TSB medium.

Tobramycin is used to treat various types of bacteria infections, particularly Gram-negative infections. It has been found to be especially effective against *Pseudonomas* species. As a member of the aminoglycosides group, tobramycin kills the bacteria in two synergistic ways, working from the outside and from the inside of bacteria. Being positively charged, or polycationic, it electrostatically binds to the negatively charged lipopolysaccharide residues in the outer membrane of the bacteria, causing disruption of the membrane integrity and its degradation [34]. Acting from the inside of the bacterial cell, it interferes with protein synthesis by inhibiting ribosomal translocation [35], affecting proofreading process and therefore causing increased rate of error in synthesis with premature termination [36]. Specifically, on the molecular level, it occupies aminoacyl transfer RNA (tRNA) binding site on the bacterial 30S and 50S ribosomes, preventing binding of tRNA to the 30S ribosome-mRNA complex and formation of the 70S complex. As a result, mRNA cannot be translated into protein, and cell death ensues.

Fig. 4 contains the main result of this paper. When the suspension of the of *P. aeruginosa* bacterial culture in a liquid medium was incubated overnight with tobramycin-coupled Fe_3O_4 NPs, it inhibited the bacterial survival, resulting in 14-fold reduction (0.1 *versus* 1.4) in the OD_{600} of the sample compared to the corresponding control cultures incubated only with ddH_2O (vehicle), and in 5-fold reduction of the original concentration of bacteria (OD_{600} 0.1 *versus* 0.5). The measured OD_{600} for the *P. aeruginosa* cultures incubated with Fe_3O_4 NPs was 1.6 (see Fig. 4), which is comparable to the OD_{600} of 1.4 for the control cultures where ddH_2O was added, indicating that the reduction in bacteria concentration is not caused by the Fe_3O_4 NPs.

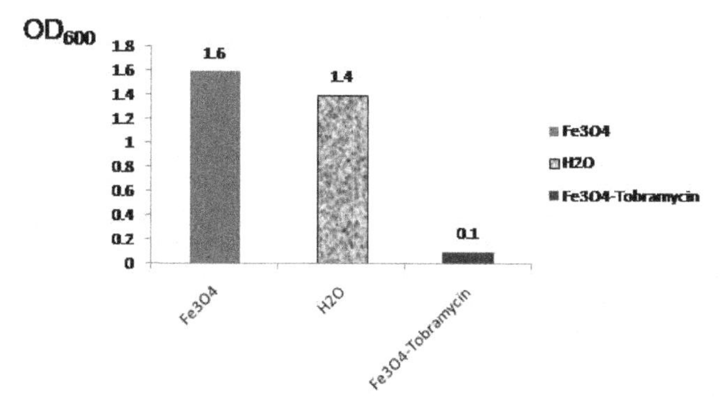

Figure 4. Tobramycin conjugated to Fe_3O_4 NPs inhibits *P. aeruginosa* growth in culture.

DISCUSSION

Since the recent expansion of the nanoscientific discoveries into the fields of biology and medicine with numerous applications in theranostics, the studies dealing with the consequences of interaction of man-made nanomaterials with living organisms and their potential toxic effects are becoming of paramount importance. However, when it comes to predicting the potential cellular permeability of *P. aeruginosa* by the tobramycin-loaded Fe_3O_4 complexes, we come to realize that very little is yet known even about the uptake of the naked NPs by the bacterial cells. What is known is that hydrophobic and hydrophilic compounds get inside the *P. aeruginosa* cell via two different routes, involving separate mechanisms. Hydrophobic substances bind to the cell membrane and penetrate it by diffusion [37], [38]. The hydrophilic compounds, such as aminoglycoside antibiotics, get inside the bacterial cell through the water-filled channels, formed by trans-outer membrane proteins, or porins. It has been demonstrated [37] that contradictory to the low outer membrane permeability based on the values of its permeability coefficient (approximately 8% that of *E. coli*), *P. aeruginosa* has a large exclusion limit with around 3000 molecular weight cut off (compared to the 500 molecular weight cut off of the *E. coli*) and allows passage of such diverse compounds as organic acids, carbohydrates, alcohols, aliphatic compounds, aromatic compounds, amino acids, and other nitrogenous compounds used as nutrient sources [39].

As for penetration of bacteria by the NPs, recent study by Morrow *et al.* [40] showed that commercially obtained CdSe/ZnS quantum dots (QDs) with two different surface chemistries, modified with either carboxyl (COOH) or polyethylene glycol (PEG), and respective effective diameters of 23 nm and 32 nm, were not internalized by the individual *P. aeruginosa* cells within a biofilm. Of interest to us is the fact that COOH-terminated QDs penetrated the biofilm matrix by diffusion more easily than PEG-terminated QDs. McQuillan in his dissertation work [41] observed that after incubation in 8 nm silver bovine serum albumin (BSA) or citrate-coated silver NPs that demonstrated a pronounced toxicity against the *E. coli*, the NPs were not found within the cytosol of the bacterial cells. On the contrary, in a similar experiment, 15 nm gold nanospheres with either coating, even though lacking toxicity against bacteria, did get inside a small number of the *E. coli* cells (~2%). In our case, it also may occur that initial disruption of the outside membrane of bacteria by tobramycin assists the subsequent penetration of NP/ tobramycin complexes into bacteria by diffusion through discontinuous and damaged cell walls.

CONCLUSIONS

Our results demonstrate that tobramycin-coupled iron oxide NPs exhibit a pronounced antibacterial effect against *P. aeruginosa*, resulting in a significant reduction of bacterial population. This finding implies that at least some of the tobramycin molecules, when bound to the surface of the NP, preserve their activity. The lack of the necessity to separate tobramycin from the NP to achieve therapeutic effect simplifies the task of improving treatment of the *P. aeruginosa* in CF patients by one step. This important finding in the clinical settings would translate into major savings in labor, time, and resources. Overall, our study shows promising preliminary results for the treatment of *P. aeruginosa* infection in CF patients using SPIONs. Further experiments will be conducted to confirm our findings and assess the statistical validity of the experimental data.

ACKNOWLEDGMENTS

This work was supported by NIH under the Grant No. 1R21HL092812-01A1 "Multifunctional Nanoparticles: Nano-Knives and Nano-Pullies for Enhanced Drug Delivery to the Lung", and by the NSF REU Site program on "Nanophotonics at the University of New Mexico", Grant No. EEC-1063142.

Leisha M. Armijo is supported by the NSF IGERT program on "Integrating Nanotechnology with Cell Biology and Neuroscience", Grant No. DGE-0549500, and by the More Graduate Education at Mountain State Alliance (MGE@MSA) AGEP program, NSF Cooperative Agreement #HRD-0450137.

REFERENCES

1. R. L. Gibson, J. L. Burns, and B. W. Ramsey, "Pathophysiology and management of pulmonary infections in cystic fibrosis", Am. J. Respir. Crit. Care Med. **168**, 918-951, 2003.
2. Cystic Fibrosis Foundation, www.cff.org, accessed on July 20, 2013.
3. M. J. Welsh, B. W. Ramsey, F. Accurso, and G. Cutting, "Cystic Fibrosis", in *The Metabolic and Molecular Basis of Inherited Diseases* (C. R. Scriver, A. L. Beaudet, W. S. Sly, and D. Valle, Eds.), 8th Ed., McGraw-Hill, New York 2001, pp. 5121–5188.

4. M. Mense, P. Vergani, D. M. White, G. Altberg, A. C. Nairn, and D. C.Gadsby, "*In vivo* phosphorylation of CFTR promotes formation of a nucleotide-binding domain heterodimer", The EMBO J. **25**, 4728 – 4739, 2006.

5. M. E. Hodson and D. M. Geddes, *Cystic Fibrosis*, Chapman and Hall Medical, London 1995.

6. J. L. Burns, B. W. Ramsey, and A. L. Smith, "Clinical manifestations and treatment of pulmonary infections in cystic fibrosis", Adv. Pediatr. Infect. Dis. **8**, 53-66, 1993.

7. N. Hoiby, "Antibiotic therapy for chronic infection of *Pseudomonas* in the lung", Annu. Rev. Med. **44**, 1-10, 1993.

8. L. Garcia-Contreras and A. J. Hickey, "Aerosol treatment for cystic fibrosis", Crit. Rev. Ther. Drug Carr. Syst. **20**, 317-356, 2003.

9. J. A. Voynow and B. K. Rubin, "Mucins, mucus, and sputum", Chest **135** (2), 505-512, Feb. 2009.

10. B. K. Rubin, "Mucus structure and properties in cystic fibrosis", Paediatric Resp. Rev. 8 (1), 4-7, March 2007.

11. H. Matsui, B. R. Grubb, R. Tarran, S. H. Randell, J. T. Gatzy, C. W. Davis, and R. C. Boucher, "Evidence for periciliary liquid layer depletion, not abnormal ion composition, in the pathogenesis of cystic fibrosis airway disease", Cell **95** (7), 1005-1015, 1998.

12. R. C. Boucher, "New concepts of the pathogenesis of cystic fibrosis lung disease", Eur. Respir. J. **23**, 146-158, 2004.

13. J. Perez-Vilar and R. C. Boucher, "Reevaluating gel-forming mucins' roles in cystic fibrosis lung disease" Free Rad. Bio. Med. **37**, 1564-1577, 2004.

14. J. Emerson, M. Rosenfeld, S. McNamara, B. Ramsey, and R. L. Gibson, "*Pseudomonas aeruginosa* and other predictors of mortality and morbidity in young children with cystic fibrosis", Pediatr. Pulmonol. **34**, 91-100, 2002.

15. M. Fegan, P. Francis, A. C. Hayward, G. H. Davis, and J. A. Fuerst, "Phenotypic conversion of *Pseudomonas aeruginosa* in cystic fibrosis", J. Clin. Microbiol. **28**, 1143-1146, 1990.

16. J. W. Costerton, P. S. Stewart, and E. P. Greenberg, "Bacterial biofilms: A common cause of persistent infections", Science **284**, 1318-1322, 1999.

17. P. K. Singh, A. L Schaefer, M. R. Parsek, T. O. Moninger, M. J. Welsh, and E. P. Greenberg, "Quorum-sensing signals indicate that cystic fibrosis lungs are infected with bacterial biofilms", Nature **407**, 762-764, 2000.

18. D. A. Cabral, B. A. Loh, and D. P. Speert, "Mucoid *Pseudomonas aeruginosa* resists nonopsonic phagocytosis by human neutrophils and macrophages", Pediatr. Res. **22**, 429-431, 1987.

19. G. B. Pier, F. Coleman, M. Grout, M. Franklin, and D. E. Ohman, "Role of alginate O acetylation in resistance of mucoid *Pseudomonas aeruginosa* to opsonic phagocytosis", Infect. Immun. **69**, 1895-1901, 2001.

20. H. Anwar, M. Dasgupta, K. Lam, and J. W. Costerton, "Tobramycin resistance of mucoid *Pseudomonas aeruginosa* biofilm grown under iron limitation", J. Antimicrob. Chemother. **24**, 647-655, 1989.

21. N. A. Hodges and C. A. Gordon, "Protection of *Pseudomonas aeruginosa* against ciprofloxacin and β-lactams by homologous alginate", Antimicrob. Agents Chemother. **35**, 2450-2452, 1991.

22. M. Whiteley, M. G. Bangera, R. E. Bumgarner, M. R. Parsek, G. M. Teitzel, S. Lory, and E. P. Greenberg, "Gene expression in *Pseudomonas aeruginosa* biofilms", Nature **413** (6858) 860-864, 25 Oct. 2001.

23. R. M. Harshey, "Bacterial motility on a surface: Many ways to a common goal", Annu. Rev. Microbiol. **57**, 249-273, 2003.

24. H. P Schweizer, "Efflux as a mechanism of resistance to antimicrobials in *Pseudomonas aeruginosa* and related bacteria: Unanswered questions", Genet. Mol. Res. **2**, 48-62, 2003.

25. F. Ratjen, G. Döring, and W. H. Nikolaizik, "Effect of inhaled tobramycin on early *Pseudomonas aeruginosa* colonisation in patients with cystic fibrosis", Lancet **358,** 983-984, 2001.

26. M. Griese, I. Müller, and D. Reinhardt, "Eradication of initial *Pseudomonas aeruginosa* colonization in patients with cystic fibrosis", Eur. J. Med. Res. **7** (2), 79-80, 21 Feb. 2002.

27. R. R. Qiao, C. H. Yang, and M. Y. Gao, "Superparamagnetic iron oxide nanoparticles: From preparations to *in vivo* MRI applications", J. Mater. Chem. **19**, 6274-6293, 2009.

28. R. T. Castaneda, A. Khurana, R. Khan, and H. E. Daldrup-Link, "Labeling stem cells with ferumoxytol, an FDA-approved iron oxide nanoparticle", J. Vis. Exp. **57**, Art. e3482, 4 Nov. 2011.

29. Y. Sahoo, A. Goodarzi, M. T. Swihart, T. Y. Ohulchanskyy, N. Kaur, E. P. Furlani, and P. N. Prasad, "Aqueous ferrofluid of magnetite nanoparticles: Fluorescence labeling and magnetophoretic control", J. Phys. Chem. B **109**, 3879-3885, 2005.

30. H. Iida, K. Takayanagi, T. Nakanishi, and T. Osaka, "Synthesis of Fe_3O_4 nanoparticles with various sizes and magnetic properties by controlled hydrolysis", J. Colloid. Interface Sci. **314**, 274-280, 2007.

31. L. M. Armijo, Y. I. Brandt, N. J. Withers, J. B. Plumley, N. C. Cook, A. C. Rivera, S. Yadav, G. A. Smolyakov, T. Monson, D. L. Huber, H. D. C. Smyth, and M. Osiński, "Multifunctional superparamagnetic nanocrystals for imaging and targeted drug delivery to the lung", *Colloidal Nanocrystals for Biomedical Applications VII* (W. J. Parak, M. Osiński, and K. Yamamoto, Eds.), SPIE International Symp. on Biomedical Optics BiOS 2012, San Francisco, CA, 21-23 Jan. 2012, Proc. SPIE **8232**, Paper 82320M (11 pp.).

32. D. J. Herman, P. Ferguson, S. Cheong, I. F. Hermans, B. J. Ruck, K. M. Allan, S. Prabakar, J. L. Spencer, C. D. Lendrum, and R. D. Tilley, "Hot-injection synthesis of iron/iron oxide core/shell nanoparticles for T2 contrast enhancement in magnetic resonance imaging", Electr. Suppl. Material (ESI), Chem. Communic. **47**, 9221-9223, 2011.

33. G. T. Hermanson, *Bioconjugate Techniques*, 2nd Ed, Academic Press 2008, p. 598.

34. S. Shakil, R. Khan, R. Zarrilli, and A. U. Khan, "Aminoglycosides versus bacteria - A description of the action, resistance mechanism, and nosocomial battleground", J. Biomed. Sci. 15 (1), 5-14, Jan. 2008.

35. L. Saiman, "Microbiology of early CF lung disease", Paediatr. Respir. Rev. **5** (Suppl A), S367-S369, 2004.

36. F. Le Goffic, M. L. Capmau, F. Tangy, and M. Baillarge, "Mechanism of action of aminoglycoside antibiotics. Binding studies of tobramycin and its 6'-n-acetyl derivative to the bacterial ribosome and its subunits", Eur. J. Biochem. **102**, 73-81, 1979.

37. H. Nikaido and R. E. W. Hancock, "Outer membrane permeability of *Pseudomonas aeruginosa*", in *The Bacteria: A Treatise on Structure and Function* (J. R. Sokatch, Ed.), Academic Press, London 1986, pp. 145-193.
38. H. Nikaido, "Nonspecific and specific permeation channels of the *Pseudomonas aeruginosa* outer membrane", in *Pseudomonas. Molecular Biology and Biotechnology* (E. Galli, S. Silver, and B. Witholt, Eds.), Am. Soc. Microbiol., Washington, DC 1992, pp. 146-154.
39. R. Y. Stanier, N. J. Palleroni, and M. Doudoroff, "The aerobic pseudomonads: A taxonomic study", J. Gen. Microbiol. **43** (2), 159-271, May 1966.
40. J. B.Morrow, C. P. Arango, and R. D. Holbrook, "Association of quantum dot nanoparticles with *Pseudomonas aeruginosa* biofilm", J. Environ. Qual. **39**, 1934-1941, 2010.
41. J. McQuillan, Bacterial-Nanoparticle Interactions, Ph.D. Dissertation, Univ. of Exeter, UK, 2010.

Mater. Res. Soc. Symp. Proc. Vol. 1617 © 2013 Materials Research Society
DOI: 10.1557/opl.2013.1176

Emission variation in CdSe/ZnS quantum dots conjugated to Papilloma virus antibodies

Juan A. Jaramillo Gómez[1], Tetyana V. Torchynska[2], Jose L. Casas Espinola[2] and Janna Douda[1]

[1]UPIITA – Instituto Politécnico Nacional, México D. F. 07340, México

[2]ESFM – Instituto Politécnico Nacional, México D. F. 07738, México

ABSTRACT

The paper presents a comparative study of the photoluminescence (PL) and Raman scattering spectra of core-shell CdSe/ZnS quantum dots (QDs) in nonconjugated states and after the conjugation to the anti-human papilloma virus (HPV), HPV 16-E7, antibodies. All optical measurements are performed on the dried droplets of the original solution of nonconjugated and bioconjugated QDs located on the Si substrate. CdSe/ZnS QDs with emission at 655 nm have been used. PL spectra of nonconjugated QDs are characterized by one Gaussian shape PL band related to the exciton emission in the CdSe core. PL spectra of bioconjugated QDs have changed essentially: the core PL band shifts into the high energy spectral range ("blue" sift) and becomes asymmetric. A set of physical reasons has been proposed for the "blue" shift explanation of the core PL band in bioconjugated QDs. The variation of PL spectra versus excitation light intensities has been studied to analyses the exciton emission via excited states in QDs. Finally the PL spectrum transformation for the core emission in bioconjugated QDs has been attributed to the electronic quantum confined effects stimulated by the electric charges of bioconjugated antibodies.

INTRODUCTION

Semiconductor nanocrystals of the II-VI group (CdSe, CdS, CdSeTe..etc), known as quantum dots (QDs), have been investigated for the different applications including the light emitting diodes, solar cells, optical amplifier media, biology and medicine [1-6]. The integration of nanotechnology with biology and medicine is expected to produce the major advances in the fluorescence resonance energy transfer (FRET) [5-7], in gene technology [8], cell tracking [5,6] and imaging [9-11]etc. The conjugation of biomolecules with core/shell QDs has been achieved using functional groups (linkers) on the QD surface [12,13], and/or with the help of electrostatic interaction between the QDs and biomolecules [14]. However the full impact of bioconjugation process on the QD optical properties and bioconjugation mechanisms are not yet completely understood. This paper presents the study of PL of CdSe/ZnS QDs covered by the amine (NH_2)-derivatized polyethylene glycol polymer in non-conjugated states and after the conjugation to biomolecules – anti-human papilloma virus (HPV), HPV 16-E7, antibodies.

EXPERIMENTAL DETAILS

Core-shell QDs commercially available [15] covered by the amine (NH_2)-derivatized polyethylene glycol (PEG) polymer were used in a form of colloidal particles diluted in a phosphate buffer saline (PBS) with a 1:200 volumetric ratio. CdSe/ZnS QDs with emission at 655 nm (1.90 eV) have been investigated. All optical measurements are performed on the dried droplets of the original solution of non-conjugated and bio-conjugated QDs located on the Si substrate. At the first PL spectra of QDs are studied in the nonconjugated state (named 655N). Then some part of QDs has been conjugated (named 655P) to the anti-papilloma virus (mouse anti-HPV 16-E7) using the 655 nm QD conjugation kits [15]. These kits contain amine-derivatized polymer coated QDs and the amine-thiol cross-linker SMCC. The protocol and details of the bioconjugation process can be found in [16]. The samples of QDs (bioconjugated and nonconjugated) in the form of a 5 mm size spot were dried on a surface of crystalline Si substrates (Fig.1) as described earlier in [17,18].

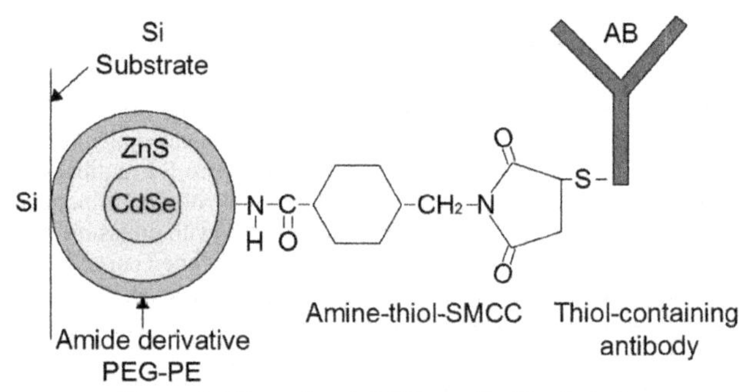

Figure 1. Bioconjugated QDs on the Si substrate

PL spectra were measured at the excitation by a He-Cd laser with a wavelength of 325 nm and a beam power of 76 mW at 300K using a PL setup on a base of spectrometer SPEX500 described in [19-21].

EXPERIMENTAL RESULTS

PL spectra of nonconjugated (655N) and bioconjugated (655P) QDs have been presented in Fig.2. In the nonconjugated state the PL spectra of QDs are characterized by one Gaussian shape PL band with the maxima at 1.90 eV (Fig.2a) related to exciton emission in CdSe cores. PL spectra have varied at the QD bioconjugation without changing of PL band intensity. Simultaneously, the peak of PL band I shifts to higher energy (1.92eV), its half width increases and the shape of PL band becomes asymmetric with the essential high energy tails (Fig.2b).

The energy shift of PL band at the QD bioconjugation to antibodies can be assigned to: i) the quantum-confined Stark effect stimulated by the charge of antibodies [22], ii) the quantum confined effect, owing to the shift of QD energy levels, stimulated by the change of electric potential at the QD surface [23], or iii) by decreasing the effective QD size in bioconjugated QDs, iv) dominated emission of excitons located at the excited QD states, v) the degradation of

CdS material at the bioconjugation [24,25], or vi) emission of the antibodies or PBS buffer at high energy UV excitation. To distinguish between these reasons the PL spectra of anti-bodies and the PBS buffer have been studied at UV excitation as well (Fig.3). As it is clear from figure 3 the PL spectrum of antibodies is characterized by the PL band in the spectral range of 2.0-3.0 eV (Fig.3, curve 1).

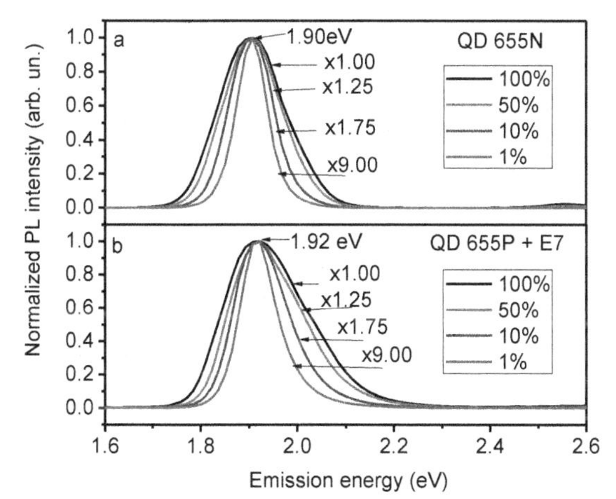

Figure 2. Normalized PL spectra of nonconjugated 655N (a) and bioconjugated 655P (b) QD samples measured at 300 K at the excitation intensities 100%, 50%, 10% and 1%, respectively. The excitation intensity of 100% corresponds to the laser power of 76 mW. Numbers at the curves (x1.00, x1.25, x1.75, x9.00) indicate on the multiplication coefficients used at the normalization of experimental PL spectra.

Note that the antibodies were kept in the PBS solution. The PL spectrum of PBS is presented in figure 3, curve 2 as well. The intensity of wide PL band in the spectral range 2.0-3.0 eV related to PBS is smaller eight-fold than the intensity of PL band related to the antibodies. It is essential to note that the intensity of PL band related to the antibodies smaller ten-fold than the PL intensity of core/shell QD emission in studied samples at the UV excitation. Thus we can conclude that the energy shift of PL spectra, revealed in the bioconjugated QDs, does not connect with emission of antibodies or the PBS buffer.

To the study the role of exciton emission localized at the QD excited states in bioconjugated QDs, the PL spectra were measured at the different excitation intensities of UV light (Fig.2). PL spectra of nonconjugated 655N QDs kept the Gaussian shape of the core PL band (1.90 eV) for all excitation intensities (Fig.2a). The full width at half maximum (FWHM) of this PL band at the first increases with excitation intensity due to rising the excited QD numbers and the QD size distribution and then, the FWHM saturates for high excitation light intensities (Fig.4, curve 1).

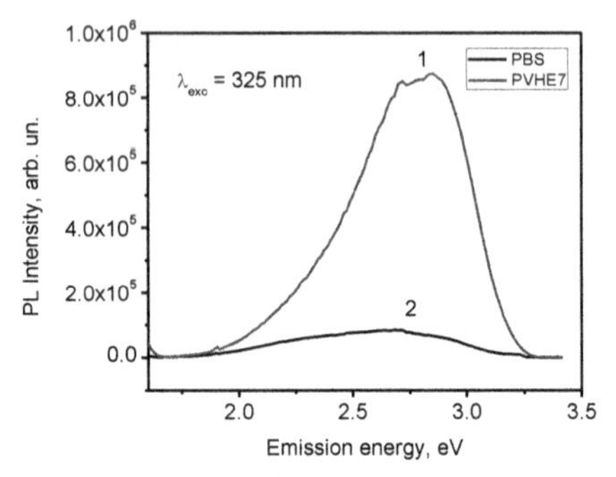

Figure 3. PL spectra of the antibodies (1) and the PBS buffer (2) without QDs.

The intensity of PL band of bioconjugated QDs increases versus excitation light intensity that is accompanied by the formation of high energy tails of the PL band (Fig.2b).

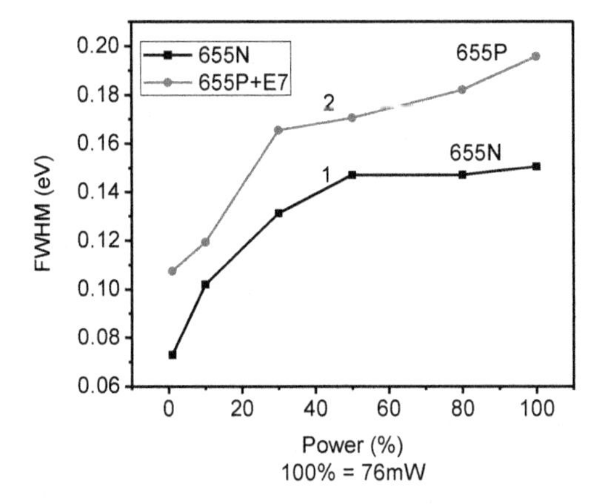

Figure 4. The FWHM of PL band in PL spectra of nonconjugated 655N (1) and bioconjugated 655P (2) QDs versus excitation light power.

The intensity of low energy side of PL band increases a little bit vs excitation intensity owing to rising the excited QD numbers and QD size distribution by the same way as in nonconjugated QDs. The FWHM of PL band in bioconjugated QDs increases monotonously versus excitation intensity without saturation (Fig.4, curve 2). Thus this study has shown that the intensity of high energy side of the PL band increases versus excitation light power due to raising the role of exciton emission via excited states in bioconjugated QDs. This effect can explain the asymmetric shape of the PL band and increasing monotonously the FWHM (Fig.4, curve 2) in bioconjugated QDs. However exciton emission via excited states cannot explain shifting (20 meV) to high energy the peak position of PL band in the PL spectrum of bioconjugated QDs.

DISCUSSION

Different reasons, as it is mentioned above, can be responsible for the shift of peak positions of the PL band in bioconjugated QDs. Experimental results presented have shown that the influence of exciton emission via excited QD states and the emission of antibodies or PBS buffer can be avoided from the consideration of the reasons of PL band shift into higher energy at the bioconjugation. Thus other effects, such as the quantum - confined Stark effect and/or the quantum confined effect, owing to the shift of QD energy levels stimulated by the change of electric potential at the QD surface, or by decreasing the effective QD size in bioconjugated QDs, have to be discussed. Note it was shown earlier that the antibody molecules are characterized by the electric charges (dipoles) that enable them to stimulate the surface enhanced Raman scattering (SERS) effects in bioconjugated QDs as well [17,18].

The Stark effect is shifting and splitting of spectral lines of atoms and molecules due to the presence of the external static electric field. The quantum-confined Stark shifts in QD ensemble emission were found early to be purely quadratic in the applied electric field, increasing with the QD sizes, and directed to the "red" spectral range due to the QD polarizability variation mainly [22]. In our experiment the QD ensemble emission has been studied as well. Thus it is possible to conclude that the "blue" energy shift of PL band in the PL spectrum of bioconjugated QDs has been not related to the quantum-confined Stark effect.

The reason for the "blue" shift of emission energy in bioconjugated QDs can be related to the change of potential barrier at the surface of QDs owing to the electric charge of bioconjugated antibodies or to decreasing the effective QD size. It is known that the position of energy levels in QDs for the strong quantum confined regime depends on the value of potential barrier [26]. As expected, the lower potential barrier reduces and the higher potential barrier increases the quantum confinement energy levels and these changes are more essential for the higher energy states [26]. Thus we can conclude that the "blue" shift of the PL band in the PL spectrum of bioconjugated QDs is related to the quantum confined effect and stimulated by the change of the surface potential barrier (or decreasing the effective QD size) at the conjugation of charged antibodies to the QD surface.

CONCLUSIONS

Photoluminescence spectra of core-shell CdSe/ZnS QDs in nonconjugated states and after the conjugation to the anti-human papilloma virus (HPV), HPV 16-E7, antibodies have been studied. It is shown that the energy shift of core exciton emission in bioconjugated QDs has been assigned to the electronic quantum confined effects stimulated by the charged antibodies.

ACKNOWLEDGMENTS

The work was partially supported by CONACYT Mexico (project 130387) and by SIP-IPN, Mexico.

REFERENCES

1. D. J. Norris, Al. L. Efros, M. Rosen, M. G. Bawendi, Phys.Rev.B, **53,** 16347 (1996).
2. M. Dybiec, G. Chomokur, S. Ostapenko, A. Wolcott, J. Z. Zhang, A. Zajac, C. Phelan, T. Sellers, G. Gerion, Appl. Phys. Lett. **90**, 263112 (2007).
3. M. Kuno, D.P. Fromm, H.F. Hamann, A. Gallagher, D.J. Nesbitt, J. Chem. Phys.**115,** 1028 (2001).
4. N. Tessler, V. Medvedev, M. Kazes, S.H. Kan, U. Banin, Science, **295,** 1506 (2002).
5. T. Jamieson, R. Bakhshi, D. Petrova, R. Pocock, M. Imani, A. M. Seifalian, Biomaterials **28**, 4717 (2007).
6. R.E. Bailey, A.M. Smith, Sh. Nie, Physica E, **25,** 1-12 (2004).
7. C.Y. Zhang, H.C. Yeh, M.T. Kuroki, T.H. Wang. Nat Mater **4**, 826–31 (2005).
8. M.Y. Han, X.H. Gao, J.Z. Su, S. Nie, Nat Biotechnol, **19,** 631 (2001).
9. T.V. Torchynska, Nanotechnology, **20**, 095401 (2009).
10. T. V. Torchynska, J. Douda, P. A. Calva, S. S. Ostapenko and R. Peña Sierra. J. Vac. Sci. &Technol. **27(2**), 836 (2009).
11. L. G. Vega Macotela, J. Douda, T. V. Torchynska, R. Peña Sierra and L. Shcherbyna, phys.stat.solid. (c), **7**, 724 (2010).
12. W. J. Parak, D. Gerion, D. Zanchet, A. S. Woerz, T. Pellegrino, Ch. Micheel, Sh. C. Williams, M. Seitz, R. E. Bruehl, Z. Bryant, C. Bustamante, C. R. Bertozzi, and A. P. Alivisatos, Chem. Mater. **14**, 2113 (2002).
13. A. Wolcott, D. Gerion, M. Visconte, J. Sun, Ad. Schwartzberg, Sh. Chen, and J. Z. Zhang, J. Phys. Chem. B, **110**, 5779 (2006).
14. A.R. Clapp, I. L. Medintz, J. M, Mauro, Br. R. Fisher, M. G. Bawendi, and H. Mattoussi, J. AM. Chem. Soc. **126**, 301-310 (2004).
15. www.invitrogen.com, http://probes.invitrogen.com/media/pis/mp19020.pdf
16. http://probes.invitrogen.com/media/pis/mp19010.pdf
17. T. V. Torchynska, J. Douda, S. S. Ostapenko, S. Jimenez-Sandoval, C. Phelan, A. Zajac, T. Zhukov, T. Sellers, J. of Non-Crystal. Solid. **354**, 2885 (2008).
18. T. V. Torchynska, A. Diaz Cano, M. Dybic, S. Ostapenko, M. Morales Rodrigez, S. Jimenes Sandoval, Y. Vorobiev, C. Phelan, A. Zajac, T. Zhukov, T. Sellers, phys. stat. sol. (c), **4**, 241 (2007).
19. T. V. Torchynska, J. Douda, and R. Peña Sierra, phys. stat. sol. (c), **6,** S143 (2009).
20. T. Torchynska, J. Aguilar-Hernandez, A.I.Diaz Cano, G. Contreras-Puente, F.G. Becerril Espinoza, Yu.V. Vorobiev, Y. Goldstein, A. Many, J. Jedrzejewski, B.M. Bulakh and L.V. Scherbina, Physica B, Condensed. Matter **308-310**, 1108-1112 (2001).
21. T.V.Torchinskaya, N.E.Korsunskaya, B.Dzumaev, B.M.Bulakh, O.D.Smiyan, A.L.Kapitanchuk, S.O.Antonov, Semiconductors, **30**, 792-796 (1996).
22. S.A. Empedocles, M.G. Bawendi, Science, **278,** 2114 (1997).
23. T.V. Torchynska, "Nanocrystals and quantum dots. Some physical aspects" in the book "Nanocrystals and quantum dots of group IV semiconductors", Editors: T. V. Torchynska and Yu. Vorobiev, American Scientific Publisher, 1-42 (2010).
24. N.E. Korsunskaya, I.V. Markevich, T.V. Torchinskaya and M.K. Sheinkman, J. Phys. Chem. Solid. **43**, 475-479 (1982).
25. N.E. Korsunskaya, I.V. Markevich, T.V. Torchinskaya and M.K. Sheinkman, phys. stat. sol (a), 1980, **60**, 565 -572 (1980).
26. D.B. Tran Thoai, Y.Z. Hu, S.W. Koch, Phys. Rev. B, **41**, 6079 (1990).

Mater. Res. Soc. Symp. Proc. Vol. 1617 © 2013 Materials Research Society
DOI: 10.1557/opl.2013.1177

Modeling of the effect of bio-conjugation to anti-interleukin-10 antibodies on the photoluminescence of CdSe/ZnS quantum dots

Tetyana V. Torchynska[1], Yuri V. Vorobiev[2], and Paul P. Horley[3]

[1]ESFM Instituto Politécnico Nacional, México, D.F. 07738, México

[2]CINVESTAV Instituto Politécnico Nacional, Querétaro, QRO 76230, México

[3]CIMAV Chihuahua / Monterrey, Chihuahua, CHIH 31109, México

ABSTRACT

Bio-conjugated CdSe/ZnS core/shell quantum dots (QDs) attract essential scientific interest due to their possible nano-medicine applications, including selective highlighting of affected tissues and targeted drug delivery to the certain type of cells. The paper is focused on the theoretical description of the blue shift observed in the luminescence spectra of CdSe/ZnS QDs upon their bio-conjugation with the anti-interleukin-10 antibodies. We propose a model that describes the ground state of the exciton confined in a quantum dot and explaining the bio-conjugation phenomenon by the change of the effective confinement volume.

INTRODUCTION

The modern developments in nano-technology allow the numerous promising applications in the field of biology and medicine, which in the future may bring an unprecedented development of nano-medicine [1, 2]. This novel technique will allow targeted drug delivery to cells with certain parameters, as well as improvements in diagnostics and localization of affected tissues in the human body. To achieve these aims, it is necessary to make a bridge between the immunity-related agents (such as cytokines) and nano-particles with pre-defined properties. The key issue for nano-particles intended for biological use is to ensure their stability, which naturally suggests formation of core-shell structures with a certain enveloping layer required to isolate a nano-particle. The core defines the desired properties of the system – it may be the material with outstanding luminescence or magnetic properties for precise detection of particle's location inside the body; on the other hand, the core can be used as a reservoir for the medicine used in targeted drug delivery applications.

In this paper, we focus on the antibody molecules to the interleukin 10 (IL-10). Interleukin-10 molecules (related to cytokines) participate in the inter-cellular communication. Cytokines are being produced at the surface of one cell and interact with the receptors of its neighbors, triggering an inter-cell response. The cytokines of interleukin family (Fig. 1) participate in the immuno-regulation and control inflammatory processes [3]. The studies earlier realized proved the important role of IL-10 in blocking inflammatory and allergic reactions [4]. We used the anti-IL-10 antibodies (antihuman IL-10, Rt IgG1, clone JES3-9D7, code RHCIL1000) to perform the bio-conjugation with CdSe/ZnS core/shell QDs using the commercially available conjugation kit [5] with QDs characterized by 655 nm emission.

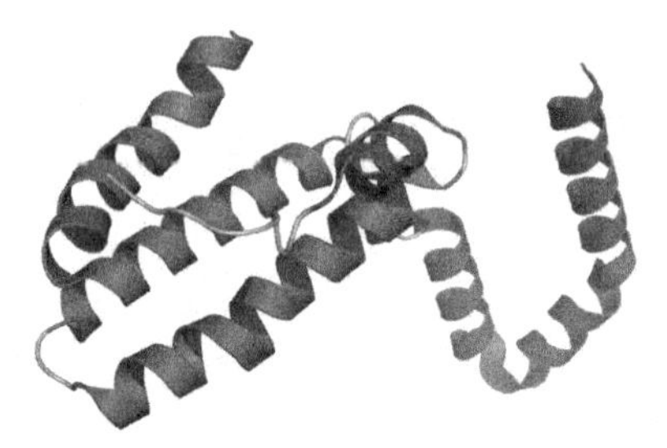

Figure 1. Protein Database (PDB) rendering of Interleukin-10 (image courtesy Wikipedia)

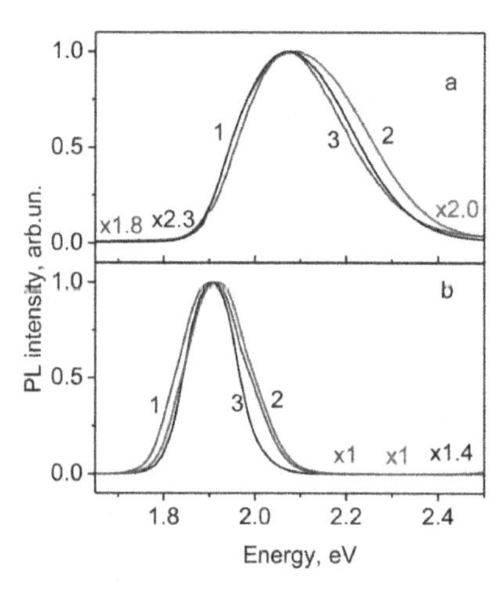

Figure 2. Comparison of photoluminescence spectra observed upon bio-conjugation of CdSe/ZnS QDs with IL-10 antibodies: a) normalized spectra for three conjugated samples showing wide asymmetric peak; b) normalized spectra for three non-conjugated samples characterized by narrow symmetric peak. Normalization coefficients are listed for each curve.

The bio-conjugated CdSe/ZnS QD systems were thoroughly studied with different experimental techniques; in particular, we used Raman scattering to reveal the influence of oxidation and compressive strains occurring upon bio-conjugation [6]. Here we would like to address the theoretical description of the photoluminescence (PL) blue-shift measured for the samples excited by the UV light (325 nm) from He-Cd laser (Figure 2). To simplify the comparison, we normalized PL spectra for their principal peak to compensate the intensity variation caused by bio-conjugation. The PL spectra for conjugated and non-conjugated QDs feature peaks with distinct appearance: the original samples are characterized with narrow peaks located approximately at 1.9 eV, while the conjugated ones feature wider asymmetrical peaks at about 2.1 eV with high-energy tail.

THEORY

The photoemission in CdSe/ZnS QDs is determined by emission of an exciton generated in the CdSe core. To find the ground state of exciton one should solve the Schrödinger's equation for a confined particle with the corresponding boundary conditions –in the most cases, assuming the impenetrable boundary [7]. The experimental results, however, for some cases do not concord with the predicted energy for the exciton peak in CdSe/ZnS QDs [8, 9], which, to our opinion, is caused by the inappropriate assumption of the strong quantum confinement in the big size CdSe core (6.4nm). The alternative way is to consider the weak confinement case by allowing the particle to tunnel through the walls of the core [10]. Yet, the consideration of tunneling probability at the boundary of nano-particle will complicate the solution of the Schrödinger equation. To keep the analytical solution relatively simple, we propose to consider mirror boundary conditions, when the walls of the quantum well are treated as mirrors "reflecting" the wave packet associated with a confined particle [11]. Thus, for each "object" point inside the quantum well it will be possible to obtain tan "image" point defined by reflection and located outside of the nano-particle. Due to the symmetry, the value of the wave function should be equal in the both points; yet, there is an important detail concerning its sign, giving two possible cases of the strong and weak quantum confinements. The former corresponds to vanishing wave function at the boundary, which is achieved by equaling the wave functions of the opposite sign to ensure the destructive interference at the boundary. The weak confinement with non-zero wave function occurs when the wave function in both points have the same signs. Solving the Schrödinger equation with the corresponding boundary conditions it is possible to obtain the wave functions Ψ and the wave numbers k, to calculate the energy of the confined particle and the probability of density distribution for the particle confined in the quantum well. As probability is proportional to the wave function squared, both the positive and negative signs of the wave function are acceptable. The proposed technique allows comparatively simple and accurate analytical solution of the Schrödinger equation for different quantum objects and was successfully tested for several geometries of nano-scale particles [11,12].

DISCUSSION

We suggest using the mirror boundary conditions for the case of CdSe/ZnS QDs conjugated to the anti-IL-10 antibodies. In the case of a spherical particle one should build the "image" points using the formulas for reflections in a spherical mirror [12], yielding the energy of the ground state

$$E_{GS_0} = E_g + \frac{h^2}{32mR^2}, \tag{1}$$

which is four times smaller than that obtainable with the impenetrable wall boundary condition. The calculations made with the CdSe parameters give the results that better fit the experimental data.

Of course, when a nano-particle becomes bio-conjugated, its energy balance changes much so that the solution found for a spherical quantum dot will give incorrect results. The problem comes from the fact that the attachment of the biomolecules to the external core will result in the generation of van der Waals forces. The extra-charge attached to the exterior shell will form a blocking electric field, which "truncates" the effective volume of the particle (Figure 3).

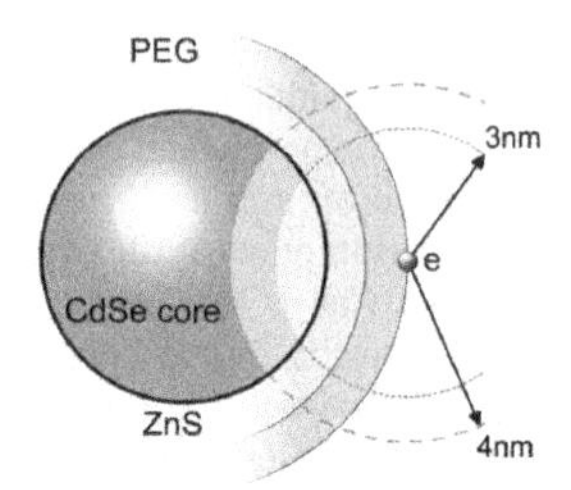

Figure 3. A schematic depiction of CdSe/ZnS QD with the 6.4nm core, encased in the ZnS and poly-ethylene glycol (PEG) protective shells. A charge generated upon bio-conjugation is shown attached to the PEG shell to the right. We consider two radii of the blocking sphere of $d = 3$ and 4 nm, respectively.

In this situation, it is natural to expect lower energy of the emitted light. At the same time, the stronger confinement will produce a blue shift of the particle emission spectrum. The question about the accurate calculation of the truncated volume is a complicated one. For a point-charge the equipotential surfaces will be spheres, which will make the blocked volume corresponding to a spherical segment. The remained effective volume will lose its spherical symmetry; we propose to approximate it with a cylinder with a height c and a circular base of the diameter a. The Schrödinger equation for such a system can be solved using the mirror boundary conditions [11], providing the ground state energy relative to the material band-gap:

$$E_{GS_C} - E_g = \frac{h^2}{2m}\left(\frac{0.268}{a^2} + \frac{0.25}{c^2}\right). \tag{2}$$

The parameters of a cylinder can be estimated for different strength of the blocking field. We consider two cases with the blocking field extending for the distance $d = 3$ and 4 nm from the charge attached to the protective PEG layer covering nano-particle (Figure 3). For these cases the parameters of the effective cylinder are: $a = 1.6\ R$, $c = 1.55\ R$ (for 4 nm blocking radius) and $a = 1.65\ R$, $c = 1.6\ R$ (for 3 nm blocking radius), where $R = 3.2$ nm is the radius of the CdSe core of the particle.

Using these formulas, one can obtain the expressions for the ground state of conjugated nano-particle with 4 nm blocking:

$$E_{GS_c} - E_g \approx \frac{h^2}{9.6mR^2}, \tag{3}$$

and also for conjugated nano-particle with 3 nm blocking distance:

$$E_{GSc} - E_g \approx \frac{h^2}{11.1mR^2}. \tag{4}$$

In a real system, the absorbed energy that triggered the formation of an exciton can be partially dissipated, so that the peak of photoluminescence spectra will occur at lower energy, producing Stokes shift E_{St}. Therefore, the total blue shift observed in the system can be defined as a ratio between peak positions of a bio-conjugated sample $h\nu_c$ versus non-conjugated sample $h\nu_0$:

$$\frac{E_{GS_c} - E_g}{E_{GS_0} - E_g} = \frac{h\nu_c + E_{St} - E_g}{h\nu_0 + E_{St} - E_g}. \tag{5}$$

Using the experimental data shown in the figure 2, one can see that $h\nu_0 = 1.87$ eV and $h\nu_c = 2.08$ eV; the typical Stokes shift is $E_{St} = 50.0$ meV [14]. Using the band gap $E_g = 1.74$ eV for the bulk CdSe at 300K is [13], one can obtain the experimental shift equal to 2.16 from equation (5). The ratios calculated with (1), (3) and (4) are 3.30 (for $d = 4$ nm) and 2.88 (for $d = 3$ nm), which is considerably close to the experimental value in view of simplifications made in treating the truncation of the nano-particle's core volume.

CONCLUSIONS

We have proposed to use the mirror boundary condition for simplifying solution of the Schrödinger equation for a spherical CdSe / ZnS core / shell QDs. It was shown that the blue shift in the photoluminescence spectrum observed in the experiment with bioconjugated QDs can be explained by the attachment of a charge to the exterior shell of nano-particle, blocking away a fraction of core's volume. Representing the truncated effective part of the core as a cylinder, we obtained a blue shift values with a considerable accuracy in relation to the experimental data. We suggest that the application of mirror boundary conditions can be an effective tool for solution of the similar problems with the different nano-particle geometry.

ACKNOWLEDGMENTS

This work was supported in part by the CONACYT of México via the research project 130387 and by the SIP-IPN.

REFERENCES

1. *Nanomedicine: Design of particles, sensors, motors, implants, robots and devices*, edited by M.J. Schulz, V.N. Shanov and Y. Yun (Artech House, Boston, 2009).
2. H.F. Tribbals, *Medical Nanotechnology and Nanomedicine* (CRC Press, Boca Raton, 2011)
3. R. de Waal Malefyt, Interleukin-10. in *Cytokines*, edited by A. Mire-Sluis and R. Thorpe (Academic Press, San Diego, 1998) pp. 151-168.

4. M.S. Grimbaldeston, S. Nakae, J. Kalesnikoff, M. Tsai and S.J. Galli, Nature Immunology **8**, 1095 - 1104 (2007).
5. <http://probes.invitrogen.com/media/pis/mp19020.pdf>
6. A.L. Quintos Vazquez, T.V. Torchynska, J.L. Casas Espinola, J.A. Jaramillo Gómez, and J. Douda. Journal of Luminescence **143**, 38–42 (2013).
7. Al. L. Efros and A.L. Efros, Soviet Physics Semiconductors **16,** 772 (1982).
8. N. Tessler, V. Medvedev, M. Kazes, S.-H. Kan, U. Banin, Science **295**, 1506-1508 (2002).
9. M. Dybec et al. Applied Physics Letters **90**, 263112 (2007).
10. D.A.B. Miller, *Quantum Mechanics for Scientists and Engineers* (Cambridge University Press, Cambridge, 2008).
11. P.P. Horley, P. Ribeiro, V.R.Vieira, J. González-Hernández, Yu.V. Vorobiev, L.G. Trápaga-Martínez, Physica E **44** 1602–1607 (2012).
12. Y.V. Vorobiev, P.M. Gorley, V.R. Vieira, P.P. Horley, J. González-Hernández, T.V. Torchynska, and A. Diaz Cano, Physica E **42**, 2264–2267 (2010).
13. C. Kittel, Introduction to Solid State Physics, 6th Ed. New York, John Wiley, 1986, p.185.
14. D.J. Norris, M.G. Bawendi, Physical Review B **53** , 16338 (1996).

Mater. Res. Soc. Symp. Proc. Vol. 1617 © 2013 Materials Research Society
DOI: 10.1557/opl.2013.1178

Microwave-Assisted Synthesis of Cadmium Sulfide Nanoparticles: Effect of Hydroxide Ion Concentration

Israel López[1] and Idalia Gómez[1]
[1]Universidad Autónoma de Nuevo León, UANL, Facultad de Ciencias Químicas, Laboratorio de Materiales I, Av. Universidad, Cd. Universitaria 66451, San Nicolás de los Garza, Nuevo León, Mexico.

ABSTRACT

Cadmium sulfide nanoparticles were synthesized by a microwave-assisted route in aqueous dispersion. The cadmium sulfide nanoparticles showed an average diameter around 5 nm and a cubic phase corresponding to hawleyite. The aqueous dispersions of the nanoparticles were characterized by UV-Vis spectroscopy, luminescence analysis, transmission electron microscopy and X-ray diffraction. The addition of sodium hydroxide solutions at different concentrations causes a red-shift in the wavelength of the first excitonic absorption peak of the cadmium sulfide nanoparticles, indicating a reduction of the band gap energy. Besides, the intensity of the luminescence of the nanoparticle dispersions was increased. However, there is a threshold concentration of the hydroxide ion above which the precipitation of the cadmium sulfide nanoparticles occurs.

Keywords: nanostructure; luminescence; optoelectronic.

INTRODUCTION

Nanomaterials have attracted attention in recent decades due to their unique physical and chemical properties. Among these materials the semiconductors are important, not only because their unconventional properties which depends on size, but also because these nanomaterials can be useful for many technological applications such as photocatalytic processes [1], solar cells [2], optoelectronic devices [3] and biologic systems [4]. Cadmium sulfide (CdS) is one of the most important II-VI semiconductors. This material exhibits a direct-band transition with band gap energy (E_g) of 2.53 eV [5]. CdS has important optoelectronic applications for optical devices based on nonlinear properties [6] and laser light emitting diodes [7].

There are many reports about synthesis of CdS nanoparticles such as spray-pyrolysis [8,9], chemical vapor deposition [10,11], solvothermal [12,13] and sol-gel [14,15]. Nevertheless, a rapid and inexpensive alternative is the microwave-assisted synthesis [5,16,17]. In general, the microwave-assisted routes offer advantages like high energy efficiency and short reactions times [18]; besides, it can be easily scaled up to industrial level.

The CdS exists in three types of crystalline structures [19]: hexagonal wurtzite, cubic zinc blend and high pressure rock-salt phase. Among these materials, the hexagonal phase is the most stable and can be found in both bulk and nanocrystalline systems whereas cubic and rock-salt phases are observed only in nanometric systems [20]. To stabilize the surface of these nanocrystals, suitable organic ions, known as capping agents, are commonly used during the synthetic process, avoiding the growth.

Many studies focus on the development of new synthetic process to obtain semiconducting nanoparticles with high luminescence quantum yield. The effect of ionic and molecular species on the luminescence of CdS nanoparticles has been also studied. Particularly, the effect of the pH during the synthetic process was reported by Henglein *et al.* [21] in 1987, and any other paper has been reported in this area. This research group studied the effect of the pH in the luminescence of CdS nanoparticles obtained by precipitation. The authors reported that hydroxide ions increase the luminescence of CdS nanoparticles by about 50%. This effect was attributed to the formation of a $Cd(OH)_2$ shell around the CdS nanoparticle, which effectively eliminates the nonradiation recombination of charge carriers.

In this work, we report the effect of hydroxide ions on the luminescence of CdS nanoparticles, also a red-shift in the wavelength of the first excitonic absorption peak was observed. However, there is a threshold concentration of the hydroxide ion above which the precipitation of CdS occurs.

EXPERIMENT

All the reactants and solvents used in this work were of analytical grade and used without any further purification. The colloidal dispersions were characterized by means of UV-Vis spectroscopy, with a Perkin Elmer Lambda 12 spectrophotometer. Luminescence analysis was carried out in a Perkin Elmer LS 55 spectrophotometer. Transmission electron microscopy (TEM) images were recorded on a FEI Titan 80-300 microscope. X-ray diffraction (XRD) pattern was carried out in a Siemens D5000 (λ = 1.5418 Å). A Thermo Scientific iCE 3000 AA spectrometer was employed for the atomic absorption (AA) spectrometry.

Two aqueous solution of concentration 30 mM were prepared, the first one of thioacetamide and the second one of cadmium chloride. These aqueous solutions were mixed in stoichiometric ratio, the resulting solution was diluted to 50 mL with a 2 mM sodium citrate solution and the pH value was fixed at 8 with a sodium hydroxide (NaOH) solution. Finally, the reaction mixture was heated in a conventional microwave oven LG-intelowave at 2.45 GHz and 1650 W for 60 s [5]. The dispersion was dried at room temperature, and the resulting solid was analyzed by XRD.

With the purpose of evaluating the hydroxide ions on the luminescence of the CdS nanoparticles, 4 mL of a NaOH solution was added into 4 mL of CdS colloidal dispersion under vigorous stirring. In order to show the hydroxide concentration effect, the NaOH solutions were prepared in six different concentrations (from 0.0050% to 1.0% w/v in NaOH). The percent of removed CdS nanoparticles was calculated by AA spectrometry.

DISCUSSION

Figure 1 (a) shows the absorbance spectrum for the colloidal dispersion of CdS nanoparticles. The E_g value was evaluated by fitting a straight line through a lineal portion of the curve to zero absorbance, and the value of the wavelength (447 nm) was converted to energy in eV units (2.77 eV). The CdS colloidal dispersion shows effects of quantum confinement, which suggest the nanometric scale of the particles. The wavelength value, λ(nm), was employed for determining the particle size, D(nm), of the CdS using the following empirical relationship [22]:

$$D = -6.6521 \times 10^{-8} \lambda^3 + 1.9557 \times 10^{-4} \lambda^2 - 9.2352 \times 10^{-2} \lambda + 13.29 \qquad (1)$$

which is valid in aqueous dispersion. The value of the particle size was estimated at 5.1 nm.

A TEM image of the nanoparticles synthesized under the conditions described is shown in Figure 1 (b). Particles with diameter around 5 nm are observed, which is in good agreement with the values obtained using the empirical relationship.

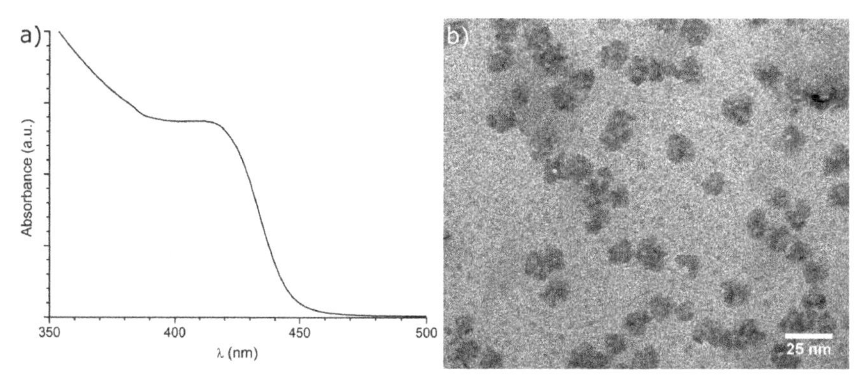

Figure 1. a) UV-Vis absorption spectrum and b) TEM image of the CdS nanoparticles.

In Figure 2 (a), the XRD pattern of CdS nanoparticles shows broad peaks indicating the presence of nanocrystals. The peaks in the XRD pattern were indexed according to the JCPDS data of the cubic structure of CdS (10-454). The XRD peaks in the pattern at 26.3°, 43.8° and 51.9° correspond to the (111), (220) and (311) planes, respectively. After addition of the hydroxide ion solution at 1.0% w/v, the CdS nanoparticles exhibited the same XRD pattern, as shown in Figure 2 (b).

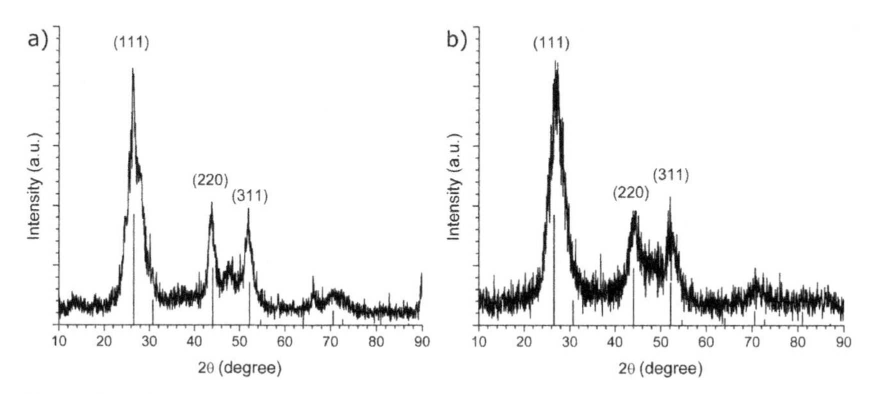

Figure 2. XRD patterns of a) early synthesized CdS nanoparticles and b) same nanoparticles after addition of the solution at 1.0% w/v in NaOH.

UV-Vis absorption spectra of the CdS nanoparticles, obtained by addition of hydroxide ion solutions at different concentrations, are shown in Figure 3. For the solutions at 0.0050% and

0.010% w/v in NaOH, a weak red-shift was observed; besides, the absorbance of the colloidal dispersions was increased by 60% and 200%, respectively. In the case of the solutions at 0.050% and 0.10% w/v in NaOH, a strong red-shift was exhibited; however, the shape of the absorption edges was significantly altered indicating a change in the electronic structure of the semiconductor, which can be attributed to the formation of a $Cd(OH)_2$ shell around the CdS nanoparticles [21]. However, the XRD analysis does not show the reflexes of this phase, which can be due to the small thickness of the $Cd(OH)_2$ shell. On the other hand, the CdS nanoparticles were removed from the water when NaOH concentration was higher than 0.50% w/v.

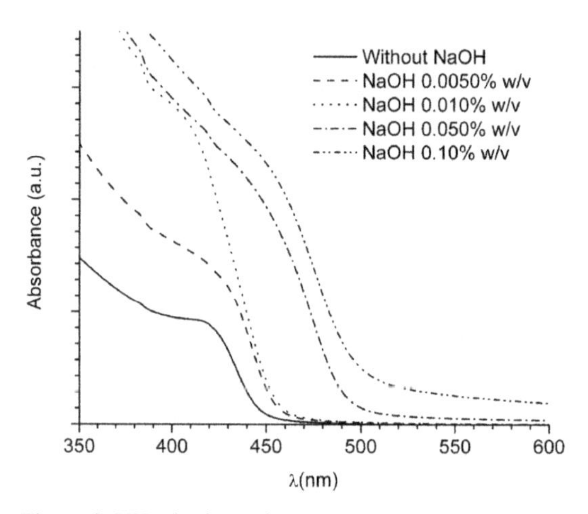

Figure 3. UV-Vis absorption spectra of the CdS nanoparticles obtained by addition of hydroxide ion solutions at different concentrations.

Figure 4 shows the luminescence spectra of the CdS nanoparticles after the addition of hydroxide ions at different concentrations. All the dispersions were excited at 450 nm. A wide emission band was observed at 600 nm, which correspond to a yellow-red color. This emission can be attributed to the transition of bound electrons from surface states to the valence band, and to the transition of Cd-interstitial donors to valence band [23]. The intensity of the luminescence was increased by 70% when the NaOH concentration was 0.010% w/v, and by 85% when this concentration was 0.010% w/v. No significant increment in intensity of luminescence was observed for higher NaOH concentrations.

Table I summarizes the results of the present research. This table presents the CdS removal percentages obtained by AA spectroscopy for the different hydroxide ion concentrations. The higher CdS removal percentage was observed when the NaOH concentration was 0.50% w/v. The limited of detection of the AA analysis corresponds to the higher CdS removal percentage; this is the reason why it was not possible to obtain a more accurate value.

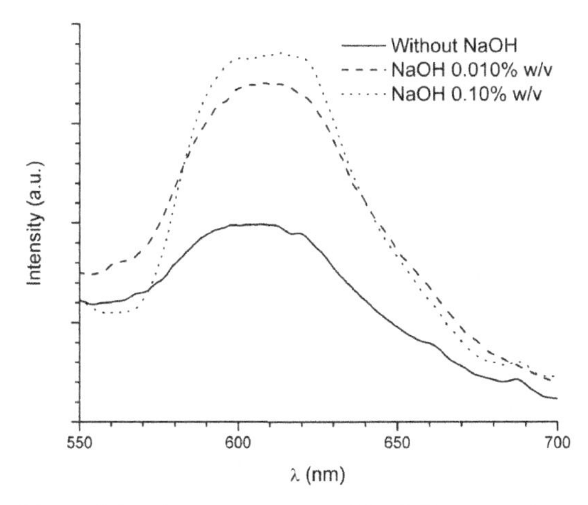

Figure 4. Luminescence spectra of the CdS nanoparticles obtained by addition of hydroxide ion solutions at different concentrations.

Table I. Optical properties and removal percentage of the CdS nanoparticles at different NaOH concentrations.

NaOH (% w/v)	**E_g (eV)**	**Relative absorbance**	**Relative intensity of luminescence**	**Removal percentage of CdS nanoparticles (%)**
0	2.77	1	1	0
0.0050	2.73	1.60	–	3
0.010	2.72	3.00	1.70	3
0.050	2.52	–	–	54
0.10	2.46	–	1.85	83
0.50	–	–	–	> 87
1.0	–	–	–	> 87

CONCLUSIONS

CdS nanoparticles with an average diameter around 5 nm, and cubic phase, were obtained by a microwave-assisted synthetic method. Absorbance of the CdS nanoparticles showed a red-shift at all the different hydroxide concentrations, the change in the shape of the absorption edge at NaOH concentrations higher than 0.050% w/v, which could be attributed to the formation of a $Cd(OH)_2$ shell around the CdS nanoparticles. An increment of 85% in the intensity of luminescence was achieved by using a 0.10% w/v in NaOH aqueous solution. Approximately, the 87% of the CdS nanoparticles were removed from the water when the NaOH concentration was higher than 0.50% w/v.

REFERENCES

1. L. Ge and J. Liu, Mater. Lett. 65, 1828 (2011).
2. Y. Hao, Y. Cao, B. Sun, Y. Li, Y. Zhang, and D. Xu, Sol. Energy Mater. Sol. Cells 101, 107 (2012).
3. Y. Xi, C. Hu, C. Zheng, H. Zhang, R. Yang and Y. Tian, Mater. Res. Bull. 45, 1476 (2010).
4. G. Wei, M. Yan, L. Ma and H. Zhang, Spectrochim. Acta A 85, 288 (2012).
5. I.A. López, A. Vázquez and I. Gómez, Rev. Mex. Fis. 59, 160 (2013).
6. M. Feng, Y. Chen, L. Gu, N. He, J. Bai, Y. Lin and H. Zhan, Eur. Polym. J. 45, 1058 (2009).
7. J. Kim, Y. Kim and H. Yang, Mater. Lett. 63, 614 (2009).
8. S.J. Ikhmayies and R.N. Ahmad-Bitar, Appl. Surf. Sci. 255, 8470 (2009).
9. N. Badera, B. Godbole, S.B. Srivastava, P.N. Vishwakarma, L.S. Chandra, D. Jain, M. Gangrade, T. Shripathi, V.G. Sathe and V. Ganesan, Appl. Surf. Sci. 254, 7042 (2008).
10. T. Zhai, Z. Gu, H. Zhong, Y. Dong, Y. Ma, H. Fu, Y. Li and J. Yao, Cryst. Growth Des. 7, 488 (2007).
11. J.K. Dongre, V. Nogriya and M. Ramrakhiani, Appl. Surf. Sci. 255, 6115 (2009).
12. A. Phuruangrat, T. Thongtem and S. Thongtem, Mater. Lett. 63, 1538 (2009).
13. A. Tang, F. Teng, Y. Hou, Y. Wang, F. Tan, S. Qu and Z. Wang, Appl. Phys. Lett. 96, 163112 (2010).
14. N.V. Hullavarad and S.S. Hullavarad, Photonics Nanostruct. 5, 156 (2007).
15. S.M. Reda, Acta Materialia 56, 259 (2008).
16. E. Caponetti, D.C. Martino, M. Leone, L. Pedone, M.L. Saladino and V. Vetri, J. Colloid Interface Sci. 304, 413(2006).
17. R. Amutha, M. Muruganandham, G.J. Lee and J.J. Wu, J. Nanosci. Nanotechnol. 11, 7940 (2011).
18. S. Das, A.K. Mukhopadhyay, S. Datta and D. Basu, Bull. Mater. Sci. 32, 1 (2009).
19. M. Pal, N.R. Mathews, P. Santiago and X. Mathew, J. Nanopart. Res. 14, 916 (2012).
20. C. Ricolleau, L. Audinet, M. Gandais and T. Gacoin, Eur. Phys. J. D 9, 565 (1999).
21. L. Spanhel, M. Haase, H. Weller and A. Henglein, J. Am. Chem. Soc. 109, 5649 (1987).
22. W.W. Yu, L. Qu, W. Guo and X. Peng, Chem. Mater. 15, 2854 (2003).
23. C.T. Tsai, D.S. Chuu, G.L. Chen and S.L. Yang, J. Appl. Phys. 79, 9105 (1996).

Mater. Res. Soc. Symp. Proc. Vol. 1617 © 2013 Materials Research Society
DOI: 10.1557/opl.2013.1179

Micro-Photoluminescence Study of Bio-conjugated CdSe/ZnS Nanocrystals

L. Borkovska[1], N. Korsunska[1], T. Stara[1], O. Kolomys[1], V. Strelchuk[1], T. Kryshtab[2], S. Ostapenko[3], G. Chornokur[3] and C. M. Phelan[3]
[1] V. Lashkaryov Institute of Semiconductor Physics of NASU, pr. Nauky 41, 03028 Kyiv, Ukraine
[2] Instituto Politécnico Nacional – ESFM, Av. IPN, Ed. 9, U.P.A.L.M., 07738 Mexico D.F., Mexico
[3] H Lee Moffitt Cancer Center and Research Institute, Tampa, Fl-33612, USA
E-mail:bork@isp.kiev.ua; kryshtab@gmail.com

ABSTRACT

The effect of bio-conjugation of CdSe/ZnS core-shell quantum dots (QDs) with Interleukin 10 (IL-10) antibodies on the aging of photoluminescence (PL) spectra of the QDs was investigated. The aging occurred upon storage of QDs for about 2 years or thermal annealing at 190 °C for up to 12 hours at atmospheric ambience and consisted in "blue" shifting the PL band position, increasing a PL band half-width and decreasing the PL intensity. The bio-conjugation is found to promote PL aging. The aging upon storage is attributed to the oxidation that decreases the QD core dimension, while the aging upon thermal annealing can be due to both oxidation and alloying of CdSe core and ZnS shell.

INTRODUCTION

To date, one of the most relevant applications of II-VI quantum dots (QDs) is in fluorescence bio-imaging and bio-sensing [1, 2]. In a liquid medium, a bio-conjugation affects mainly the intensity of QD emission and so the dependence of QD photoluminescence (PL) intensity on concentration of attached target molecules is used for QD application as protein sensors [1]. However, other spectroscopic confirmations of successful bio-conjugation (specifically, a spectral shift of emission maximum) can improve the sensitivity of immunofluorescent assays.

We have founded earlier that, if a liquid solution of bio-conjugated QDs is dropped on a silicon substrate and then dried at room ambience for several days, a shift of the PL band position to the short-wavelength region (blue shift) occurs [3, 4]. In contrast, a spectral position of the PL band in non-conjugated reference QD samples does not change at the same conditions. The shift in bio-conjugated QDs strongly increased under thermal treatment at elevated temperatures [4, 5] or during long-term storage at room temperature at air ambience [6]. In these cases the changes in color of the QD emission seen by the naked eye can be reached.

In this paper, the mechanisms of large blue shifts in bio-conjugated QDs are studied by the micro-PL technique.

EXPERIMENTAL DETAILS

The CdSe/ZnS core-shell QDs embedded in a silica shell [7] were conjugated to the IL 10 (antihuman IL 10, rat IgG2a) molecule from Serotec using a commercially available QD conjugation kit from Invitrogen Inc. [8]. The bio-conjugation was confirmed with the agarose gel electrophoresis technique [3]. A principal PL maximum of the QDs dispersed in a buffer solution was at 585- 590 nm that corresponded to average diameter of core/shell nanocrystals of about 4.4 nm [7]. The samples were produced by deposition of a ~10 μl droplet of buffered QD solution on a crystalline Si substrate. Before initial PL study the spots were dried in the atmospheric ambience at room temperature for 1 h. The samples were thermally annealed at ambient atmosphere at 190 ^{0}C for up to 12 hours in a temperature stabilized oven or stored at room temperature at the atmospheric ambience for about 2 years.

The micro-PL spectra were measured at room temperature using T-64000 Horiba Jobin-Yvon spectrometer, equipped with a cooled charge-coupled device (CCD) detector, or using a SPEX 500M spectrometer, equipped with a cooled photomultiplier tube coupled with a lock-in amplifier. The 457.9 nm line of an Ar-Kr ion laser or a 325 nm of HeCd laser were used as the excitation sources with the laser power density of 100 and 2 W/cm^2, respectively. Laser beam was focused at the 3 μm diameter spot.

RESULTS

The micro-PL spectra of both conjugated (C-QD) and non-conjugated (NC-QD) samples measured before and after thermal annealing for 2 and 4 hours are shown in Fig.1a, b. The PL spectra of thermally treated C-QD sample as compared with that of NC-QD demonstrate the next transformations: (i) a shifting of the PL band position to shorter wavelengths (a blue shift), (ii) the increasing of the PL band full width at a half maximum (FWHM) and (iii) the decreasing of the PL intensity.

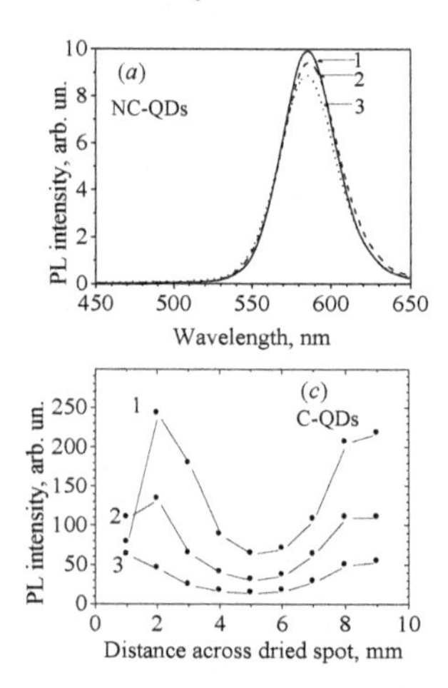

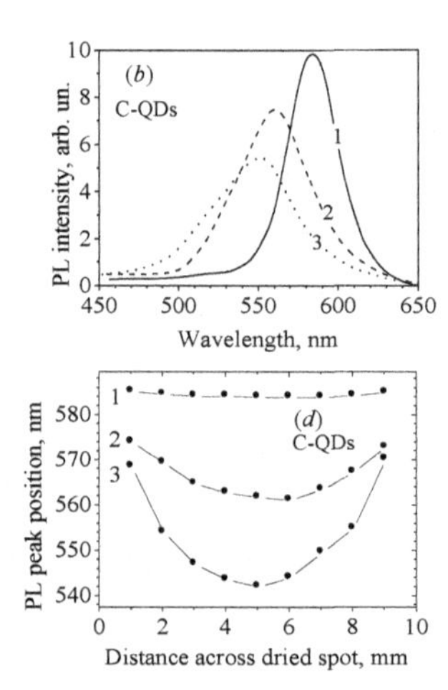

Figure 1. Micro-PL spectra of NC-QD (a) and C-QD (b) samples, as well as the PL intensity (c) and peak position (d) line scans across the C-QD sample before (1) and after thermal annealing at 190 ^{0}C for 2 hours (2) and 4 hours (3), λ_{exc}=325 nm.

Our previous study of the similar QD-samples annealed at 190 C for 2 hours has shown that in addition to above mentioned changes the conventional PL spectrum of the C-QDs shows also a larger contribution of defect-related band [5]. This band is centered at 700-750 nm and is more pronounced at low temperatures when thermal quenching of the PL intensity is strongly suppressed. In the present micro-PL study, this band is found as a shoulder on the long-wavelength edge of the excitonic band and its intensity is more that 10 times lower than that of the excitonic band.

A mapping of the PL spectra across the dried spot with the C-QDs revealed a radial gradient profile for both the PL intensity and a blue shift magnitude (Fig.1c, d). The higher PL intensity and lower blue shift magnitude are found at the ring area of the spot periphery, while the reduced PL intensity and increased blue shift magnitude are observed in the central part of the spot. This effect was found to arise after the initial 30 minutes of drying of buffered QD solution on a Si substrate [3]. Fig.1 shows that thermal annealing increases a radial gradient of a blue shift magnitude.

Analysis of the micro-PL spectra of the C-QD samples annealed for various time periods shows that with the increase of annealing time a blue shift magnitude and FWHM of the excitonic band increase, but the PL intensity decreases (Fig. 2, triangles). It is also found, that upon the annealing qualitatively the same changes take place in the PL spectra of NC-QDs, but their scale is of much lower magnitude (Fig. 2, circles). It should be noted, that the spectra analyzed were measured in the central part of the spot.

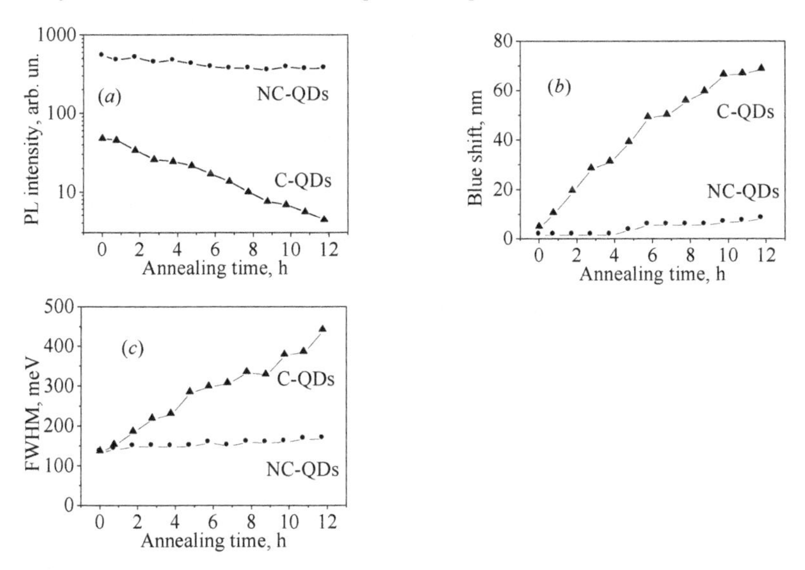

Figure 2. The change of the PL intensity (a), blue shift magnitude (b) and FWHM (c) due to thermal annealing at 190 ^{0}C vs annealing time for non-conjugated (circles) and conjugated to IL-10 (triangles) CdSe/ZnS QDs. λ_{exc}=325 nm.

It is found that the PL spectra of the QD samples change in the same manner when the QDs deposited on a Si substrate are stored at room temperature for a long time period. In the samples stored for about 2 years we found the PL intensity decreased by over two orders of magnitude, the PL band blue shifted and FWHM increased (Fig. 3a). The changes are observed for both NC-QD and C-QD samples, being larger for conjugated QDs. In the QD samples stored for about 2 years a radial gradient profile for the PL intensity and a blue shift magnitude was clearly observed. A mapping of the PL spectra in stored QD samples across the dried spot revealed strong correlation between the PL characteristics: the larger was a blue shift magnitude, the lower was PL intensity and the larger were both the FWHM of the excitonic band and relative intensity of defect-related band (Fig. 3). All PL characteristics were obtained from the fitting of the PL spectra with two Gaussian lines corresponding the excitonic and defect-related bands.

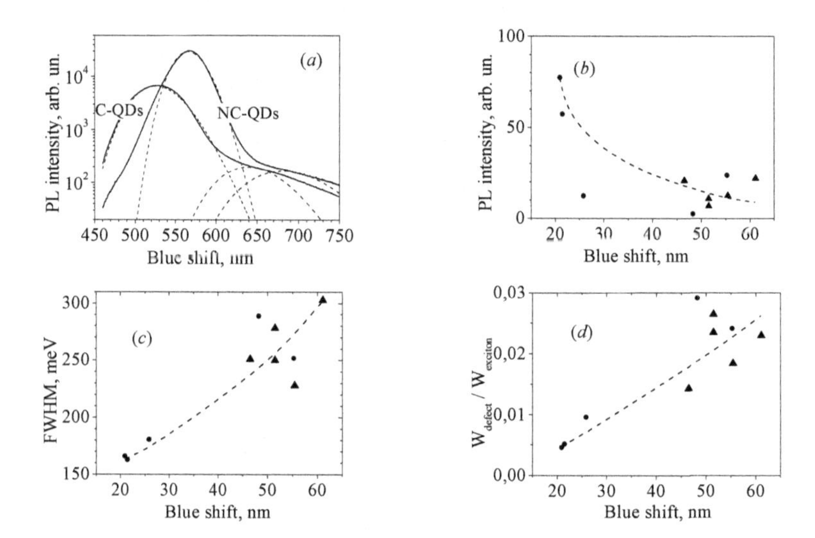

Figure 3. Micro-PL spectra (a) of conjugated (C-QDs) and non-conjugated (NC-QDs) QDs stored for 2 years and their fitting with two Gaussian lines (dashed lines). Integrated PL intensity (b), FWHM (c) and the ratio of the defect-related to excitonic PL band intensities (d) via blue shift magnitude in the NC-QDs (circles) and C-QDs (triangles) stored for 2 years. Dashed lines are drawn for better eye guide. λ_{exc}=457.9 nm.

DISCUSSION

Thus, both the thermal treatment and long-term storage of QD samples dried on a Si substrate produce similar changes in the PL spectra: (i) blue spectral shift, (ii) PL decrease, (iii) FWHM increase and (iv) the increase of relative contribution of defect-related band in the PL spectrum. These features can be assigned to “aging” process, which will be discussed below. Since the changes in the PL characteristics correlate with each other, they can be caused by the same mechanism. The changes are found to be larger in C-QDs, so the bio-conjugation seemingly promotes the PL aging.

A blue shift effect for C-QDs dried on a Si substrate for several days [3, 4] was explained by an extra hydrostatic compressive stress produced in consolidated solution of a droplet. In the annealed and long-term stored samples, an estimated magnitude of applied hydrostatic pressure to produce observed blue shift of 60 nm (or about 240 meV) is about 53 kbars. This value is of the same order as the pressures for wurtzite-rocksalt transition in bulk CdSe (27-30 kbars) and CdSe/ZnS QDs (69 kbars) [9]. It is difficult to imagine the cause of such large strains in dried QD solution. Moreover, our previous X-ray diffraction study of long-term stored CdSeTe/ZnS QDs dried on a Si substrate have shown much smaller elastic deformations than those mentioned above [6]. So, we have to rule out compressive strains as the main "aging" mechanism.

A blue spectral shift of QD luminescence band may be caused by increasing of QD core band gap due to ZnCdSe alloy formation as a result of Zn diffusion from a ZnS shell. The effect of alloying was observed in CdSe/ZnS QDs [10] and CdSe/ZnSe nanorods [11] upon thermal treatment at 220-270^{0}C and accompanied by blue shifting of PL peak position, decreasing of a quantum yield and broadening of PL band [11]. In the samples studied, 240 meV blue shift of PL band can easy be explained by partial intermixing of ZnS shell (of about 0.7 nm [7]) and CdSe core (no more than 4.1 nm [7]), taking into account larger band gap of ZnS (~3.6 eV) as compared to CdSe (~1.74 eV). The alloying of CdSe core and ZnS shell can explain also the change of other PL characteristics. It seems that alloying is more probable process for thermally annealed samples than for the samples stored at room temperature.

Alternatively, the PL shift with aging can be interpreted as a reduction in the QD core diameter due to oxidation. Oxidation occurs when QD samples are placed in the environments containing oxygen and proceeds under continuous irradiation or in the dark [12]. Oxidation of CdSe and CdSe/ZnS QDs has been found to result in a blue shift of the PL peak, increased FWHM and decreased PL intensity [12]. The latter is obviously caused by the creation of non-radiative defects on the QD surface. The degradation of CdSe/ZnS interface can also promote the generation of radiative defects on the QD surface and increasing the intensity of defect-related luminescence. The changes observed in the PL spectra of oxidized QDs [12] are similar to those found in our samples subjected to thermal annealing or long-term storage. Hence, we hypothesize that oxidation of QDs contribute to aging of the PL spectra occurred upon thermal annealing and long-term storage. It seems that the silica layer covered the core-shell nanocrystal does not prevent QDs from oxidation since it is known to be thin and porous that provides direct access of oxygen to ZnS layer.

An additional argument in favor of an oxidation hypothesis follows from the result of our X-ray diffraction study of long-term stored CdSeTe/ZnS QDs conjugated to Anti-Caveolin 1 antibodies [6]. It has been found that the diameter of core in stored bio-conjugated QDs is ~1.5 nm smaller than that in non-conjugated [6]. For studied silica coated CdSe/ZnS QDs the decrease in CdSe core diameter from 4.1 to 2.7 nm results in the shift of emission peak from yellow to green [7]. Hence, the decrease in QD core diameter by ~1-1.5 nm can easy explain 240 meV "blue" shift of PL peak position.

CONCLUSIONS

In conclusion, we studied the transformation of the micro-PL spectra of non-conjugated and bio-conjugated CdSe/ZnS QDs dried on a Si substrate and stored at room temperature or thermally annealed. A blue spectral shift in conjunction with the increased FWHM, decreased PL intensity and increased contribution of defect-related band in the PL spectrum was observed. A bio-conjugation is found to promote these transformations. The oxidation of QDs is proposed to be the main reason of blue shift that occurs upon long-term storage or thermal annealing, while the alloying of CdSe core and ZnS shell is supposed to contribute also in blue shift produced by thermal annealing.

ACKNOWLEDGMENTS

This work was partially supported by the National Academy of Sciences of Ukraine through the program "Nanotechnologies and nanomaterials" (Grant No 2.2.1.14) and the project "Physical and Physical-Technological Aspects of Fabrication and Characterization of Semiconductor Materials and Functional Structures for Modern Electronics" (Grant No III-41-12).

REFERENCES

1. K.E. Sapsford, T. Pons, I.L. Medintz and H. Mattoussi, *Sensors* **6**, 925 (2006).
2. H. Mattoussi, G. Palui, and H. B. Na, *Advanced Drug Delivery Reviews* **64**, 138 (2012).
3. G. Chornokur, S. Ostapenko, Yu. Emirov, N. E. Korsunska, T. Sellers, and C. Phelan, *Semicond. Sci. Technology* **23**, 075045 (2008)
4. M. Dybiec, G. Chornokur, S. Ostapenko A. Wolcott, J. Z. Zhang A. Zajac, C. Phelan, T. Sellers, and D. Gerion, *Appl. Phys. Letter* **90**, 263112 (2007).
5. L. V. Borkovska, N. E. Korsunska, T. G. Kryshtab, L. P. Germash, E. Yu. Pecherska, S. Ostapenko, and G. Chornokur, *Semiconductors* **43**, 775 (2009).
6. T.G. Kryshtab, L.V. Borkovska, O. Kolomys, N.O. Korsunska, V.V. Strelchuk, L.P. Germash, K.Yu. Pechers'ka, G. Chornokur, S.S. Ostapenko, C.M. Phelan, and O.L. Stroyuk, *Superlattices and Microstructures* **51**, 353 (2012).
7. D. Gerion, F, Pinaud, S.C. Williams, W.J. Parak, D. Zanchet, S. Weiss, and A. P. Alivisatos, *J. Phys. Chem. B* **105**, 8861 (2001).
8. http://www.invitrogen.com
9. H.M. Fan, Z.H. Ni, Y.P. Feng, X.F. Fan, J.L. Kuo, Z.X. Shen, and B.S. Zou, *Appl. Phys. Lett.*, **90**, 021921 (2007).
10. X. Xia, Z. Liu, G. Du, Y. Li, and M. Ma, *J. Phys. Chem. C* **114**, 13414 (2010).
11. H. Lee, P.H. Holloway, H. Yang, L. Hardison, and V.D. Kleiman, *J. Chem. Phys.* **125**, 164711 (2006).
12. A. Y. Nazzal, X. Wang, L. Qu, W. Yu, Yu. Wang, X. Peng, and M. Xiao, *J. Phys. Chem. B* **108**, 5507 (2004).

Mater. Res. Soc. Symp. Proc. Vol. 1617 © 2013 Materials Research Society
DOI: 10.1557/opl.2013.1180

Colloidal indium sulfide quantum dots in water: synthesis and optical properties

Mariia V. Ivanchenko, Oleksandra E. Rayevska, Oleksandr L. Stroyuk, and Stepan Ya. Kuchmiy
L.V. Pysarzhevsky Institute of Physical Chemistry of National Academy of Sciences of Ukraine,
31 Nauky av., 03028, Kyiv, Ukraine

ABSTRACT

Colloidal β-In_2S_3 quantum dots stabilized in water by a number of polymers or sodium polyphosphate and mercaptoacetate were synthesized. An increase in the stabilizer content was found to result in a decrease in the average dot size from 20–30 to 5–10 nm and formation of a narrow absorption band centered at 290 nm. The position and spectral width of the band were found to be independent on stabilizer concentration, synthesis temperature and molar In:S ratio. The band was assumed to belong to a molecular cluster smaller than 1 nm which is a precursor for formation of larger regular indium sulfide quantum dots.

INTRODUCTION

Semiconductor nanocrystals smaller than the Bohr exciton diameter $2a_B$ for a given compound, the so-called quantum dots (QDs), belong to the most intriguing objects of the modern chemistry and find numerous applications in light-emitting devices, luminescent biomarkers, solar cells and photocatalytic systems [1]. Of special interest are the QDs with a large a_B, such as PbS (a_B = 18 nm), PbSe (a_B = 46 nm) [1], In_2S_3 (a_B > 30 nm) [2], etc. For these semiconductors the exciton becomes spatially confined already in comparatively large particles with the size d of tens of nanometers, while at decreasing d to a few nm we can observe evolution of the electron properties of the semiconductor at extreme exciton confinement. Vigorous research of lead chalcogenide QDs inspired by their unique properties such as IR photoluminescence and the multi-exciton generation phenomenon [1] stimulated progress in synthesis protocols allowing to stabilize PbS and PbSe QDs in various solvents and tailor QD size and surface chemistry. At the same time, the synthetic chemistry of In_2S_3 QDs, specifically in the form of stable colloids in polar solvents as well as strategies of influencing the average QD size are studied fragmentarily. The present paper reports synthesis of stable aqueous colloids of indium(III) sulfide using a number of water-soluble polymers and low-molecular-weight agents capable of forming complexes with In(III) such as sodium polyphosphate and mercaptoacetate. For the latter two a detailed study of synthesis conditions on the optical properties of colloidal indium sulfide QDs was performed and means of influencing the In_2S_3 QD size were found.

EXPERIMENTAL DETAILS

Indium(III) chloride, $Na_2S \times 9H_2O$, 50 w.% aqueous solution of polyethyleneimine (molecular mass M = 50 KDa), polyvinyl alcohol (M = 100 KDa), polyvinylpyrrolidone (M = 360 KDa), gelatin from porcine skin, polyethylene glycol (M = 6 KDa), sodium polystyrenesulfonate, sodium polyphosphate, sodium mercaptoacetate, $NaBH_4$ and NaOH were supplied by Sigma Aldrich. Colloidal In_2S_3 NCs were synthesized at room temperature via interaction between stoichiometric amounts $InCl_3$ and Na_2S in aqueous solution of a stabilizer. In a typical procedure, to produce 10.0 mL of 1×10^{-3} M colloidal solution, 5.0 mL of aqueous solution of 0.004 M $InCl_3$ and a stabilizer (typically 0.06 M sodium polyphosphate or 0.02 M sodium mercaptoacetate) were mixed at vigorous refluxing with 5.0 mL of aqueous solution of 0.006 M Na_2S. Absorption spectra were registered using a Specord 220 spectrophotometer. The hydrodynamic size of In_2S_3 NCs was determined by the dynamic light scattering technique using a Malvern Zetasizer Nano setup. Transmission electron microscopy (TEM) experiments were carried out on a Selmi S-100 microscope with an accelerating voltage of 100 kV.

RESULTS AND DISCUSSION

Interaction between $InCl_3$ and Na_2S in aqueous solution with no stabilizers present results in instant deposition of a yellow precipitate. In the presence of water-soluble polymers – gelatin, polyvinyl alcohol, polyvinylpyrrolidone, polyethyleneglycol, polyethyleneimine and sodium polystyrenesulfinate, as well as some low-molecular-weight compounds such as sodium polyphosphate (SPP) or sodium mercaptoacetate (SMA) in similar conditions stable colloidal solutions form with the color and optical properties depending on the nature and concentration of a stabilizer.

Figure 1 shows absorption spectra of indium sulfide colloids synthesized in the presence of several polymer agents (a), SPP (b), and SMA (c). Solutions stabilized by polymers as well as by SPP and SMA at a low content (Fig. 1b,c, curves 4,5) exhibit structureless bands with an edge at 340 – 420 nm depending on the stabilizer nature. When plotted in the coordinates of Tauc equation for indirect electron transitions $(\alpha\times h\nu)^{1/2}$ vs. $h\nu$ [1] ($h\nu$ – quantum energy, α – absorption coefficient) they manifest a linear section in the range of $h\nu > 3$ eV. From the intersection of the tangent to this section and the x axis the indirect transition energy can be determined which in the first approximation equals to the band gap E_g of colloidal indium sulfide particles. The band gaps for different stabilizers are given in Table I in a descending order. The highest band gap, 3.46 eV, was achieved in case of polyethyleneimine, the lowest – for the low-molecular-weight stabilizers (SPP, SMA). In all cases the band gap of colloidal particles is considerably higher that E_g of the bulk indium(III) sulfide, ~2 eV [a1, a4]. The fact indicates that the In_2S_3 particles behave as quantum dots exhibiting considerable spatial electron confinement.

Among the tested range of stabilizers for a more detailed investigation of the QD structure and relationships between the synthesis conditions and QD properties two low-molecular-weight stabilizers, SPP and SMA, were chosen as their presence does not hinder studies of the QDs by means of the dynamic light scattering spectroscopy and electron microscopy. According to the bright-field TEM (Fig. 2a) the In_2S_3 QDs synthesized at a lowest concentration of SPP or SMA, 1×10^{-3} M (corresponding spectra are given by the curves 5 in Fig. 1b,c), have a size of 20 – 40 nm. Similar results were obtained by the dark-field TEM (Fig. 2b) indicating crystalline nature of the QDs.

The electron diffraction patterns of indium sulfide QDs (Fig. 2b, inset) shows a series of concentric reflexes corresponding to the lattice distances of 2.9 Å, 2.06 Å, 1.9 Å, 1.4 Å, 1.2 Å, 1.05 Å, 0.8 Å, and 0.7 Å, characteristic for the tetragonal β-In_2S_3 [3,4]. Besides, a separate reflexes belonging to α-In_2S_3, 1.9 Å, 1.55 Å, 1.1 Å, and to InS (2.8 Å) [3] can be observed. These are the most intense reflexes for α-In_2S_3 and та InS and the absence of less intense signals indicates that the phases are present as traces.

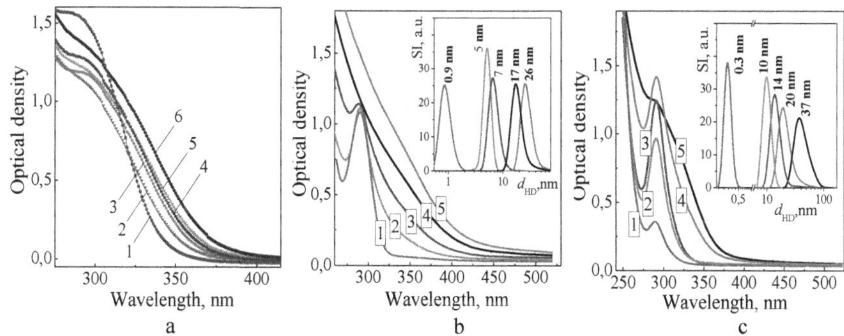

Figure 1. (a) Absorption spectra of 1×10^{-3} M indium sulfide colloids stabilized by polyethyleneimine (curve 1), gelatin (2), polyvinylpyrrolidone (3), sodium polystyrenesulfonate (4), polyvinyl alcohol (5), and polyethylene glycol (6). The stabilizer content is 0.5 w.%. (b,c) Absorption spectra and hydrodynamic size (insets) of indium sulfide QDs stabilized by sodium polyphosphate (b) and sodium mercaptoacetate (c). SI – scattering intensity. [SPP] = 3×10^{-2} M (curves 1), 1×10^{-2} M (2), 5×10^{-3} M (3), 1×10^{-3} M (4), and 5×10^{-4} M (5). [SMA] = 5×10^{-2} M (1), 2×10^{-2} M (2), 1×10^{-2} M (3), 5×10^{-3} M (4), and 1×10^{-3} M (5). [In_2S_3] = 1×10^{-3} M, cuvette – 1.0 mm.

Table I. Band gap E_g of In_2S_3 QDs stabilized by various agents

Sample	Stabilizing agent	E_g (eV)
1	Polyethyleneimine	3.46
2	Gelatin	3.29
3	Polyvinylpyrrolidone	3.25
4	Sodium polystyrenesulfonate	3.21
5	Polyvinyl alcohol	3.16
6	Polyethyleneglycol	3.10
7	Sodium mercaptoacetate	3.11
8	Sodium polyphosphate	2.74

Note: the band gap values determined for the colloids containing [In_2S_3] = 1×10^{-3} M, 0.5 w.% of the stabilizers # 1–6, 5×10^{-3} M of the stabilizer #7, and 1×10^{-3} M of the stabilizer #8. The band gap determination error is 0.01 eV.

The average hydrodynamic size of In_2S_3 QDs synthesized at a minimal SPP concentration is 26 nm (Fig. 1b, inset, curve 5) which agrees with the TEM data. Colloidal In_2S_3 QDs stabilized by a minimal amount of SPA have a somewhat larger hydrodynamic size d_{HD} = 37 nm (Fig. 1c, inset, curve 5). In both cases an increase in the stabilizer content results in a blue shift of the absorption band edge and simultaneous decrease of the hydrodynamic QD size. For example, increasing SPP concentration by an order of magnitude – from 1×10^{-3} M to 1×10^{-2} M results in a decrease of d_{HD} from 26 to 5 nm (Fig. 1b, inset, curve 2). In case of SMA, when the stabilizer content is elevated from 1×10^{-3} M to 2×10^{-2} M d_{HD} lowers from 37 to 10 nm (Fig. 1c, inset, curve 2). The data indicate that the blue shift of the fundamental absorption band edge of In_2S_3 QDs does originate from escalation of the exciton confinement due to the QD size reduction resulting in an increase of the interband electron transition energy.

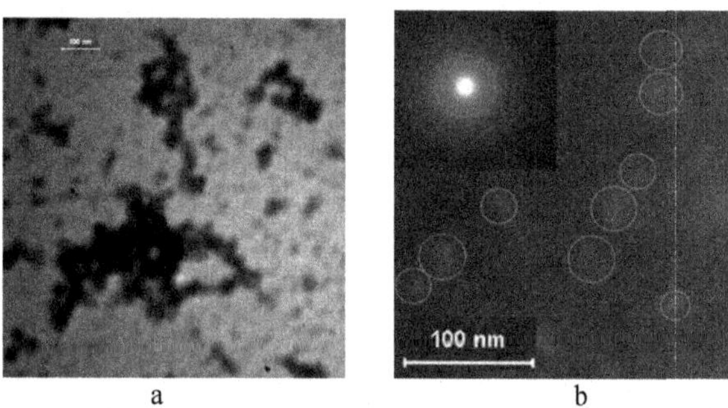

Figure 2. Bright-field TEM (a), dark-field TEM (b), and electron diffraction pattern (inset in (b)) of indium sulfide QDs stabilized by sodium polyphosphate.

Figure 1b shows that an increase of SPP concentration, besides the blue shift of the absorption threshold of In_2S_3 QDs, results also in growth of a new narrow band with a sharp maximum at λ = 290 nm (curves 1–3). In case of SMA the peak is present in the absorption spectrum already at the lowest (in the range studied) stabilizer concentration (Fig. 1c). The peak position is the same both for SPP and SMA and does not depend on their concentration in colloidal solution. According to the dynamic light scattering measurements, the average hydrodynamic size of particles in the solutions prepared at the highest stabilizer concentration – 3×10^{-2} M for SPP and 5×10^{-2} M for SMA, does not exceed 1 nm (Fig. 1b,c, insets, curves 1). From the fact, as well as lack of dependence between the peak position and the stabilizer content one can conclude that the band centered at 290 nm belongs not to the indium sulfide QDs but rather to some molecular species.

It is well-known that the most effective means of influencing the average size of metal sulfide QDs stabilized in water by polymers and various molecular agents are those allowing to influence a ratio of the rates of primary nuclei formation and nuclei growth [1]. For example, lowering of the synthesis temperature or a large excess of one of the ions forming the QDs typically results in reduction of the average QD size and a blue shift of the absorbance threshold of the final colloidal solutions. It was found that in case of indium sulfide these factors do not affect the spectral parameters of the band centered at λ = 290 nm. At variation of the synthesis temperature in quite a broad range – from 7 to 83 °C the band retains its position and width unchanged. Introduction of an excess of sodium sulfide results in an increase of the band intensity but again without changes in its spectral parameters (Fig. 3a). In case of the above-discussed dependences of the optical properties of indium sulfide QDs on the SPP or SMA concentration we also basically deal with an excess of sulfide-ions because the stabilizers bind In(III) and reduce its amount in the system. On the contrary, introduction of an excess of In(III) relatively to the amount necessary for the complete binding of the sulfide ions results in a decrease of the intensity of the band at $\lambda = 290$ nm the more pronounced the higher is the excess $InCl_3$ concentration (Fig. 3b).

Two conclusions can be made from the results: (i) by the character of relationship between the synthesis conditions and the properties of the band centered at 290 nm most probably it belongs no to the QDs but rather to some molecular cluster; (ii) a prerequisite for the band formation and growth is the presence of sulfide ions which act as a strong reducing agent. Arguments in favor to the second conclusion were obtained by synthesizing colloidal solutions in the presence of another strong reductant – sodium borohydride. In the presence of $NaBH_4$ the characteristic band at 290 nm can be observed already at a low SPP concentration of 1×10^{-3} M.

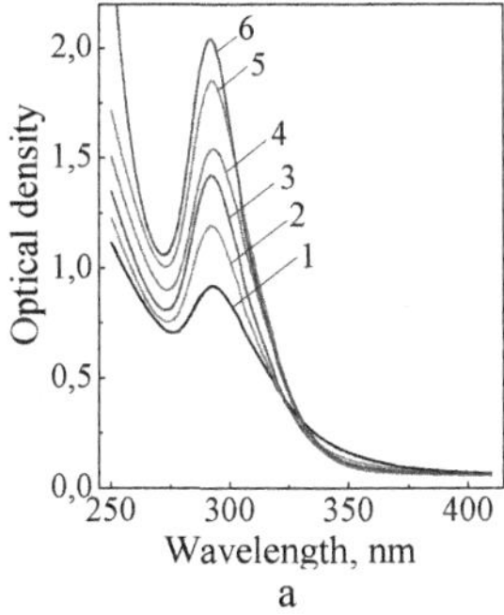

a

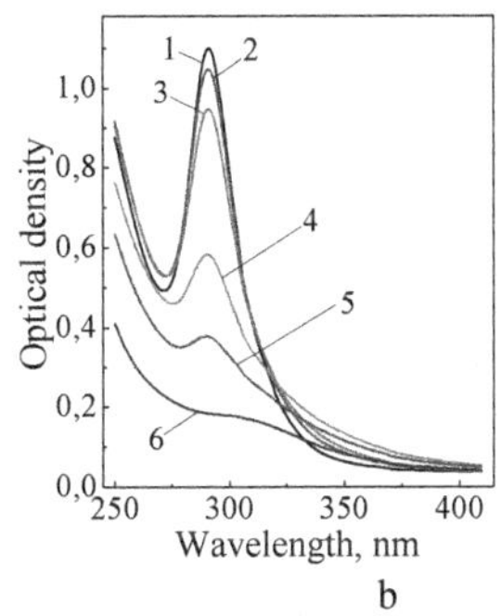

b

Figure 3. Absorption spectra of colloidal indium sulfide synthesized (a) at $[In_2S_3] = 1\times10^{-3}$ M with no Na_2S excess (curve 1) and a Na_2S excess of 5×10^{-4} M (2), 1×10^{-3} M (3), 1.5×10^{-3} M (4), 3×10^{-3} M (5), and 1×10^{-2} M (6); (b) at $[In_2S_3] = 1\times10^{-3}$ M with no $InCl_3$ excess (curve 1) and a $InCl_3$ excess of 5×10^{-4} M (2), 1×10^{-3} M (3), 2×10^{-3} M (4), 3×10^{-3} M (5), and 5×10^{-3} M (6).

It should be noted that almost no reports can be found on the nature of the band centered at $\lambda = 290$ nm in the absorption spectra of colloidal indium sulfide, though it was observed for In_2S_3 QDs stabilized not only in water [3,5], but also in acetonitrile [5] and hexane [4]. The authors of [5], though not going into details about the nature of this band, noted that most probably it belongs to molecular indium sulfide clusters similar to the well-known "magic-size" cadmium chalcogenide clusters. Analysis of the reports on the coordination compounds containing In_xS_y clusters as structural fragments showed that absorption spectra of $Ni(bpy)_3]_3[H_4In_{10}S_{20}]\times bpy\times 2EG\times H_2O$ (bpy – bipyridyl, EG – ethyleneglycol) complex with the pyramidal $In_{10}S_{20}$ fragments actually exhibit a band centered at 290 nm originating from the electronic excitation of such structural blocks [6]. Combining the reported data and the present experimental results we may assume that the band with the maximum at 290 nm, appearing at certain conditions in the absorption spectra of colloidal indium sulfide, belongs to a molecular cluster smaller than 1 nm. The structure of the cluster and the exact valence state of indium atoms constituting the cluster are now unknown and will be the subject of a separate study.

Although the influence of the basic synthesis parameters on the spectral properties of the band centered at 290 nm differs from that typical for the colloidal metal sulfide QDs, the evolution of the band at the post-synthesis ageing resembles the behavior of the colloidal systems. In particular, prolonged ageing of SPP- or SMA-stabilized indium sulfide colloids at room temperature or a much faster ageing at 80–90 °C results in a red shift of the absorption maximum and edge indicating gradual transformation of the discussed molecular clusters into larger species and regular QDs. One may assume that the molecular clusters exhibiting the absorption band at 290 nm serve as "building bricks" in formation of larger In_2S_3 QDs.

CONCLUSIONS

Aqueous indium sulfide colloids stabilized by a number of water-soluble polymers (gelatin, polyvinyl alcohol, polyvinylpyrrolidone, polyethyleneimine, polyethyleneglycol, sodium polystyrenesulfonate, etc.) as well as sodium polyphosphate and mercaptoacetate were synthesized. The colloids were shown to contain tetragonal β-In_2S_3 quantum dots with the band gap considerably higher than that of bulk indium sulfide due to the spatial exciton confinement. In case of the quantum dots stabilized by sodium polyphosphate and sodium mercaptoacetate it was shown that an increase in the stabilizer concentration results in reduction of the average dot size from 20–30 to 5–10 nm and formation of a narrow absorption band with a maximum at 290 nm. The position and spectral width of the band were found to be independent on the stabilizer concentration, synthesis temperature and the molar In:S ratio, but changing at the post-synthesis ageing of the colloidal solutions. It was assumed that the band at 290 nm belongs to a molecular cluster smaller than 1 nm with a core of indium sulfide. The cluster acts as a building "brick" at formation of larger regular In_2S_3 quantum dots.

ACKNOWLEDGMENTS

The authors acknowledge LLC NanoMedTech (Kyiv, Ukraine) for their kind help in accessing TEM and DLS facilities. The work was financially supported by a Joint Project of National Academy of Sciences of Ukraine and Siberian Branch of Russian Academy of Sciences (Project # 07-03-12(U)).

REFERENCES

1. Semiconductor nanocrystal quantum dots: synthesis, assembly, spectroscopy and applications, ed. by A. Rogach, Springer-Verlag GmbH, Vienna, 2008.
2. W. Chen, J.-O. Bovin, A.G. Joly, S. Wang, F. Su, G. Li, *J. Phys. Chem. B* **108**, 11927-11934 (2004).
3. D.K. Nagesha, X. Liang, A.A. Mamedov, G. Gainer, M.A. Eastman, M. Giersig, J.-J. Song, T. Ni, N.A. Kotov, *J. Phys. Chem. B* **105**, 7490-7498 (2001).
4. K.H. Park, K. Jang, S.U. Son, *Angew. Chem. Int. Ed.* **45**, 4608-4612 (2006).
5. P.V. Kamat, N. Dimitrijević, R.W. Fessenden, *J. Phys. Chem.* **92**, 2324-2329 (1988).

6. Y.-P. Zhang, X. Zhang, W.-Q. Mu, W. Luo, Y.-Q. Bian, Q.-Y. Zhu, J. Dai, *Dalton Trans.* **40**, 9746-9751 (2011).

Mater. Res. Soc. Symp. Proc. Vol. 1617 © 2013 Materials Research Society
DOI: 10.1557/opl.2013.1181

Photoluminescence and Structural Properties of CdSe Quantum Dot-Polymer Composite Films

L. Borkovska[1], N. Korsunska[1], T. Stara[1], V. Bondarenko[1], O. Gudymenko[1], O. Stroyuk[2], O. Raevska[2] and T. Kryshtab[3]

[1] V. Lashkaryov Institute of Semiconductor Physics, NASU, pr. Nauky 41, 03028 Kyiv, Ukraine

[2] L. Pysarzhevsky Institute of Physical Chemistry, NASU, pr. Nauky 31, 03028 Kyiv, Ukraine

[3] ESFM - Instituto Politécnico Nacional, Av. IPN, Ed.9 U.P.A.L.M., 07738 Mexico D.F.Mexico

ABSTRACT

Thermal stability of the luminescent properties of CdSe and CdSe/ZnS quantum dots (QDs) in polymer films of gelatin and polyvinyl alcohol (PVA) is studied. Thermal annealing of the films at the air ambience at 100 °C is found to result in two effects in the photoluminescence (PL) spectra: (i) an enhancement of the PL intensity and (ii) a red spectral shift of the PL bands. The first effect is observed in both QDs-gelatin and QDs-PVA composites, while the second one - in the QDs-gelatin only. The passivation of CdSe QDs with ZnS shell reduces the effects. The enhancement of the PL intensity is supposed to be due to the decrease of nonradiative defect density. The red shift is explained by dissociation of coordination bonds between surface Cd atoms and amino-groups of gelatin. This dissociation decreases the PL intensity too. This effect competes with the effect of PL enhancement and is supposed to be responsible for non-monotonous dependence of the PL intensity versus annealing time in the QDs-gelatin composite.

INTRODUCTION

The progress achieved in fabrication of semiconductor quantum dot (QD)-polymer composites made an impact in many applications including optotelectronics, photonics, analytical chemistry and bioengineering [1- 3]. The challenges in present-day applications of semiconductor QDs are their stability and biocompatibility. The release of Cd^{2+} ions from the core of II-VI QDs (CdSe, CdS, CdTe) in response to ultraviolet radiation has been shown to be the main cause of their cytotoxicity [2, 3]. The gelatin, - natural nontoxic, water-soluble and biodegradable polymer, was proposed as a surface capping agent for CdTe [4], CdSe and CdS [5] QDs. It can also be used as a stabilizing agent during the synthesis of colloidal QDs [5].The possibility to produce CdTe and CdSe QDs embedded in gelatin nanoparticles was demonstrated [6, 7]. The CdTe QDs–gelatin system showed inherent stability against photo-oxidation damage and salt effect [6], as well as biocompatibility and the reduced toxicity [4]. However, thermal stability of QD-gelatin system was not investigated in detail.

Here, we present the results of our study of the effect of thermal annealing at 100 °C on the photoluminescence (PL) of composite polymer films with embedded CdSe and CdSe/ZnS QDs. The polymers studied were gelatin and polyvinyl alcohol (PVA). The latter contained a small amount of gelatine which was used as a stabilizing agent during the synthesis of the QDs.

EXPERIMENTAL DETAILS

CdSe and CdSe/ZnS QDs embedded in polymer film of gelatin or PVA deposited on a glass substrate were studied. CdSe QDs were produced by the reaction of Na_2SeSO_3 and $CdCl_2$ in aqueous solution of gelatin [5]. The gelatin was used as a stabilizer and a host matrix agent during the synthesis and purification of initial colloidal CdSe QDs. Passivation of CdSe QDs with ZnS shell was achieved by addition of measured volumes of $Zn(NO_3)_2$ solutions followed by stirring for 5-10 min and subsequent addition of an equimolar amount of sodium sulphide solution. ZnS amount was varied by increasing molar concentration ratio [ZnS]:[CdSe] in the solution from 0 to 2. The aqueous QDs-gelatin solution contained $1x10^{-3}$ M QDs (in terms of CdSe concentration) and 10 w. % gelatin, while QDs-PVA solution contained $1x10^{-3}$ M QDs, 10 w. % PVA and 0.5 w. % gelatin. In the last case QDs were purified as a gelatin gel by dialysis at 5 °C and then diluted with PVA with HCl acid added to the solution to maintain homogeneity of the polymer mixture and optical transparency of the films. The average diameter of CdSe QDs was about 2.3-2.4 nm as estimated from spectral position of the absorption maximum (at about 2.5-2.45 eV at 300 K) in the optical absorption spectrum [8].

The films were annealed at 100 °C for a period of time in the range of 5-210 minutes in an atmospheric ambience. Before annealing the films were removed from the glass substrate. The PL and PL excitation spectra were studied at room temperature. The PL was excited with a light of a halogen lamp passed through a grating monochromator and recorded using a prism monochromator equipped with a photomultiplier and an amplifier with synchronous detector. X-ray diffraction (XRD) study was realized using a D-8 ADVANCE BRUKER one-crystal X-ray diffractometer operating with the Cu K_{α} - radiation.

RESULTS

The XRD patterns of the composite films of QDs-gelatin and QDs-PVA (Fig. 1) show the intense diffraction peaks around 2θ =14° and 2θ = 19.5° of the amorphous phase of gelatin and PVA matrix, respectively [9, 10]. The signals of much lower intensity found in the XRD pattern of the QDs-PVA composite are evidently caused by polymer matrix of partially crystalline nature [10]. The peaks from crystalline CdSe phase are not observed apparently because of a low QD concentration in the composites (less than 1 % w/v).

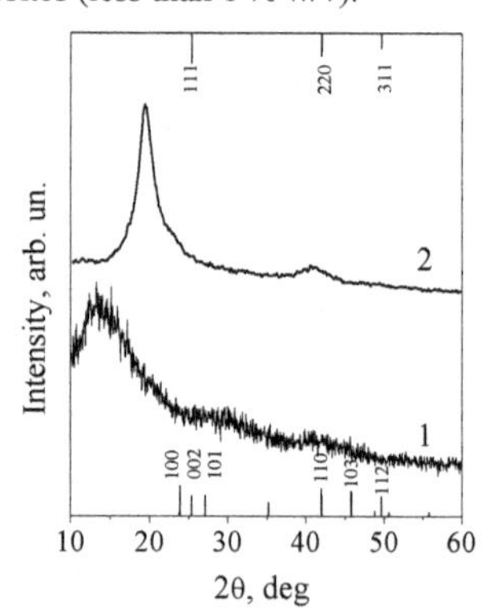

Figure 1. X-ray diffraction pattern of CdSe QDs –gelatin (1) and CdSe QDs - PVA (2) composites. The line spectrum indicates the reflections for the bulk hexagonal and cubic CdSe.

The PL spectra of the composite films are characterized by two bands denoted as I_1 and I_2 (Fig. 2a). Usually, the high energy band in the PL spectra of colloidal QDs is ascribed to the exciton radiative recombination or radiative recombination of carriers via shallow levels of defects in the QDs (band-edge PL), while the low energy band is attributed to radiative recombination of carriers via deep levels of surface defects in QDs [11]. In the PL spectra of the QDs-gelatin composite, the band-edge PL dominates, while in the PL spectra of the QDs-PVA composite the defect-related band contributes mainly. A large contribution of deep trap emission in the PL spectra of composites studied is not surprising owing to small size of the QDs. This is often observed in bare CdSe QDs of small size passivated by organic capping groups and is explained by large surface-to-volume ratio and insufficient surface defect passivation by organic ligands [12]. In the case of CdSe-PVA composite an insufficient passivation of QD surface defects by functional groups of PVA should be assumed. Besides, an addition of HCl in the composite is found to promote the decrease of the band-edge emission too.

The passivation of CdSe nanocrystals with ZnS shell results in an increase of the PL intensity and a shift of the PL bands to lower energies (Fig. 2a) as usually observed for CdSe/ZnS QDs [12, 13]. The shift is explained by partial tunnelling of the electron wavefunction into the shell [12]. The shift magnitude is somewhat larger for the I_1 band. As the ZnS quantity increases, the shift for the I_1 band increases up to 100 meV for [ZnS]:[CdSe]=2 in the QDs-PVA composite (Fig. 2b).

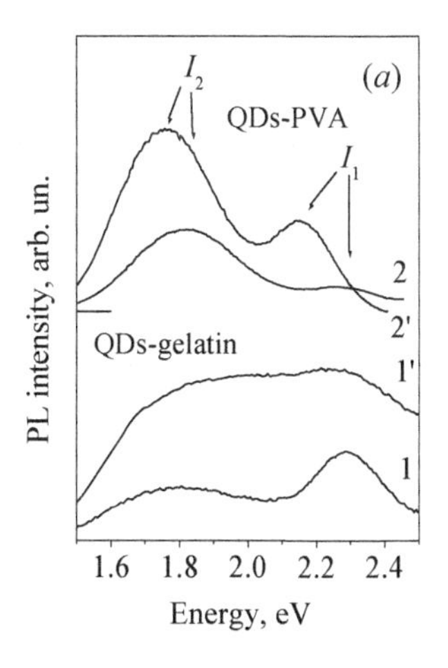

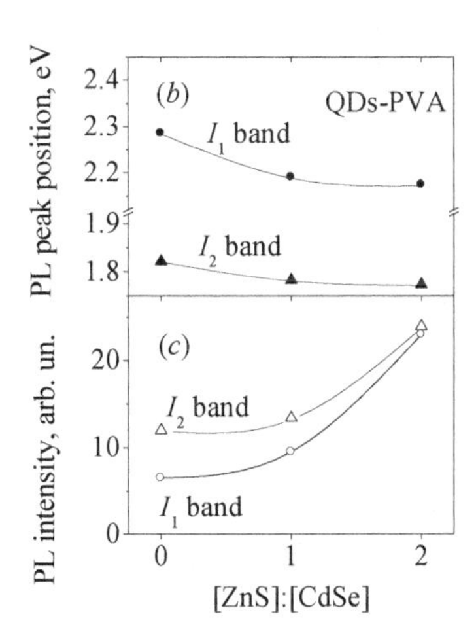

Figure 2. Photoluminescence spectra (a) of CdSe QDs–gelatin (1), CdSe/ZnS QDs–gelatin (1'), CdSe QDs-PVA (2), CdSe/ZnS QDs-PVA (2') composite films. Spectral positions (b) and intensities (c) of the PL bands in CdSe/ZnS QDs-PVA composite in dependence on ratio of [ZnS]:[CdSe] in starting solution. T=300 K, λ_{exc}=470 nm.

The effect of thermal annealing on the PL and PL excitation spectra of the QDs-gelatin composite is found to be slightly different from that of the QDs-PVA (Fig. 3). The PLE spectra presented in Fig. 3 are detected in the maximum of the I_2 band. The PLE spectrum of the I_1 band (not shown) was similar to that of the I_2 band, but due to its strong overlapping with the I_1 band it was not examined here.

In the PL spectra of the QDs-gelatin, the PL intensity increases and the PL bands shift to lower energies upon the annealing (a red shift effect) (Fig. 3a, curves 1, 2). In different samples annealed for about 1 hour, the shift magnitude ranges between 10 and 85 meV. In the PL excitation spectra, a step at about 2.4-2.5 eV associated with light absorption by the ground states in the QDs shifts to lower energies too (Fig. 3a, curves 3, 4). However, the spectral shift in the PLE spectra is smaller and is in the range of 10-20 meV. In contrast, for thermally treated QDs-PVA composite no spectral shifts are observed in both the PL and PL excitation spectra, although the PL intensity increases (Fig. 3b).

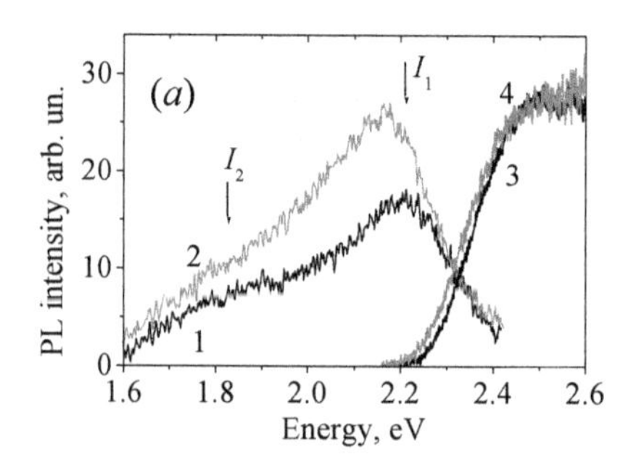

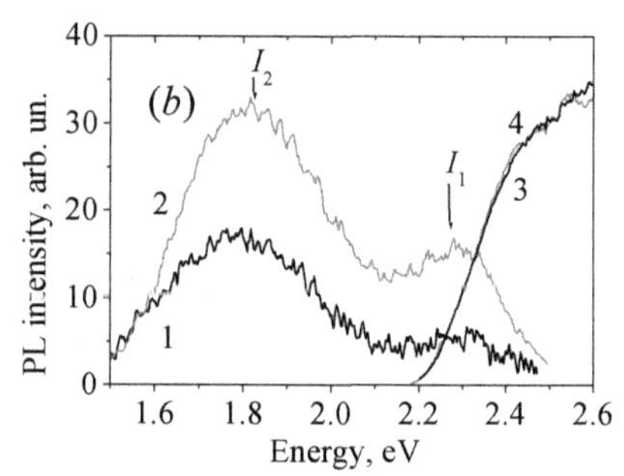

Figure 3. Photoluminescence spectra (curves 1, 2) and PL excitation spectra of I_2 band (curves 3, 4) for CdSe QDs-gelatin (a) and CdSe QDs-PVA (b) composites before (curves 1, 3) and after (curves 2, 4) thermal annealing at 100 °C for 1 h. T=300 K, λ_{exc}=470 nm.

The changes in the PL spectra caused by thermal treatment depend on the duration of annealing (Fig.4). In the QDs-gelatin composite, the PL intensity increases at first and then drops down, while its spectral position progressively shifts to lower energies. At the same time in the QDs-PVA composite, the PL intensity increases at first and then saturates. Though the PL bands in the QDs-PVA composite do not shift during the first hour of annealing, a small red shift of the PL bands is found for long annealing times. Passivation of CdSe QDs with ZnS shell decreases the effect of annealing on the PL spectra, and the higher is [ZnS]:[CdSe] ratio, the smaller is the effect.

The changes in the PL spectra occurred upon the annealing are found to be partially reversible. Storage at room ambience of annealed QDs-gelatin composites for a week results in full recovery of the PL peak position as well as in a partial recovery of the PL band intensity.

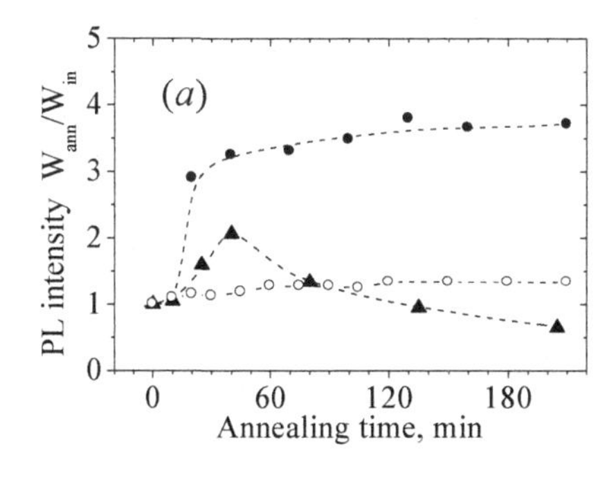

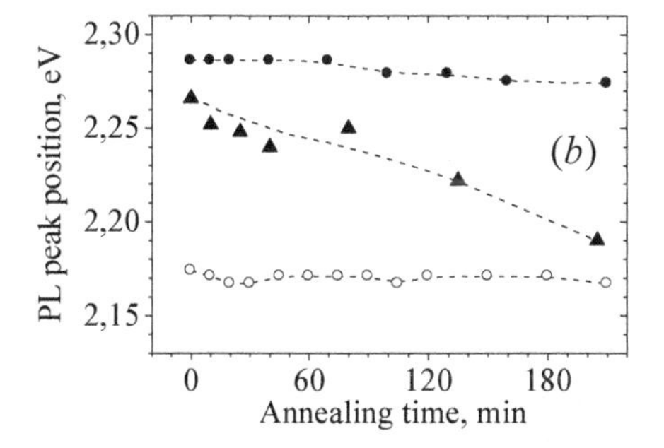

Figure 4. Relative change of the intensity (a) and spectral position (b) of the I_1 band vs annealing time for CdSe QDs-gelatin (triangles), CdSe QDs-PVA (solid circles) and CdSe/ZnS QDs-PVA (open circles) composites. The annealing temperature is 100 ^{0}C. Dashed lines are drawn for a better eye guide.

DISCUSSION

Thus, thermal annealing of the QDs-polymer composites studied affects both the intensity and spectral position of QD-related PL bands. Since passivation of CdSe QDs with ZnS shell decreases the effects produced by annealing as well as the effect character depends on the type of polymer matrix, it can be supposed that the main reason of observed PL transformations is the change of surface defect passivation in the QDs.

The enhancement of the PL intensity observed for both QDs-gelatin and QDs-PVA composites can be caused by a decrease of a density of nonradiative defects due to their passivation upon the annealing. In the untreated polymer, there is a certain amount of hydrogen bonds between neighboring units of a backbone chain (intra-molecular bonds) and between adjacent macromolecules of polymer (intermolecular bonds). The annealing of polymer at elevated temperatures stimulates breaking of hydrogen bonds between the macromolecules as well as between neighboring parts of a macromolecule, splitting of polymer chains and formation of free radicals. At elevated temperatures various transformations of free radicals, including chain-transfer reactions between free macro-radicals, give rise to a partial cross-linking of polymer chains [14-16]. It can be supposed that some of these processes promote the passivation of surface defects in QDs.

The effect of spectral shift of the PL bands observed in the QDs-gelatin composite can be caused by the change of exciton confinement energy in the QDs, since it occurs simultaneously with the shift of the QD absorption maximum in the PL excitation spectra. The shifts in both the emission and absorption spectra have been observed for CdTe and CdSe colloidal QDs upon surface ligand exchange [17, 18]. The red shift was explained either by nanoparticle aggregation in solution and efficient energy transfer from the smaller to the larger QDs in the ensemble [18] or by the extension of exciton wave-function to the outer ligands [19]. The red shift of the PL band and the PL intensity quenching observed in thermally annealed CdSe QDs embedded in polymethylmethacrylate were ascribed to the loss of ligand surfactants and coalescence of the adjacent QDs [20]. In our case a reversible character of a red shift effect accompanied by the PL decrease at prolonged annealing allows supposing that rather not QD coalescence, but the detachment of ligands causes a red shift effect.

Specifically, gelatin can passivate surface Cd atoms with amino- and carboxy-groups, as well as with –NH–CO– fragments of the polymer chains. Thermal annealing can cause a dissociation of some of these low-energy coordination bonds, for example, the ones with amino-groups. In fact, it has been shown that a number of amines are reversibly bound to the surfaces of macroscopic CdSe single crystal causing an increase in the PL [21]. In CdSe nanocrysals, an exchange of trioctylphosphine oxide (TOPO) with amines produced an order of magnitude increase in PL quantum yield and a blue shift in the PL spectra [17]. So, we can hypothesize that the dissociation of coordination bonds between surface Cd atoms and amino-groups of gelatin upon thermal annealing results in the decrease of PL intensity and the red shift of PL bands. Competition of this process with the effect of PL enhancement can explain non-monotonous dependence of the PL intensity versus annealing time for CdSe QDs embedded in gelatin matrix (Fig. 4a). A decrease of confinement energy in the QDs can be caused by redistribution of electronic density in QDs due to depassivation of QD surface atoms by amino-groups. Small red shifts observed in the QDs-PVA composite at long annealing times can also be caused by the presence of small amount of gelatin around the QDs.

CONCLUSIONS

The PL and structural properties of the QDs-gelatin and QDs-PVA composite films have been investigated. It is shown that CdSe and CdSe/ZnS core-shell QDs synthesized in aqueous solution of gelatin can be successfully transferred into PVA matrix. It is found that PL spectra of the QDs-PVA composite as compared with that of the QDs-gelatin show increased contribution of defect-related band and change in a different way upon thermal annealing. The annealing of the QDs-gelatin induces shifting the QD-related PL bands to lower energies (the red shift) and non-monotonic changing in the PL intensity, while thermal treatment of the QDs-PVA produces increasing the PL intensity without changing the PL band position. It is supposed that PL increasing is caused by the change of surface defect passivation in the QDs stimulated by rearrangement of polymer chains. The red shift and the PL intensity decrease are explained by the dissociation of coordination bonds between the surface Cd atoms and amino-groups of gelatin molecules.

ACKNOWLEDGMENTS

This work was partially supported by the project "Physical and Physical-Technological Aspects of Fabrication and Characterization of Semiconductor Materials and Functional Structures for Modern Electronics" (Grant N III-41-12).

REFERENCES

1. A. J. Nozik, M. C. Beard, J. M. Luther, M. Law, R. J. Ellingson, J. C. Johnson, *Chem. Rev.* **110**, 6873 (2010).
2. H. Mattoussi, G. Palui, H. B. Na, *Advanced Drug Delivery Reviews* **64**, 138 (2012).
3. M.A. Walling, J.A. Novak, J.R.E .Shepard, *Int. J. Mol. Sci.* **10**, 441 (2009).
4. S.J. Byrne, Y. Williams, A. Davies, S.A. Corr, A. Rakovich, Y.K. Gun'ko, Y.R. Rakovich, J.F. Donegan, Y. Volkov, *Small* **3**, 1152 (2007)
5. A.E. Raevskaya, A.L. Stroyuk, S. Ya. Kuchmiy, *J. Colloid Interface Sci.* **302**, 133 (2006).
6. Y. Wang, H. Chen, C. Ye, Y. Hu, *Materials Letters* **62**, 3382 (2008).
7. L. Chen, A. Willoughby, J. Zhang, *Luminescence* (2013) Doi: 10.1002/bio.2505.
8. S. Baskoutas, and A.F. Terzis, *J. Appl. Phys.* **99**, 013708 (2006).
9. I. Yakimets, N. Wellner, A. C. Smith, R. H. Wilson, I. Farhat, J. Mitchell, *Polymer* **46**, 12577 (2005).
10. R. Jayasekara, I. Harding, I. Bowater, G.B.Y. Christie, G.T. Lonergan, *Polymer Testing* **23**, 17 (2004).
11. A.R. Kortan, R. Hull, R.L. Opila, M.G. Bawendi, M.L. Steigerwald, P.J. Carroll, L.E. Brus, *J. Am. Chem. Soc.* **112**, 1327 (1990).
12. B.O. Dabbousi, J. Rodriguez-Viejo, F.V. Mikulec, J.R. Heine, H.Mattoussi, R. Ober, K.F. Jensen, M.G. Bawendi, *J. Phys. Chem.* **101**, 9463 (1997).
13. A. V. Baranov, Yu. P. Rakovich, J. F. Donegan, T. S. Perova, R. A. Moore, D. V. Talapin, A. L. Rogach, Y. Masumoto, I. Nabiev, *Phys. Rev. B* **68**, 165306 (2003).
14. P.V. Kozlov, and G.I. Burdygina, *Polymer* **24**, 651 (1983).
15. E. Esposito, R. Cortesi, C. Nastruzzi, *Biomaterials* **17**, 2009 (1996).
16. I.V. Yannas, and A.V. Tobolsky, *Nature* **215**, 509 (1967).
17. D. V. Talapin, A. L. Rogach, A. Kornowski, M. Haase, H. Weller, *Nano Lett.* **1**, 207 (2001).
18. S. F. Wuister, I. Swart, F. van Driel, S. G. Hickey, C. de Mello Donega, *Nano Lett.* **3**, 503 (2003).
19. C.-T. Yuan, W.-C. Chou, D.-S. Chuu, W. H. Chang, H.-S. Lin, R.-C. Ruaan, *J. Med. and Biol. Eng.* **26**, 131 (2006).
20. L.Hu, H. Wu, L. Du, H. Ge, X. Chen, and N. Dai, *Nanotechnology* **22**, 125202 (2011).
21. G. C. Lisensky, R. L. Penn, C. J. Murphy, and A. B. Ellis, *Science* **248**, 840 (1990).

Metamaterials and others

Mater. Res. Soc. Symp. Proc. Vol. 1617 © 2013 Materials Research Society
DOI: 10.1557/opl.2013.1182

Spatial resonances of the Cherenkov emission in dispersive metamaterials

GennadiyBurlak and Erika Martinez-Sanchez
Center for Research on Engineering and Applied Sciences, Autonomous State University of Morelos, Cuernavaca, Mor.Mexico

ABSTRACT

We systematically study the Cherenkov optical emission by a nonrelativistic modulated source crossing 3D dispersive metamaterial. It is found that the interference of the field produced by the modulated source with the periodic plasmonic-polariton excitations leads to the specific interaction in the frequency range where the dispersive refractive index of a metamaterial is negative. Such resonance considerably modifies the spatial structure of the Cherenkov fieldand the reversed Cherenkov emission. In our study parameters of metamaterial and modulated source are fixed while the frequency spectrum of the plasmonic excitations is formed due to the fields interplay in the frequency domain.

INTRODUCTION

Nanophotonics is the study of the behaviour of light on the nanometer scale with involving the interaction of light with nano-structures. Novel optical properties of materials results from their extremely small size that have a variety of applications in nanophotonics and plasmonics. The investigations of optical negative-index [1] metamaterials (NIM) using the nanostructured metal-dielectric composites already have led to both fundamental and applied achievements that have been realized in various structures [2]-[33]. The main applications of negative index metamaterials (or left-handed materials (LHM)) are connected with a remarkable property: the direction of the energy flow and the direction of the phase velocity are opposite in NIM that results unusual properties of electromagnetic waves propagating in these mediums.

Cherenkov radiation by a charged source that moves in a left-handed material and has not the intrinsic frequency has been studied in number of works [17]-[23]. Both experimental and theoretical frameworks are investigated; see review [18] and references therein. However in some important cases a moving particle has the intrinsicfrequency [27] ω_0, for instance the ion oscillating at the transition frequency ω_0. In this paper the Cherenkov optical radiation in 3D metamaterials by a nonrelativistic modulated source having the intrinsic frequency ω_0 with an emphasis on the dispersive properties of the medium is numerically studied. We performed the FDTD simulations with the use of the material parameters at various modulating frequencies ω_0, however without references to the operational frequency range.

THEORY

In metamaterials, it is necessary to treat electromagnetic wave interactions with a metal ingredient using a dispersive formulation that allows correct description of the internal electron dynamics. In this paper we exploit the Drude model that became widely used for modelling in complex materials where for a range of frequency the negative refraction index n is expected. The Maxwell equations read

$$\nabla \times \mathbf{E} = -\mu_0\mu_h \frac{\partial \mathbf{H}}{\partial t} - \mathbf{J}_m - \sigma_m \mathbf{H}, \quad (1)$$

$$\nabla \times \mathbf{H} = \varepsilon_0\varepsilon_h \frac{\partial \mathbf{E}}{\partial t} + q\mathbf{v}_0 f(r,t)\cos(\omega_0 t) + \mathbf{J}_e + \sigma_e \mathbf{E} \quad (2)$$

where a radiating particle (bunch) has modulating frequency ω_0 , $\mathbf{J}_e$ is the electrical current and $\mathbf{J}_m$ is the magnetic current which obey the following material equations

$$\dot{\mathbf{J}}_e + \gamma_e \mathbf{J}_e = b_e \mathbf{E},\ \dot{\mathbf{J}}_m + \gamma_m \mathbf{J}_m = b_m \mathbf{H}, \quad (3)$$

hereγ_e and γ_m are the electrical and magnetic collision frequencies respectively, $b_e = \varepsilon_0\omega_{pe}^2, b_m = \mu_0\omega_{pm}^2$, ω_{pe}, and $\omega_{pm}$are frequencies of electric and magnetic plasmons respectively, $\sigma_e$and $\sigma_m$are conductivities,$\varepsilon_h, \mu_h$are dielectric and magnetic functions of the host medium respectively [26], [27]. For metals such as aluminum, copper, gold, and silver, the density of the free electrons is on the order of$10^{23} cm^{-3}$. The typical value ω_{pe} is $\omega_{pe} \approx 2 \cdot 10^{16} s^{-1}$([28], p.44). In a metamaterial with fishnet structure [14] we consider the charge particle (charge q) moving with a uniform velocity parallel to x direction: $\mathbf{v}_0 \parallel \hat{\mathbf{e}}_x$ and the bunch density is defined by the Gaussian as $f(r,t) = W^{-3}\exp\{-\frac{[(x-v_0t)^2+y^2+z^2]}{W^2}\}$, where W is the width; at $W \to 0$ such a distribution is simplified to the isotropic point-source distribution $f(r,t) \to \pi^{\frac{3}{2}}\delta(x - v_0t)\delta(y)\delta(z)$. Further for simulations we use dimensionless variables, normalized with the use: the vacuum light velocity $c = (\varepsilon_0\mu_0)^{-0.5}$ and the typical spatial scale for nanooptics objects $l_0 = 75nm$. With such normalizationthe above indicated metal plasma frequency becomes $\omega_{pe} = 5$. The electrical and magnetic fields are renormalized with the electrical scale $E_0 = ql_0\varepsilon_0$ and magnetic scale $H_0 = (\varepsilon_0/\mu_0)^{0.5}E_0$ accordingly. Some metamaterials exhibit anisotropic properties with tensor permittivity and permeability. To seek for simplicity in this paper we concentrated in the isotropic geometry. Modelling anisotropic medium is a straightforward extension of this model, see details in Refs. [18], [24].
The idea of our simulations is as follows. In optical experiments normally we can refer only to the parameters of material ($\gamma_e, \sigma_e, \omega_{pe}$and $\gamma_m, \sigma_m, \omega_{pm}$). It is of significant interest to consider Cherenkov emission produced by a particle with modulating frequency ω_0, proceeding from simple principles, using the only material parameters and without of references to the operational frequency band. In this situation the frequency spectrum of internal excitations ω must be left as a free parameter that has to be defined from simulations by a self-consistent way. For 3D dispersion material such a problem becomes too difficult for analytical consideration. Therefore in this paper the standard numerical algorithms in the time domain [29] were used, for details see Refs.[11], [33].

We consider general 3D case in Cartesian coordinates since such a geometry normally is used on the optical investigations [14]. We examine a spatially averaged metamaterial composition: nanostructured metal-dielectric composites (fishnet), similarly that was used in the experiment [14]. In this case the spatial average scale is less then the infrared and visual wavelengths, so we can deem that the material (dielectric and magnetic) dispersion is allowed by the Drude model and the role of the active dielectric ingredient is reduced to a compensation of losses due to the metal ingredient. In our simulations numerical grid $L^3, L = 100, 120, 150$, for more details see Refs.[11], [33].

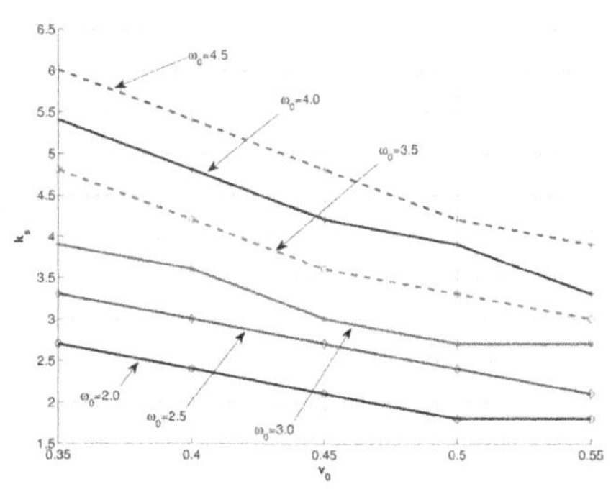

Figure 1.The dependence of the spectral peak position (wavenumber ks) of the field E_x on the particle velocity $\boldsymbol{v_0}$ for different$\boldsymbol{\omega_0}$.

In our approach the following steps have been used: (i) First, we calculated the time-spatial field dynamics that is raised by the crossing radiating particle for different modulating frequencies $\boldsymbol{\omega_0}$, Eqs.(1)-(3). (ii) In second step we apply the Fourier analysis for the time and spatial dependencies of the field calculated in the first step in order to reveal the dynamics of internal excitations.

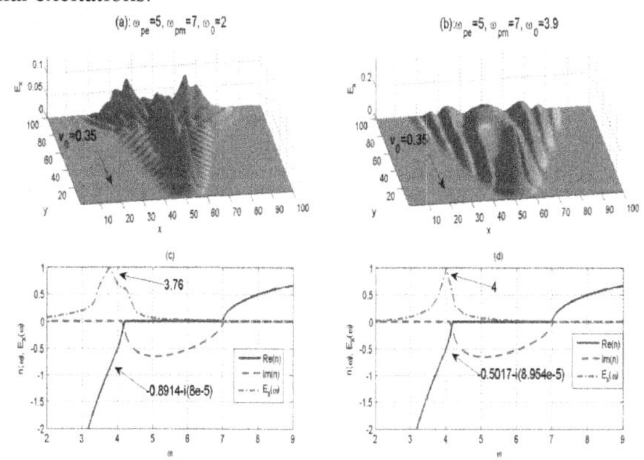

Figure 2. Snapshots of fields E_x(r,t) and the complex refractive indices n($\boldsymbol{\omega}$) in NIM for velocity$\boldsymbol{v_0} = \mathbf{0.52}$and different modulated frequencies $\boldsymbol{\omega_0}$. (a) Field E_x for $\boldsymbol{\omega_0}$=2corresponds to the plasmonic-polariton excitation (PPE) at the peak frequency $\boldsymbol{\omega_c}$=3.776 where Re(n)=-0.872 indicated in (c); (b) field E_x for $\boldsymbol{\omega_0}$=3.9(resonant structure) that generate PPE at the peak frequency $\boldsymbol{\omega_c} = \mathbf{3.839}$where n($\boldsymbol{\omega_c}$)=-0.672-i8.21 · 10^{-5}.

The following dimensionless parameters were used in our simulations: $\omega_{pe} = 5, \omega_{pm} = 7$, $\varepsilon_h = 1.44$, $\mu_h = 1, \gamma_e = \gamma_m = 10^{-7}$, $W = 4$, $q_2 = 2$. We also varied the particle velocity v_0 for different ω_0 to study regimes of the Cherenkov emission. Our observation of the spatial resonancefor Cherenkov emissionis shown in Fig. 2.

DISCUSSION

Finally we study the frequency spectra of the Cherenkov emission for different values of $\omega_0$andv_0. In the frequency domain the dispersive permittivity and permeability have the following form ([29] Chapter 9) $\varepsilon(\omega) = \varepsilon_h - \omega_{pe}^2/(\omega^2 + i\gamma_e\omega)$and $(\omega) = \mu_h - \omega_{pm}^2/(\omega^2 + i\gamma_m\omega)$. For such $\varepsilon(\omega)$ and $\mu(\omega)$ the complex refraction index for the NIM metamaterial can be written as [32]

$$n(\omega) = \sqrt{|\varepsilon(\omega)\mu(\omega)|}e^{i[\phi_\varepsilon(\omega)+\phi_\mu(\omega)]/2} \tag{4}$$

In this case a peak frequency $\omega = \omega_c$ has to be substituted into $\varepsilon(\omega), \mu(\omega)$ and then in $n(\omega)$ Eq.(4). Fig. 2 (left and right panels respectively) show the structure of field $E_x(\mathbf{r}, t)$ and the complex refractive indices $n(\omega)$ in NIM for the cases different particle modulating frequencies ω_0 and velocity $v_0 = 0.52$. We observe from Fig. 2 (a) that the field structures corresponds to the inverse Cherenkov emission. Fig. 2 (c) shows that for$\omega_0 = 2$ the plasmon-polaritons excitations (PPE) are generated at the peak frequency$\omega_c = 3.77$where the complex refractive index is $\omega_c = -0.872 - i8.105 \cdot 10^{-5}$. For larger frequency$\omega_0 = 3.9$ the field E_x has monotonic spatial structure, see Fig. 1 (b) that allows exciting PPE at the peak frequency $\omega_c = 3.839$ where $n(\omega_c) = -0.672 - i8.21 \cdot 10^{-5}$. In both cases$Re(n) < 0$, thus, the optical waves have negative phase velocity that corresponds to reverse Cherenkov emission in NIM.

CONCLUSIONS

We numerically studied the Cherenkov optical emission by a nonrelativistic modulated source crossing 3D dispersive metamaterial. It is found that the resonant interaction of the field produced by the modulated source with the spectrum of the periodic plasmonic-polariton excitations leads to considerable change of spatial structure for the Cherenkov emission. The spatial resonance arises due toDoppler effect around of wave resonances in k-domain.The field acquires monotonic shape in the frequency range where the dispersive refractive index of metamaterial is negative and the reversed Cherenkov radiation is generated. This effect opens new interesting possibilities in various applications metamaterials in nanophotonics with the potential for creating and control light confining structures to considerably enhance light-matter interactions.

ACKNOWLEDGMENTS

The work is partially supported by CONACyT (Mexico) grant 169496.

REFERENCES

1. V. G.Veselago, Sov. Phys. Usp., 10, 509 (1968). [Usp.Fiz.Nauk 92, 517 (1967)].
2. V. M. Shalaev,Nature Photonics, 1, 41 (2007).
3. C. M.Soukoulis, and M. Wegener, Nature Photonics, 5, 523 (2011).
4. O.Hess, J. B. Pendry, S. A. Maier, R. F. Oulton, J. M. Hamm, K. L. Tsakmakidis, Nature Materials, 11, 573 (2012).

5. Huanyang Chen, C. T. Chan , Ping Sheng, Nature Materials, 9, 387 (2010).
6. J. A.Gordon, and R. W. Ziolkowski, Opt. Express, 16, 6692 (2008).
7. G. W.Milton, New Journal of Physics, 12, 033035 (2010).
8. V.Podolskiy, A. Sarychev, V. Shalaev, Optics Express, 11, 735 (2003).
9. V.Shalaev, M. WenshanCai, Uday K. Chettiar, Hsiao-Kuan Yuan, A. K. Sarychev, V. P. Drachev, A. V. Kildishev, Opt. Lett., 30,3356 (2005).
10. G.Burlak, A. D-de-Anda, R. Santaolaya Salgado, J. Perez Ortega, OpticsCommun., 283, 3569 (2010).
11. G. Burlak, PIER, 132, 149 (2012).
12. G. Burlak, V. Rabinovich, SIGMA, 8, 096(2012).
13. G. Burlak, A. D-de-Anda, JAMOP, Article ID 217020, 1-13 (2011).
14. ShuminXiao, V. P. Drachev, A. V. Kildishev, Xingjie Ni, K. UdayChettiar, Hsiao-Kuan Yuan, V. M. Shalaev, Nature, 466, 735 (2010).
15. Subimal Deb, S. Dutta Gupta, J. Phys., 75, 5 (2010).
16. Cherenkov, P. A.Dokl. Akad. Nauk, 2, 451 (1934).
17. Averkov, Yu. O., V. M. Yakovenko, Phys. Rev. B, 79, 193402 (2005).
18. Z. Duan, B. I. Wu, S. Xi, H. S. Chen, M. Chen, PIER, 90, 75 (2009).
19. Sheng Xi, Hongsheng Chen, Tao Jiang, Lixin Ran, JiangtaoHuangfu, Bae-Ian Wu, Jin Au Kong, Min Chen, Phys. Rev. Lett., 103, 194801 (2009).
20. Averkov, Yu. O., A. V. Kats, and V. M. Yakovenko, Phys. Rev. B, 72, 205110 (2005).
21. Jun Zhou, ZhaoyunDuan, Yaxin Zhang, Min Hu, Weihao Liu, Ping Zhang, Shenggang Liu, Nuclear Instruments and Methods in Physics Research Section A, 654, 475 (2011).
22. Z. Duan, Y. S. Wang, X. T. Mao, W. X. Wang, and M. Chen, PIER,121, 215(2011).
23. Zhu, Lei, Fan-Yi Meng, Fang Zhang, Jiahui Fu, Qun Wu, Xu Min Ding, and Joshua Le-Wei Li,PIER, 137, 239 (2013).
24. ZhaoyunDuan, Chen Guo, Min Chen, Opt.Express, 19, 13825 (2011).
25. V. L. Ginzburg,Phys. Usp. 39, 973, (1996).
26. J. D.Jackson, *Classical electrodynamics*, (John Willey and Sons, 1975).
27. K. E.Oughstun,*Electromagnetic and Optical Pulse Propagation 2: Temporal Pulse Dynamics in Dispersive*, (Attenuative Media, Springer Series in Optical Sciences, Springer, 2009).
28. P.Yeh,*Optical waves in Layered Media*,(John Wiley and Sons, New York, 1988).
29. A.Taflove, S. C. Hagness, *Computational Electrodynamics: The Finite-Difference Time-Domain Method*(Artech House, Boston, 2005).
30. J. Schneider, Understanding the Finite-Difference Time-Domain Method, <www.eecs.wsu.edu/ ~schneidj/ufdtd>, (2010).
31. G.N.Afanasiev, *Vavilov-Cherenkov and Synchrotron Radiation, Fundamental Theories of Physics*(Kluwer Academic Publishers, 2004).
32. R. W.Ziolkowski, Phys. Rev. E, 63, 046604 (2001).
33. G. Burlak, E. Martinez-Sanchez, PIER, 139, 277 (2013).

Mater. Res. Soc. Symp. Proc. Vol. 1617 © 2013 Materials Research Society
DOI: 10.1557/opl.2013.1183

Plasmon-polariton signature in the transmission and reflection spectra of one-dimensional metamaterial heterostructures

E. Reyes-Gómez[1], S. B. Cavalcanti[2], and L. E. Oliveira[3]
[1] Instituto de Física, Facultad de Ciencias Exactas y Naturales, Universidad de Antioquia UdeA, Calle 70 No. 52-21, Medellín, Colombia
[2] Instituto de Física, Universidade Federal de Alagoas, Maceió-AL, 57072-970, Brazil
[3] Instituto de Física, Univ. Estadual de Campinas- Unicamp, Campinas - SP, 13083-859, Brazil

ABSTRACT

The transmission and reflection properties of a meta-stack composed of a periodic AB arrangement of an air(A)/metamaterial(B) bilayer is presented, with the multi layered system embedded between two semi-infinite layers of the A material. For oblique incidence, a finite projection along the growth direction of the electric or magnetic field of the incident wave associated with the TM or TE modes, respectively, leads to a coupling of the photon modes with the bulk electric or magnetic metamaterial plasmons, in each layer of the meta-stack. This field-matter coupling gives rise to plasmon-polariton modes and signatures of electric or magnetic longitudinal bulk-plasmon polariton modes in the transmission, as well as in the reflection properties of the meta-stack, by means of a plasmon-polariton gap. Such features survive even in the case of a single bilayer and experimental observation should be, therefore, easily achieved.

INTRODUCTION

The advent of metamaterials [1-3] has opened a new era for light control, and has also given a considerable thrust to the recent area of plasmonics that deals with the generation, propagation and detection of plasmon polaritons (PPs). PPs are collective electronic and magnetic excitations generated by a resonant electromagnetic field. A typical example of a metamaterial is a dispersive left-handed material (LHM) that exhibits negative refraction. Studies on the propagation of light obliquely incident in one-dimensional (1D) stacks of a periodic arrangement characterized by the repetition of a unit double-layer cell composed of a conventional material (A-layer:air) and a dispersive LHM (B-layer) - henceforth referred to as meta-stacks – have reported unusual features. For example, non-Bragg gaps, such as the null-average ($< n >= 0$) refractive index gap [4-6], and longitudinal PP modes [7, 8] of both magnetic (in a TE configuration) and electric (in a TM configuration) nature have been reported in such meta-stacks. Disordered and quasiperiodic meta-stacks have also exhibited essentially the same features and additional ones [9,10]. The present work is concerned with a systematic theoretical study of the transmission and reflection spectra of periodic meta-stacks. Calculated results are obtained within the transfer-matrix approach, and the meta-stack transmission and reflection coefficients are straightforwardly evaluated. The paper is organized as follows. In Sec. II we present the theoretical framework of this study. Numerical results and discussion are presented in Sec. III, and in Sec. IV we conclude.

THEORETICAL FRAMEWORK

Let us consider a meta-stack defined as a finite repetition of a bilayer composed of building blocks A and B whose widths, electric permittivities, and magnetic permeabilities are given by a and b, ε_A and ε_B, and μ_A and μ_B, respectively. The multilayered system is supposed to be sandwiched between two semi-infinite layers of material A. We have considered the slabs A as non-dispersive, and for which $\varepsilon_A = \mu_A = 1$ (air). Slabs B have been assumed with two different dispersion responses. The slab B was first assumed with Drude-like metamaterial dispersion responses, where the electric permittivity and magnetic permeability are given by [2]

$$\varepsilon_B(\omega) = \varepsilon_0 - \omega_e^2/[\omega\,(\omega + i\,\gamma_e)] \qquad (1)$$

and

$$\mu_B(\omega) = \mu_0 - \omega_m^2/[\omega\,(\omega + i\,\gamma_m)], \qquad (2)$$

respectively. Frequencies associated with the bulk electric and magnetic plasmon modes are $\nu_e = \omega_e/(2\pi\,\varepsilon_0^{1/2})$ and $\nu_m = \omega_m/(2\pi\,\mu_0^{1/2})$ respectively. The damping constants γ_e and γ_m account for the absorption in the active medium B. Secondly, we take for the slab B a more realistic dispersive response, such as a split-ring resonator (SRR) metamaterial. In such case, the electric permittivity is still given by Eq. (1), but the magnetic permeability is given by

$$\mu_B(\omega) = \mu_0 - F\omega^2/(\omega^2 - \omega_m^2 + i\,\gamma_m\omega) \qquad (3)$$

where $F < \mu_0$ is a positive parameter determined by the geometry of the split ring [3], and ω_m is the magnetic resonance frequency. The magnetic plasmon frequency is then related to the resonance frequency through the expression $\nu_m = \omega_m\,\mu_0^{1/2}/[2\pi\,(\mu_0 - F)^{1/2}]$.

The transfer-matrix formalism [5] may be used to evaluate both the transmission ($T = |t^2|$) and reflection ($R = |r^2|$) coefficients [10] of the meta-stacks, in the cases of TE or TM modes.

RESULTS AND DISCUSSION

In Fig. 1, we display the calculated results, for incidence angle $\theta = \pi/12$, for both the reflection ($R = |r^2|$) and transmission ($T = |t^2|$) coefficients, respectively, as functions of frequency ($\nu = \omega/2\pi$), in the case of TE modes, obtained by stacking m bilayers AB. We choose the slabs A as non-dispersive, with $\varepsilon_A = \mu_A = 1$, and consider slabs B as metamaterials with Drude-like responses for both electric permittivity and magnetic permeability [cf. Eqs. (1) and (2)], with $\varepsilon_0 = 1.21$ and $\mu_0 = 1$, and $\omega_e = \omega_m = 6\pi$ GHz. Theoretical calculations are performed neglecting absorption effects, and results are shown for slabs with equal widths ($a = b = 12$ mm). The results in Fig. 1 were performed with the same parameters for the electric and magnetic responses, for layers A and B, as in the 1D infinite periodic meta-stacks in the study by Reyes-Gómez *et al* [7]. An excellent agreement between the present calculated reflection and transmission spectra, (in Figs. 1(e) and 1(f) - obtained for $m = 60$ double slabs AB) and the results obtained for the corresponding infinite meta-stack via the photonic dispersion relation [cf. Fig. 3 (a) of Reyes-Gómez *et al* [7]], is expected: for oblique incidence, a finite projection of the electric (TM) or magnetic field (TE) of the incident wave along the growth direction leads to a coupling of the photon modes with the longitudinal bulk-like electric or magnetic metamaterial plasmons in each layer of the stack. Notice [Figs. 1(e) and 1(f)] that the $< n(\nu_0) >= 0$ gap is around $\nu_0 = 1.95$ GHz, the magnetic longitudinal bulk-like PP gap opens up at $\nu = 3$ GHz, and the Brillouin-zone-boundary Bragg gap is at $\nu \cong 6.5 - 7$ GHz. It is clear [Figs. 1(a) and 1(b)] that the signature of the magnetic longitudinal bulk-plasmon polariton modes in the transmission and reflection spectra of the meta-stack survive even in the case of a single (m = 1) AB bilayer.

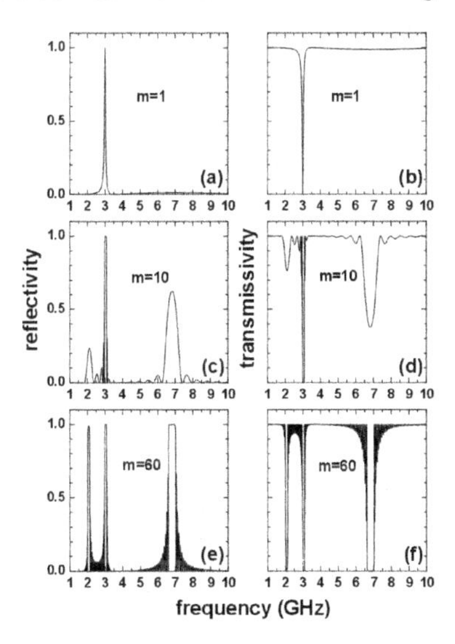

FIG. 1: Reflection- and transmission-coefficient spectra of TE modes for finite meta-stacks obtained by the stacking of m bilayers AB. Results are for incidence angle $\theta = \pi/12$. Slabs A (width a = 12 mm) are considered as non-dispersive, with $\varepsilon_A = \mu_A = 1$, whereas dispersion of slabs B (width b = 12 mm) is considered via Eqs. (1) and (2), with $\varepsilon_0 = 1.21$, $\mu_0 = 1$, and $\omega_e = \omega_m = 6\pi$ GHz. No absorption is considered.

As one could argue that the Drude-like model used for the magnetic response of the metamaterial layers B is too naive, we have also performed the calculations for the reflection and transmission spectra of the meta-stack with a more realistic model for the frequency dependence of the magnetic permeability of the slabs B. As mentioned before, we have then considered a SRR response for the metamaterial slabs B, where the electric permittivity is still given by the Drude model [cf. Eq. (1)] and the magnetic permeability is given by Eq. (3). Results for the reflection (reflection ($R = |r^2|$) and transmission ($T = |t^2|$) coefficients (TE modes), for a meta-stack of m bilayers AB, are depicted in Fig. 2. As before, notice [cf. Figs. 2(e) and 2(f)] that the magnetic longitudinal bulk-like PP gap opens up at $\nu_m = \omega_m \mu_0^{1/2}/[2\pi (\mu_0 - F)^{1/2}] = 3{:}46$ GHz, and the Brillouin-zone-boundary Bragg gap is at $\nu \cong 7 - 7.5$ GHz. Also, one should notice that the features (value of 1 for the reflection and 0 for the transmission) associated with the magnetic longitudinal bulk-like PP modes in the transmission and reflection spectra of the SRR meta-stack survive even in the case of a single (m = 1) AB SRR bilayer [Figs. 2(a) and 2(b)]. One should mention that $< n(\nu_0) >= 0$ occurs at $\nu_0 = 3.1$ GHz. However, one should stress that no absorption is considered for the results in Fig. 2, and the magnetic-permeability response $\mu_B(\nu)$ diverges at

$\nu = 3.0$ GHz, rendering the model theoretical results for the magnetic permeability of slabs B not reliable around the $\nu = 3.0$ GHz low-frequency limit. Again, although not shown here, similar results are also obtained in the case of TM waves.

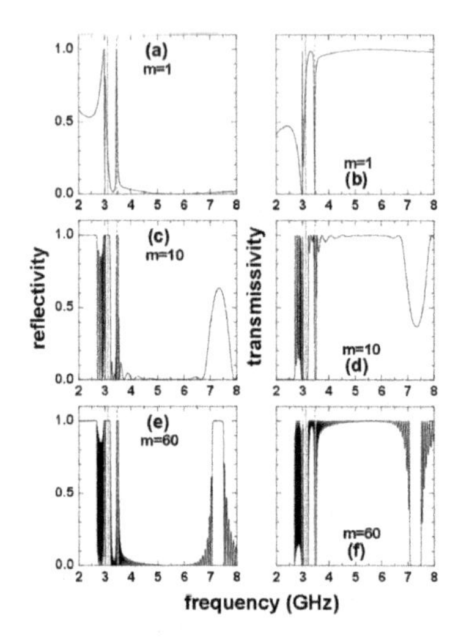

FIG. 2: Frequency-dependence of the reflection and transmission coefficients of TE modes for finite meta-stacks of m bilayers. Parameters of slabs A are taken as in Fig. 1, whereas slabs B follow the SRR dispersion [cf. Eqs. (1) and (3)] with $\varepsilon_0 = 1.21$, $\omega_e = 8\,\pi$ GHz, $\mu_0 = 1$, $\omega_m = 6\,\pi$ GHz, and $F = 0.25$. Widths of the slabs A and B are $a = 12$ mm and $b = 12$ mm, respectively. The incidence angle is $\theta = \pi/12$, and no absorption is considered. In all panels, vertical dotted lines correspond to the frequency at which the geometrical average of the heterostructure refractive index vanishes, whereas vertical dashed lines correspond to the bulk magnetic-plasmon frequency of slabs B.

We have also investigated the effects of absorption on the reflection and transmission spectra of the SRR meta-stack. Such effects may be appropriately introduced by modifying the electric-permittivity and magnetic-permeability responses of the metamaterial dispersive slabs B: absorption effects are phenomenologically taken into account by introducing electric/magnetic damping constants [cf. Eqs. (1) and (3)]. Calculated results, in the range $\nu = 2\text{-}4$ GHz, for the reflection and transmission spectra are displayed in Fig. 3 with absorption modeled via phenomenological $\gamma_e = 10^{-2}\,\omega_e$ and $\gamma_m = 10^{-2}\,\omega_m$ damping parameters. It is unambiguously clear that moderate absorption effects do not change the essential findings, i.e., that photon-plasmon coupling results in the formation of PP modes and also, gives rise to signatures of electric and magnetic longitudinal bulk-like PP modes in the transmission and reflection spectra of the meta-stack. Such features [at $\nu_m = \omega_m\,\mu_0^{1/2}/[2\pi\,(\mu_0 - F)^{1/2}] = 3.46$ GHz] survive even in the case of a single AB bilayer. Experimental observation should be, therefore, easily achieved.

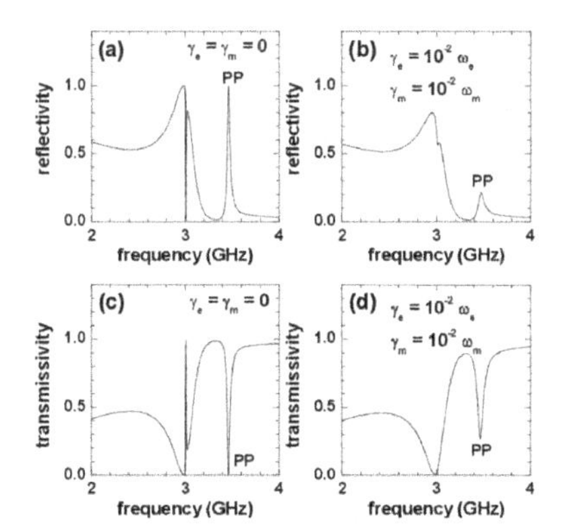

FIG. 3: Reflection and transmission spectra (TE waves), for incidence angle $\theta = \pi/12$, for a single AB bilayer in the vicinity of the magnetic-plasmon frequency of the bulk metamaterial B. Calculations were performed for both non-absorptive and SRR absorptive B metamaterials. Parameters are the same as in Fig. 2.

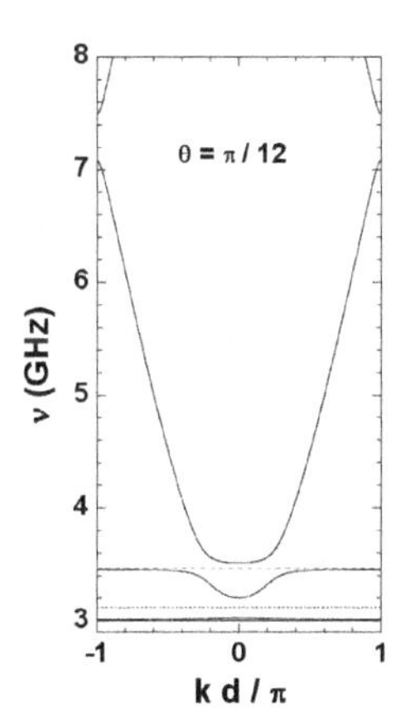

FIG. 4: Dispersion relation corresponding to an infinite meta-stack, with period AB, composed of non-dispersive slabs A and SRR dispersive metamaterial slabs B. Parameters are the same as in Fig. 2. Calculations were performed for TE modes with an incidence angle angle $\theta = \pi/12$. The horizontal dashed line in panel corresponds to the bulk magnetic-plasmon frequency of slabs B, whereas the horizontal dotted line corresponds to the frequency value at which the geometrical average of the heterostructure refractive index vanishes.

The calculated dispersions, for TE modes, in the case of an infinite periodic SRR metamaterial superlattice are displayed in Fig. 4. Null transmission results in Fig. 2 should be compared with the corresponding gaps in Fig. 4. Of course, results for large values of m double slabs AB should corroborate the results obtained for the corresponding infinite SRR metamaterial superlattice via the dispersion relation in Fig. 4. What is remarkable, however, is that the PP feature associated with the magnetic longitudinal bulk-like PP modes at $\nu_m = 3.46$ GHz survives for the single bilayer AB. Similar results corresponding to electric longitudinal bulk-like PP modes are also obtained in the case of TM waves.

CONCLUSIONS

In conclusion, we have performed a systematic study of the transmission and reflection spectra of periodic meta-stacks. Photon-plasmon coupling results in the formation of PP modes and gives rise to signatures of the electric or magnetic longitudinal bulk-like PP modes in the transmission and reflection spectra of the meta-stack. Such features survive even in the case of a single AB bilayer, and experimental observation should be, therefore, easily achieved. We would like to stress that, in the context of nanophotonics, plasmonics based on PP propagation is a very promising area of research. Of course, nanomaterials involving magnetic active LHMs should reveal many unexpected features to be used in nanoelectronics as well as in nanomagnetonics in the quest for nanodevices.

ACKNOWLEDGMENTS

The present work was partially financed by the Scientific Colombian Agency CODI - University of Antioquia and Brazilian Agencies CNPq and FAPESP (Proc. 2012/51691-0).

REFERENCES

[1] V. G. Veselago, *Sov. Phys.Usp.* **10**, 509 (1968).
[2] N. I. Zeludhev, *Science* **328**, 582 (2010), and references therein.
[3] F. S. S. Rosa, D. A. R. Dalvit, and P. W. Milonni, *Phys. Rev. Lett.* **100**, 183602 (2008).
[4] J. Li, L. Zhou, C. T. Chan, and P. Sheng, *Phys. Rev. Lett.* **90**, 083901 (2003); H. Jiang, H. Chen, H. Li, Y. Zhang, and S. Zhu, *Appl. Phys. Lett.* **83**, 5386 (2003); I. V. Shadrivov, A. A. Sukhorukov, and Y. S. Kivshar, *Appl. Phys. Lett.* **82**, 3820 (2003); H. Daninthe, S. Foteinopoulou, and C.M. Soukoulis, *Photon. and Nanostruc. – Fund. and Appl.* **4**, 123 (2006).
[5] S. B. Cavalcanti, M. de Dios-Leyva, E. Reyes-Gómez, and L. E. Oliveira, *Phys. Rev. B* **74**, 153102 (2006); *ibid., Phys. Rev. E* **75**, 026607 (2007).
[6] E. Silvestre, R. A. Depine, M. L. Martínez-Ricci, and J. A. Monsoriu, *J. Opt. Soc. Am. B* **26**, 581 (2009); J. Schilling, *Nature Photon.* **5**, 449 (2011).
[7] E. Reyes-Gómez, D. Mogilevtsev, S. B. Cavalcanti, C. A. A. Carvalho, and L. E. Oliveira, *Europhys. Lett.* **88**, 24002 (2009).
[8] C. A. A. de Carvalho, S. B. Cavalcanti, E. Reyes-Gómez, and L. E. Oliveira, *Phys. Rev. B* **83**, 081408(R) (2011); *ibid., Superl. and Microstruct.* **54**, 96 (2013).
[9] D. Mogilevtsev, F. A. Pinheiro, R. R. dos Santos, S. B. Cavalcanti, and L. E. Oliveira, *Phys. Rev. B* **84**, 094204 (2011); E. Reyes-Gómez, S. B. Cavalcanti, and L. E. Oliveira, *J. Phys.: Cond. Matter* **25**, 075901 (2013).
[10] E. Reyes Gómez, N. Raigoza, S. B. Cavalcanti, C. A. A. de Carvalho, and L. E. Oliveira, *Phys. Rev. B* **81**, 153101 (2010); J. R. Mejía-Salazar, N. Porras-Montenegro, E. Reyes-Gómez, S. B. Cavalcanti, and L. E. Oliveira, *Europhys. Lett.* **95**, 24004 (2011).

Mater. Res. Soc. Symp. Proc. Vol. 1617 © 2013 Materials Research Society
DOI: 10.1557/opl.2013.1184

Optimized nanomagnetic system for spintronic applications

Mishel Morales Meza[1], Paul P. Horley[1] and Alexander Sukhov[2],
[1]CIMAV Chihuahua / Monterrey, 120 Av. Miguel de Cervantes,
Chihuahua, CHIH 31109, México
[2]Institut für Physik, Martin-Luther Universität Halle-Wittenberg
Halle (Saale), 06099, Germany

ABSTRACT

Magnetic properties at nano-scale provide a whole spectrum of new phenomena that can be beneficial for spintronic devices characterized with ultra-short response time, high sensitivity to magnetic field and miniature size. The properties and stability of a magnetic system can be enhanced by creating ordered arrays of ferromagnetic nano-particles. Here we report a considerable reduction of coercitivity for a magnetic array using triangular, square and hexagonal particle arrangement. The reduction of coercitivity can be explained by fine-tuning of dipole-dipole interaction between magnetic particles, which is to large degree influenced by the number of nearest neighbors and distance between the particles.

INTRODUCTION

The discovery of the giant magnetoresistance [1, 2] paved a way for significant developments in nano-scale magnetic devices, leading to unprecedented increase in the density of information storage [3]. Yet, the small device size makes it susceptible to the destabilizing action of the temperature, which may degrade the required magnetic configuration. One of solutions providing more robustness to a spintronic device can be found in synchronization of several such devices – which may represent, for example, spin valves [4]. A spin valve is a multi-layer device formed with thick ferromagnetic polarizer layer and thin analyzer layer, separated with a non-magnetic spacer. Polarized layer is used to define spin orientation for the carriers passing through it, so that the dynamics of the analyzer layer can be effectively controlled with an applied magnetic field and a torque generated by spin-polarized current. Therefore, we considered ferromagnetic particles shaped as cylinders with diameter of 12 nm and height of 2.2 nm to mimic the geometry of analyzer layers of spin valves. These particle dimensions agree by the order of magnitude with particle size constrains when the magnetization rotation occurs in a uniform manner [5], permitting to use macrospin approximation [6]. With the current advances of nano-scale etching and lithography, we consider that it would be feasible to create the arrays of ferromagnetic nano-particles of the desired size and configuration. We considered several types of grids formed by regular polygons covering the surface – triangles, hexagons and squares. The working hypothesis was that the different number of neighbors – as well as distance between them – will be one of the factors defining the magnetic response of the system. As triangular grid has six nearest neighbors (while hexagonal and square ones have three and four nearest neighbors, respectively) it was expected that the particles arranged into triangular grid will be characterized with larger magnetic stiffness.

THEORY

The magnetization dynamics of a macrospin is governed with the Landau-Lifshitz-Gilbert equation [7], augmented with a viscous damping term. The Gilbert form of the equation reads

$$\frac{d\mathbf{m}_i}{dt} = -\gamma\mu_0(\mathbf{m}_i \times \mathbf{H}_{EFF}) + \frac{\alpha}{M_S}\left(\mathbf{m}_i \times \frac{d\mathbf{m}_i}{dt}\right), \quad (1)$$

with gyromagnetic ratio γ, vacuum permeability μ_0, damping coefficient α, saturation magnetization M_S, normalized magnetization of the i-th particle $\mathbf{m}_i$ and effective field $\mathbf{H}_{EFF}$ (which is different for each nano-particle). The Gilbert form of equation can be re-written to obtain an ODE with a derivative in its left-hand-side:

$$\frac{d\mathbf{m}_i}{dt} = -\frac{\gamma\mu_0}{1+\alpha^2}\{\mathbf{m}_i \times \mathbf{H}_{EFF} + \alpha[(\mathbf{m}_i \cdot \mathbf{H}_{EFF})\mathbf{m}_i - \mathbf{H}_{EFF}]\}. \quad (2)$$

This equation can be numerically solved using well-developed and stable tools such as Heun and Runge-Kutta methods. The effective field acting on each particle consists of several terms:

$$\mathbf{H}_{EFF} = \mathbf{H}_{EXT} + \mathbf{H}_{ANI} + \mathbf{H}_{DEM} + \mathbf{H}_{DDI}\ . \quad (3)$$

Here $\mathbf{H}_{EXT}$ is the applied external field, $\mathbf{H}_{ANI}$ is the crystalline field responsible for easy-axis anisotropy, $\mathbf{H}_{DEM}$ is the demagnetizing field (approximation of thin infinite magnetic film) and $\mathbf{H}_{DDI}$ is the long-range dipole-dipole interaction field given by the formula:

$$\mathbf{H}_{DDI} = -\frac{VM_S}{4\pi}\sum_{i\neq j}\left[\frac{\mathbf{m}_i}{r_{ij}^3} - 3\frac{(\mathbf{m}_i \cdot \mathbf{r}_{ij})\mathbf{r}_{ij}}{r_{ij}^5}\right], \quad (4)$$

with particle's volume V and distance between particles' centers r_{ij}. The calculation of dipole-dipole interaction requires considerable CPU power because of the need to calculate interaction of every particle with all others. However, as particle positions are fixed, it is possible to pre-calculate the coefficients related to inter-particle distance, which allows for a certain gain in calculation time.

The response of particle system to varying external magnetic field can be obtained by numeric solution of the equation (2), allowing construction of hysteresis curve using the averaged magnetization value for the entire particle ensemble. The coercitivity corresponds to the threshold field required to reverse magnetization of the system; the remanence gives the magnitude of magnetization after external field is removed. Intending to use particle arrays in non-volatile magnetic memory modules, it will be important to keep the re-writing field as small as possible (to avoid disturbance of the neighboring arrays), while the remanence should be large enough to provide good signal-to-noise ratio for room-temperature use. Therefore, one can formulate the optimization conditions as determining grid structure required to minimize coercitivity and keep remanence at considerably high values.

DISCUSSION

In our calculations, we used the parameters of cobalt reported from the experimental studies [4]. As for the system of non-overlapping ferromagnetic particles the only type of magnetic interaction will be dipole-dipole field given by equation (4), we decided to study the influence of system geometry on the coefficients of this formula. The key question was the quantity of neighbor groups for a particle and the distance at which they are located. It is obvious that nearest neighbors for the case of triangular, square and hexagonal grid will be six, four and three, respectively. The situation is not so straightforward when one talks about the other neighbor particle groups. To illustrate this, we plotted histograms presenting the number of neighbor particles versus distances (Fig. 1, left-side panels). To simplify the comparison, the distances x_i were normalized over the grid parameter d.

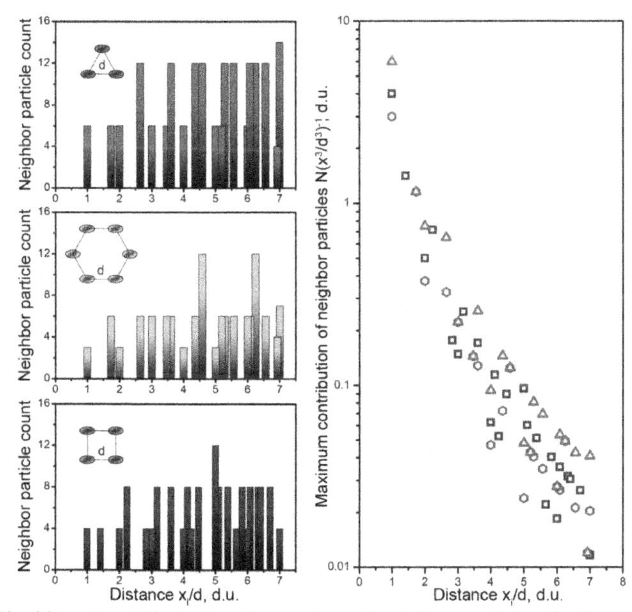

Figure 1. The histograms illustrating the number of neighbor particles for three types of array grids considered: triangular (upper left), hexagonal (center left) and square (bottom left). The right panel shows the maximum contribution of each neighbor group into the dipole-dipole interaction field. The shapes correspond to the grid types considered.

As one can see from the figure, the triangular grid has larger number of neighbors per group coming in numbers of six and twelve. This will mean that even when the distance to the corresponding neighbor group is large, the number of the neighbors may make a considerable contribution. For a hexagon grid, the most common numbers of particles in neighbor groups are three and six, with only two peaks of twelve particles located past four grid periods. The square grid is in somewhat better position with its predominant four and eight particles per neighbor group. It is important to note that the distance between the particles r_{ij} enters the formula (4) as an inverted cube; therefore, it was thought useful to make a plot illustrating this quantity as the maximum possible contribution of the corresponding neighbor group (Fig. 1, right panel). As one can see from the plot, triangular grid is characterized with major coefficient values for all the particle distances studied. Hexagonal grid produces the lowest values, which suggest that it should be the most magnetically soft out of the three particle systems studied.

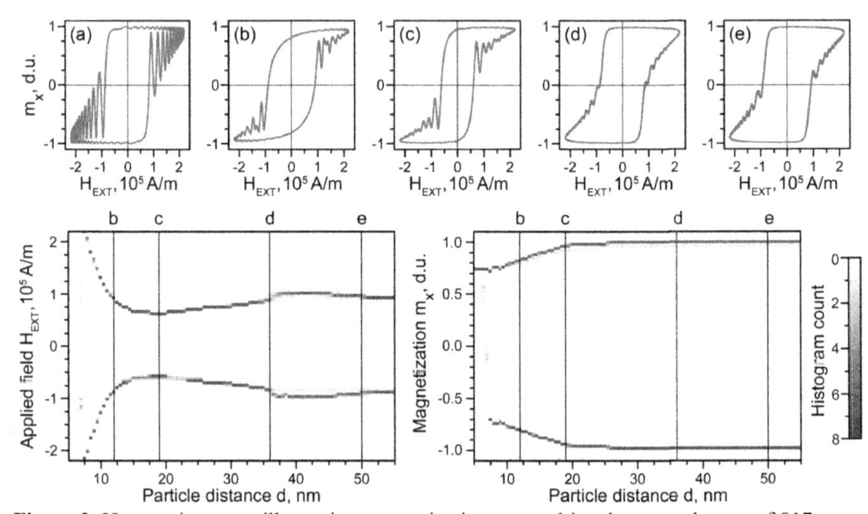

Figure 2. Hysteresis curves illustrating magnetization reversal in a hexagonal array of 817 nano-particles (upper panels) compared hysteresis of a single particle (panel a). The hysteresis curves were calculated for field frequency 0.5GHz and grid parameters: b) $d = 12$ nm, c) $d_{OPT} = 19$ nm, d) $d = 36$ nm and e) $d = 50$ nm. Coercitivity and remanence histograms are shown below (lower left and lower right panels, respectively).

Therefore, one can expect that hexagonal particle array can be reversed with the lowest magnetic field. However, as the intensity of dipole-dipole interactions vary with d, it is important to study the influence of hysteresis parameters over inter-particle distance. The corresponding results are presented in Figure 2. The upper panels present the most characteristic hysteresis curves compared to the hysteresis of the single particle (Fig. 2a). Two bottom panels display the histogram for coercitivity and remanence for eight full cycles of magnetic field. We propose to use histograms as a convenient method to study parameters of hysteresis curves [8]. When the proper hysteresis is achieved, there would be two peaks in a histogram for a fixed value of d. The absence of hysteresis is signaled by a single peak. The repetitivity of the hysteresis curve can be also deduced from the plot: if the values of coercitivity and remanence vary from cycle to cycle, they will fall into the different bins of a histogram, which will produce lower and wider peaks. If the curves are repetitive, the histograms will be characterized with narrow peaks with the height corresponding to the number of field cycles observed.

The hysteresis curve for a single particle features numerous oscillations corresponding to precessional magnetization relaxation. The coercitivity of the single particle H_{C0} coincides with the value obtained for the system where the distance between the centers of nano-particles is equal to their diameter (Fig. 2b); the remanence in this case is 0.85. For the smaller d the value of coercitivity grows abruptly reflecting increase of magnetic interactions for partially overlapping particles. In this paper we focus on disperse particle system, for which grid parameter should be larger than particle size d = 12 nm. As one can see, the increase of inter-particle distance decreases H_C, reaching a minimum value of 60.3 kA/m (about 70% of H_{C0}) at optimal distance d_{OPT} = 19 nm (Fig. 2c). It is remarkable that the remanence value remains considerably high – 0.97 – offering good perspectives for system stability against the thermal noise. For grid parameters larger than d_{OPT} one can observe increasing coercitivity that eventually exceeds H_{C0} (Fig. 2d). The histogram peaks become blurred because the oscillations corresponding to magnetization relaxation crosses the line $m_x = 0$ several times. The characteristic inter-particle distance at which dipole-dipole interaction field loses its dominant character in (4) can be detected by a jump of coercitivity values at d = 36 nm. Further increase of d leads to coercitivity decrease, reaching single-particle value of H_{C0} at d = 50 nm (Fig. 2e) meaning that for such large inter-particle distances H_{DDI} becomes negligible.

CONCLUSIONS

We report considerable improvement of magnetization dynamics in nano-particle arrays allowing achieving a 30% decrease of coercitivity value for the arrays of ferromagnetic particles in comparison to single-particle case. It was also shown that coercitivity and remanence histograms are useful tools, presenting information about hundreds of hysteresis cycles in a clear and concise way that considerably simplifies further analysis.

ACKNOWLEDGMENTS

This work was supported in part by the CONACYT of México via basic science project #129269 and the German Research Foundation (Grant No. SU 690/1-1).

REFERENCES

1. P. Grünberg, R. Schreiber, Y. Pang, M.B. Brodsky, H. Sowers, Phys. Rev. Lett. **57**, 2442 (1986).
2. M.N. Baibich, J.M. Broto, A. Fert, F. Nguyen Van Dau, E. Petroff, P. Eitenne, G. Creuzet, A. Friederich, J. Chazelas, Phys. Rev. Lett. **61**, 2472 (1988).
3. I. Kaitsu, R. Inamura, J. Toda, and T. Morita, Fujitsu Sci. Tech. J. **42**, 122 (2006).
4. S.I. Kiselev, J.C. Sankey, I.N. Krivorotov, N.C. Emley, R.J. Schoelkopf, R.A. Buhrman, D.C. Ralph, Nature **425**, 380 (2003).
5. M. Beleggia, M. De Graf, J. Magn. Magn. Mater. **285**, L1 (2005).
6. J. Xiao, A. Zangwill, and M.D. Stiles, Phys. Rev. B **72**, 14446 (2005).
7. L.D. Landau and E.M. Lifshitz, Phys. Z. Sowjetunion **8**, 153 (1935).
8. P.P. Horley, A. Sukhov, C.-L. Jia, E. Martínez, and J. Berakdar, Physical Review B **85**, 054401 (2012).

Mater. Res. Soc. Symp. Proc. Vol. 1617 © 2013 Materials Research Society
DOI: 10.1557/opl.2013.1185

Polarization Control of Optically Pumped Terahertz Lasers

Gabriela Slavcheva[1] and Alexey V. Kavokin[2]

[1] Blackett Laboratory, Imperial College London and Centre for Photonics and Photonic Materials, University of Bath, U.K.
[2] Spin Optics Laboratory, St. Petersburg State University, Russia and School of Physics and Astronomy, University of Southampton, U. K.

ABSTRACT

Two-photon pumping of excited exciton states in semiconductor quantum wells is a tool for realization of ultra-compact terahertz (THz) lasers based on stimulated optical transition between excited *2p* and ground *1s* exciton state. We show that the probability of two-photon absorption by a *2p*-exciton is strongly dependent on the polarization of both photons. Variation of the threshold power for THz lasing by a factor of 5 is predicted by switching from linear to circular pumping. We calculate the polarization dependence of the THz emission and identify photon polarization configurations for achieving maximum THz photon generation quantum efficiency.

INTRODUCTION

Excitons in nanoscale semiconductor materials exhibit low-energy excitations in the range of the exciton binding energy, analogous to inter-level excitations in atoms, yielding infrared and terahertz (THz) transitions. Thus intra-excitonic transitions between excited exciton ladder states represent a natural system for generating THz radiation and coherence.

Towards the goal of development of new compact and efficient coherent THz sources recently a new scheme of microcavity polariton-triggered THz vertical cavity surface emitting laser (VCSEL) has been proposed [1], whereby the *2p* dark quantum well (QW) exciton state is pumped by two-photon absorption (TPA) using *cw* laser beam, and subsequently decays to *1s* exciton-polariton state emitting THz radiation.

The inverse process of a strong few-cycle THz pulses resonantly-driven intra-excitonic $|1s\rangle \rightarrow |2p\rangle$ transition in a QW has been experimentally observed and investigated [2, 3]. Generation of incoherent p-type exciton population and population inversion by single-photon pulsed excitation of *1s* or *2s* exciton resonances in a QW has been predicted in [4].

By contrast, within the framework of the proposed polariton-triggered THz laser scheme, we shall be interested in an optical *cw* pumping of the *2p* exciton state by two-photon absorption mechanism. Although there has been a considerable body of theoretical work on two-photon absorption in QWs (see e.g. [5]), the polarization selection rules in a bosonic THz laser remain largely unknown. In the present study we demonstrate polarization control of THz emission and of the quantum efficiency for THz photon generation. We employ crystal symmetry point group theoretical methods [6] to calculate the polarization dependence of the excitonic TPA and intra-excitonic *2p* to *1s* THz transition radiative decay rates in GaAs/AlGaAs QWs. This enables us to calculate the polarization dependence of the quantum efficiency for THz photon generation and thus identify maximum efficiency regimes of operation.

THEORY

The optically pumping scheme to a *2p* exciton state by two photons, each of half the energy of the *2p* exciton state, is shown in figure 1(a).

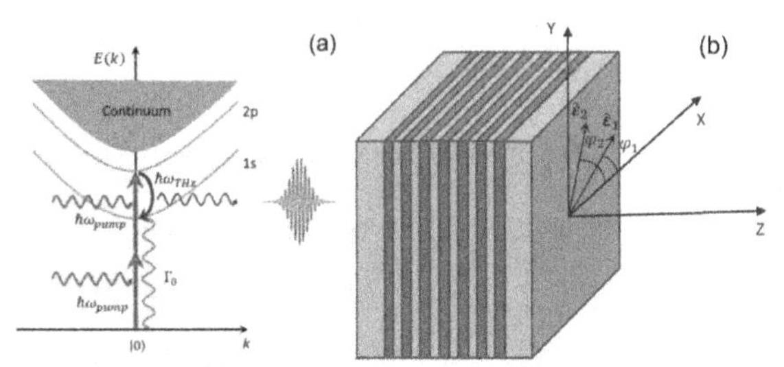

Figure 1. (a) Energy-level diagram of two-photon transitions to *2p* exciton states in a QW. 1s: ground, 2p: excited (dark) bound state; $\hbar\omega_{pump} = E_{2p}/2$; ω_{THz}: center frequency of the emitted THz pulse; Γ_0: ground (*1s*) state exciton spontaneous emission rate; (b) excitation geometry.

Two-photon *p*-exciton absorption probability calculation

The wave function of a quasi-2D (Q2D) exciton at the Γ point can be written as:

$$\psi_\lambda(\boldsymbol{r}_e,\boldsymbol{r}_h)=\frac{v_0}{\sqrt{S}}U_\lambda^{\alpha\beta}(\rho)\phi_c^\alpha(z_e)\phi_v^{\beta*}(z_h)u_{ck}(\boldsymbol{r}_e)u_{vk}^*(\boldsymbol{r}_h)e^{i\boldsymbol{k}_{||}\cdot\boldsymbol{R}_{||}} \tag{1}$$

where v_0 is the unit-cell volume, S is the QW area, **R** is the centre-of-mass (c.o.m.) co-ordinate, $\boldsymbol{r}=\boldsymbol{r}_e-\boldsymbol{r}_h=(\boldsymbol{\rho}_e-\boldsymbol{\rho}_h,z)$; α,β: subband indices and $\phi_{c(v)}^\alpha$ is the α-subband envelope function of the conduction (valence) band; $U_\lambda^{\alpha\beta}$ is the 2D exciton envelope function; $\lambda=(n,m)$ is the 2D exciton quantum number (n=1,2,..., $|m|<n$); u_{ck}, u_{vk} are the periodic parts of the Bloch wave function for conduction/valence bands; $\boldsymbol{k}_{||}\sim\boldsymbol{0}$ is the exciton c.o.m. wave vector.

The TPA probability for allowed-allowed transitions at the Γpoint is given by:

$$W_{TPA}=\frac{2\pi}{\hbar}\sum_{if}|V_{if}|^2 S_f(E) \tag{2}$$

where S_f is the final density of states and the momentum matrix element is given by:

$$V_{fi}=\frac{e^2}{m^2c^2}A_1A_2\sum_l\left[\frac{\langle f|\varepsilon_1\cdot p|l\rangle\langle l|\varepsilon_2\cdot p|i\rangle}{E_l-E_i-\hbar\omega_2}+\frac{\langle f|\varepsilon_2\cdot p|l\rangle\langle l|\varepsilon_1\cdot p|i\rangle}{E_l-E_i-\hbar\omega_1}\right] \tag{3}$$

reflecting the order of absorption of the photon with polarization vector $\varepsilon_{1,2}$, energy $\hbar\omega_{1,2}$ and vector potential $A_{1,2}$. Since the TPA is a two-step process, one should sum over all intermediate states $|l\rangle$ with energy E_l. Two types of matrix elements enter Eq. (3).

The first matrix element has been calculated in the 3D case [7] and for Q2D case reads:

$$\langle l|\varepsilon_\mu\cdot\boldsymbol{p}|i\rangle=\sqrt{S}\Psi_\lambda^*(0)\langle c|\varepsilon_\mu\cdot\boldsymbol{p}|v\rangle=v_0U_\lambda^{\alpha\beta}(0)\phi_c^\alpha(z_e)\phi_v^{\beta*}(z_h)\langle c|\boldsymbol{p}|v\rangle \tag{4}$$

where $\Psi_\lambda(\boldsymbol{0})$ is the relative motion exciton wave function at $\boldsymbol{r}=\boldsymbol{0}$ and the interband momentum matrix element is given by: $\langle c|\boldsymbol{p}|v\rangle=\frac{1}{v_0}\int_{cell}d^3r\,u_c^*(\boldsymbol{r})\frac{\hbar}{i}\nabla u_{v\hat{z}}(\boldsymbol{r})$.

The second matrix element is between hydrogenic-type exciton states:

$$\frac{1}{m}\langle f|\varepsilon_v.\boldsymbol{p}|l\rangle = \frac{1}{\mu_\xi}\int d^3r\ \Psi_\delta^{\alpha\beta*}(\boldsymbol{r})(\hat{\varepsilon}_v.\boldsymbol{p})\Psi_\lambda^{\alpha\beta}(\boldsymbol{r}) \tag{6}$$

and can be calculated introducing reduced Coulomb Green's function [8] according to:

$$G(\boldsymbol{\rho},\boldsymbol{0}) = \frac{U_\lambda^{\alpha\beta}(\rho)U_\lambda^{\alpha\beta*}(0)}{E_\lambda - \Omega} = \frac{1}{2\pi}e^{-2\rho/k_\mu a_B^*}\left[-ln\left(\frac{4\rho}{k_\mu a_B^*}\right) - \gamma + 3 - 4\left(\frac{\rho}{k_\mu a_B^*}\right)\right] \tag{7}$$

where $\Omega = -E_G + \hbar\omega_\mu < 0$ and E_G is the band gap energy; a_B^*:3D exciton Bohr radius, γ: Euler's constant, $k_\mu^2 = \frac{E_B}{E_G - \hbar\omega_\mu}$, $E_B = \frac{\hbar^2}{2\bar{\mu}_\xi a_B^2}$: exciton binding energy; $\bar{\mu}_\xi$: reduced exciton mass.

Taking 2D hydrogen atom wave function as excitonic relative motion wave function we obtain for the two-photon optical transition to *2p* exciton states (n=1,m=±1) matrix element:

$$V_{fi} = \frac{v_0^2}{\sqrt{5}}\frac{e^2}{mc^2}A_1A_2\frac{4\hbar I_{\alpha\beta}}{3\pi i\sqrt{3\pi}}\frac{1}{E_B a_B^{*2}}J_{eff,2}\langle c|\boldsymbol{p}|v\rangle \tag{9}$$

The polarization dependence is derived considering invariance of the interband matrix element $M = \langle c|\boldsymbol{p}|v\rangle$ under cubic (O_h) crystal point symmetry group transformations [9]:

$$J_{eff,2}^2 = \frac{1}{2}(\hat{\varepsilon}_1\times\hat{\varepsilon}_2)^2\left|J_{p,2}(k_1) - J_{p,2}(k_2)\right|^2 + \frac{1}{2}[1 + (\hat{\varepsilon}_1.\hat{\varepsilon}_2)^2]\left|J_{p,2}(k_1) + J_{p,2}(k_2)\right|^2 = \frac{C_1^2}{2}\left\{(\hat{\varepsilon}_1\times\hat{\varepsilon}_2)^2|k_1^2 - k_2^2|^2 + [1 + (\hat{\varepsilon}_1.\hat{\varepsilon}_2)^2]|k_1^2 + k_2^2|^2\right\} \tag{10}$$

where $J_{p,2}(k_\alpha) = C_1 a_B^{*3}k_\alpha^3$ and $C_1 = -\frac{9(143 + 36\,ln\,(2/3))}{2048}$.

Using subband envelope wave functions of an infinite QW with well width L_z for equal pumping frequency: $\hbar\omega_1 = \hbar\omega_2 = \frac{E_{2p}}{2}$ we obtain the TPA probability for *2p* final states:

$$W_{2p}^{(2)} = \frac{K_{TPA}}{27\pi^3}\frac{C_1^2}{2}I_{\alpha\beta}^2 M^2\left(\frac{\hbar^2}{a_B^{*4}}\right)S_{2p}^{c1,hh1}\frac{16\,E_B^2}{(2E_G - E_{2p})^2}[1 + ((\hat{\varepsilon}_1.\hat{\varepsilon}_2)^2)] \tag{11}$$

where $I_{\alpha\beta}$ is the overlap integral of the subband envelope functions and $K_{TPA} = \frac{128\,\pi\,e^4 A_1^2 A_2^2 v_0^2}{\hbar\bar{\mu}_\xi^2 m^2 c^4 S L_z^2 E_B^2}$.

Polarisation dependence of the intra-excitonic *2p→1s* transition probability

We are interested in the optical transition matrix element between initial two-fold degenerate state Ψ_δ, $\delta = (n = 2, m = \pm 1) = (2, p_{x(y)})$ and final state Ψ_λ, $\lambda = (n = 1, m = 0) = (1,\ s\ )$. The matrix element is of the second type and for normal incidence ($\hat{\varepsilon}\perp\hat{z}$) reads:

$$M_{if} = \frac{v_0^2}{S}\frac{m}{\bar{\mu}_\xi}I_{\alpha\beta}^2\int d^2\rho\ U_\delta^{\alpha\beta*}(\rho)(\hat{\varepsilon}.\hat{\boldsymbol{\varrho}})\frac{\partial U_\lambda^{\alpha\beta}(\rho)}{\partial\rho} \tag{12}$$

Substituting excitonic relative motion wave functions by 2D hydrogen wave functions for *1s* and *2p* states and taking the infinite QW subband envelope wave functions, we obtain:

$$W_{2p\to 1s}^{(2)} = \frac{27}{512}K_{OPA}^{THz}A_{THz}^2\frac{\hbar^2}{\pi^2 a_B^{*2}}I_{\alpha\beta}^4 S_{1s}(E)\Phi^2(\varphi) \tag{13}$$

where the one-photon emission coefficient is given by: $K_{OPA}^{THz} = \frac{16\pi}{\hbar}\left(\frac{e}{\bar{\mu}_\xi c}\right)^2 \frac{v_0^2 A_{THz}^2}{L_z^4 S}$ and the THz radiation field vector potential $A_{THz} = \left(\frac{2\pi c I_{THz}}{n\omega_{THz}^2}\right)^{1/2}$, with I_{THz} : THz radiation intensity. The angular dependence is given by: $\Phi(\varphi) = \cos\varphi e^{\mp i\varphi}$, where φ is the polar angle of the THz photon. For linear ($\hat{\varepsilon}||\hat{x}$) polarization of the emitted THz photon, $\Phi^2(\varphi) = 1$; for linear ($\hat{\varepsilon}||\hat{y}$), $\Phi^2(\varphi) = 0$ and therefore there is no THz emission, and for circularly polarized THz photon, $\Phi^2(\varphi) = 1/2$, the corresponding THz emission rate is half of the one for x-linear polarization.

Quantum efficiency of THz photon generation

The quantum efficiency of THz photon generation η, is given by:

$$\eta = \frac{T}{P} \sim \left(\frac{G}{g}\right)^2, \tag{14}$$

where T is the THz photon generation rate, P is the two-photon pumping rate; G and g are the oscillator strengths of the $2p \rightarrow 1s$ and TPA transitions, respectively, given by:

$$G^2 = \frac{6\pi c^3 \varepsilon \bar{\mu}_\xi S}{\omega_{THz}^2 n^3 e^2} W_{2p\rightarrow 1s}^{(2)}, \qquad g^2 = \frac{6\pi c^3 \varepsilon \bar{\mu}_\xi S}{\omega_{2p}^2 n^3 e^2} W_{2p}^{(2)}, \tag{15}$$

where $\omega_{THz} = \frac{E_{2p} - E_{1s}}{\hbar}$; $\omega_{2p} = \frac{E_{2p}}{\hbar}$. Substituting the expressions for the probabilities of the respective transitions obtained above, after some algebra one obtains for infinite QW:

$$\eta = \frac{144 m^2 n^2 a_B^{*2} S\, I_{\alpha\beta}^2}{e^2 \hbar^2 L_z M^2 \left(143 + 36\, ln(2/3)\right)^2} \frac{E_{2p}^4 \left(2E_G - E_{2p}\right)^2}{\left(E_{2p} - E_{1s}\right)^3} \left(\frac{S_{1s}^{c1,hh1}}{S_{2p}^{c1,hh1}}\right) P\left(\hat{\varepsilon}_1, \hat{\varepsilon}_2, \hat{\varepsilon}\right) \tag{16}$$

where the polarization dependence is given by:

$$P\left(\hat{\varepsilon}_1\, \hat{\varepsilon}_2, \hat{\varepsilon}\right) = \frac{\cos^2\varphi}{1 + \left(\cos\varphi_1 \cos\varphi_2 + \cos(\varphi_1 + \delta_1)\cos(\varphi_2 + \delta_2)\right)^2} \tag{17}$$

DISCUSSION

The polarization dependence of the two-photon absorption probability to $2p$ exciton states is given in figure 2.

We suggest adding an external THz cavity at the VCSEL output that will filter out the linear polarization of the emitted THz radiation, fixing the direction of our reference co-ordinate system x-axis. Maximum (*5*-fold) increase of the TPA rate with respect to *YY* polarization is achieved for linearly $XX, X\bar{X}, \bar{X}X, \bar{X}\bar{X}$ polarized photons (figure2 (a)). The TPA rate can increase by a factor of *3* for linearly-circularly or circularly-linearly polarized photons (b,c); by a factor of *2* for both circularly polarized (d,e), by a factor close to *5* (but < 5) for elliptically polarized (f). Thus changing polarization from *YY* to *XX* one can vary the lasing threshold by a factor of *5*.

Maximum quantum efficiency $\eta = 1$ is found along *YY*-lines for co-linearly polarized photons (figure 3), for *Y*-polarized first (second) photon (linear-circular and circular-linear polarization), or along diagonal lines for co- and counter- circular-circular polarization. The plots for circularly polarized THz radiation are scaled down by a factor of 2 (not shown), resulting in $\eta = 0.5$. In addition, unconditional maximum quantum efficiency is achieved for counter-X ($\bar{X}$) linearly polarized pumping photons and linearly polarized THz emission (not shown).

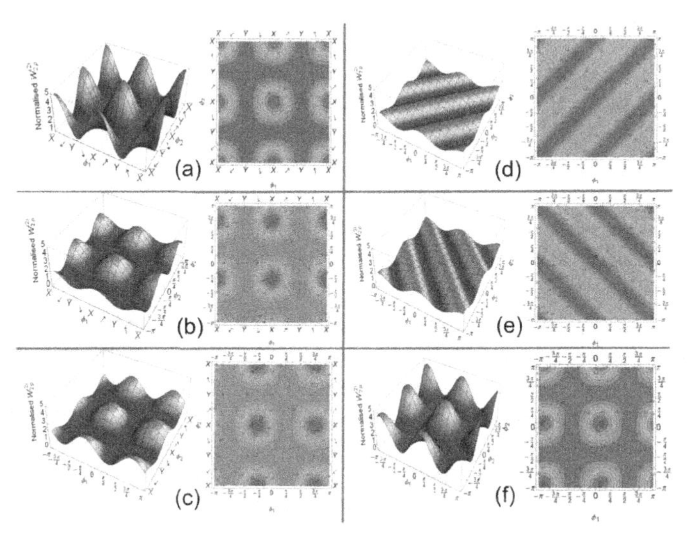

Figure 2. 3D surface and contour plots of the excitonic TPA vs φ_1, φ_2 at specific phase shifts δ_1, δ_2; (a) co-linearly polarized photons $\delta_1 = 0, \pm\pi$; $\delta_2 = 0, \pm\pi$; (b) 1st linear-2nd σ^+polarized $\delta_1 = 0$; $\delta_2 = \frac{\pi}{2}$; (c) 1st σ^+polarized -2nd linearly polarized $\delta_1 = \frac{\pi}{2}$; $\delta_2 = 0$; (d) $\sigma^+ - \sigma^+$or $\sigma^- - \sigma^-$co-circularly polarized $\delta_1 = \pm\frac{\pi}{2}$; $\delta_2 = \pm\frac{\pi}{2}$; (e) $\sigma^+ - \sigma^-$or $\sigma^- - \sigma^+$ counter-circularly polarized $\delta_1 = \pm\frac{\pi}{2}$; $\delta_2 = \mp\frac{\pi}{2}$; (f) co-left-elliptically polarized $\delta_1 = -\frac{\pi}{6}$; $\delta_2 = -\frac{\pi}{8}$.

To verify these predictions experimentally, we suggest pumping the QW structure by two laser beams with the same frequency but different polarization (figure 4). The delay line provides phase shift π to obtain counter-linearly polarized beams. The intensity and polarization of the emitted THz light could be measured as a function of intensities and polarizations of the two pumping beams. As reference experiments the intensity of THz emission with one pump beam switched off can be measured. Comparing the spectra obtained with both beams switched on with those with only the first/second beam switched on, the signal generated by absorption of the two photons coming from different beams and having different polarizations can be extracted.

CONCLUSIONS

We have developed a theory of the two-photon excitonic absorption to p-exciton states in QWs and calculated the polarization dependence of TPA probability, using crystal symmetry point group theory. Our model predicts variation of the lasing threshold by a factor of 5 by switching pumping photons polarization from co-linearly polarized along the p-state to co-linearly polarized in an orthogonal direction. Conditions for achieving maximum quantum efficiency for different polarizations of the pumping photons are identified, thereby opening routes for polarization control of the THz VCSEL and a whole new range of applications.

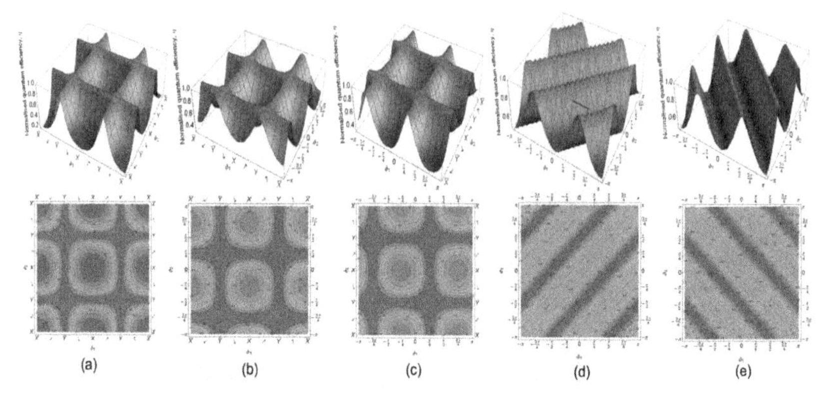

Figure 3. 3D surface (top) and contour (bottom) plots of the normalized quantum efficiency of THz photon generation against φ_1, φ_2 at different phase shifts δ_1, δ_2 at linear $\varphi = 0, \pm\pi$ polarization of the emitted THz radiation: (a) co-linearly polarized photons; (b) 1st linear – 2nd circularly polarized photon; (c) 1st circular – 2nd linearly polarized; (d) $\sigma^+ - \sigma^+$or $\sigma^- - \sigma^-$co-circularly polarized photons; (e) $\sigma^+ - \sigma^-$or $\sigma^- - \sigma^+$ counter-circularly polarized photons.

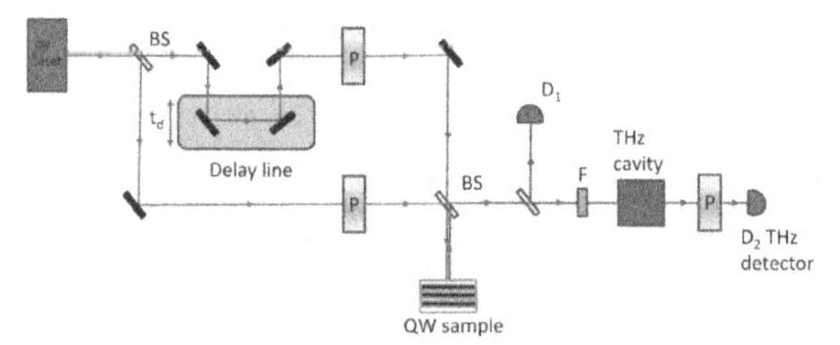

Figure 4. Suggested experimental setup for polarization control of THz VCSEL; BS: beam splitter, P: polariser, F: band-pass filter, D1: visible light detector, D2: THz detector.

REFERENCES

1. A. Kavokin, I. Shelykh, T. Taylor, and M. Glazov, Phys. Rev. Lett. **108**, 197401 (2012).
2. R. Huber, B. A. Schmid, R. A. Kaindl, and D. S. Chemla, Phys. Stat. Sol. (b) **245**, 1041 (2008);R. Huber et al., Phys. Rev. B **72**, 161314(R) (2005).
3. J. L. Tomaino et al., Phys. Rev. Lett. **108**, 267402 (2012).
4. M. Kira and S.W. Koch, Phys. Rev. Lett. **93**,076402 (2004).
5. A. Pasquarello and A. Quattropani, Phys. Rev. B **42**, 9073 (1990).
6. E. L. Ivchenko and G. E. Pikus, *Superlattices and Other Heterostructures* (Springer-Verlag, Berlin, 1997).
7. R. J. Elliott, Phys. Rev. **108**, 1384 (1957).
8. R. Zimmermann, Phys. Stat. Sol. (b), **146**, 371 (1988).
9. G. D. Mahan, Phys. Rev. **170**, 825 (1968); V. Heine, *Group theory in quantum mechanics*, (Dover Publications, 1993).

Mater. Res. Soc. Symp. Proc. Vol. 1617 © 2013 Materials Research Society
DOI: 10.1557/opl.2013.1186

Experimental Characterization of frequency-depend electrical parameters of On-Chip Interconnects

Diego M. Cortés Hernández, Mónico Linares Aranda, Reydezel Torres Torres
Instituto Nacional de Astrofísica, Óptica y Electrónica (INAOE)
Luis Enrique Erro No. 1, Sta. María Tonanzintla, Puebla, C.P. 72000, México
dmcortes@inaoep.mx, mlinares@inaoep.mx, reydezel@inaoep.mx

ABSTRACT

An exhaustive analysis of the frequency-dependent series resistance associated with the on-chip interconnects is presented. This analysis allows the identification of the regions where the resistance curves present different trending due to variations in the current distribution. Furthermore, it is explained the apparent discrepancy of experimental curves with the well-known square-root-of-frequency models for the resistance considering the skin-effect. Measurement results up to 40 GHz show that models involving terms proportional to the square root of frequency are valid provided that the section of the interconnect where the current is flowing is appropriately represented.

INTRODUCTION

In each generation of manufacturing process for integrated circuits, the metal layers used for interconnection purposes is increased [1].Moreover, due to the high-data rates presented in current electronic systems, many of these interconnects are routed in a very complex way and are operated at very-high (GHz range) frequencies. Hence, these interconnects must be treated as transmission lines (TL).

Traditionally, IC developers were particularly concerned about the series resistance (R) and shunt capacitance (C) presented by a TL, as well as for the number of lumped-circuit stages required to represent a line of certain length within a given bandwidth [2],[3]. This motivated research oriented to determine and represent these parameters as a function of geometry, frequency (f), and fabrication materials [4]–[6]. In several papers the capacitance (C) and the losses represented by a parallel conductance (G) show a weak dependence on f when the interconnect presents a ground shield [7], [8]. In contrast, R, which is associated with the metal losses occurring along the interconnect is strongly dependent on f and it is expected to be proportional to the square root of f when the skin depth is comparable or less than the thickness of the metal layer [9]. Nevertheless, this ideal trending of R is not evident in the vast majority of the experimental results obtained for on-chip interconnects [7], [10]. Many square-root-of-f models have been developed, however these models use fitting parameters to compensate for effects that are still not well explained. For instance, assuming that the series resistance that the interconnect presents at low frequencies introduces a f-independent parameter that is observed even at very high frequencies [5], [9]. On the other hand, as f increases the RC model lacks of accuracy, especially when representing relatively long interconnects such as those used for global networks (e.g. clock distribution) [11]. This paper presents a characterization of the series *Resistance* impacted by f-dependent effects.

THEORETICAL FRAMEWORK

It is known that once the operation frequency of a TL is so high that the metal skin depth (δ) is smaller than the thickness of the metal layers used to form the signal plane t_s and ground plane t_g (See Figure 1), R become dependent on f. Particularly, for TL transmitting signals in quasi-TEM mode it is expected that R is proportional to $\sqrt{f}$, [9]. Although this behavior is clearly observed in packaging technologies at microwave frequencies, it is not evident for IC interconnects. This is due to the fact that the dimensions of the metal layers forming the IC interconnects are much smaller than in packaging technologies, which originates that R vary with f in a more complicated way since each metal region forming the interconnects is impacted by the skin and proximity effects within different f-ranges.

In the Figure 1, a typical stack-up defining the different metal layer levels within an IC is shown. In this figure, the ground plane is located at the Metal-3 level. The ground plane has a width w_g and shields the upper interconnects from the silicon semiconductor substrate. In this figure, h is the thickness of the SiO_2 layer separating the ground plane from the interconnect on top, which presents a width w.

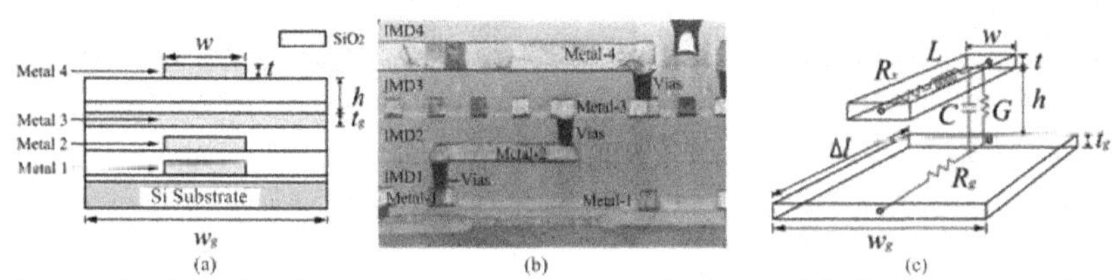

Figure 1. Typical interconnects on an integrated circuit: (a) Ideal cross section, (b) Actual cross section, (c) Lumped model

Figures 1c shows a section of an interconnect with length $l = \Delta l$ that can be represented by means of an electrical *RLGC* lumped model. Although the *R, L, G,* and *C* elements have been widely studied in the literature, the dependence of R on f is simplified most of the times assuming a $\sqrt{f}$ variation with constant proportionality coefficient. Further examination of this assumption is the subject of study in this paper as explained hereafter.

Resistance

In accordance to Figure 1c, the total series resistance of the interconnect can be defined as:

$$R = R_s + R_g \quad (1)$$

where R_s is the resistance associated with the signal trace or signal plane, whereas R_g corresponds to the ground plane.

Due to the fact that these two resistive contributions to R are located at different regions of the structure, the configuration of the electromagnetic fields will affect the distribution of the current flow in different ways as f changes, such as is shown in the Figure 2

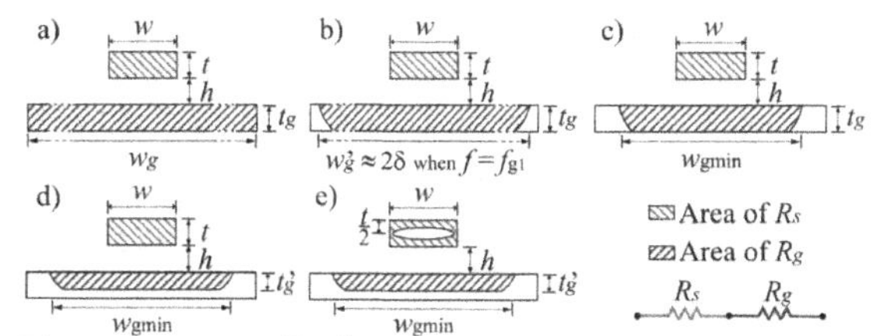

Figure 2. Current distribution for different *f*-regions.

Starting the analysis at f so low that δ is much bigger than t and t_g, both R_s and R_g can be analyzed assuming DC conditions. For the case of R_s, since the current can be considered to uniformly flow within the cross section of the signal plane (Figure 2a), the following equation can be used:

$$R_s = R_{s0} = \frac{1}{\sigma wt} \qquad (2)$$

where σ is the conductivity of the metal.

For the case of R_g, although the contribution of R_g at low frequencies is weak (the corresponding value in a conventional design is usually negligible ($R_{g0} \approx 0$) it can be approximately calculated as:

$$R_g = R_{g0} = \frac{1}{\sigma w_g t_g} \qquad (3)$$

Thus, in accordance to (1)–(3), the series resistance can be defined as (See Figure 3):

$$R_I = R_{s0} + R_{g0} \qquad (4)$$

According to the Figure 3, R remains constant until f reaches a value f_{g1} that makes the resistance to vary proportionally to $\sqrt{f}$. This first transitional frequency is associated with the reduction of the effective width of the cross section where the current is flowing within the ground plane (w_g') as f increases, which is explained as follows. At low f, the transverse fields surrounding the signal trace reach the top of the ground plane near below it and also laterally penetrate the whole plane since δ is very big. However, when f increases to a value at which $\delta \approx w_g/2$ (see Figure 2b) the TL operates in region II, where w_g' becomes proportional to δ. In this case, since δ is inversely proportional to $\sqrt{f}$, when substituting w_g'$= \delta/2$ for w_g in (3) It is clear that R_g increases with $\sqrt{f}$. This effect has been previously analyzed by considering that the current flows through the ground path that presents the smallest impedance, and it is referred to as the current crowding effect [12],[13]. Hence, considering that within this *f*-region δ is still bigger than the dimensions of the signal plane (which is a reasonable assumption in most practical technologies), R_s in (1) remains constant, and R presents the following mathematical form:

$$R_{II} = R_{s0} + k_1\sqrt{f} \qquad (5)$$

The upper *f*-limit of region II defines the starting frequency f_{gsat} at which a plateau is observed in the R versus f curve (region III in Figure 3). This plateau is only observed if w_{gmin} is bigger than the thickness of the ground plane and also than the width and thickness of the signal plane. Within this frequency range, defined as region III, R can approximately be represented by:

$$R_{III} = R_{s0} + R_{gsat} \qquad (6)$$

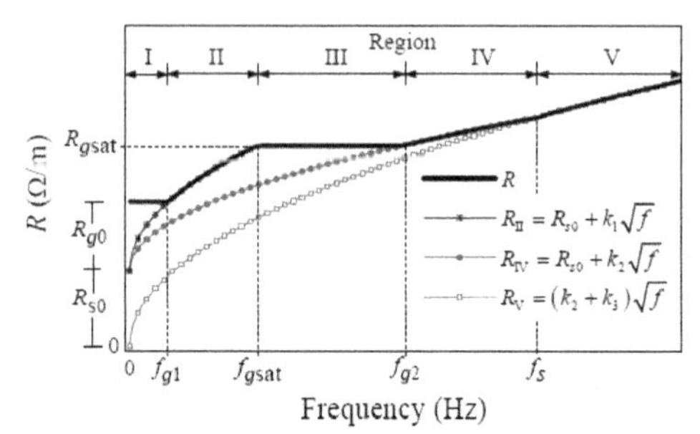

Figure 3. Plot showing the regions presenting different variation of the R versus f curve.

When the skin depth reaches a value that equals the thickness of the ground plane ($\delta \approx t_g$), a reduction in the vertical distribution of the current is observed in this plane as f increases. This confinement of the current concentration on top of the ground plane generates another region at which R_g proportionally increases with $\sqrt{f}$ (Figure 2d), this behavior defines the region IV in Figure 3, and it is represented by:

$$R_{IV} = R_{s0} + k_2\sqrt{f} \qquad (7)$$

Observe that in region IV, due to the configuration of the fields, once the cross section area for which the current is flowing is reduced as f increases, most of the current is confined to the upper section of the ground plane. Conversely, for the signal plane the current is confined in comparable proportions along the perimeter of its cross section (Figure 2e). Approximately, R_s starts to become f-dependent at $f = f_s$ (i.e. when $\delta \approx t/2$), which defines the onset frequency for region V, which is given by following expression:

$$R_V = k_2\sqrt{f} + k_3\sqrt{f} = k_4\sqrt{f} \qquad (8)$$

EXPERIMENTAL VALIDATION

Several microstrip lines with different lengths, widths and heights from the ground plane were designed and fabricated to verify the validity of the analysis presented in this paper. The microstrip lines were made of aluminum in the Metal-4 level and present widths of 2μm and 4μm, and lengths of 400μm and 1000μm. These lines present a ground shield at the Metal-1 level, the thickness of the SiO_2 dielectric for these lines is h = 3 μm. In all cases, the signal plane presents a nominal thickness t = 925 nm, whereas the metal thickness of the ground shields is t_g = 665nm. Figure 4 shows the fabricated lines.

The propagation constant γ and characteristic impedance Z_c were determined from the experimental data of each pair of microstrip lines, presenting identical cross section and varying only in length. Thus, once the γ and Z_c are obtained from measured data, the experimental series resistance versus frequency curves are respectively determined as R=Re($\gamma \cdot Z_c$) and L =Im($\gamma \cdot Z_c$)/2πf. This allows the direct inspection of these parameters to identify the different operation regions.

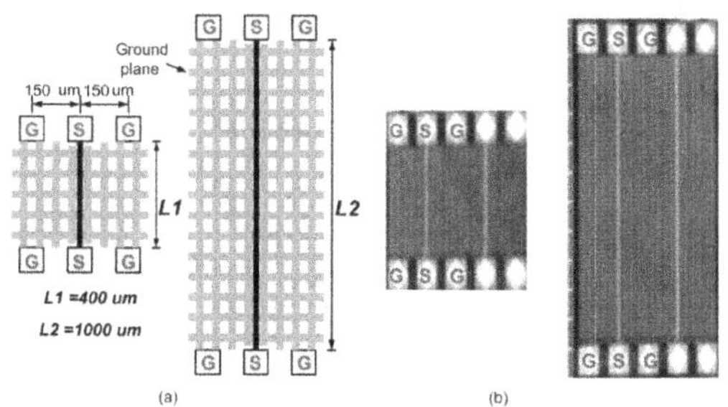

Figure 4. Fabricated microstrip lines: (a) Layout and (b) A micrograph

RESULTS AND DISCUSSION

Figure 5 shows the data for microstrip lines for the signal plane and the ground plane in Metal-1 level. As f increases, region II can be observed in the plot of Figure 5 by noticing a linear increase of R with $\sqrt{f}$. This region is observed in both lines within the same frequency range. Moreover, $k1$ takes approximately the same value to represent R_{II} for both lines (i.e., 0.06 and 0.063, respectively), which means that the lateral skin effect occurring in the ground plane increases R roughly in the same proportion independently of the line width. This behavior continues up until region III becomes apparent. In this case, a plateau is observed in the R versus $\sqrt{f}$ curves, meaning that the lateral skin effect in the ground plane reached saturation since the current is confined below the signal plane. This region remains until the skin depth equals the thickness of the ground plane, where the resistance increases again proportionally with $\sqrt{f}$ in region IV. Notice that, similarly as k_1 in region II, k_2 in Figure 5 presents approximately the same value. Besides, another interesting observation is that in regions I, II, III, and IV, R_s remains constant (i.e. $R_s = R_{s0}$) since the skin depth is not comparable with the thickness of the signal plane. Thus, R_s becomes frequency dependent until the skin depth is smaller than one half of the signal plane thickness as explained in Section II. This occurs in region V, where R is represented by a line that passes through the origin since R_s and R_g are frequency dependent and there is no justification to add a constant to the model of R. R_V is typically the model used to represent the frequency dependent resistance of transmission lines. The experimental transition frequencies values are presented in Table I. When determining the parameters to represent R in the different operation regions, k_1, k_2, and k_4 are the coefficients that allow the representation of the corresponding frequency dependence within different frequency ranges. Table II summarizes the different parameters obtained for R.

CONCLUSIONS

An exhaustive analysis for identifying the different operation regions affecting the frequency dependence of the series resistance of on-chip interconnects was carried out. From this analysis it was possible to associate physical phenomena to the different variations occurring in this parameter at frequencies of gigahertz. Moreover, the separate models for the different regions allow the representation of experimental resistance data in an accurate and consistent way. This analysis then allows the implementation of physical models for the accurate representation of the series parasitics occurring in interconnects used in ICs.

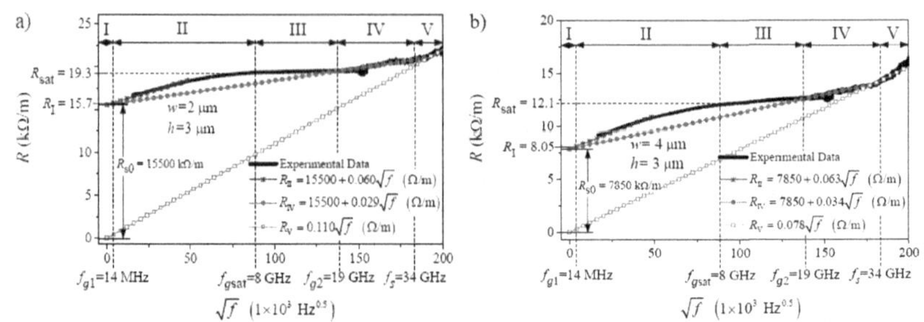

Figure 5. Experimental Resistance curves up to f = 40 GHz for the microstrip lines: (a) w=2μm, (b) w=4μm

Table I. Transition frequencies obtained for the fabricated lines.

Structure	f_{g1} (MHz)	f_{sat} (GHz)	f_{g2} (GHz)	f_s (GHz)
w=2μm, h=3μm	14.2	8	16.4	18.6
w=4μm, h=3μm	13.7	8	16.4	18.6
w=2μm, h=1μm	9.8	8	15.2	20
w=4μm, h=1μm	8.2	8	15.2	20

Table II. Resistance parameters

Ground plane	Level Metal-1 h=3μm		Level Metal-3 h=1μm	
Trace width / Parameter	*2μm*	*4μm*	*2μm*	*4μm*
R_{s0} (kΩ m^{-1})	15.50	7.85	15.40	7.15
R_{g0} (kΩ m^{-1})	0.20	0.20	0.20	0.20
R_{gsat} (kΩ m^{-1})	19.3	12.10	21.10	13.10

REFERENCES

1. T. Quémerais, L. Moquillon, J. M. Fournier, and P. Benech. *IEEE Trans. on Microwave Theory and Techniques*, **58**, 2426 (2010).
2. T. Sakurai, *IEEE J. Solid-State Circuits*, **SC-18**, 418 (1983).
3. Y. I. Ismail E. G. Friedman, and J. L. Neves, *IEEE Trans. on VLSI Systems*, **7**, 4 (1999).
4. J. H. Kim, D. Oh, and W. Kim, IEEE *Trans. Adv. Packag.*, **33**, 4, 857 (2010).
5. J. Brinkhoff K.et. al., *IEEE Trans. Microwave Theory Tech.*, **56**, 12, 2954 (2008).
6. T. Makita I. Tamai, and S. Seki, *IEEE Trans. on Electron Devices*, **58**, 3 (2011).
7. H. Heng-Ming L. Tai-Hsin, and H. Chan-Jung, *IEEE J. on Emerging and Selected Topics in Circuits and Systems*, **2**, 2 (2012).
8. D. Zeng, H. et.al., *Custom Integrated Circuits Conference*, 1-4 (2010).
9. J. Zhang, et.al., *IEEE Trans. Electromagn. Compat.*, **52**, 1, 189 (2010).
10. H. Y. Cho, J. K. Huang, C. K. Kuo, S. Liu, and C. Y. Wu,, *IEEE Trans. Electron Devices*, **56**, 12, 3160 (2009).
11. O. Gonzalez, et al. *Analog ICs and Signal Processing,* **71**, 221 (2012).
12. B. Young. 1st ed. New Jersey, Prentice Hall PTR, 2001.
13. H. Johnson and M. Graham, *"High-Speed Signal Propagation: Advanced Black Magic,"* New Jersey, Prentice Hall PTR, 2003.

Mater. Res. Soc. Symp. Proc. Vol. 1617 © 2013 Materials Research Society
DOI: 10.1557/opl.2013.1187

Low Frequency Admittance Measurements in the Quantum Hall Regime

Carlos Hernández[1] and Christophe Chaubet[2]
[1]Departamento de Física, Universidad de los Andes, A.A. 4976,
Bogotá D.C., Colombia.
[2]Laboratoire Charles Coulomb L2C, Université Montpellier II, Pl. E. Bataillon,
34095 Montpellier Cedex 5, France.

ABSTRACT

In this paper we present an ac-magneto-transport study of a two-dimensional electron gas (2DEG) in the quantum Hall effect (QHE) regime, for frequencies in the range [100Hz, 1MHz]. We present an approach to understand admittance measurements based in the Landauer-Buttiker formalism for QHE edge channels and taking into account the capacitance and the topology of the cables connected to the contacts used in the measurements. Our model predicts an universal behavior with the a-dimensional parameter $R_H C\omega$ where R_H is the 2 wires resistance of the 2DEG, C the capacitance cables and the angular frequency, in agreement with experiments. For a specific configuration, we measure the electrochemical capacitance of the quantum Hall edge channels as predicted by Christen and Büttiker.

INTRODUCTION

The quantum Hall effect (QHE) is widely used by national laboratories thanks to its great stability in dc [1,2]. The accuracy of measurements has allowed establishment of the international standard for the electrical resistance unit (Ohm). The QHE plateaus are perfectly quantized in dc (up to 10^{-9}) but in ac the quantization has never been better that 10^{-7} [3]. The quantized Hall resistance of a QHE sample is usually found to be current and frequency dependent. That lack of precision in metrological studies, is caused by ac-losses [4,5]. One knows that these losses are linked to the charge distribution on the edges of the 2DEG, and that a good ac-quantization of the Hall resistance can be achieved in gated samples, by biasing a gate [6,7]. For ungated samples, ac-losses are explained by a "Polarization Model" and a "Capacitive Model" [4,5], however the origin of these ac-losses is not yet understood and other approaches have been explored. Desrat et al [8] have interpreted their measurements using the theory proposed by T. Christen and M. Büttiker [9, 10] based on the Büttiker, Prêtre and Thomas [11] theory of finite frequency transport that can be used to predict the frequency dependence of the conductor's admittance. However this interpretation was very controversial because it did not take into account the parasitic capacitances of the experimental setup [12].

EXPERIMENTAL DETAILS

We have performed a study of the longitudinal admittance of two-dimensional electron gases. The samples used for the experimental measures were GaAlAs/GaAs heterojonctions (Table 1), of metrological quality, obtained with the collaboration of the "Laboratoire National de Métrologie et d'Essais (LNE)", french national laboratory of metrology. The samples were entirely processed in the Philips laboratory of Limeil Breivannes, as a Hall bar having six independent lateral voltage contacts in addition to the source and drain contacts (Fig. 1.(a)). Details of the epitaxial growth as well as sample characteristics are described elsewhere [13,14]. We have first performed dc-measurements using a standard configuration to obtain magnetoresistance and quantum Hall resistance. For impedance measurements we have used an Agilent 4294A precision impedance analyzer, using different two point configurations (Fig.1.(b)).

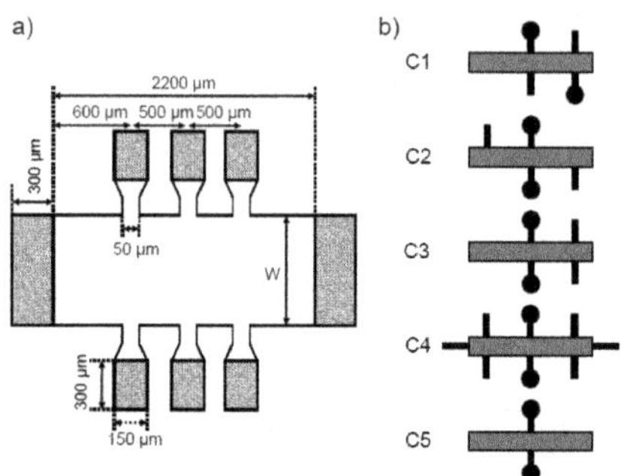

Figure 1. a) The shape of the Hall bar. All samples have same geometry, and same dimensions except for W (200µm or 400µm). b) Measurement configurations. Solid black lines represent the ohmic contacts which are wire-bonded to the TO8 sample holder (thus connected to coaxial cables). Black circles represent the two contacts which are connected to the impedance analyzer. In config.4 all contacts are wirebonded and so connected to coaxial cables; in config.5, only two.

All the wiring of the experimental system is made with coaxial cables whose characteristics are well known. We thus obtain directly the admittance as a function of frequency in the range [100Hz-1MHz], and we could make a great number of measurements for different values of the magnetic field in the range [0-12T]. All experiments were performed in a cryostat at 1.5K.

The experimental results for Hall effect in dc were excellent as expected for metrological quality samples [15]. To analyze the admittance in alternative current, we have plotted conductance and susceptance, as a function of the frequency for different filling Hall plateaus.

The two-points admittance in alternative current, $Y = G + jS$ where G is the conductance and S the susceptance, was measured using different measurement configurations, (Fig. 1.(b)): C1 (PL175 W = 200µm), C2 (PL175 W = 400µm), C3 (PL175 W = 200µm), C4 (LEP514 W = 400µm) and C5 (PL174 W = 400µm), with an Agilent 4294A impedancemeter at a fixed value of the magnetic field, with a polarization voltage of 100 mV (this is low enough to guarantee the linear regime). The first observation concerns the influence of the topology of contacts connections on the conductance (real part of the admittance) measured at $\nu = 2$. Fig.2 presents the results obtained with different configurations. In dc, one measures $2e^2/h$, which is the measured value at low frequency for any configuration. At high frequency, one observes a decrease of the conductance, more or less pronounced, depending on the configuration. Results for C1 and C2 are superimposed (see Fig.2). We remark that there exists a frequency domain where the conductance is negative for configuration C3 and C4. Finally, in configuration C5 the conductance is constant and independent on the frequency in a wide range.

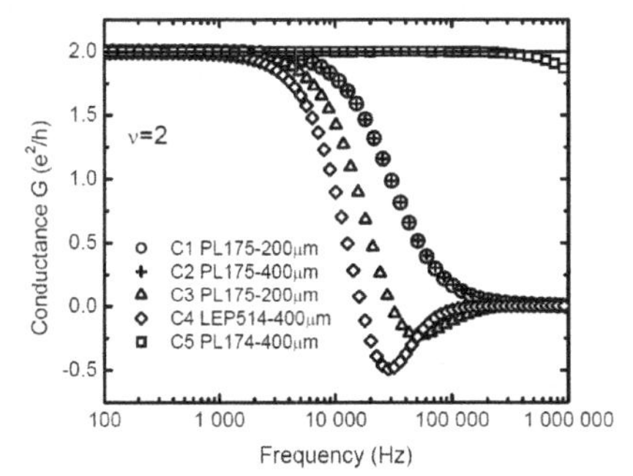

Figure 2. Conductance as a function of the frequency for all configurations. Results for C1 and C2 are superimposed

DISCUSSION

In order to understand our admittance measurements, we present an approach based on the Landauer-Buttiker formalism for QHE edge channels [11]. For this approach we will consider all the coaxial cables connected to sample contacts, including those not connected to the instrument. Those unconnected cables are represented by their capacitance connected to earth, because the guard of all cable is grounded. We do not consider for the moment the sample electrochemical capacitance [9,10], because it is much smaller than the cable capacitance.

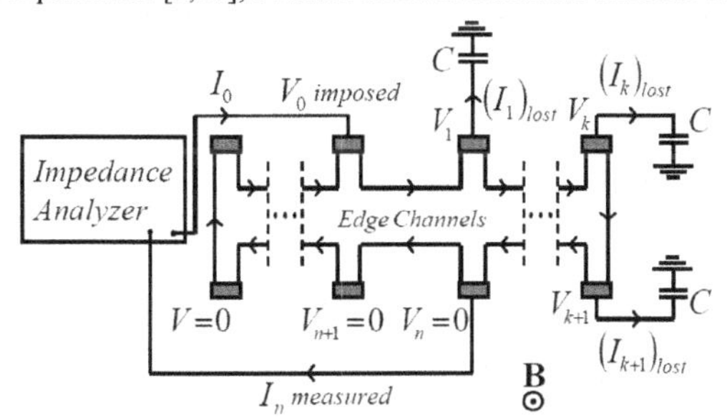

Figure 3. Simplified equivalent circuit of the Hall bar showing chiral edge states connecting contacts. C is the capacitance of a coaxial cable. V_k is the potential of contact k. V_0 is the current imposed by the impedance analyser, and I the measured current. $V_{k>n} = 0$ because contact n is grounded. Contacts $k > n$ are also connected to a capacitance C, but this capacitance is short-circuited because $V_{k>n} = 0$; it is not represented on this figure for clarity. B is the magnetic field.

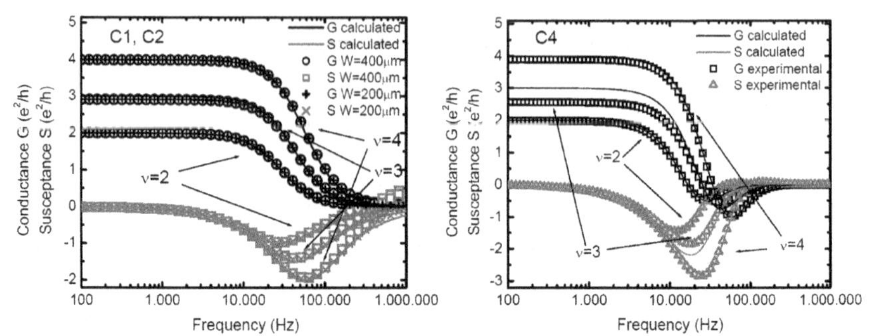

Figure 4. Experimental and calculated conductance and susceptance, for configuration C1, C2 (PL175 200μm and PL175 400μm) and C4 (LEP514 400μm). As predicted, the results for C1 and C2 are identical. The value C=425pF was used in all calculations.

We number the contacts of the sample by following the direction of the edge states: contacts 0 and n are at fixed potentials V_0 and $V_n = 0$ imposed by the connection to the impedance analyzer. Other contacts (1 to n-1) are connected to the capacitance of unplugged coaxial cables, and their respective potentials are not imposed. Contacts connected by the edge states flowing out from $V_n = 0$ (thus located after V_n) are all at the same potential (earth) and do not enter the calculations (Fig. 3).

On the quantum Hall plateaus we apply the Landauer-Buttiker formalism, in which edge states connects adjacent contacts to the same chemical potential. In that case for contacts 0 and n we have,

$$I_0 = \frac{\nu e^2}{h}\left(V_n - V_0\right) \quad (1)$$

Using the equations for the capacitance connections and edge channels, we can calculate as well the current losses for other contacts. We find for contact k,

[illegible] (2)

And finally, the admittance for the system is:

[illegible] (3)

For Configuration C3 (Fig.4), we have n=3 (because V3=0) and there will be losses at contacts 1 and 2. For configuration C1, we have n=2 (V_2=0) and there will be losses only at contacts 1. In this configuration, contact 3 is at the same potential than contact 2 and does not enter in the calculations. Then we can deduce than configurations C1 and C2 are equivalent with n=2. For C3 configuration, n=3; for C4, n=4.

For multiple contact configurations, C1, C2, C3 and C4 we have used the capacitance value C = 425pF to fit the admittance. This value is close to the measured capacitance of a single coaxial line C_m = 480pF.

Our calculations reproduce experimental results with very good accuracy for frequencies up to 100kHz and for all configurations (Fig.4). For ν=2, theoretical results correspond with experimental data for all samples. For ν=3 and the other even numbers filling factors, some differences are observed but it can be explained by the quality of quantization of Hall plateaus and the quality of the thermalization of low potential contacts with earth.

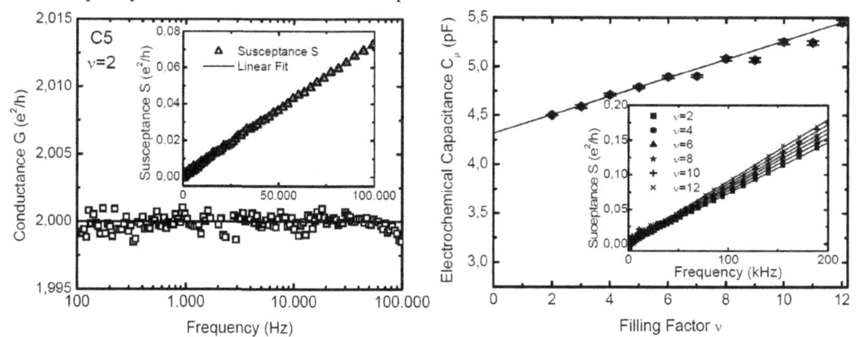

Figure 5. Left: experimental and theoretical value of the conductance for configuration C5 (PL174). Insert: the susceptance exhibits a linear dependence with frequency. Right: electrochemical capacitance as a function of the filling factor. Insert: the susceptance has a linear dependence with the frequency; Cμ is proportional to the slope.

For two contacts configuration C5 shown in Fig.5, the conductance value does not depend on the frequency, in accordance with the theoretical approach (because $n = 1$). The measure stays close to $2e^2/h$ until 100kHz. However the susceptance have a linear behavior unpredicted by the previous formula and the capacitance of the cables plays a minor role in the measurements. In those conditions we can measure the electrochemical capacitance of the sample. According to Büttiker [9,10] the admittance $G_{\alpha\beta}(\omega)$ gives the linear current response at contact α resulting from a voltage oscillation at contact β. The admittance matrix is written:

$$G_{\alpha\beta}(\omega) = G^0_{\alpha\beta} - i\omega E_{\alpha\beta} \qquad (4)$$

where $G^0_{\alpha\beta}$ is the d.c. conductivity matrix and $E_{\alpha\beta}$ the emittance matrix of the Hall bar. Those matrix are 8 × 8 square matrix in case of a 8-terminal Hall bar. In our experimental conditions, the measured admittance is thus $G_{\alpha\beta}(\omega) = G^0 - i\omega\, C\mu$, with $C\mu = (E_{00} + E_{01} + E_{02} + E_{04})$. This is in accordance with results of Fig.5: the conductance is independent of the frequency and equal to $G^0 = 1/R_H = 2e^2/h$, the imaginary part of the admittance is proportional to the frequency (Fig.5) with a proportionality constant obviously positive. With our sign convention this corresponds to a positive capacitance.

We have measured the susceptance as a function of the frequency at all possible filling factor. For each filling factor we measure the slope and obtain a positive Cμ, an apparent electrochemical capacitance. The values of Cμ are reported in Fig.5 for several values of the filling factor. We observe a clear linear variation of this capacitance with the filling factor. This shows that we observe here the sample electrochemical capacitance, composed of a relation between the geometrical capacitance matrix of the sample and its density of states, able as well to accumulate charges. We add that we certainly also measure some unwanted parasitic capacitance caused by the environment (most probably the sample holder, the gold wires, or a bad compensation of external cables because of non equal lengths). The dependence in the filling factor proves the role of the density of states in the electrochemical capacitance. To go further in the analysis and the measurements of the sample geometrical capacitance, more compact sample geometries are suitable, as those considered in Ref. [10] (having only two or four ohmic contacts). As a last remark, we observe that experimental points for filling factors $\nu = 7, 9, 11$ are slightly below the line. This is compatible with the absence of minima at those filling factors in the Shubnikov-de Haas d.c. oscillations: the quantization is not perfect for those odd filling factors.

CONCLUSIONS

In this work we have proposed a new approach to understand ac-losses in the QHE regime. Our predictions are in great agreement with experimental data, and an universal dependence on the a-dimensional parameter $RC\omega$ is found. Our model predicts a strong dependence upon the pairs of contacts that are used. Our model suggests that a two terminal configuration might help the realization of a quantum resistance standard in ac. In this two cables configuration we could measure the electrochemical capacitance as a function of the filling factor .

REFERENCES

1. J. Melcher, P. Warnecke, R. Hanke, IEEE Trans. Instrum. Meas., 42, 292-294 (1993).
2. F. Delahaye, Metrologia, 31, 367-373 (1994).
3. F. J. Ahlers, B. Jeanneret, F. Overney, J. Schurr and B. M. Wood, Metrologia, 46, R1-R11 (2009).
4. B. Jeanneret and F. Overney, IEEE Trans. Instrum. Meas., 56, 431-434 (2007).
5. J. Schurr, F. J. Ahlers, G. Hein, J. Melcher, K. Pierz, F. Overney and B. M. Wood, Metrologia, 43, 163-173 (2006).
6. J. Schurr, F. J. Ahlers, G. Hein and K. Pierz, Metrologia, 44, 15–23 (2007).
7. J. Schurr, J. Kucera, K. Pierz and B. P. Kibble, Metrologia 48, 47 (2011)
8. W. Desrat, D.K. Maude, L.B. Rigal, M. Potemski, J.C. Portal, L. Eaves, M. Henini, Z.R. Wasilewski, A. Toropov, G. Hill and M.A. Pate, Phys. Rev. B 62, 12990 (2000)
9. M. Büttiker, J. Phys. Condens. Matter 5, 9361(1993)
10. T. Christen and M. Büttiker, Phys. Rev. B 53, 2064 (1996)
11. M. Büttiker, A. Prêtre and H. Thomas, Phys. Rev. Lett. 70, 4114 (1993)
12. J. Melcher, J. Schurr, F. Delahaye and A. Hartland, Phys. Rev. B 64, 127301 (2001)
13. Y. M. Meziani, C. Chaubet, S. Bonifacie, A. Raymond, W. Poirier, and F. Piquemal, J. Appl. Phys. 96, 404 (2004).
14. F. Piquemal, G. Genev`es, F. Delahaye, J. P. Andr´e, J. N. Patillon, and P. Frijlink, IEEE Trans. Instrum. Meas. 42, 264 (1993).
15. C. Hernández, C. Chaubet, Rev. Mex. Fis. 55, 432-436 (2009).

Appendix:
Volume 1534
(Symposium 6B)

Papers in the Appendix were published in electronic format as Volume 1534.

IV group semiconductors

Mater. Res. Soc. Symp. Proc. Vol. 1534 © 2013 Materials Research Society
DOI: 10.1557/opl.2013.291

Si Quantum Dot Structures and Some Aspects of Applications

Lyudmula V. Shcherbyna[1] and Tetyana V. Torchynska[2]

[1]V. Lashkarev Institute of Semiconductor Physics at National Academy of Science, Kiev, Ukraine
[2]ESFM – Instituto Politécnico Nacional, México D.F. 07738, México

ABSTRACT

This paper presents briefly the history of emission study in Si quantum dots (QDs) in the last two decades. Stable light emission of Si QDs and NCs was observed in the spectral ranges: blue, green, orange, red and infrared. The analysis of recombination transitions and the different ways of the emission stimulation in Si QD structures, related to the element variation for the passivation of surface dangling bonds, as well as the plasmon induced emission and rare earth impurity activation, have been discussed.

The different applications of Si QD structures in quantum electronics, such as: Si QD light emitting diodes, Si QD single union and tandem solar cells, Si QD memory structures, Si QD based one electron devices and double QD structures for spintronics, have been presented. The different features of poly-, micro- and nanocrystalline silicon for solar cells, that is a mixture of both amorphous and crystalline phases, such as the silicon nanocrystals (NCs) or QDs embedded in a α-Si:H matrix, as well as the thin film 2-cell or 3-cell tandem solar cells based on Si QD structures have been discussed as well. Silicon NC based structures for non-volatile memory purposes, the recent studies of Si QD base single electron devices and the single-electron occupation of QDs as an important component to the measurement and manipulation of spins in quantum information processing have been analyzed as well.

INTRODUCTION

The significant worldwide interest directed toward the silicon-based light emission for integrated optoelectronics is related to the complementary metal-oxide semiconductor compatibility and the possibility to be monolithically integrated with very large scale integrated (VLSI) circuits. This makes Si NC and QD structures very attractive for the fabrication of optoelectronics and microelectronics devices [1]. In the field of Si NCs and QDs the structure of small Si nanoclusters with a number of Si atoms $n<10$ is well understood, but the transition to the bulk silicon, which is realized via QD and NC particles, is still a subject of intensive experimental and theoretical works.

It was shown that improved optical properties of Si NCs smaller than 5 nm (QDs) are due to a combination of two effects: the stimulation of electron and hole radiative recombination rate due to the increased overlap of electron and hole wave functions confined in the quantum dots, as well as the reduction of recombination rate via nonradiative defects at the QD (NC) surface [2-4]. Optical properties of Si QDs are sensitive to the NC sizes, due to the quantum confined effect in QDs, and to defect and surface effects owing the large surface to volume ratio in small crystallites. Decreasing surface defect emission can be achieved using the different types of QD passivation. At the same time some questions concerning the real correlation of PL peak positions, caused by exciton recombination inside of Si QDs, with their sizes, the nature of other PL bands, as well as the relative contributions of different elementary PL bands into the total PL spectrum are not clear yet [1].

Emission of Si NCs and QDs

Stable light emission of Si QDs and NCs is observed in the spectral range: blue (2.64–3.0 eV), green (2.25 eV), orange (2.05 eV), red (1.70–1.80 eV) and infrared (1.2–1.6 eV) [1–11]. These PL bands were attributed: to exciton recombination in Si QDs [3, 4, 6, 7, 11-13], the carrier recombination through defects inside of Si NCs [12] or via oxide related defects at the Si/SiOx interface [1, 5, 8, 9, 11,13]. Due to the strong competition between the exciton recombination inside of Si NCs and the defect-related recombination at the Si/SiOx interface, the new effect: emission controlled by the hot carrier ballistic transport to the Si/SiOx interface, has been revealed in Si NCs [1, 14-16]. Note, the photoluminescence related to the Si/SiOx interface and controlled by the hot carrier ballistic transport depends strongly on the Si NC size as well [1, 17, 18].

The surface properties of silicon NCs and QDs, and their interplay with quantum confined effects are still unclear even for porous Si that was intensively studied in the last two decades. It is essential that many scientific groups have reported that the PL energy in air does not increase more than 2.1 eV even when the Si NC size falls down below 3 nm [5, 6, 18-21]. The stabilization of emission energy at 2.1 eV during the oxidation of Si QD surface in the first approach is attributed to the change of recombination mechanism: from the one dealing with free exciton emission in QDs with optical gap enhanced due to the quantum confinement, to the carrier recombination via the oxygen-related interface states or via oxide-related defects inside of the silicon oxide on the NC surface [1, 5, 6, 18-21].

The other approach deals with an idea that the surface oxidation process and appearance of double Si=O bonds causes a significant modulation or even closing an optical band gap of Si QDs with respect to that in hydrogen saturated nanoclusters [22]. The authors suggested that when a Si QD is passivated by hydrogen, the recombination is via free excitons and follows the quantum confinement model for all sizes. If the QD is passivated by oxygen, a stabilized electronic state may be formed on the double Si=O bond. The trapped electron state is a p-state localized on the Si atom of the Si=O bond and the trapped hole state is a p-state localized on the oxygen atom [22]. As a result in big QDs (3-5nm) the recombination is via free excitons and PL energy follows the quantum confinement model. In QDs with the size 1.5-3.0 nm the recombination involves a trapped electron and a free hole and the variation of PL energy is slowed down, since the energy of trapped electron state is size independent. In last case the PL energy remains constant even if the NC size decreased [22].

The first results of *ab initio* calculation of the effect of different surface passivants on the optical gap of Si QDs with diameters up to 2 nm were presented in [23]. The authors used the density function theory (DFT), based on the local density approximation (LDA), the generalized gradient approximation (GGA) using a pseudopotential plane wave approach, and quantum Monte Carlo (QMC) techniques [23]. After performing DFT geometry optimizations and HOMO-LUMO gap calculations, QMC calculations were carried out to verify that the DFT trends for the HOMO-LUMO gap as a function of size and surface chemistry are predictive [23]. In the case of oxygen, the optical gap reduction computed as a function of QD sizes provided a consistent interpretation of several recent experiments.

It was shown that the molecular energy levels of $Si_{35}H_{35}X$ complexes, where X is: H, OH, Cl, or F, are nearly identical, with a 0.1 to 0.2 eV variation on the gap. Thus the effect on the gap of single bonded passivants such as a hydroxyl group and the elements H, F and Cl is negligible. Three passivants forming double bonds with the surface (O, S, and CH_2), have similar energy levels (very different from the single bonded passivants). The effect of double bonded passivants such as S and CH_2 is similar to that of oxygen. A large closing of the LDA gap from 3.4 to 1.8 eV occurs when a double bonded sulfur atom is used to passivate the surface [23]. Thus it was demonstrated that specific surface chemistry must be taken into account in order to quantitatively explain the optical properties of Si QDs.

The quantum yield of Si QD structures has been generally reported equal to a few percents or less [24, 25]. Recently it was shown that a further improvement of the optical properties can be achieved via a careful passivation of the Si QD surface [26,27]. There is information that Si QD ensemble quantum yields of 23% [28] and 30% [29] have been achieved by terminating the surface of Si QDs with covalently bonded organic molecules. Other approaches for the Si QD emission improvement related to: using the silicon nitride matrix, the interaction with surface plasmons excited in metallic nanoparticles, the join of QDs with rare earth atoms or with photonic structures, have been discussed in [1].

Si QD based light emitting devices

The investigations related to the design of LEDs based on porous Si appeared just after the discovery of the room temperature PL of PSi [30]. Electroluminescence (EL) of these devices strongly depended on annealing conditions in different ambient: pure nitrogen or diluted oxygen. Under direct bias, EL was detected at voltages of ~2 V and current densities of ~1 mA/cm^2 [30]. This type of LED suffers from a low external quantum efficiency of less than 10^{-3}-1% [30, 31].

An interesting technical design for the enhancement of light emission efficiency using a periodic micron-scale rugged surface pattern on the Si-QD LED structures with the SiNx matrix was studied in [32]. Si-QD LEDs with the *a*-SiN_x /ITO/*n*-Si in *a*-SiN_x / *p*-Si structure were prepared (Fig.1a). Micron-scale surface patterns were fabricated on the top layer (Fig.1b) of Si-QD LEDs to increase the extraction of light from the active layer [32].

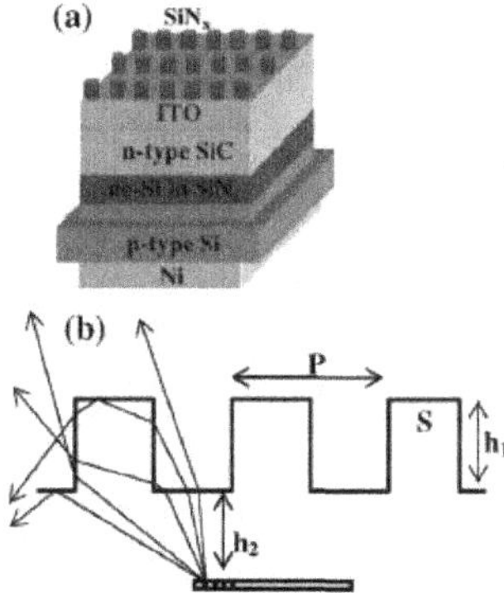

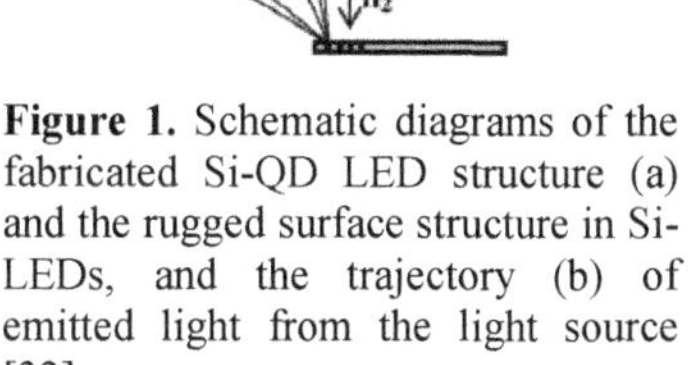

Figure 1. Schematic diagrams of the fabricated Si-QD LED structure (a) and the rugged surface structure in Si-LEDs, and the trajectory (b) of emitted light from the light source [32].

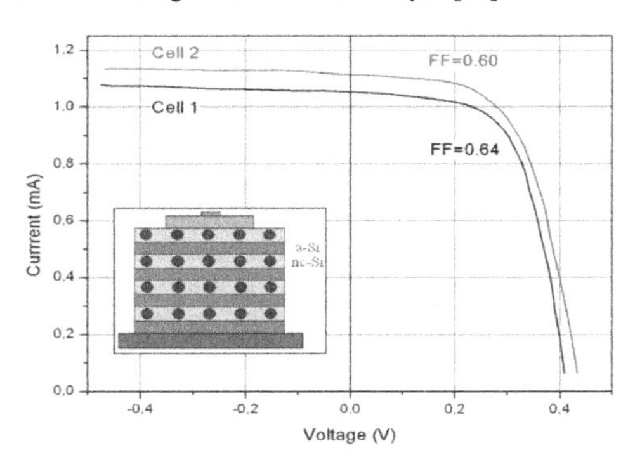

Figure 2. Illuminated I–V curves for two solar cells made with superlattice structures. The cell 1 had amorphous layer grown for 60s, and the cell 2, for 180 s. Insert represents the design of superlattice cell employing alternating layers of amorphous Si (a-Si) and nanocrystalline Si (nc-Si) [39].

The authors [33] proposed to couple the radiative QD dipoles to metal (Ag) nanoparticles located on the surface of SiOx layer with Si QDs that provide an opportunity to enhance the Si QD radiative decay rate. It was shown that Si QDs exhibited up to sevenfold luminescence enhancement in the presence of Ag nano-island arrays at the emission frequencies corresponding to the collective dipole plasmon resonance frequency of the Ag island array. Enhanced radiative emission was attributed to electromagnetic coupling between the semiconductor QD active dipole emitters and surface plasmons in the metal particles [33].

The fabrication of extremely efficient Si-based LEDs on the base of MOS structures with Er implanted in the thin gate oxide was discussed in [34]. Devices exhibited strong electroluminescence with the peak at 1.54 μm at 300 K and with 10% external quantum efficiency have been discussed [34]. MOS structures with Tb-and Yb-doped SiO_2 gates were prepared as well with EL at 540 nm (Tb) and 980 nm (Yb) at 300 K and an external quantum effciency of 10% and 0.1%, respectively [34].

Si QD based solar cells

Nanocrystalline silicon (*nc*-Si:H) films, that are silicon NCs or QDs embedded in amorphous hydrogenated silicon (α-Si:H) matrix, are considered as a promising candidate for the low cost and high efficient solar cells. During last thirty years α-Si:H became one of the most mature photovoltaic material. Significant progress has been made in the area of multi-junction α-Si:H solar cells (tandem), and the stabilized efficiencies of 11.2 and 10.2% have been reported for small-area cells and large-area panels, respectively [35]. Besides α-Si:H films other Si-based materials could be also used for solar cell applications. In dependence on specific tasks, polycrystalline silicon (poly-Si:H) [36, 37], microcrystalline silicon (μc-Si:H) [38] or nanocrystalline silicon (nc-Si:H) (Fig.2) [39, 40] were considered as alternative materials. The problem of the collection of incoming light could be solved by increasing of the absorption ability of layers. In this case the creation of active layers on the combination of α-Si:H and nc-Si:H allow to enlarge the wavelength range of absorption light and to increase the efficiency of solar cells.

Tandem solar cells permit to increase efficiency due to a multi-band gap approach. The radiative efficiency limit for a single junction Si crystalline cell is 29% and it increases to 42.5% and 47.5% for 2-cell and 3-cell tandem cells, respectively [41]. For an AM1.5 (air mass 1.5) solar spectrum the optimal band gap of the top solar cell for a 2-cell tandem is in the range 1.7 - 1.8 eV with a crystalline Si bottom cell. In the case of 3-cell tandem the optimal band gaps are 1.5 eV and 2.0 eV for the middle and upper cells, respectively [41]. The authors [41] have considered the thin film 2-cell tandem based on Si QD structures, which permits to increase solar cell efficiency and to satisfy the thin film low cost deposition, required for third generation photovoltaic devices [41].

Other type of Si QD solar cells based on the p-i-n structure: n++- type polysilicon/non-doped oxygen containing Si QD superlattice (Si-QDSL)/p-type hydrogenated amorphous silicon/ aluminum electrode on a quartz substrate have been considered in [42]. It is shown that the open-circuit voltage, fill factor and photocurrent density in these solar cells are 518 mV, 0.51 and 14.3 mA/cm2, respectively, and the conversion efficiency of 3.8 % can be achieved [42].

Si NC based memory devices

Silicon NC based structures are studied recently for non-volatile memory purposes [43]. This is the metal-insulator-semiconductor (MIS) structure containing semiconductor nanocrystals in the insulator layer, where NCs serve as charge storage media replacing the floating gate in conventional memory field effect transistors (FETs) [44].

The main goal of the wide research activity in the field of MIS structures with embedded semiconductor NCs is to overcome the difficulties of floating gate non-volatile memory devices (used e.g. in flash memories) [43]. The reduction of lateral dimensions requires the reduction of voltage level used, and finally the thickness of insulator layers in memory FETs. One of the problems in floating gate FETs is that through defects of tunnel oxide with reduced thickness the whole amount of stored charge can be lost [43]. The other problem of floating gate FETs is the drain turn-on effect [43]. It is connected with strong capacitive coupling between the drain and floating gate, and between the source and floating gate [43]. These disadvantages of floating gate FETs can be decreased by replacing a floating gate with separated semiconductor NCs [43], which are electrically isolated.

Si QD based single electron devices

Single electron tunneling and charging effects in Si QDs as well as their potential applications in single electron devices have been extensively studied, not only as physical phenomena in nanostructures, but also as operating principles for future integrated circuits [45]. To make the nanodevices operating at room temperature, the dot size has to be scaled down to several nanometers to guarantee the Coulomb blockade energy and quantum confinement energy to be larger than the thermal vibration energy.

Recently, the single electron effects for the individual Si QDs have been measured by using the scanning tunneling microscopy (STM) [46]. The studied structures of Si QDs with the QD density 2 x 10^{12} cm^{-2} were elaborated by low-pressure chemical vapor deposition on the 1.2 nm thick SiO_2 layer grown on a Si substrate. A double barrier junction (insert in Fig. 4) was created: a first barrier at the gap between the tip and a Si QD, and a second one, between the QD and a substrate separated by a 1.2 nm thick silicon dioxide layer [46]. I-V characteristics show an exponential behavior, as it is expected for a single tunnel barrier. Features appear when the I-V measurements are performed with the tip located just above a Si QD (Fig. 3, curve a, b).

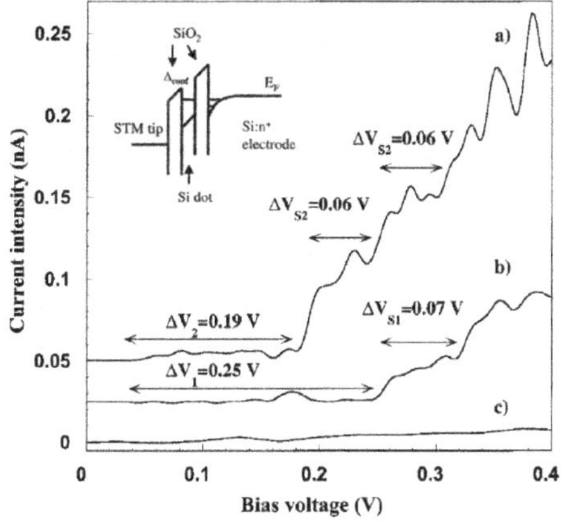

Figure 3. I-V characteristics taken on (a) 4.7 nm Si NCs at 30 K, (b) 4.5 nm Si NCs at 300 K, and (c) 1.2 nm tunnel oxide without Si NCs. Insert present band structure of the double tunnel barrier junction formed by STM tip above Si NCs and 1.2 nm thick oxides [59].

It is exhibited (i) a threshold voltage around 0.2 V and (ii) staircase structures on which the current variations are superposed. Discrete current jumps of 0.05 and 0.02 nA are observed that were interpreted in terms of Coulomb effects [46].

The peaks observed in the I-V curves (Fig. 3) were attributed to resonant tunneling between the coupled states in electrodes and in NCs. The value of Coulomb gap ΔV_{cb}, extracted from the experimental I-V characteristics of NCs with different sizes from the range of 3.8 nm up to 6.0 nm, decreases from 0.30 down to 0.10 V [46]. Presented experimental results related to the properties of a single NC (or QD) located between two barriers from thin oxides, or the large array of Si NCs embedded in thick oxide. There is other interesting and perspective area of the research dealing with electron transport in few lateral quantum dots in series. In [47] it has been shown that double QD systems are real candidates for developing the solid state quantum bits.

Spintronics with QDs

The great number of spin-related investigations in quantum confined nanostructures with long spin relaxation (dephasing) times demonstrates that the spin of electrons offers new mechanisms for the information processing and transmission [48,49]. In quantum confined structures the dephasing times are microseconds and the distance for spin transportation coherently approaches micrometers [48]. The miniaturized spintronics is developing towards single spins, so called as single-spintronics, which is the interplay between spin dependent transport and single electron physics.

The number of electrons inside a QD is possible to control due to the electrostatic repulsion between the electrons. Strong magnetic fields due to the Zeeman effect separate the states with opposite spins and perform a condition for tunneling out the QD for only one type of spin. Electron spins in silicon, in particular, are excellent candidates for quantum information processing and for the spintronics application. The charge of electrons is manipulated by charged gates easily, and the spin degree of freedom is well isolated from the charge fluctuations that lead to good spin qubit stability or quantum coherence properties.

Additionally silicon is characterized by small spin-orbit coupling and the existence of a nuclear spin-zero ^{28}Si isotope permits to achieve long single spin coherence times in silicon QDs. The fabrication and measurement of a top-gated QD occupied by a single electron in a Si/SiGe heterostructure have been presented in [50]. It was shown that the transport through QDs was directly correlated with a charge sensing from an integrated point contact that permitted to confirm the single-electron QD occupancy [50]. One of the main problems for the realization of quantum computation with QDs deals with the formation of a single-qubit gate. The monitoring of single spins in semiconductors has not been realized yet.

CONCLUSIONS

The history of emission study in Si quantum dots (QDs) in the last two decades have been presented together with the analysis of Si QD recombination transitions and the different ways of the emission stimulation in Si QD structures, related to the element variation for the passivation of surface dangling bonds, the stimulation by plasmon induced emission, the activation by the rare earth impurities etc... The different applications of Si QD structures in quantum electronics, such as: Si QD light emitting diodes, Si QD single union and tandem solar cells, Si QD memory structures, Si QD based one electron devices and double QD structures for spintronics, have been discussed as well.

ACKNOWLEDGMENTS

The authors would like to thank the CONACYT (project 130387) and SIP-IPN, Mexico, for the financial support.

REFERENCES

1. T.V. Torchynska, " Nanocrystals and quantum dots. Some physical aspects" in the book "Nanocrystals and quantum dots of group IV semiconductors", Editors: T. V. Torchynska and Yu. Vorobiev, American Scientific Publisher, 2010, 1-42.
2. S. Furukawa, and T. Miyasato, Jpn. J. Appl. Phys., Part 2 27, L2207, (1988).
3. A.G. Cullis, and L. T. Canham, Nature, 335, 335 (1991).
4. V. Lehmann, U. Gosele, Appl. Phys. Lett. 58, 856 (1991).
5. S.M. Prokes, Appl. Phys. Lett. 62, 3244 (1993).
6. Y. Kanemitsu, Sh. Okamoto, Phys. Rev. B 56, R1696 (1997).
7. S. Schuppler, S.L. Friedman, M.A. Marcus, D.L. Adler, Y.H. Xie, F.M. Ross et al. Phys. Rev. B 52, 4910 (1995).
8. T.V.Torchynska, L.I. Khomenkova, N.E. Korsunska, M.K. Sheinkman, Physica B. Conden. Mat. 273-274, 955-958 (1999).
9. T.V. Torchinskaya, J. Aguilar Hernandez, L. Schacht Hernandez, G. Polupan, Y. Goldstein, A. Many, J. Jedrzejewski, A. Kolobov, Microelect. Engineer. 66, 83-90 (2003).
10. K.A. Jeon, J.H. Kim, J.B. Choi, K.B. Han, S.Y. Lee, Mater. Sci. Eng. B 23 (2003) 1017.
11. T.V. Torchynska, A.Vivas Hernandez, M. Dybiec, Y.Emirov, I. Tarasov, S. Ostapenko, Y. matsumoto, phys. stat.solid. (c). 2 (6), 1832-1836 (2005).
12. T.V. Torchynska, A. Diaz Cano, M. Dybic, S. Ostapenko, M. Mynbaeva, Physica B, Conden. Mat. 376-377, 367-369 (2006).
13. M. Dybiec, S. Ostapenko, T. V. Torchynska, E.V. Lozada, Appl. Phys. Lett. 84 (25), 5165-5167 (2004).
14. T.V. Torchynska, M. Morales Rodriguez, F.G. Becerril-Espinoza, N.E. Korsunskaya, L.Yu. Khomenkova, L.V. Shcherbyna, Phys. Rev.B 65, 115313 (2002).
15. T.V. Torchynska, J. Appl. Phys. 92, 4019 (2002).
16. T. Torchynska, J. Aguilar-Hernandez, A.I.Diaz Cano, G. Contreras-Puente, F.G. Becerril Espinoza, Yu.V. Vorobiev, Y. Goldstein, A. Many, J. Jedrzejewski, B.M. Bulakh and L.V. Scherbina, Physica B, Conden. Mat. 308-310, 1108-1112 (2001).
17. T.V. Torchynska, Opto-electronics Review, 2, 121-130 (1998)
18. N.E. Korsunskaya, T.V. Torchinskaya, B.R. Dzhumaev, L.Yu. Khomenkova, B.M. Bulakh, Semiconductors, 31, 773 (1997).
19. Y. Kanemitsu, H. Uto, Y. Masumoto, T. Matsumoto, T. Futagi, H. Mimura, Phys. Rev. B, 48, 2827 (1993).
20. G. G. Qin, X. S. Liu, S. Y. Ma, J. Lin, G. Q. Yao. X. Y. Lin, K. X. Lin, Phys. Rev. B, 55, 12876 (1997).
21. T. V. Torchynska, N. E. Korsunskaya, L. Yu. Khomenkova, B. R. Dzhumaev, S. M. Prokes, Thin Solid Films, 381/1, 88 (2001).
22. M. V. Wolkin, J. Jorne, P. M. Fauchet, G. Allan, C. Delerue, Phys. Rev. Lett. 82, 197 (1999).
23. A. Puzder, A.J.Williamson, J. C. Grossman, G. Galli, Phys. Rev. Lett.88, 097401 (2002).
24. K.A. Littau, P.J. Szajowski, A.J. Muller, A.R. Kortan, and L.E. Brus, J. Phys. Chem. 97, 1224 (1993).
25. J.P. Wilcoxon, G.A. Samara and P.N. Provencio, Phys. Rev. B 60, 2704 (1999).
26. G. Ledoux, J. Gong, F. Huisken, O. Guillois and C. Reynaud, Appl. Phys. Lett. 80, 4834 (2002).
27. T.V. Torchynska, "Si and Ge quantum dot structures" in the book "Nanocrystals and quantum dots of group IV semiconductors", Editors: T. V. Torchynska and Yu. Vorobiev, American Scientific Publisher, 2010, 42-84.
28. J.D. Holmes, K.J. Ziegler, C. Doty, L.E. Pell, K.P. Johnston, B.A. Korgel, J. Am. Chem. Soc. 123, 3743 (2001).
29. R.M. Sankaran, D. Holunga, R.C. Flagan, K.P. Giapis, Nano Lett. 5, 531 (2005).

30. L. Tsybeskov, S.P. Duttagupta, K.D. Hirschman, P.M. Fauchet, Appl. Phys. Lett. 68, 2058 (1996).
31. M.E. Castagna, A. Muscara, S. Leonardi, S. Coffa, L. Caristia, C. Tringali, S. Lorenti, J. Lumin. 121, 187 (2006).
32. K.H. Kim, J.H. Shin, H.M. Park, Ch. Huh, T.Y. Kim, K.S. Cho, J.Ch. Hong, G.Y. Sung, Appl. Phys. Lett. 89, 191120 (2006).
33. J. S. Biteen, N. S. Lewis and H. A. Atwater, Appl. Phys. Lett. 88, 131109 (2006).
34. M.E. Castagna, S. Coffa, M. Monaco, L. Caristia, A. Messina, R. Mangano, C. Bongiorno, J. Phys. E 16, 547 (2003).
35. L. Yu. Khomenkova, "Si nanocrystals and quantum dots embedded in amorphous Si matrix", in the book "Nanocrystals and quantum dots of group IV semiconductors", Editors: T. V. Torchynska and Yu. Vorobiev, American Scientific Publisher, 2010, 85-112.
36. R. Iiduka, A. Heya, H. Matsumura, Solar Energy Materials and Solar Cells, 48, 279 (1997).
37. H. Meiling, A.M. Brockhoff, J.K. Rath, R.E.I. Schropp, J. Non-Cryst. Solids, 227–230, 1202 (1998).
38. A.V. Shah, J. Meier, E. Vallat-Sauvain, N. Wyrsch, U. Kroll, C. Droz, U. Graf, Solar Energy Materials and Solar Cells, 78, 469 (2003).
39. V. L. Dalal, A. Madhavan, J. Non-Cryst. Solids, 354, 2403 (2008).
40. M.A. Green, Third Generation Photovoltaics: Advanced Solar Energy Conversion, Springer, 2003.
41. G. Conibeer, M. Green, E.-Ch. Cho, D. König, Y.-H. Cho,T. Fangsuwannarak, G. Scardera, E. Pink, Y. Huang, T. Puzzer, Sh. Huang, D. Song, Ch. Flynn, S. Park, X. Hao, D. Mansfield Thin Solid Films, 516, 6748 (2008)
42. Shigeru Yamada, Yasuyoshi Kurokawa, Shinsuke Miyajima, Akira Yamada, and Makoto Konagai, Proc. IEEE 35th PVSC, No. 5617097, 766 (2010).
43. Zs.J. Horváth, P. Basa, "Nanocrystal memory structures" in the book "Nanocrystals and quantum dots of group IV semiconductors", Editors: T. V. Torchynska and Yu. Vorobiev, American Scientific Publisher, 2010, 225-252.
44. P. Normand, E. Kapetanakis, P. Dimitrakis, D. Skarlatos, K. Beltsios, D. Tsoukalas, et al. Nucl. Instr. and Meth. B, 216, 228 (2004).
45. I. Antonova, "Electrical properties of semiconductor nanocrystals and quantum dots in dielectric matrix", in the book "Nanocrystals and quantum dots of group IV semiconductors", Editors: T. V. Torchynska and Yu. Vorobiev, American Scientific Publisher, 2010, 149-187.
46. T. Baron, P. Gentile, N. Magnea, P. Mur, Appl. Phys. Lett., 79, 1175 (2001).
47. W.G. van der Wiel, S. De Franceschi, J.M. Elzerman, T. Fujisawa, S. Tarucha, L. Kouwenhoven, Rev. Mod. Phys. 75, 1, (2003).
48. J.M. Kikkawa, I.P. Smorchkova, N. Samarth, D.D. Awschalom, Science 277, 1284 (1997).
49. V. Cerletti, W.A. Coish, O. Gywat and D. Loss, Nanotechnol. 16, R27 (2005).
50. C.B. Simmons, M. Thalakulam, N. Shaji, L.J. Klein, H. Qin, R.H. Blick, Appl. Phys. Lett. 91, 213103 (2007).

Mater. Res. Soc. Symp. Proc. Vol. 1534 © 2013 Materials Research Society
DOI: 10.1557/opl.2013.292

Quantum mechanical description of Electronic transitions in Cylindrical nanostructures, including pores in semiconductors

Yuri V. Vorobiev[1], Pavel Horley[2] and Jesus González-Hernández[2]
[1]CINVESTAV-Querétaro, Libramiento Norponiente 2000, Fracc. Real de Juriquilla, CP 76230 Querétaro, QRO., México.
[2]CIMAV Chihuahua/Monterrey, Avenida Miguel de Cervantes 120, CP 31109, Chihuahua, CHIH., México

ABSTRACT

Cylindrical nanostructures NS (namely, nanowires and pores) with rectangular or circular cross-section are analyzed using Mirror Boundary Conditions (MBC) in solution of the Schrödinger equation. The MBC are formulated as equivalence of the module of electron's Ψ-function in an arbitrary point inside the NS and its images in NS's walls treated as mirrors. Thus the two types of MBC – odd (OMBC) and even (EMBC) could be applied, when Ψ-functions in real point and its images are equated with the opposite or the same sign correspondingly. The first case gives zero value of Ψ-function at NS's boundaries and therefore corresponds to the strong quantum confinement, whereas the second (non-zero Ψ-function at the boundary) gives a weak confinement. The analytical expressions for energy spectra of electron in a NS found for all cases examined contain no adjustable parameters, and show reasonable agreement with experimental data found in the literature.

INTRODUCTION

Nanostructures (NS) of different kind have been actively studied during the last two decades, both theoretically and experimentally; a special interest was focused on quasi-one-dimensional NS such as nanowires, nanorods and elongated pores that can not only modify the main material's parameters, but can introduce totally new characteristics like optical and electrical anisotropy, birefringence etc. All these elongated NS can be approximated as cylinders with different shape of cross-section.

Theoretical treatment of NSs is based on the solution of the Schrödinger equation, usually within the effective mass approximation [1-4], although for small NS its application could be questioned. An important element of this description is the boundary conditions; the traditional *impenetrable wall* conditions (i) are not realistic, and (ii) in many cases could not be written in simple analytical form making the analysis quite difficult. Recently [5-7] we introduced a new - mirror boundary conditions (**MBC**) assuming that electron in a NS is reflected by NS's walls acting as mirrors; it is obvious that this assumption favors the effective mass approximation.

Additional advantage of this approach is the possibility to treat pore as "inverted" NS – a void surrounded by semiconductor material – considering the "reflection" of particle wave function from the pore's boundary. Thus, essentially the same solution of the Schrödinger equation (and the energy spectrum) will describe a pore and NS of the same geometry and size.

In our approach, the boundary condition equalizes absolute values of the particle's Ψ-function in an arbitrary point inside the NS and the corresponding image point with respect to a mirror-reflective wall. Thus, depending on the sign of the equated Ψ values, even and odd mirror boundary conditions could exist. For the case of **odd mirror boundary conditions** (**OMBC**, Ψ-functions in real point and its images are equated with the opposite sign) the incident and reflected waves de Broglie cancel each other at the boundary, so this case is equivalent to that of impenetrable walls with zero Ψ-function at the boundary (''strong'' confinement case).

However, many experimental data ([4], for example) show that it is not always so – there is a possibility that a particle may penetrate the barrier, returning later into the confined volume. Thus, the wave function will not vanish at the boundary, and the system can be considered as a ''weak'' confinement. This case corresponds to **even mirror boundary conditions (EMBC)**: Ψ-function in real point and its images have the same sign. Below we analyze solutions of the Schrödinger equation for several structures, using mirror boundary conditions of both types, and making comparison of the energy spectra obtained with experimental data found in the literature.

THEORY

We start with the simplest case that is easily treated on the basis of traditional approach – a NS shaped as **rectangular prism with square base** (the sides $a = b$ correspond to the axes "x" and "y", the side $c > a$ is oriented along "z" direction). Assuming, as usual, the potential inside the NS equal to zero and using the separation of variables, we look for the solution of the stationary Schrödinger equation $\Delta\Psi + k^2\Psi = 0$ (where $k^2 = 2mE/\hbar^2$, m being particle's effective mass) as the product of plane waves in both directions along all the coordinate axes

$$\Psi = \prod_j \Psi_j(x_j) = \prod_j (A_j \exp(ik_j x_j) + B_j \exp(-ik_j x_j)) \tag{1}$$

The **even mirror boundary conditions (EMBC)** for this case are (see [7])

$$\Psi(x,y,z) = \Psi(-x,y,z) = \Psi(x,-y,z) = \Psi(x,y,-z) = \Psi(2a-x,y,z) = \Psi(x,2b-y,z) = \Psi(x,y,2c-z) \tag{2}$$

Application of them to the solution (1) gives the final solution

$$\Psi(x,y,z) = A\cos k_x x \cos k_y y \cos k_z z \tag{3}$$

The wave vector components are $k_x a = \pi n_x$, $k_y b = \pi n_y$ and $k_z c = \pi n_z$

It gives the energy spectrum:

$$E = \frac{h^2}{8m}\left(\frac{n_x^2}{a^2} + \frac{n_y^2}{b^2} + \frac{n_z^2}{c^2}\right) = \frac{h^2}{8m}\left(\frac{n_x^2 + n_y^2}{a^2} + \frac{n_z^2}{c^2}\right) \tag{4}$$

The **odd mirror boundary conditions (OMBC)** will be obtained from (2) by taking the sign "–" for the first function. The solution then will be

$$\Psi(x,y,z) = B\sin k_x x \sin k_y y \sin k_z z \tag{5}$$

The wave vector components will again be given by relations above leading to the same spectrum (4). Using the traditional **impenetrable walls** boundary conditions, the solution in the form (5) will be obtained (it coincides with the solution for OMBC case because the Ψ-function is zero at the boundary), with the same wave vector and the energy spectrum. So we see that for this simple geometry of NS, the energy spectrum is the same for both type of **MBC** (although the solutions are not the same), and is the same as the spectrum obtained with impenetrable walls. In [7] we demonstrated that for NS of spherical shape the energy spectrum found with EMBC (weak confinement) is different from that corresponding to impenetrable walls conditions.

From (4) it is seen that the energy spectrum of prismatic (cylindrical) NS is a sum of the spectrum of the 2-dimensional cross-sectional NS and the one-dimensional wire of length c. In similar manner, the spectrum for cylinders with the other cross-section shape could be built, for example, using solutions for 2-dimensional triangular or hexagonal structures analyzed previously [5, 6]. Below we give analysis of the cylindrical NS with circular cross-section.

We consider **cylindrical NS** with a circle of diameter a in cross-section and the length c; for treatment made in traditional approach see [8, 9]. Here the variable separation can be used, in cylindrical coordinates:

$$\Psi\,(r,\varphi,z) = AF(r)\exp(ip\varphi)[B\exp(ikz) + C\exp(-ikz)],\ p - \text{integers } (0, \pm1, \pm2,\ldots). \tag{6}$$

We note that the value of p defines the angular momentum: $L = p\hbar$.

In the case of weak confinement (EMBC), we use the condition of mirror reflection from the base, which gives $B = C$, so that the function becomes

$$\Psi\,(r,\varphi,z) = AF(r)\exp(ip\varphi)\cos kz. \tag{6A}$$

Strong confinement (OMBC) gives $B = -C$, so there will be $\sin kz$ instead of $\cos kz$ in (7A).

The radial function $F(r)$ s the solution of the radial equation

$$\frac{d^2F(r)}{dr^2} + \frac{1}{r}\frac{dR}{dr} + \left(k^2 - \frac{p^2}{r^2}\right)F(r) = 0 \tag{7}$$

It is Bessel's differential equation in the variable kr, and its solution is represented by cylindrical Bessel function of the integer order $|p|$: $J_{|p|}(kr)$, here $k = \hbar^{-1}(2mE_n)^{1/2}$, m is particle's effective mass, E_n – quantized kinetic energy corresponding to the motion in 2-dimensional circular NS. The total energy will include that of the motion along axis z: $E = E_n + E_z$.

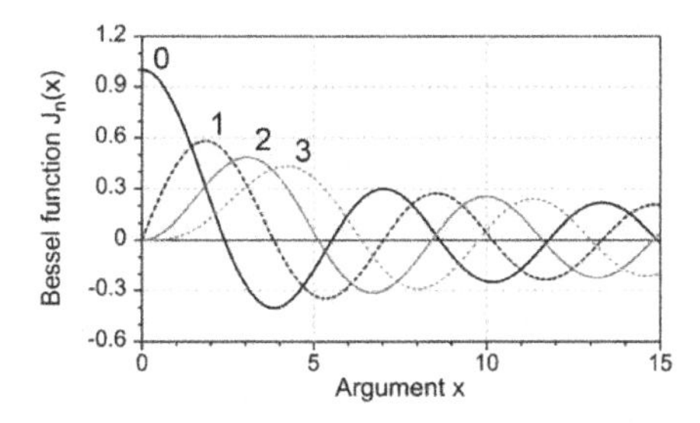

Figure 1. Cylindrical Bessel functions $J_n(x)$. Curve numbers correspond to order n.

The energy E_n (defined by the values of k) is determined by the boundary conditions. In traditional treatment (impenetrable boundary), the F-function value at the boundary is zero, thus the energy values are determined by the roots (nodes) of the cylindrical Bessel function (see Fig. 1 where the function's order is denoted by n, and Table 1). The same situation will be in case of the wall-mirrors, if the **OMBC** are applied: again, zero of the function at the boundary, and the nodes $q_{|p|i}$ define the energy.

If the **EMBC** are used, situation is different since the function values close to the wall have to be equal from both sides: thus the boundary corresponds to the extremes of this function (we have strictly proved this point in analysis of spherical QDs [7]).

The Table 1 gives several values of the Bessel function argument (kr) corresponding to the function nodes ($q_{|p|i}$) and extremes ($t_{|p|i}$), for function order $|p|$ equal to 0, 1, 2 and 3.

Table 1. Argument values at nodes and extremes of cylindrical Bessel function.

$\|p\|$	$q_{\|p\|1}$	$t_{\|p\|1}$	$q_{\|p\|2}$	$t_{\|p\|2}$	$q_{\|p\|3}$	$t_{\|p\|3}$	$q_{\|p\|4}$	$t_{\|p\|4}$
0	2.4	**0**	5,5	**3.713**	8.5	**7.10**	11.6	**10.15**
1	0	**1.625**	3.7	**5.375**	6.9	**8.55**	10.25	**11.6**
2	0	**2.92**	5.11	**6.775**	8.4	**10.0**	11.65	**13.15**
3	0	**4.325**	6.4	**8.1**	9.85	**11.4**	13.2	**14.2**

At the boundary $r = a/2$, therefore the corresponding value of k is $2q_{|p|i}/a$ for OMBC and $2t_{|p|i}/a$ for EMBC. Then the energy spectrum for circular motion is:

$$E_n = \frac{2\hbar^2}{ma^2} s_{|p|i}^2 = \frac{h^2}{2\pi^2 ma^2} s_{|p|i}^2 \tag{8}$$

Here the defining parameter $s_{|p|i}$ is $q_{|p|i}$ for OMBC case (strong confinement) and $t_{|p|i}$ for EMBC (weak confinement).

The quantization along "z" axis for both boundary conditions will be the same as in the previous case: $E_z = \frac{h^2}{8m}\frac{n_z^2}{c^2}$, with the total energy

$$E = \frac{h^2}{2m}\left(\frac{s_{|p|i}^2}{\pi^2 a^2} + \frac{n_z^2}{4c^2}\right) \tag{9}$$

In the case when conditions EMBC are applicable, the ground state energy GS will be determined by $t_{11} = 1.625$:

$$E_{GS} = (h^2/2m)(0.268/a^2 + 1/4c^2). \tag{10}$$

If the OMBC are correct, the GS will be determined by the smallest q-value of 2.4:

$$E_{GS} = (h^2/2m)(0.584/a^2 + 1/4c^2). \tag{10A}$$

The expressions (9-10A) could be used for analysis of optical processes in NSs: (10, 10A) give the blue shift in exciton ground state energy, if instead of m the reduced exciton mass is used. Using (9) it is possible to obtain the energies for the higher excites states.

For long enough NSs, pores in particular, when the value of c is sufficiently large, the second term in energy does not affect the GS. Thus the solution for cylindrical (circular) NS based on even mirror boundary conditions EMBC (weak confinement) gives the GS shift due to quantum confinement 2.2 times smaller (that is $(2.4/1.625)^2$) than that corresponding to strong confinement (OMBC and traditional approach with *impenetrable wall* conditions); in case of spherical QD [7], the difference was 4 times. It looks reasonable that in case of strong confinement, the blue shift is essentially larger than that for weak confinement.

The applicability of the OMBC or EMBC conditions (strong or weak confinement in our sense) ought to be determined by the conditions at the NS's boundary, first of all by the probability of electron's tunneling. One could expect that in case of isolated NS, a strong confinement (OMBC) approximation should be correct, whereas for NS surrounded by another solid or liquid media (such was the situation with core-shell QDs [7]; the pores in semiconductor could be similarly treated), a weak confinement with EMBC should be the better approximation.

DISCUSSION

There are many publications related to semiconductor nanorods (nanowires) and cylindrical pores, like arrays of cylindrical pores in sapphire [10], ZnO nanorods grown within these pores [11], In_2O_3 nanowires etc. However, the major part of these papers study relatively large NSs (30 nm or more); it is easy to see from (10, 10A) that the blue shift in these cases will be quite small (order of 0.01 eV or less, both for weak and strong confinement). Nevertheless, we found some publications referring to small nanorods for the confinement effect to be seen.

In [12], CdS nanorods with diameter of 5 nm and length 40 nm in liquid crystal were studied: the actual subject was optical anisotropy caused by the alignment of the nanorods, and to observe the anisotropy, the authors studied polarization of photoluminescence due to electron-hole recombination. The spectral maximum of luminescence corresponded to 485 nm (2.56 eV), which is by 0.14 eV above the band gap of the bulk CdS. The exciton reduced mass in CdS [14] $\mu = 0.134\ m_o$, and from (10) the blue shift is 0.12 eV, in reasonable agreement with experiment. Since the CdS NSs were surrounded with liquid crystal, we use the EMBC approximation.

Another example of cylindrical QDs is presented by circular organic molecules, like coronene $C_{24}H_{12}$ (Fig. 2); here $c << a$, and the second term in (9-10A) is very large even at $n_z = 1$ and does not affect the optical properties (transitions between states with different n_z correspond to ultraviolet). Thus the spectrum is defined by the first term in (9, 10), as in a long cylinder.

In [14] optical properties of coronene molecules in THF (tetrahydrofuran) solution were studied. Since the molecules are not isolated, we can expect that the weak confinement (EMBC case) will be appropriate. Strong absorption lines were registered at photon energies of 4.1 – 4.3 eV, with the weaker absorption down to 3.5 eV. To apply our consideration, the diameter a of corresponding cylinder (circle) must be calculated; it is diameter of the circle containing all 12 C atoms (Fig. 2) at the largest distance from the molecule's center.

The C-C bond length (side of a hexagon) in coronene is $d = 1.4$ Å, that gives $a = d\sqrt{28} =$ 0.741 nm. Taking in (10) m as free electron mass and using only the first term, we obtain the ground state energy $E_{GS} = 0.73$ eV. The higher energy states (9) are defined by the values of $s_{|p|i}$ $= t_{|p|i}$ equal to 2.92, 3.713, 4.30. The corresponding energies are 2.353, 3.805 and 5.1 eV, and the transition energies 1.62, 3.1 and 4.37 eV. The first of them is beyond the studied spectral range; the two other reasonably well account for the absorption observed. If we treat the case using the OM BC, instead of $t_{|p|i}$ in (9, 10A), the $q_{|p|i}$ -values should be used: 2.4, 3.7, 5.11, 5.5 etc. It gives the transition energies of 2.19, 5.62 and 6.76 eV that do not agree with the experiment.

Another paper [15] is devoted to studying of coronene-like nitride molecules (Fig. 3) with composition $N_{12}X_{12}H_{12}$ (X being B, Al, Ga or In). Depending on X, the bond length is changing thus giving the different values of a. The authors give the transition energies between ground state and the first excited state (HOMO-LUMO transition E_{HL}). For these isolated molecules, the strong confinement case (OMBC) will be appropriate; the Table 2 gives the bond length and E_{HL} values from [15] together with calculated values of a, and the transition energies ΔE found using our expression (9) with q-values in place of s. We see that our ΔE values are reasonably close to the E_{HL} values. The treatment based on EMBC give large discrepancies between E_{HL} and ΔE.

Figure 2. Coronene molecule

Figure 3. Coronene-like nitrides $N_{12}X_{12}H_{12}$ (X Pink, N blue, H white)

Table 2. The lowest transition energies in coronene-like molecules (see text).

Material	*d*, Å	*a*, nm	ΔE, eV	E_{HL}, eV [15]
BN	1.44	0.762	6.351	5.18
AlN	1.79	0.95	4.11	4.59
GaN	1.84	0.974	3.88	3.94
InN	2.06	1.09	3.1	2.33

CONCLUSIONS

Theoretical treatment of cylindrical nanostructures NSs, made with application of two types of mirror boundary conditions MBC in solution of the Schrödinger equation (even EMBC and odd OMBC) give simple and reasonable account of the optical properties of NSs. The energy spectra are defined by NS's shape and dimensions; EMBC corresponds to weak confinement and describes the cases when NS is embedded in another media (like pore), and the OMBC are more appropriate in treatment of isolated NSs where strong confinement exists.

REFERENCES

1. Al.L. Éfros, A.L. Éfros, Sov. Phys. Semicond. 16 (7), 772 (1982).
2. S.V. Gaponenko. Optical Properties of Semiconductor Nanocrystals, Cambridge University Press, Cambridge, 1998.
3. J.L. Liu, W.G. Wu, A. Balandin, G.L. Jin, K.L. Wang:, Appl. Phys. Lett. 74, 185 (1999).
4. B.O. Dabbousi, J. Rodriguez-Viejo, F.V. Mikulec, J.R. Heine, H. Mattoussi, R. Ober , K.F. Jensen, M.G. Bawendi, J. Phys. Chem. B. 101, 9463 (1977).
5. V.R. Vieira, Y.V. Vorobiev, P.P. Horley, P.M. Gorley, Phys. Stat. Sol. C. 5, 3802 (2008).
6. Y.V. Vorobiev, V.R. Vieira, P.P. Horley, P.N. Gorley, J. González-Hernández, Science in China Series E: Technological Sciences. 52, 15 (2009).
7. Y.V. Vorobiev, P.P. Horley, V.R. Vieira, Physica E. 42, 2264 (2010).
8. R.W. Robinett, Am. J. Phys. 64(4), 440 (1996).
9. L.A. Mel'nikov, A.V. Kurganov, Tech. Phys. Lett. 23(1), 65 (1997).
10. J. Choi, Y. Luo, R.B. Wehrspohn, et al., J. Appl. Phys. 94(4), 4757 (2003).
11. M.J. Zheng, L.D. Zhang, G.H. Li, W.Z. Shen, Chemical Physics Letters 363, 123 (2002).
12. K-Ju. Wu, K-Ch. Chu, Ch-Yu. Chao, Y.F. Chen, et al., Nano Letters. 7(1), 1908 (2007).
13. J. Singh, Physics of semiconductors and their heterostructures. McGrow-Hill, 1993.
14. J. Xiao, H. Yang, Z. Yin, J. Guo, et al., Journal of Materials Chemistry. 21, 1423 (2011).
15. E. Chigo Anota, M. Salazar Villanueva, et al., Physica Status Solidi C. 7, 2252 (2010).

Mater. Res. Soc. Symp. Proc. Vol. 1534 © 2013 Materials Research Society
DOI: 10.1557/opl.2013.293

Effect of Single-Wall Carbon Nanotubes Layer on Photoelectric Response of Au/Si Photovoltaic Structures

Nicholas L. Dmitruk[1], OlgaYu. Borkovskaya[1], Tetyana.S. Havrylenko[1], Sergey V. Mamykin [1], VladimirR. Romanyuk [1] and ElenaV. Basiuk [2]
[1]V. Lashkaryov Institute of Semiconductor Physics, National Academy of Sciences of Ukraine, Kyiv, 03028, Ukraine
[2]Centro de Ciencias Aplicadas y Desarrollo Tecnológico Universidad Nacional Autónoma de México, Circuito Exterior C. U., 04510, México D.F. Mexico

ABSTRACT

The effect of the deposited on Si substrate Single-wall carbon nanotube (SWCNT) nanolayers on optical, photoelectric and electrical properties of Au/n-Si structures has been investigated. Highly purified SWCNTs were prepared by the arc-discharge method. The significant enhancement of the photocurrent (increasing to the long-wave range) and the photoconversion efficiency was found for structures with SWCNT and its mechanism was analyzed with taking into account optical and electrical characteristics of structures.

INTRODUCTION

Single-wall carbon nanotubes (SWCNT) as well as the other low-dimensional carbon materials have attracted great interest for the applications in nanoelectronics and optoelectronics due to their unique physical properties [1]. Specifically, SWCNTs are proposed for the photovoltaic efficiency enhancement in organic photovoltaic devices due to the introduction in the polymer photoactive layer [2] or using them as conductive transparent electrode [3, 4] that can exceed the performance of indium tin oxide (ITO) material. The enhancement of the short-circuit photocurrent and photoconversion efficiency was found recently for the metal/n-Si structures with the fullerene C60 intermediate layer [5]. Unlike the fullerens, SWNTs are known to have characteristic electronic structure of spike like density of states (the van Hove singularities) due to the one-dimensional structure [6], exhibiting either semiconducting or metallic behavior. The energy gaps between spikes are determined by chiral indexes and diameters of SWNTs, the values and distribution of which are dependent on the method and parameters of SWNT synthesis [6, 7]. In its turn, the optical and electrical properties of SWCNT thin films deposited on semiconducting or dielectric substrates are greatly affected by both the methods of deposition and solvents (or surfactants) used for untangling bundles of SWCNTs and also by preliminary substrate processing, which improves adhesion of SWCNT. In this work we investigate the effect of the deposited SWNT nanolayers on the Si substrate on optical, photoelectric and electrical properties of Au/n-Si structures that can be perspective for photovoltaic applications in solar cells.

EXPERIMENTAL DETAILS

— Highly purified SWCNTs, synthesized by the arc-discharge method, were deposited on n-Si and on glass substrates.

— To improve adhesion of SWCNT to a substrate, the well-sonicated SWCNT-ethanol mixture was deposited on Si and glass surfaces modified with poly (vinil pyridine) (PVP)-ethanol solution, earlier used for metal nanoparticles immobilization [8]. Both dip- and drop-coating methods of SWCNT deposition were used, iterated some (5÷7) times. In the last case SWCNTs and bundles of them form the more dense net but the less uniform layer (Fig.1a)

— The photovoltaic structures were fabricated by evaporation of semitransparent Au layer through the opaque mask on both the pristine or PVP-modified Si surfaces and ones with SWCNT nanolayers prepared by dip- or drop-coating with SWCNT-ethanol mixture. The Au barrier contacts for Au/SWCNT/Si structures have both circular form with diameters of diodes equal to 0,5 mm and a form of grid used for solar cells. The In ohmic contact was fabricated on the back side of Si plate.

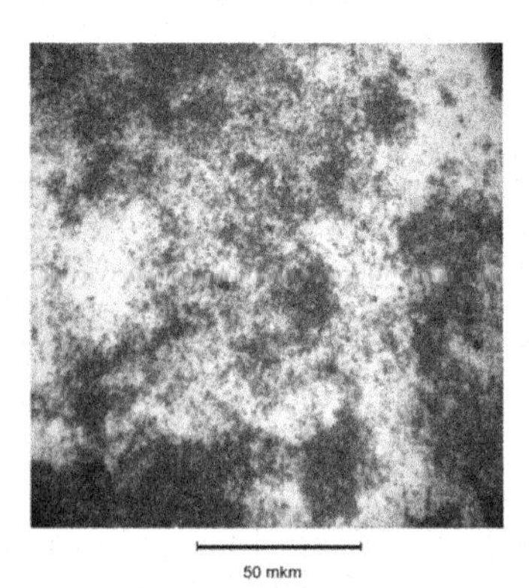

Figure 1. a) Image of SWCNT layer deposited by drop coating on glass.

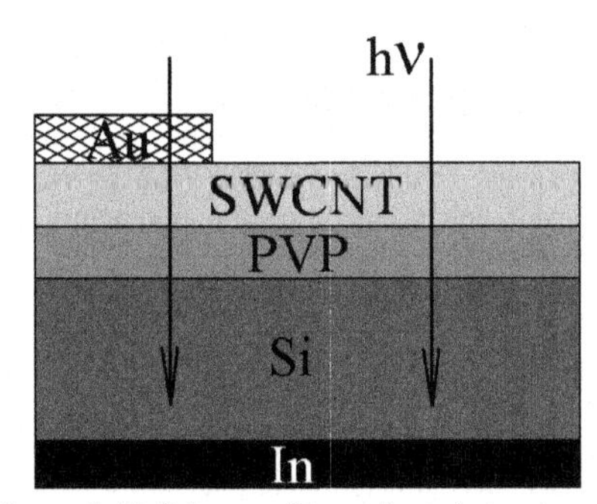

Figure 1. b) Scheme of investigated structures.

EXPERIMENTAL RESULTS AND DISCUSSION

Optical properties

The light transmittance spectra were measured for SWCNT films on glass substrates in the 0.4-1.88 μm spectral range (0.66-3.1 eV) by using Si and Ge calibrated photodetectors (Fig.2a). Fig.2b demonstrates the optical absorption spectra for SWCNT films deposited by dip (1) and drop (2) coating after the subtraction of background absorption due to the π plasmons and impurities. Some peaks are seen which can be related to transmission between the electronic densities of states singularities of SWCNT. S_{11} and S_{22} correspond to the first and second energy gaps between spikes of semiconducting tubes, and M_{11} corresponds to metallic ones with almost the same diameters. Evaluation of SWCNT diameters distribution according to Kataura graphics [6] showed that they grouped around 1,09÷1,15 and 0,82÷0,9 nm.

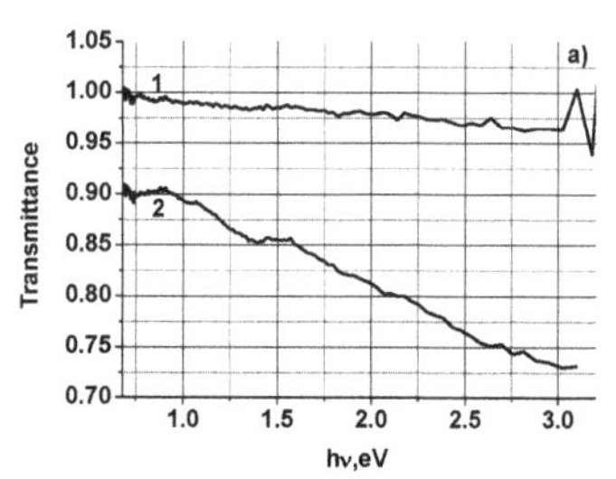

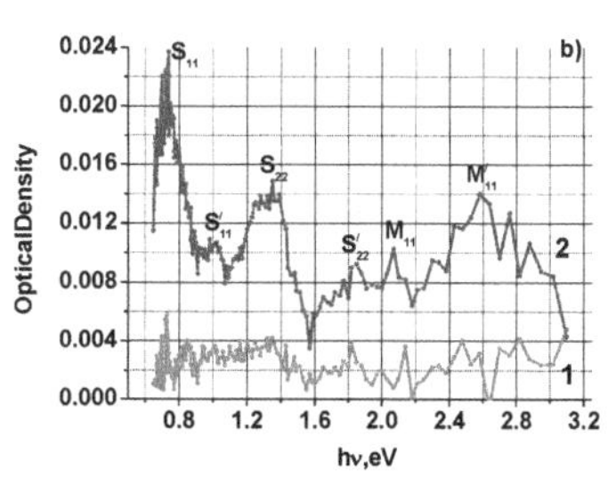

Figure 2. a) The light transmittance spectra of SWCNT films deposited by dip- (1) and drop (2) coating on the glass substrate.
b) The optical density spectra for the same films obtained from optical absorption spectra after subtraction of background absorption.

To estimate the change of the light transmittance spectra for the Si substrate due to SWCNT layers, both under the Au electrode and outside of it, the spectra of refractive index (n) and extinction coefficient (k) for SWCNT were determined from the SWCNT/Si reflectance spectra measured for p- and s-polarized light at a number of angles of incidence.

The satisfactory agreement of experimental transmittance spectra for the drop-deposited SWCNT layer on glass substrate with the calculated one taking into account determined n and k spectra and the effective SWCNT layer thickness (Fig.3a) allowed to calculate the transmittance spectra for the investigated Au/Si structures (Fig.3b). It is seen that the introduction of SWCNT layer somewhat decreased the light transmittance into Si substrate.

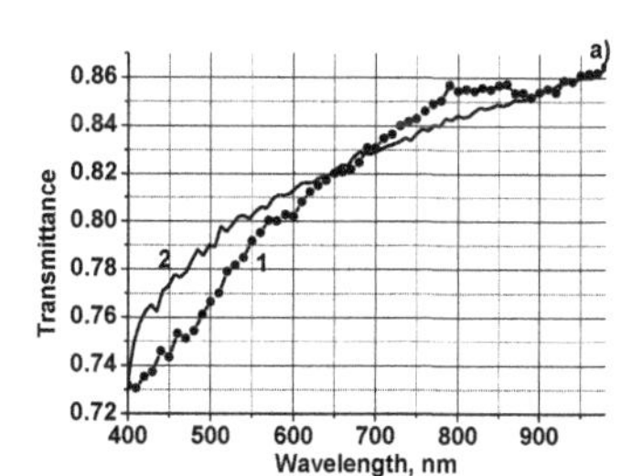

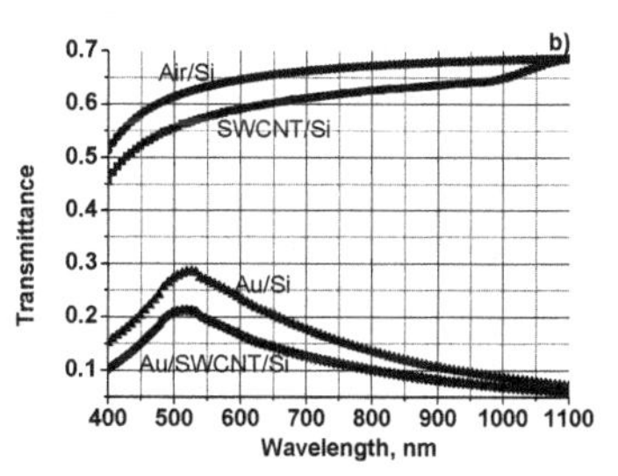

Figure 3. a) Experimental (1) and calculated (2) transmittance spectra of drop-coated SWCNT layer on glass.
b) The calculated spectra of light transmittance into Si substrate through the layers: SWCNT, Au, and Au/SWCNT.

Photoelectric and electric properties

The spectra of the short-circuit photocurrent for the investigated structures are shown in Fig.4a. It is seen that introduction of thin PVP layer decreases the photocurrent value while the deposition of SWCNT film drives to photocurrent enhancement that increasing with wavelength increase. (Fig.4b)

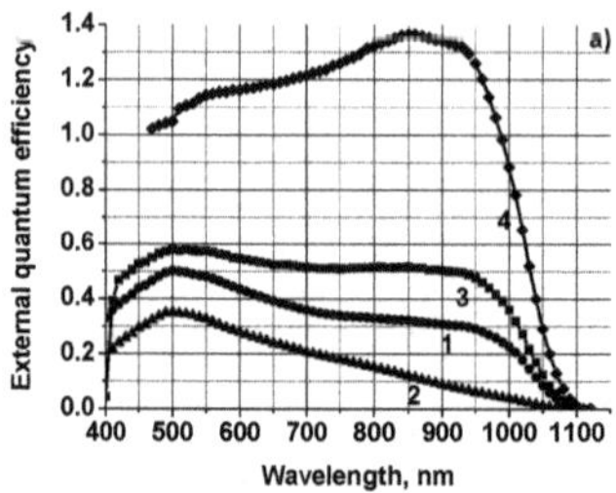

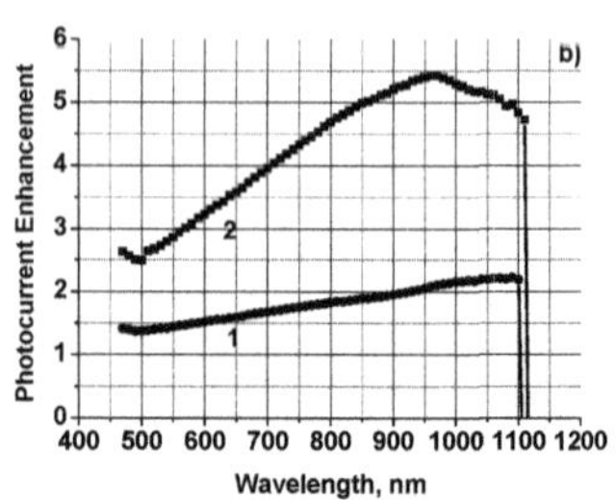

Figure 4 a) Spectra of the short-circuit photocurrent expressed as the external quantum efficiency normalized to the area of diode for Au/Si (1), Au/PVP/Si (2), Au/SWCNT(dip)/PVP/Si (3), Au/SWCNT(drop)/PVP/Si (4).
b) The photocurrent enhancement spectra due to introducing of SWCNT(dip) (1) and SWCNT(drop) (2) layer into Au/n-Si photodiode structure.

Analogical changes are observed for the light current voltage characteristics, measured under simulated AMO illumination (Fig.5).

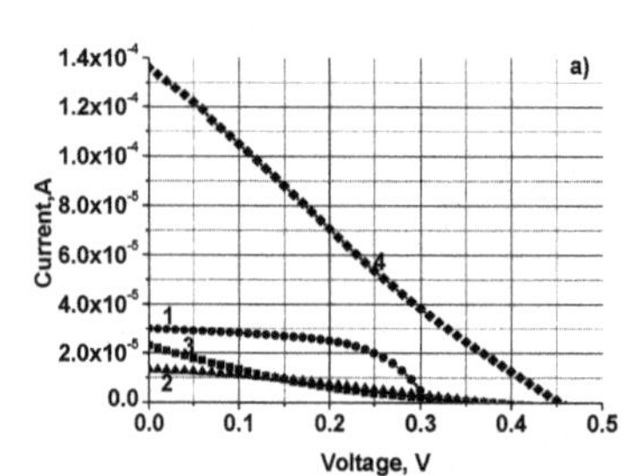

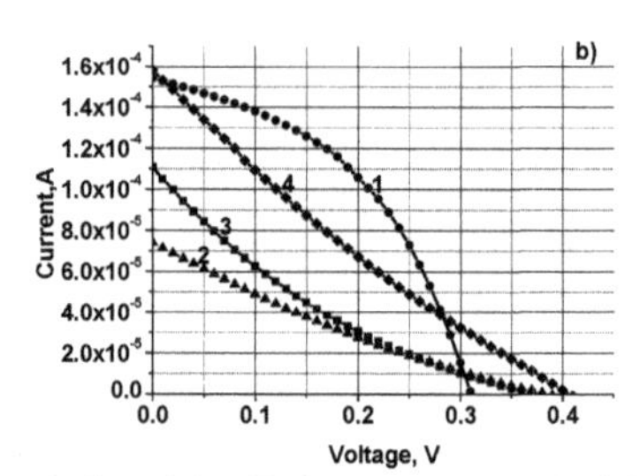

Figure 5. The light current-voltage characteristics of the diode structures (a) and the structures of the grid form (b) at AMO simulated illumination: Au/Si (1), Au/PVP/Si (2), Au/SWCNT(dip)/PVP/Si (3), Au/SWCNT(drop)/PVP/Si (4).

Besides the photocurrent enhancement the increase of the open-circuit voltage, but decrease of the fill factor, caused by introducing of thin dielectric PVP layer, are found. As the analysis of dark current-voltage characteristics (Fig.6) shows this is accompanied with the series resistance increase related to PVP layer, decreasing after the SWCNT layer deposition.

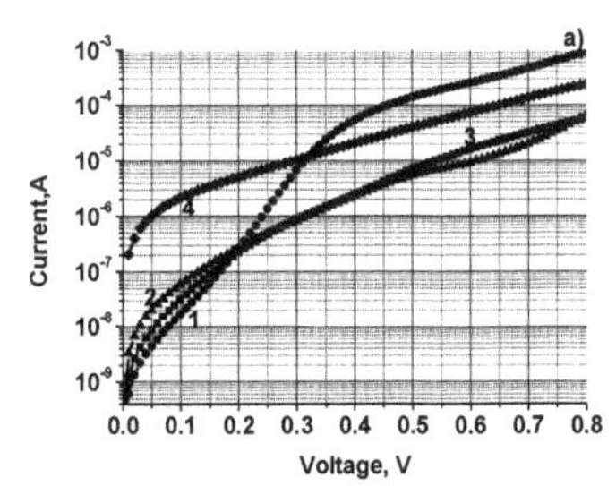

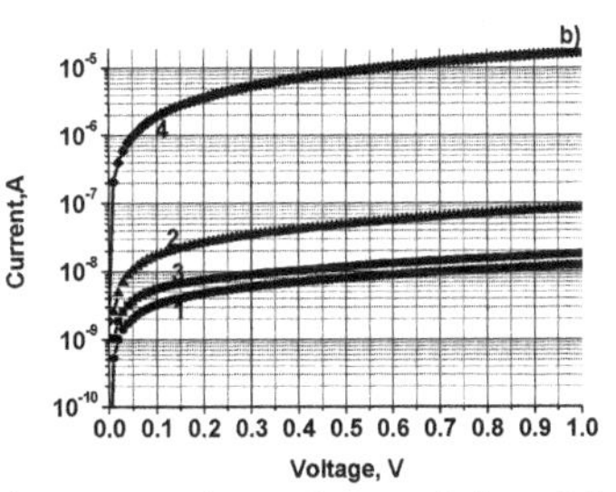

Figure 6. The forward (a) and reverse (b) dark current-voltage characteristics of diode structures: Au/Si (1), Au/PVP/Si (2), Au/SWCNT(dip)/PVP/Si (3), Au/SWCNT(drop)/PVP/Si (4).

Finally, the obtained cell efficiency increase due to SWCNT layer introducing in Au/Si photovoltaic structures was less (by about a factor of two) compared to the photocurrent enhancement (up to a factor of 4 at AMO illumination). The effect of the photocurrent enhancement is essentially determined by the optical properties of SWCNT/Si structure, dependent on the SWCNT nanolayer thickness and homogeneity. No essential changes of them (and of the electrical properties of Au/SWCNT/Si structure also) with time and temperature treatment (100°C,30 min) have been observed.

CONCLUSIONS

— Single-wall carbon nanotube layers, deposited on the n-Si substrate modified with PVP, caused the significant enhancement of photocurrent in the Au/n-Si photodiode structures, increasing in the long-wave range.

— The increase of the photoconversion efficiency due to the SWCNT interlayer in Au/n-Si structures was caused by both the photocurrent enhancement and the open circuit voltage increase that is likely to be due to PVP (dielectric) nanolayer introducing. The greatest effect of the photocurrent enhancement (up to a factor of 4 at AMO illumination) was observed for structures with SWCNT deposited by drops.

— Analysis of optical properties of the structures with SWCNT showed that the photocurrent enhancement effect was caused mainly by increasing the photocarrier collection area (lateral photoeffect), considerably exceeding the effect of transmittance decrease [5].

REFERENCES

1. Dresselhaus M.S., Dresselhaus G., Avouris P., Eds. Carbon Nanotubes: Synthesis, Structure, Properties and Application, 1st ed .; Springer-Verlag: Heidelberg, Germany, Vol.**80** (2001).
2. Kymakis, E.; Alexandrou, I.; Amaratunga, G.A.J. "High open-circuit voltage photovoltaic devices from carbon-nanotube-polymer composites". *Journal of Applied Physics* **93** (3), 1764–1768 (2003).
3. van de Lagemaat, J.; Barnes, T.M.; Rumbles, G.; Shaheen, S.E.; Coutts, T.J.; Weeks, C.; Levitsky, I.; Peltola, J.; Glatkowski, P.,. "Organic solar cells with carbon nanotubes replacing In2O3:Sn as the transparent electrode". *Applied Physics Letters* **88** (23), 233503–1–3 (2006).

4. Rowell, M. W.; Topinka, M.A.; McGehee, M.D.; Prall, H.-J.; Dennler, G.; Sariciftci, N.S.; Liangbing Hu; Gruner, G.,. "Organic solar cells with carbon nanotube network electrodes". *Applied Physics Letters* **88** (23), 233506–1–3 (2006).
5. N.L. Dmitruk, O.Yu. Borkovskaya, I.B. Mamontova , S.V. Mamykin, D.O. Naumenko, E.V. Basiuk (Golovataya-Dzhymbeeva), E. Alvazez-Zauco *Proc. of 22nd European Photovoltaic Solar Energy Conference*; 3-7 September (2007).
6. H. Kataura, Y. Kumazawa, Y. Maniwa, I. Umeru, S. Suruki, Y.Ohtsuka, Y. Achiba Synth. *Met.* **103**, 2555-2558 (1999).
7. O. Jost, A.A. Gorbunov, W. Pompe, T. Pichler, R. Friedlein, M. Knupfer, M. Reibold, H.-D. Bauer, L. Dunsch, M. S. Golden, *J.Fink. Appl. Phys. Lett.* **75** 2217-2219 (1999).
8. Malynych S., Luzinov I., Chumanov G. *J. Phys. Chem.* **B106**, 1280-1285, (2002).

Mater. Res. Soc. Symp. Proc. Vol. 1534 © 2013 Materials Research Society
DOI: 10.1557/opl.2013.294

Charging behavior of MNOS and SONOS memory structures with embedded semiconductor nanocrystals - Computer simulation

K. Z. Molnár[1], P. Turmezei[1], and Zs. J. Horváth[1,2]
[1]Óbuda University, Kandó Kálmán Faculty of Electrical Engineering, Institute of Microelectronics and Technology, Budapest,Tavaszmező u. 15-17, H-1084 Hungary
[2]Hungarian Academy of Sciences, Research Centre for Natural Sciences, Institute for Technical Physics and Materials Science, Budapest, P.O. Box 49, H-1525 Hungary

ABSTRACT

The effects of the oxide and nitride thicknesses and of the presence of semiconductor nanocrystals are studied on the charging behaviour of MNOS and SONOS non-volatile memory structures by the calculation of electron and hole tunnelling probability into the nanocrystals or to the nitride conduction or valence band, respectively, and by the simulation of memory hysteresis behaviour. The results are discussed in terms of the actual shape of potential barrier yielding different tunnelling current mechanisms.

It is concluded for both MNOS and SONOS structures that the optimal tunnel oxide thickness for the charging behaviour is about 2 nm. Low nitride thickness decreases the efficiency of the injected charge due to its loss via the blocking layer, yielding narrower memory window. The presence of nanocrystals enhances the charge injection resulting in the better performance, but for the structures with thin tunnel oxide layer (below 3-4 nm) only, and if the nanocrystals are located close to the Si substrate at the oxide/nitride interface. The results of simulations are in agreement with the experimental results obtained in MNOS structures with Si or Ge nanocrystals.

INTRODUCTION

It was obtained in our recent works that the presence of Si or Ge nanocrytals (NCs) enhanced the charge injection properties of MNOS (metal-nitride-oxide-silicon) structures [1-3]. For better understanding the experimental results, the tunnelling probability of electrons and holes into the nanocrystals and to the conduction or valence band of the nitride layer, respectively, has been calculated for MNOS structures with and without NCs [4] using WKB approximation [5]. Using these probabilities the actual memory hysteresis characteristics were simulated to understand the role of NCs and the effect of layer thicknesses [6]. In this paper the results of calculations are briefly analysed and discussed. Although the calculations have been performed for MNOS structures, the obtained dependences are valid for SONOS (silicon-oxide-nitride-oxide-silicon) structures as well, as the presence of a capping oxide layer does not influence on the charge injection directly.

THEORY

On the basis of WKB approximation [5], the tunneling probability of an electron through a potential barrier with an arbitrary shape can be expressed as

$$P = \frac{\exp\left(-\frac{2}{\hbar}\int_{x_1}^{x_2}\sqrt{2m^*[U(x)-E]}dx\right)}{\left\{1+\frac{1}{4}\exp\left(-\frac{2}{\hbar}\int_{x_1}^{x_2}\sqrt{2m^*[U(x)-E]}dx\right)\right\}^2} \quad (1)$$

where P is the tunneling probability, x_1 and x_2 are the coordinates, where the electron enters and leaves the potential barrier (turning points), m^* is the effective mass of the electron, $\hbar$ is the Planck constant (divided by 2π), $U(x)$ is the potential energy as a function of coordinate, and E is the electron energy. If the probability is much less than unity, Eq. (1) turns into the form:

$$P = \exp\left(-\frac{2}{\hbar}\int_{x_1}^{x_2}\sqrt{2m^*[U(x)-E]}dx\right) \quad (2)$$

The integration of energy (Eq. (2)) yields simple analytical expressions for MNOS structures [5].

The tunneling probability of electrons and holes has been studied as a function of the oxide thickness and of the size of NCs. For both the electron and hole, effective masses $m^*=0.42m_0$ has been used, where m_0 is the free electron mass. No image force lowering, band bending at the silicon surface, and resonant tunneling have been considered.

The charge injected via the thin oxide layer is captured and stored in traps in the nitride layer close to the oxide/nitride interface. The net charge captured in the nitride layer during a voltage pulse can be calculated by the integration of the difference of the current flowing into the structure via the oxide layer and that of flowing out of the structure via the nitride layer, over the charging pulse width τ [11]:

$$\Delta\sigma_3 = \int_0^{\tau}(J_{ox} - J_n)dt \quad (3)$$

The oxide current can be expressed as [5]:

$$J_{ox} = C_{FN}E_{ox}^2 P \quad (4)$$

where C_{FN} is a constant, E_{ox} the electric field in the oxide and P the tunneling probability via the potential barrier.

The current via the nitride layer has been obtained by experiment [7]:

$$J_n = J_{PF} + J_{EX} \quad (5)$$

where J_{PF} is the Poole-Frenkel current [5,7,8]:

$$J_{PF} = C_{PF1}\cdot E_{n2}\cdot \exp\left(C_{PF2}\cdot\sqrt{E_{n2}}\right) \quad (6)$$

where C_{PF1} and C_{PF2} are parameters depending on the insulator properties and E_{n2} is the electric field in the nitride layer (between the charge centroid and the metal) and J_{EX} is an excess current obtained at low biases [7]:

$$J_{EX} = C_{EX1} \cdot \exp(C_{EX2} \cdot E_{n2}) \quad (7)$$

where C_{EX1} and C_{EX2} are parameters obtained from experiments [7].

The integration has been performed for the each charging pulse amplitude by Eqs. (1-7). The actual barrier height values and C_{FN} have been taken from Ref. [5], while the parameters for the nitride current from Ref. [7]. The actual electric fields in the layers were obtained on the basis of the Poisson equation and of the Gauss law [6,9]. The hysteresis curves were begun to be calculated from zero pulse amplitude and zero flat-band voltage. The voltage pulse amplitude was increased or decreased by 1 V step by step.

For structures without NCs the effect of oxide thickness has been studied in the range of 0-10 nm for structures with a total thickness of nitride layer of 30 nm and with constant depth of charge centroid of 6 nm. The effect of nitride thickness has been studied in the range of 10-40 nm for structures without NCs (d_{ox}=2 nm). Two different cases were studied. In one case the charge centroid was located at a fixed distance of 6 nm from the oxide/nitride interface, in the other case it was located at the 1/5 part of the total nitride thickness. The pulse width used for simulation was 10 ms [6].

The effect of NCs located at the SiO_2/Si_3N_4 interface was studied for oxide thickness range 0.1-5 nm for a nitride thickness of 30 nm and charge centroid of 6 nm. A single step tunneling process was considered via the NCs to the nitride conduction/valence bands without capture and termalization of charge carriers in the nanocrystals, assuming a continuous energy state distribution in them.

Although the calculations have been performed for MNOS structures, the obtained tunneling probabilities are valid for SONOS structures as well. The presence of a capping oxide layer does not influence on the electric field in the tunnel oxide layer, an equivalent nitride thickness should be simply considered. In the case of simulation of memory hysteresis, the presence of the top oxide layer affects the current flowing out from the structure via the control layer. As the height of the hysteresis depends mainly on the ratio of currents flowing via the tunnel and control layers, the obtained height of hysteresis is not valid for a SONOS structure, but the character of its dependence and the width of the hysteresis are valid.

RESULTS AND DISCUSSION

Fig. 1 presents the two extremes of electron tunneling probability for MNOS structures with or without embedded semiconductor nanocrystals as a function of the oxide thickness and the electric field in the oxide layer. It presents the tunneling probability to the conduction band of the nitride layer for MNOS structures with NCs, if they are located deep in the nitride layer, deeper than the tunneling length, and also for MNOS structures without NCs. The other extreme is the tunneling probability to the nanocrystals located at the SiO_2/Si_3N_4 interface. It is just the tunneling probability via the oxide layer. The tunneling probability via NCs or traps [10] located within the tunneling depth, is between these two extremes.

The tunneling probability to the conductance band exhibits maximum as a function of oxide thickness, which depends on the actual electric field. The higher the electric field the thinner the oxide layer for the maximum probability. The tunneling probability is higher for higher electric fields. The increase of tunneling probability in the presence of a thin oxide layer in comparison with MNS structures without oxide layer is due to a high potential drop on the oxide layer caused by its lower dielectric constant: the area of potential barrier above the electron energy is smaller. At higher oxide thicknesses and/or electric fields the charge injection mechanism changes to direct tunneling to the nitride conductance band via the oxide layer. In this case the tunneling probability decreases with further increase of oxide thickness. At high oxide thickness the tunneling probability saturates at a certain level. At this point the potential drop on the oxide layer just equals to the barrier height at the SiO_2/Si_3N_4 interface, and the current mechanism changes to the Fowler-Nordheim tunneling of electrons to the oxide conduction band. For this injection mechanism, the oxide thickness does not influence the tunneling probability. So, these results indicate that the optimal charging behavior of MNOS structures without NCs can be expected for an oxide thickness of 2-3 nm.

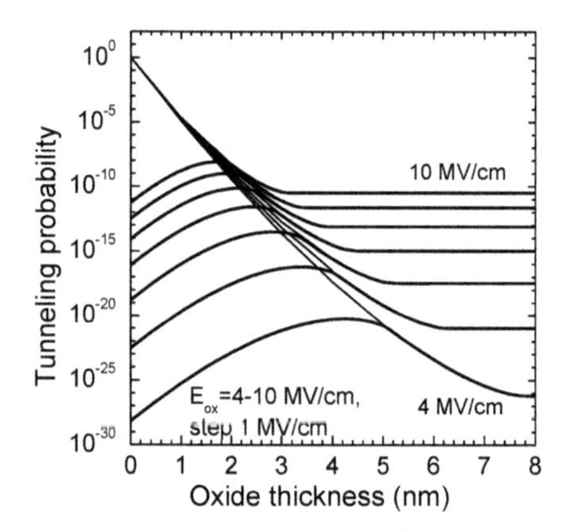

Figure 1: The two extremes of electron tunneling probability to MNOS structures with (thin lines) or without (thick lines) embedded semiconductor nanocrystals as a function of the oxide thickness and the electric field in the oxide layer.

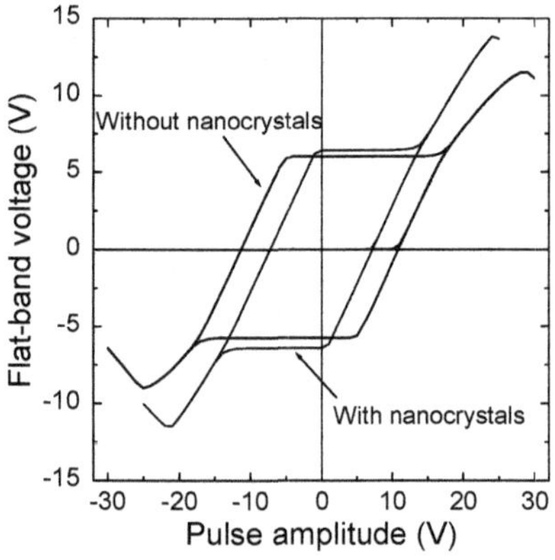

Figure 2: Simulated memory hysteresis curves exhibiting the effect of presence of nanocrystals. The oxide and nitride thicknesses are 2 nm and 30 nm, respectively, the depth of charge centroid is 6 nm, the pulse width is 10 ms.

The tunneling probability via the oxide layer (direct tunneling to NCs located at the SiO_2/Si_3N_4 interface) does not depend strongly on the electric field, but exhibits fast decrease with increasing the oxide thickness. For a given electric field the two extremes merge at a certain value of oxide thickness, when the modified Fowler-Nordheim tunneling to the nitride conductance band via the oxide and nitride layer changes to direct tunneling via the oxide layer. So, the presence of nanocrystals enhances the charge injection at thin oxide layers only, and if NCs are located close to the SiO_2/Si_3N_4 interface (within the tunneling length).

Similar results were obtained for hole tunneling, but the maximum probabilities in MNOS structures without NCs were obtained for thinner oxides about 1.5-2.5 nm.

Two simulated memory hysteresis curves are presented in Fig. 2 that demonstrate the effect of existence of nanocrysals at the SiO_2/Si_3N_4 interface. It is considered that the captured charge is stored in defects at a depth of 6 nm from the SiO_2/Si_3N_4 interface. The hysteresis loop can be characterized by its height and width. The height is connected to the maximum charge that can be injected and stored in the structure, while the width is related to the electric field accruing in the oxide layer at the end of the charging pulse. This value is considered as a threshold oxide field necessary to the charge injection at a given charging pulse duration [11]. It is seen in Fig. 2 that the memory hysteresis is thinner and higher for the structure containing NCs indicating enhanced charge injection and storage behavior.

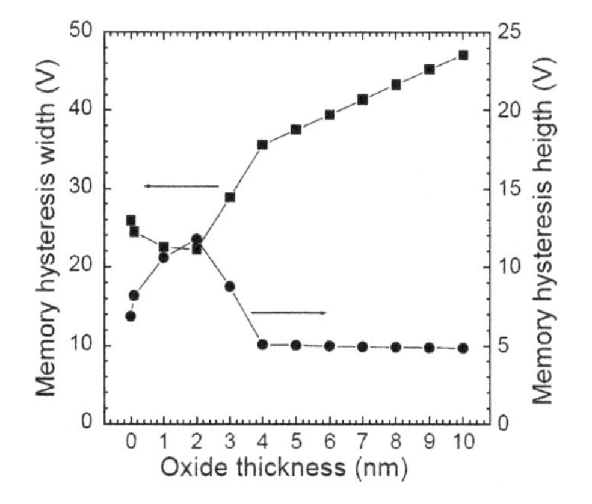

Figure 3: The width (squares) and height (circles) of memory hysteresis curves simulated for MNOS and SONOS structures without nanocrystals as a function of oxide thickness.

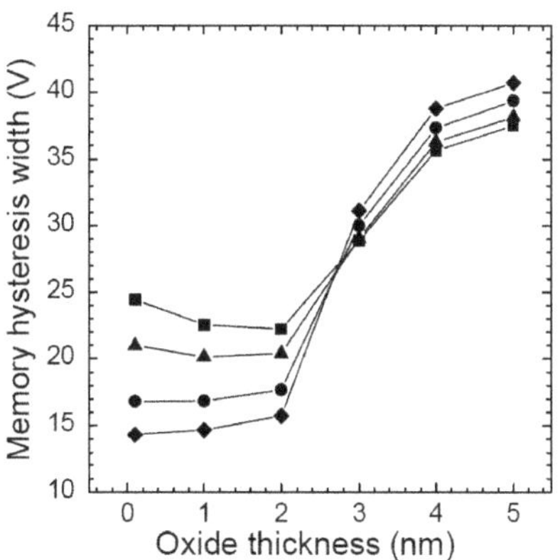

Figure 4: The width of memory hysteresis simulated for structures without (sqares) and with nanocrystals at the SiO_2/Si_3N_4 interface with nanocrystal layer thickness of 1 nm (triangles), 3 nm (circles) and 5 nm (diamonds).

Fig. 3 presents the width and height of the hysteresis curves simulated for different oxide thicknesses without NCs. The width of memory hysteresis exhibits minimum at an oxide thickness of 2 nm, while its height has maximum at the same value indicating the best charging behavior for this oxide thickness. There is an abrupt increase of the width of hysteresis between 2 nm and 4 nm, and a less abrupt increase above 4 nm. This indicates that the dominating injection mechanism is the modified Fowler-Nordheim tunneling to the nitride conductance band via the oxide and nitride layers in MNOS structures without NCs with oxide thickness below 4 nm, while above 4 nm the direct tunneling via the oxide layer dominates.

Fig. 4 presents the effect of the presence of NCs on the width of memory hysteresis as a function of oxide thickness. It is seen that NCs enhances the charging behavior for thin oxide layers only below 3 nm, as it has been concluded on the basis of tunneling probabilities [4]. The larger the nanocrystals size, the higher their effect. The increasing width of the hysteresis above an oxide thickness of 3 nm is due to the voltage drop on NCs yielding lower electric field in the oxide layer for the same charging pulse amplitude.

The effect of nitride thickness on the memory hysteresis was studied for two different cases, as mentioned above. If the charge centroid was kept at a 1/5 part of the total nitride thickness, the shape of hysteresis curves for each thickness was similar, both the horizontal and vertical dimensions of the loops decreased proportional to the decreasing nitride thickness. But in the case of a fixed depth of charge centroid of 6 nm, the height of the hysteresis decreased faster than its width. This case is closer to reality. So, in real cases the memory window shrinks faster with decreasing nitride thickness, than the charging pulse amplitude decreases. Therefore, a thin nitride layer decreases the efficiency of the injected charge. One should make a compromise between the low charging pulse amplitude and the memory window width [6].

CONCLUSIONS

The effect of the oxide and nitride thicknesses and of the presence of semiconductor nanocrystals has been studied on the charging behaviour of MNOS and SONOS non-volatile memory structures by the calculation of electron and hole tunneling probability to the nanocrystals or to the nitride conduction or valence band, respectively, and by the simulation of memory hysteresis behavior. The tunneling probabilities have been calculated by WKB approximation. The memory hysteresis behavior was simulated by the integration of the difference of the current flowing into the structure via the tunnel oxide layer and that of flowing out of structure via the blocking layer, over the charging pulse width. The results have been discussed in terms of the actual shape of potential barrier yielding different tunneling current mechanisms.

It has been obtained for both MNOS and SONOS structures that the tunnel oxide thickness, which is optimal for charge injection, is about 2 nm, while low nitride thickness decreases the efficiency of the injected charge due to its loss via the blocking layer, yielding narrower memory window. The presence of NCs enhances the charge injection resulting in better performance, but for structures with thin tunnel oxide layer (below 3 nm) only, and if the nanocrystals are located close to the Si substrate at the oxide/nitride interface, i.e., when they are within the tunneling depth. The results of simulations are in agreement with the experimental results obtained in MNOS structures with Si or Ge nanocrystals.

REFERENCES

1. Zs. J. Horváth, P. Basa, T. Jászi, A. E. Pap, L. Dobos, B. Pécz, L. Tóth, P. Szöllősi, K. Nagy, J. Nanosci. Nanotechnol., **8**, 812 (2008).
2. Zs. J. Horváth, P. Basa, Mater. Sci. Forum, **609**, 1 (2009).
3. Zs. J. Horváth, P. Basa, in: Nanocrystals and Quantum Dots of Group IV Semiconductors, (Eds. T. V. Torchynska, Yu. V. Vorobiev), American Scientific Publishers, 2010, p. 225.
4. Zs. J. Horváth, K. Z. Molnár, Gy. Molnár, P. Basa, T. Jászi, A. E. Pap, R. Lovassy, P. Turmezei, Phys. Stat. Sol. (C), **9**, 1370 (2012); doi 10.1002/pssc.201100668
5. K. I. Lundström, C. M. Svensson, IEEE Trans. El. Dev., **ED-19**, 826 (1972).
6. K. Z. Molnár, Zs. J. Horváth, Proc. 7th IEEE Int. Symp. Applied Computational Intelligence and Informatics SACI2012, May 24–26, 2012, Timisoara, Romania, p. 365.
7. P. Basa, Zs. J. Horváth, T. Jászi, A. E. Pap, L. Dobos, B. Pécz, L. Tóth, P. Szöllősi, Physica E, **38**, 71 (2007.)
8. D. Frohman-Bentchkowsky, M. Lenzlinger, J. Appl. Phys. **40**, 3307 (1969).
9. Zs. J. Horváth, V. Hardy, J. Nanosci. Nanotechnol., **8**, 834 (2008).
10. H. E. Maes, R. J. Overstraeten, J. Appl. Phys. **47**, 664 (1976).
11. Zs.J.Horváth, Solid-State Electron. **23**, 1053 (1980).

Mater. Res. Soc. Symp. Proc. Vol. 1534 © 2013 Materials Research Society
DOI: 10.1557/opl.2013.295

Iron Silicide Nanostructures Prepared by E-Gun Evaporation and Annealing on Si(001)

György Molnár[1], László Dózsa[1], Zofia Vértesy[1] and Zsolt J. Horváth[1,2]
[1]Institute of Technical Physics and Materials Science, Research Centre for Natural Sciences, HAS, Budapest, H-1525 P.O. Box 49, Hungary,
[2]Óbuda University, Kandó Kálmán Faculty of Electrical Engineering, Institute of Microelectronics and Technology, H-1084 Budapest,Tavaszmező u. 15-17, Hungary

ABSTRACT

Iron silicide nanostructures were grown on Si(001) by strain-induced, self-assembly method. E-gun evaporated iron particles were deposited both on to room temperature and high temperature Si substrates, and were further annealed in situ. The initial Fe thickness was in the 0.1-6 nm range and the annealing temperatures varied between 500 and 850°C. The phases and structures formed were characterized by the reflection high energy electron diffraction and by scanning electron microscopy. The electrical characteristics were investigated by I-V and C-V measurements, and by deep level transient spectroscopy. The size distributions of the formed iron silicide nanostructures were not homogeneous but, were oriented in perpendicular directions on Si(001). Higher temperature annealing resulted in the increased particles size and faceting. Electrical characteristics showed the critical defect concentration related to Fe.

INTRODUCTION

New generation thin film solar cells have to use environmentally friendly, non toxic and abundantly available chemical elements [1]. One of the potentially candidates is semiconducting β-$FeSi_2$, for this material the 23% theoretical efficiency in solar cells has been predicted. Efforts have been made to produce iron silicide based photovoltaic devices, since both in its thin film and nanoparticle shapes have potential applications in the photovoltaic technology [2-4]. Terasawa and coworkers suggested a composite β-$FeSi_2$/Si film for solar cells use, where iron silicide nanoparticles are embedded in silicon. In this case photocarriers are generated in the iron silicide particles, which has high photoabsorption coefficient, while carrier transport happens in silicon. This kind of material may result an excellent, new solar cell as a consequence of its high photoabsorption coefficient and high carrier mobility [5].

β-$FeSi_2$ is an indirect semiconductor, although in epitaxial configurations it shows a direct band gap on silicon substrate due to lattice distortions [6,7]. During solid phase thin film reactions, a number of phases of the Fe-Si equilibrium phase diagram have been found on Si substrates [8-11]. The mostly Fe-rich iron silicide is Fe_3Si (DO_3 type), with cubic lattice. Two types of iron monosilicides might be present in thin film form. The first monosilicide phase is ε-FeSi with cubic structure and the second phase is cesium-chloride type cubic FeSi. The iron disilicides might appear with three different crystal structures. The high temperature, metastable, tetragonal α-$FeSi_2$ phase might be present in thin film formed on Si substrates. The cubic γ-$FeSi_2$ phase is also metastable. At the end, the stable β-$FeSi_2$ has orthorhombic structure. All of the above mentioned phases, including metastable ones, might be epitaxially stabilized on the surface of Si substrates [12].

The most effective physical method of nanostructure preparation is the self-assembly, that have been observed besides compound and group IV semiconductors in a wide range of material and substrate combinations [13]. The strain induced self-assembled growth is a basic physical procedure of preparation of the nanoscale objects. During the growth of strained layers, the film often remains planar up to a critical thickness that depends on the lattice mismatch of the film and the substrate. Above that critical thickness, three dimensional dislocation free islands may form [14]. This phenomenon is the Stransky-Krastanov transition, which is an important way of self-assembled formation of quantum dots and wires. This type of growth may occur during the growth of epitaxial silicides.

The motivation of this study is to compare the formation of iron silicide nanostructures prepared by (i) iron evaporation onto Si substrate, which is kept at room temperature (RT) with subsequent annealing and by (ii) reactive deposition epitaxy method (RDE), where the iron particles are evaporated onto heated Si substrates. This research field may contribute to gain new knowledge in design of the morphology of iron silicides, and for practical side to make new steps towards more effective environmentally friendly solar cells.

EXPERIMENT

Pieces of (001) oriented Si (p-type, 12-20 Ωcm) wafers were used as substrates. Before loading the samples into the oil free evaporation chamber their surfaces were etched in diluted HF. Prior to evaporation Si wafers were annealed in situ for 5 min at 850°C. Iron ingots of 99.9% purity were evaporated using an electron gun, at an evaporation rate of 0.01-0.03 nm/s, at a pressure of $3*10^{-6}$ Pa. The film thickness was measured by vibrating quartz. The temperatures were monitored by small-heat-capacity Ni-NiCr thermocouples. The initial Fe thicknesses were in the 0.1-6.0 nm range and the annealing temperatures varied between 500 and 850°C.

Samples were annealed by two methods: (i) Fe evaporation onto room temperature (RT) substrates with subsequent annealing for 60 minutes at 850 °C, (ii) reactive deposition epitaxy method (RDE), where the substrates were heated during Fe evaporation at different temperatures. The RDE grown samples were post annealed after the end of the deposition in situ in the vacuum chamber. The total annealing time (deposition + post annealing time) for the RDE grown samples was 60 minutes too. So, the thermal budget was the same for the differently prepared samples at a given temperature.

The growth processes were tracked in situ by reflection high energy electron diffraction (RHEED) at 10 keV beam energy during the sample preparation. The morphologic features of the nanostructures were characterized by scanning electron microscopy (SEM). The electrical characteristics were investigated by current-voltage (I-V), and capacitance-voltage (C-V) measurements and the defects were measured by deep level transient spectroscopy (DLTS).

DISCUSSION

The whole sample preparation process was followed up by in situ RHEED measurements. The azimuthal orientation of the 10 keV e-beams was along Si <110> for all of the samples. In Fig. 1(a-e) can be seen the RHEED images of the two iron silicide samples, which were taken during sample preparation, where the deposited mass equivalent Fe thickness was 0.1 nm, and the heat treatment was carried out at 850°C for 60 minutes for both of the samples. The only difference is the deposition mode: The first column of images belongs to an RT deposited and then annealed sample and the second column to an RDE deposited and then further annealed sample for comparison. In Fig. 1(a-b) can be seen the RHEED images of the cleaned and annealed Si(001) substrate showing 2x1 reconstruction for both of the samples. Fig. 1(c-d) shows state of the surface after 0.1 nm iron evaporation. In case of the RT deposited sample (Fig. 1(c)) the original Si surface reconstruction disappeared and the lines weakened, as consequence of very thin Fe covering. While, the RDE deposited sample (Fig. 1(d)) shows promptly appeared new lines belonging to iron silicide phase formed on the hot substrate instantaneously. RHEED images of Fig. 1(e-f) show the state of the surfaces after annealing, where both images show a new, ordered surface with epitaxial character. The misfit differences of Si(001) and of the three iron disilicide phases, mentioned in the introduction section, are within two percent [11] that is why the RHEED cannot differentiate between them.

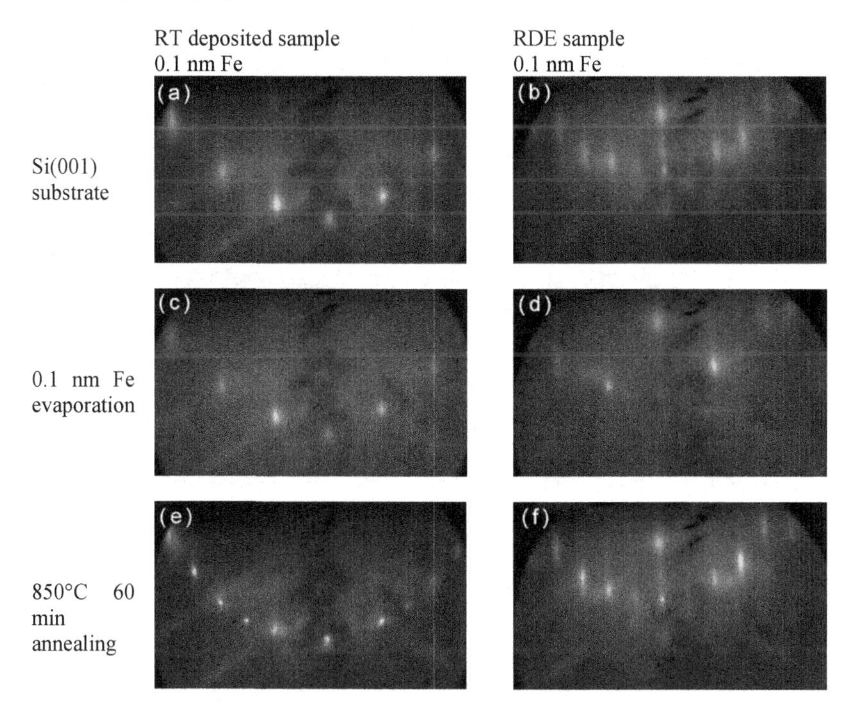

Figure 1. RHEED images of iron silicide nanostructure formation. The first column belongs to an RT deposited sample and the second to an RDE ones. (a-b) Si substrate, (c-d) after 0.1 nm Fe deposition, (e-f) after 850°C 60 min annealing

The particular phase identification of iron disilicide nanostructures by transmission electron microscopy selected area electron diffraction was presented for similar samples in a previous paper [15].

SEM images of iron silicide nanostructures are presented in Figs. 2(a-f) as a function of the annealing temperature. The first column of images shows the RT deposited and then annealed sample, the second column shows the RDE deposited and then further annealed sample for comparative study. The film thickness was 0.1 nm for each sample, and the heat treatments were carried out at 500, 600, and 850°C for 60 minutes. All of the samples show aggregated iron silicide nanostructures. The size and the distribution of the islands depend on the temperature and on the type of the annealing. At 500°C annealing; the density of the silicide nanostructures is higher for RT sample compared to RDE one (Fig. 2(a-b)). At 600°C annealing; besides the small nanostructures, bigger size aggregates appeared (Fig. 2(c-d)).

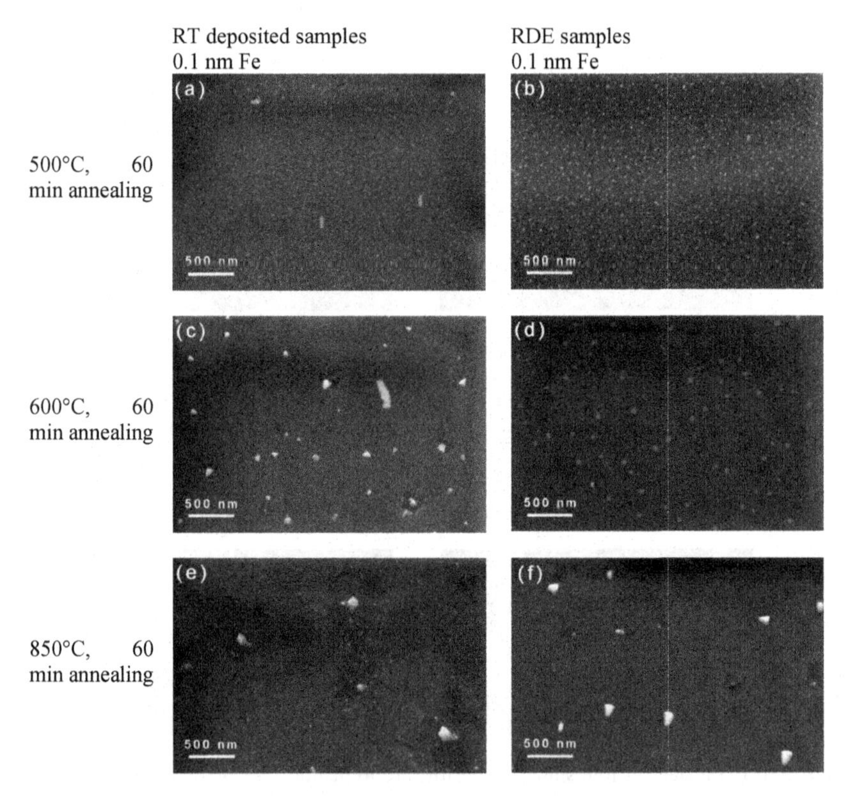

Figure 2. SEM images of iron silicide nanostructures formed at RT (first column) and RDE (second column) depositions at different temperatures for comparison. Samples annealed at (a-b) 500°C, (c-d) 600°C, (e-f) 850°C, for 60 minutes.

In case of RT sample the characteristic shape of big aggregates is the triangular, while for RDE sample the circle. At 850°C annealing the RT sample shows randomly shaped structures and the RDE sample shows perpendicularly ordered, triangular objects (Fig. 2(e-f)).The local environment of these big objects is depleted as a consequence of Ostwald ripening phenomena, where the bigger islands grow further at the cost of the smaller ones [16]. In case of thicker samples - up to 6 nm initial Fe thickness - the size of aggregated object grows continuously, reaching the 1 micrometer lateral dimensions.

According to electrical characteristics, the Fe related defects are dominant in all samples. Fig.3(a-c). The current-voltage and capacitance-voltage characteristics show these defects in about 1-2 μm depth from the surface. The doping concentration determined from the C-V characteristics decreases near the surface. The deep level defects compensate the doping of the starting wafer. The I-V and C-V values have significant scatter in different junction. In some of the junctions the leakage current is dominate at reverse bias. We assume it is due to the rough silicides/silicon interface morphology and to very large defect concentration in the vicinity of the interface.

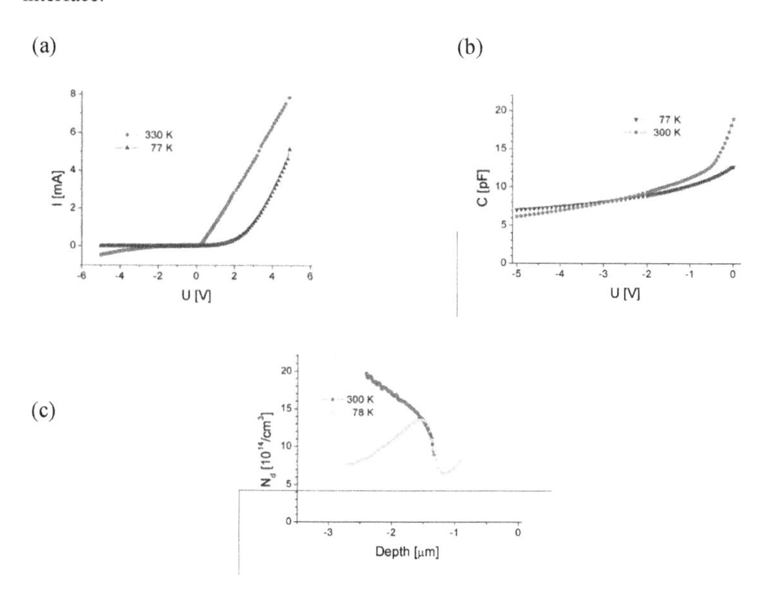

Figure 3. (a) Current-voltage plot, (b) capacitance–voltage plot, (c) depth profile of main defects of a 1 nm thick sample measured at temperatures indicated in the plots.

CONCLUSIONS

Iron silicide nanoislands were grown on Si(001) substrate by RT deposited and then annealed and by RDE deposited and then further annealed methods for comparison. The shape of the nanostructures varied from circular to triangular and quadratic depending on the initial Fe thickness and on the annealing method and temperature. The size distribution of the formed iron silicide nanoobjects was not homogeneous, but they were oriented in perpendicular directions on Si(001). Higher temperature annealing resulted in increased particles size and faceting. The RT and RDE growth mode resulted in similar nanostructures; no sharp difference was detected between them. Electrical characteristics show the critical problem for application, which is the large defect concentration related to Fe. In case of successful engineering of iron silicide nanostructures they might be used potentially as environmentally friendly semiconductors for more effective solar cells.

ACKNOWLEDGMENTS

The authors would like to thank the support of OTKA Grant No. 81998.

REFERENCES

1. F. Alharbi, J. D. Bass, A. Salhi, A. Alyamani, H. C. Kim, and R. D. Miller, Renew. Energ. **36**, 2753 (2011).
2. M. Shaban, K. Nakashima, W. Yokoyama, and T. Yoshitake, Jap. J. Appl. Phys. **46**, L667 (2007).
3. G. K. Dalapati, S. L. Liew, A. S. W. Wong, Y. Chai, S. Y. Chiam, and D. Z. Chi, Appl. Phys. Lett. **98**, 013507 (2011).
4. T. Buonassisi, A. A. Istratov, M. A. Marcus, B. Lai, Z. Cai, S. M. Heald, and E. R. Weber, Nature Materials **4**, 676 (2005).
5. S. Terasawa, T. Inoue, and M. Ihara, Solar Energy Materials & Solar Cells **93**, 215 (2009).
6. D. B. Migas, and L. Miglio, Phys. Rev. B **62**, 11063 (2000).
7. K. Yamaguchi, and K. Mizushima, Phys. Rev. Lett. **86**, 6006 (2001).
8. K. A. Mäder, H. von Känel, and A. Baldereschi, Phys. Rev. B **48**, 4364 (1993).
9. H. von Känel, K. A. Mäder, E. Müller, N. Onda, and H. Sirringhaus, Phys. Rev. B **45**, 13807 (1992).
10. N. Jedrecy, A. Waldhauer, M. Sauvage-Simkin, R. Pinchaux, and Y. Zheng, Phys. Rev. B **49**, 4725 (1994).
11. P. Villars, and L. D. Calvert, *Pearson's Handbook of Crystallographic Data for Intermetallic Phases*, Vol. 3, (American Society for Metals, Metals Park, OH 1985) p. 2232.
12. G. Molnár, L. Dózsa, G. Pető, Z. Vértesy, A. A. Koós, Z. E. Horváth, and E. Zsoldos, Thin Solid Films **459,** 48 (2004).
13. A. L. Barabási, Appl. Phys. Lett. **70**, 2565 (1997).
14. J. Tersoff, and F. K. LeGoues, Phys. Rev. Lett. **72,** 3570 (1994).
15. N. Vouroutzis, T. T. Zorba, C. A. Dimitriadis, K. M. Paraskevopoulos, L. Dózsa and G. Molnár, J. Alloys Compounds **448**, 202 (2008).
16. M. Zinke-Allmang, Thin Solid Films **346**, 1 (1999).
17. K. Wünstel, and P. Wagner, Appl. Phys. A **27**, 207 (1982).

Mater. Res. Soc. Symp. Proc. Vol. 1534 © 2013 Materials Research Society
DOI: 10.1557/opl.2013.296

Exciton - Light coupling in SiC nanocrystals

Miguel Morales Rodriguez[1] and Georgiy Polupan[2]
[1] PCIM-Universidad Autónoma Metropolitana, Un. Azcapotzalco, 02200, Mexico, DF,
[2]ESIME-National Polytechnic Institute, Mexico D.F. 07738,

ABSTRACT

The paper presents the results of the SiC:N nanocrystal characterization using the methods of photoluminescence (PL) at low temperatures and X-ray diffraction (XRD). Photoluminescence study of porous SiC:N (PSiC) layers with different PSiC thicknesses reveals the intensity stimulation for the high energy PL bands. The early investigation of temperature dependences of the high energy PL bands had shown that these PL bands related to free exciton emission in the different SiC polytypes. The SiC polytypes in the original n-type SiC:N wafers and in porous SiC layers were confirmed by XRD study. The intensity enhancement of exciton-related PL bands in big size (50-250nm) SiC NCs is attributed to the realization of the exciton week confinement and exciton-light coupling in SiC NCs. The numerical simulation of exciton radiative recombination rates for the different exciton emissions has been done using a model of exciton – light coupling in SiC NCs. The experimental and numerically calculated results have been compared and discussed.

INTRODUCTION

The large band gap of SiC polytypes: from 2.36 eV in 3C-SiC up to 3.33 eV in 2H-SiC, makes SiC NCs as a perspective compound for the blue and ultraviolet (UV) light emitter diodes (LEDs) and full-color displays [1,2]. The porous SiC structures prepared on the n- and p-type bulk SiC substrates of different polytypes (3C-, 6H-, 4H- SiC) were studied intensively in the 90th mainly as the light emitting materials [3-9]. The emission intensity of SiC can be enhanced significantly when the crystallite size diminishes to the nanometer scale [5-7]. The mechanisms of intensity enhancement for variety of photoluminescence (PL) bands in the spectral range of 1.9-3.7 eV in PSiC are under discussion up to now.

This paper presents the results of porous SiC characterization using X-ray diffraction (XRD) and photoluminescence spectroscopy techniques as well as the numerical simulation of radiative recombination rate for the different exciton related PL bands versus SiC NC sizes that has been performed using the model of exciton – light coupling in SiC NCs.

EXPERIMENTAL DETAILS

Porous SiC layers were formed by surface anodization of n-type 6H-SiC substrates doped with nitrogen with resistivity of 0.052 Ω cm and orientation (0001) in 3% aqueous solution of HF at the dc current densities 8 (#1), 16 (#2), 24 (#3) and 48 (#4) mA/cm^2 and an etching duration of 10 min without external illumination (Table 1). The X-ray diffraction experiments were made using the XRD equipment model of D-8 advanced (Bruker Co.) with K_α line from the Cu source (λ=1.5406Å).

Table 1. Parameters of PSiC samples

Sample number	Anodization current density, mA/cm^2	Thickness, μm	Size of SiC NCs, nm
#1	8	2.1	170-200
#2	16	4.3	130-160
#3	24	6.2	80-120
#4	48	12.1	40-70

Photoluminescence (PL) was measured using SPEX 500 spectrometer coupled with a photomultiplier at the temperature of 20K under the excitation by 325nm He-Cd laser line with up to 30 mW power. A thickness of the porous layers (Table 1) and their morphology were estimated earlier using planar and cross sectional SEM images, obtained on XL-FEG (FEI-Sirion) [7].

EXPERIMENTAL RESULTS

XRD signals corresponding to the reflection from the crystal planes of 6H-SiC and of the inclusions of 4H- and 15R-SiC polytypes have been obtained (Fig.1).

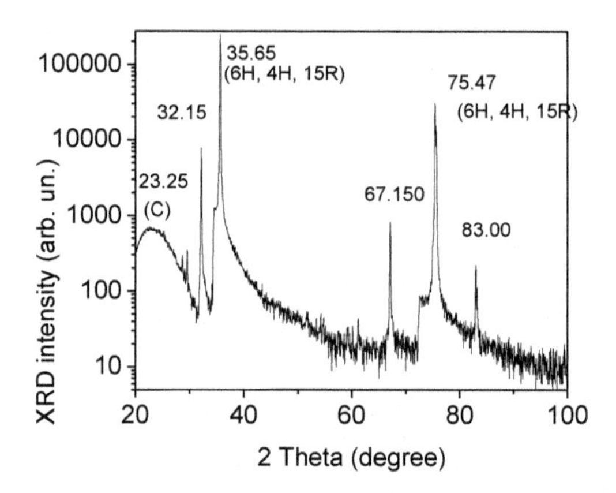

Figure1. XRD diagrams for the PSiC sample #4.

The detail investigation of the most intensive XRD peaks has shown that these peaks are a superposition of the reflections at the angles of 2Θ: (i) 35.646°, 35.699°, 35.720° and 35.780, corresponding to the planes (006) of 6H-SiC, (004) of 4H-SiC as well as (0015) and (015) of 15R-SiC, and (ii) 75.382, 75.492°, 75.604° and 75.683°, corresponding to the planes (0012) and (204) of 6H-SiC, (008) of 4H-SiC and (0030) of 15R-SiC in the bulk SiC, respectively. Note with increasing PSiC porosity and decreasing the size of SiC NCs, XRD peaks shift toward low angles. In addition, a broad band with the peak in the angle range of 2Θ = 23.3-25.15° has been revealed. The intensity of this band increases with enlargement of the PSiC layer porosity and thicknesses (Fog.1). This band is attributed to amorphous graphite, which appears at the surface of PSiC in a process of electrochemical etching.

PL spectra of studied PSiC layers are presented in figure 2. The PL spectra are obviously a superposition of strongly overlapping PL sub-bands and could be numerically deconvoluted on a set of six individual components in a spectral range of 1.80 – 3.40 eV (Fig.3).

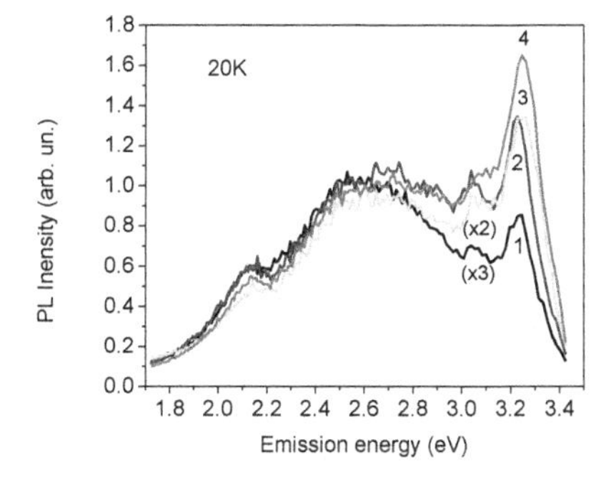

a

Figure 2 Normalized PL spectra of the PSiC samples #1 (1), #2 (2), #3 (3) and #4 (4).

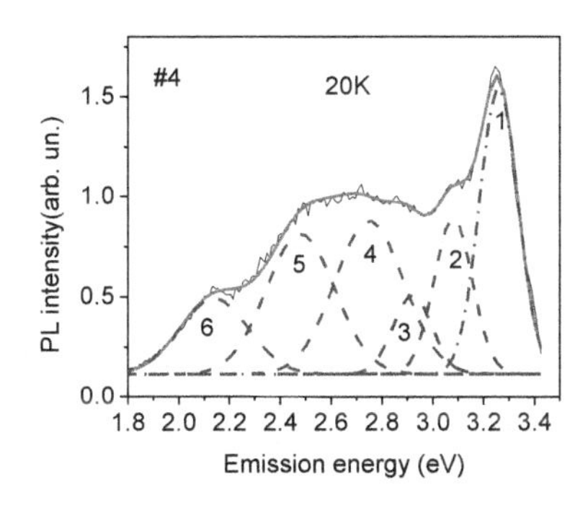

b

Figure 3. The deconvolution example of PL spectrum on six elementary PL bands.

All PL spectra include the low energy PL bands with the peaks at 2.16, 2.50 and 2.76 eV and a half-width of 300 - 500 meV, as well as the high energy PL bands with maxima at 2.90 (428 nm), 3.05 (407 nm) and 3.25 (382 nm) eV and half-widths 180-200 meV (Fig.3). Note that the intensity of the high energy PL bands (3.05 and 3.25 eV) increases noticeable with decreasing the SiC NC sizes in comparison with the intensity of low-energy PL bands (Fig.2).

The high energy PL bands (2.90, 3.05 and 3.25 eV) have a narrower half-width, which allows assigning them to exciton recombination in NCs of the different SiC polytypes. It is known that the free exciton PL bands in the SiC polytypes exhibit peaks at 3.26 eV (4H-SiC), 2.86 eV (indirect) and 3.00 eV (direct) in (6H-SiC), and 2.99eV (15R-SiC) at 4.2K [7]. Note, in n-type 6H-PSiC doped with nitrogen we can see the donor–bound exciton PL band (2.76 eV) at low temperature as well. The early investigation of temperature dependences of the high energy PL bands (2.90, 3.05 and 3.25 eV) have shown that these PL bands related to the free excitons in the SiC polytypes (4H-SiC, 6H-SiC and 15R-SiC) [10].

The intensity stimulation of the high-energy PL bands in PSiC can be attributed to exciton recombination rate increasing due to the realization of the weak exciton confinement in SiC NCs when the size of NCs decreases. The theoretical explanation of such type effects was proposed earlier in [11,12], where two weak exciton confinement regimes were discussed. The first one deals with NCs of the size “bigger” than the Bohr exciton radius, but remains “smaller” than the wavelength of emitted light in a material. The second weak confinement regime related to the exciton–light coupling with the formation of polariton, which is not important for small NCs, but becomes strong if the NC size approaches to the wavelength of light in a material. The exciton recombination rate Γ^{QD}, for the assumption of exiton-light coupling, was considered earlier as the product of the photon and exciton eigenstates in [11,12]. To get an analytical formula for the radiative recombination rate Γ^{QD}, the NC distribution function was approximated by a Gaussian function [12]. For this approximation the analytical form for Γ^{QD} were obtained [11, 12]:

$$\Gamma^{QD} = \frac{\sqrt{2\pi}}{12}\omega_{LT}(\frac{2\pi}{\lambda_o})^3\langle r\rangle^3\exp(-8\varepsilon_B\frac{\pi^2\langle r\rangle^2}{\lambda_o^2}) \quad (1)$$

where k is the wave vector of light in a material with the high frequency dielectric constant, ε_B, that equal to 6.52 for SiC, ko = $2\pi/\lambda o$ is the wave vector of light at an exciton resonance frequency [12]. On the stage of numerical calculation the radiative recombination rates (Fig.4) for three types of excitons with the PL peaks at 2.90 (428 nm), 3.05 (407 nm) and 3.25 (382 nm) eV have been estimated using Eq. 1 at the variation of NC sizes. As it is clear from figures 4 the recombination rates approach the maximum in the exciton-light coupling model at the SiC NC radii of 20, 22 and 25 nm for the excitons with the peaks at 382, 407 and 428 nm, respectively.

Figure 5 presents the variation of the integrated PL intensities in PSiC for the exciton emissions with the peaks at 382, 407 and 428 nm versus NC sizes estimated from the experimental PL spectra. The maximum PL intensity is detected at the NC size of 50 nm (NC radii equal to 25 nm) that corresponds very well to the estimated NC size for the maximum recombination rate in the exciton-light coupling model. But PL intensity decreasing with the enlargement of NC sizes passes slowly, than has been predicted in this model.

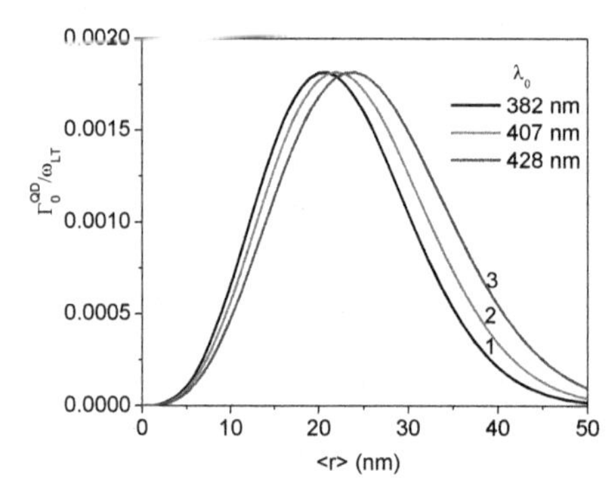

Figure 4. The numerical calculation of the recombination rates of excitons related to the PL bands with the peaks at 382 (1), 407 (2) and 428 (3) nm versus NC radii.

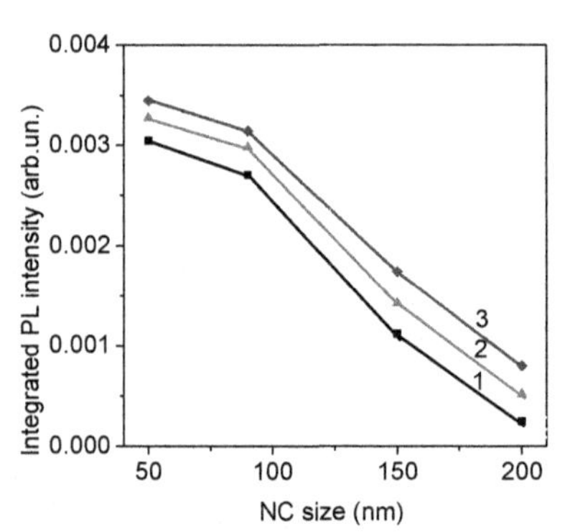

Figure 5. Integrated PL intensities of PL bands with the peaks at 382 (square), 407 (triangle) and 428 (rhomb) nm in studied PSiC samples with the different NC sizes.

The intensity of exciton related PL bands in SiC NCs can be presented as [13-15]:

$$W = \frac{C_0}{\tau_R} = (1-c)\frac{G\tau_R^{-1}}{\tau_R^{-1} + \tau_{NR}^{-1}} \qquad (2)$$

where, G is the rate of the exciton generation by excitation light and τ_R, τ_{NR} are the radiative and nonradiative recombination times in SiC NCs, respectively, and c is the porosity of SiC material. Note that in real PSiC samples an ensemble of SiC NCs has been excited by the excitation light [16-18]. In PSiC with high porosity, c, the NC size is small and owing to the high surface to volume ratio the nonradiative recombination rate, τ_{NR}^{-1}, is significant (high surface recombination) [19-21]. Additionally, due to the high porosity (60-80%) the volume of SiC material, (1-c), excited by light is small. Both of these factors lead to decreasing the PL intensity in the real PSiC samples.

On the other side in PSiC with small porosity (20-40%) the volume of SiC material, (1-c), excited by light is higher, the number of excited SiC NCs enlarges, and the rate of surface nonradiative recombination, τ_{NR}^{-1}, is smaller, that lead to increasing the PL intensity in the studied PSiC layers [20, 21]. All these factors can be the reasons why the PL intensity in PSiC decreases slowly (Fig.5), than it is predicted for the variation of the recombination rate, τ_R^{-1}, versus NC sizes in the proposed exciton-light coupling model (Fig.4).

CONCLUSIONS

Photoluminescence study of PSiC layers with the different PSiC thicknesses and SiC NC sizes has shown the stimulation of the PL intensity for the exciton PL bands with versus NC size decreasing. The PL intensity enhancement for the exciton-related PL bands is attributed to the weak quantum confined effect related to the exciton recombination rate increasing as a result of exciton-light coupling in SiC NCs of different polytypes.

ACKNOWLEDGMENTS

The authors thank G. Gasga from ESFM-IPN, Mexico for the measurement of XRD spectra. The work was partially supported by CONACYT Mexico (projects 130387), as well as by SIP-IPN, Mexico.

REFERENCES

1. J. Y. Fan, X. L. Wu, P. K. Chu, Progress in Materials Science **51,** 983 (2006).
2. J. S. Shor, L. Bemis, A. D. Kurtz, I. Grimberg, B. Z. Weiss, M. F. MacMillian, W. J. Choyke, J. Appl. Phys. **76,** 4045 (1994).
3. M. Morales Rodriguez, A. Díaz Cano, T. V. Torchynska, J. Palacios Gomez, G. Gomez Gasga, G. Polupan and M. Mynbaeva , J. of Materials Science: Materials in Electronics, 19, n. 8-9, 682-686 (2008).
4. A. O. Konstantinov, A. Henry, C. I. Harris, E. Janzen, Appl. Phys. Lett. **66,** 2250 (1995).
5. N.E. Korsunska, I.V. Markevich, T. V. Torchynska, M.k. Sheinkman, phys. stat. solidi.(a), **60** (2), 565-572 (1980).

6. G. Polupan a, T.V. Torchynska, Thin Solid Films, **518,** S208–S211(2010),
7. T.V. Torchynska, G. Polupan, Superlattice and Microstructure, **45**, 222-227 (2009).
8. V. Petrova-Koch, O. Sreseli, G. Polisski, D. Kovalev, T. Muschik, F. Koch, Thin Solid Films **255,** 107 (1995).
9. O. Jessensky, F. Muller, U. Gosele, Thin Solid Films **297,** 224 (1997).
10. T V Torchynska, A. Diaz Cano, J.A. Yescas Hernandez, Yu.V. Vorobiev, L.V. Shcherbyna, phys.stat.solid.(c). **8**, No. 6, 1974–1977 (2011).
11. 11, B. Gil, A. V. Kavokin, Appl. Phys. Lett. 81, 748 (2002).
12. S. V. Gupalov, E. L. Ivchenko, A. V. Kavokin, J. Exp. Theor. Phys. 86, 388 (1998).
13. M.K. Sheinkman, N.E. Korsunskaya, I.V. Markevich, T.V. Torchinskaya, J. Physic. Chemist. Solids, 43 (5), 475-479 (1982).
14. N.E. Korsunskaya, I.V. Markevich, T.V. Torchinskaya and M.K. Sheinkman, J. Phys. C. Solid St.Phys. 13, 2975 -2982 (1980).
15. N.E. Korsunskaya, I.V. Markevich, T.V. Torchinskaya and M.K. Sheinkman, phys. stat. sol (a), 60, 565 -572 (1980).
16. M. Morales Rodriguez, A. Díaz Cano, T. V. Torchynska, G. Polupan and S. Ostapenko, J. Non-Cryst. Solids, **354,** 2272-2275 (2008).
17. T.V.Torchynska, J. Aguilar Hernandez, L. Sanchez Hernandez, G. Polupan, Y. Goldstein, A. many etc. Microelectronic Engineering, **66,** (1-4), 83-90 (2003).
18. T. V. Torchynska, M.K. Sheinkman, N.E. Korsunska, Physica B, Condenced Matter **273-274**, 955-958 (1999).
19. T. V. Torchynska, J. Aguilar Hernandez, A.I Diaz Cano, G. Contreras Puente, F. G. Becerril-Espinoza, Yu. Vorobiev, Yu.Goldstein etc., Physica B, Conden. Matter, **308-310,** 1108-1112 (2001).
20. T. V. Torchynska, A. Diaz Cano, M. Dybic, S. Ostapenko, M. Mynbaeva, Physica B **376-377,** 367 (2006).
21. T.V.Torchynska, A.L. Quintos Vazquez, G. Polupan, Y. Matsumoto, L. Khomenkova, L. Shcherbyna, J. Non Crystal. Solids, **354**,(19-25), 2186-2189 (2008).

Mater. Res. Soc. Symp. Proc. Vol. 1534 © 2013 Materials Research Society
DOI: 10.1557/opl.2013.297

Features of Electronic Transport in relaxed $Si/Si_{1-X}Ge_X$ Heterostructures with High doping level

Lev K. Orlov[1,2], A. A. Mel'nikova[2], Mikhail L. Orlov[1], Natal'ya A. Alyabina[3], Natal'ya L. Ivina[3], V. N. Neverov[4], and Zsolt J. Horváth[5,6]
[1]Institute for Physics of Microstrucrures, Russian Academy of Sciences, Nizhny Novgorod, Russia
[2]Nizhny Novgorod State Technical University, Nizhny Novgorod, Russia
[3]Nizhny Novgorod Lobachevsky University, Nizhny Novgorod, Russia
[4]Institure for Physics of Metals, Russian Academy of Sciences, Yekaterinburg, Russia,
[5]Óbuda University, Kandó Kálmán Faculty of Electrical Engineering, Institute of Microelectronics and Technology, Budapest, Hungary
[6]Institute of Technical Physics and Materials Science, Research Centre for Natural Sciences, Hungarian Academy of Sciences, Budapest, Hungary

ABSTRACT

The low-temperature electrical and magnetotransport characteristics of partially relaxed $Si/Si_{1-x}Ge_x$ heterostructures with two-dimensional electron channel ($n_e \geq 10^{12}$ cm^{-2}) in an elastically strained silicon layer of nanometer thickness have been studied. The detailed calculation of the potential and the electron distribution in layers of the structure was carried out to understand the observed phenomena. The dependence of tunneling transparency of the barrier separating the 2D and 3D transport channels in the structure was studied as a function of the doping level, the degree of blurring boundaries, layer thickness, and degree of relaxation of elastic stresses in layers of the structure. Tunnel characteristics of the barrier between the layers were manifested by the appearance of a tunneling component in the current-voltage characteristics of real structures. Instabilities, manifested during the magnetotransport measurements using both weak and strong magnetic fields are explained by the transitions of charge carriers from the two-dimensional into three-dimensional state, due to interlayer tunneling transitions of electrons.

INTRODUCTION

The electronic properties of $Si/Si_{1-x}Ge_x$ transistor heterocompositions with two-dimensional transport channels in different layers of the heterostructure are discussed in the literature for quite a long time. However, the greatest success in the area of real transistor devices promoting in millimeter waves, is obtained mainly in strained planar heterostructures containing quantum layers of the $Si_{1-x}Ge_x$ solid solution with electron or hole conductance [1]. The smaller number of papers is devoted to studying the properties of relaxed heterostructures with an electron transport channel in the silicon layers. Earlier the main efforts were concentrated on the analysis of the structural defect features in the grown heterocompositions [2] and on low-temperature magnetotransport measurements of two-dimensional electrons in the elastic - strained layers of Si [3]. Problems arising from the formation of conductive Si channels with high-mobility electron gas in the surface layer of the solid solution $Si_{1-x}Ge_x$, are connected with the necessity of introduction a plastic deformation area in the structure. This area should be controlled by the density of extended defects due to the heterojunction formed in the vicinity of three-dimensional network of misfit dislocations [4]. The formation takes place between the buffer layer of a solid solution and the silicon substrate. At the same time due to the uncontrolled multiplication of misfit dislocations, it is difficult to increase the percentage of germanium in the solid solution layers (above 30 at.%), if the depth of the quantum well formed in the layers of the $Si/Si_{1-x}Ge_x$ heterostructure is increased.

Increasing of the doping level in the $Si/Si_{1-x}Ge_x$ transistor heterostructures leads to the specific behavior of transport characteristics of these structures. These features are associated with the fundamental properties of the appearing two-dimensional layer [3], and with quite a noticeable influence of the charge of free carriers' on the potential of the quantum well and adjacent barrier layers. The last fact was the main motivation for this work to discuss the regularities observed in experiments using partially relaxed $Si_{1-x}Ge_x/Si/Si_{1-x}Ge_x$ transistor heterostructures.

EXPERIMENT

The n^+-SiGe/n-Si/p-SiGe double heterostructures with the two-dimensional silicon transport channel, with a thickness of about 10 nm (see Figure 1) studied in the present work, were characterized by the value of the surface carrier concentration $n_{cr} > 10^{12}$ cm^{-2}, and a relatively low values of electron mobility in the system ($\mu_e \approx 3000$-5000 cm^2/Vs at liquid helium temperature) [5]. The details of sample preparation are presented elsewhere [5]. The samples were studied using magnetoresistance, Hall and current-voltage measurements.

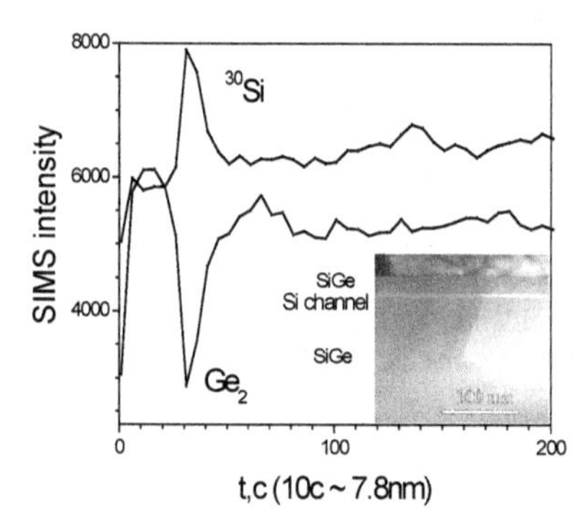

Figure 1. The distribution of Si and Ge alloy components along the thickness of the SiGe/Si/SiGe structure with the Si transport channel $d_{Si} \approx 8$ nm. The inset shows the electron microscopic image of Sample 412.

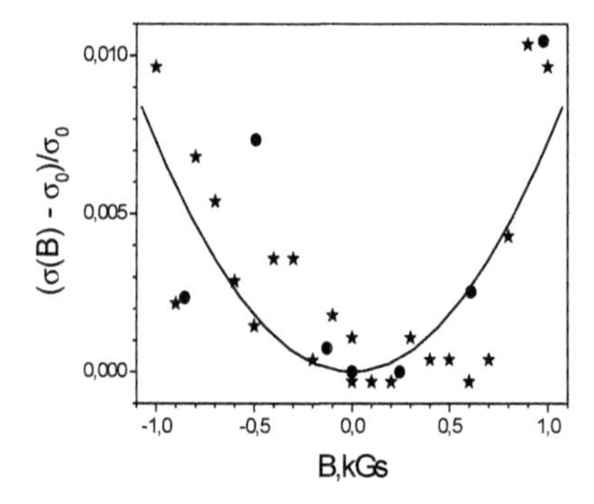

Figure 2. The magnetic-field dependence of correction to conductivity $\Delta\sigma$ for Sample 414 (T = 4.2 K). The conductivity of the structure is equal to $\sigma_0 = 0.0056\ \Omega^{-1}$ at B = 0. Symbols - experimental data, line - approximating parabola.

RESULTS AND DISCUSSION

We have previously shown that the electrons in the Si channel exhibit two-dimensional properties in a weak magnetic field even at relatively low mobility, partly related to the participation of several groups of carriers from the different layers in transport. These properties are expressed, in particular, in the appearance of a negative magnetoresistance in the system described by the logarithmic type corrections to the conductivity and associated with the manifestation of weak localization [5].

The typical form of the magnetic-field dependence of the correction to conductivity, $\Delta\sigma = \sigma(B) - \sigma_0$, where $\sigma_0 = \sigma(B = 0)$, observed in a weak magnetic field at low temperature about 5K (for Sample 414 exhibiting the highest conductivity) is shown in Figure 2. A part of the results, represented by symbols – stars, were obtained on the samples being cut in the form of a square probe with alloyed contacts (in the low-temperature experiment with the use of inserts), placed in liquid helium. Experimental data, represented by symbols – points, were obtained in a specialized liquid dilution cryostat (Oxford-instrument) on the bridge structures carved in the shape of a Greek cross. At zero magnetic field at liquid helium temperature the conductivity of the Sample 414 was $\sigma_0 = 0.0056\ \Omega^{-1}$. The line in Figure 2, approximating the experimental points, is a polynomial curve, which shows a quadratic dependence of observed correction to conductivity $\Delta\sigma = \sigma(B) - \sigma_0$ on the magnetic field: $\Delta\sigma/\sigma_0 \approx 0,00726\ B^2$.

The observed form of the dependence of correction to conductivity versus the magnetic field points to the fact that in these heterostructures with high (> 10^{12} cm^{-2}) surface electron density the magnetoresistance in weak magnetic field may be associated with a variety of mechanisms, including the existence of few groups of charge carriers in a system. These groups can be the electrons distributed among different energy valleys of the conduction band of silicon layers as well as charge carriers in the top barrier layer of the structure.

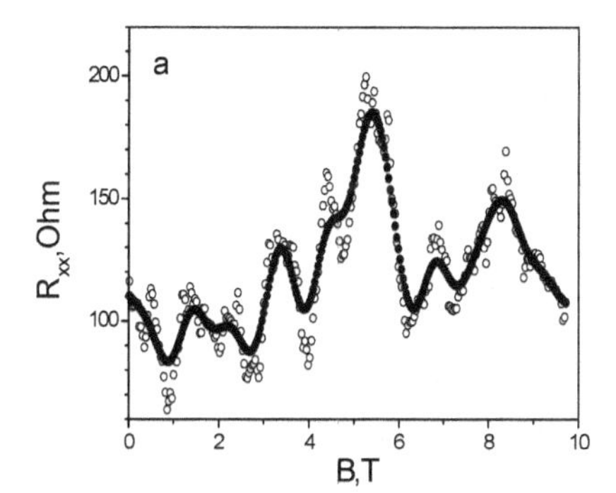

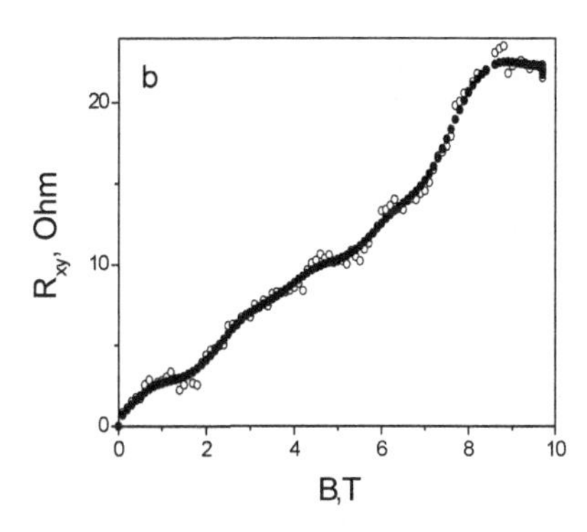

Figure 3. Magnetoresistance (a) and Hall (b) characteristics of Sample 414 as a function of magnetic field. Open symbols represent the averaged experimental data. Solid symbols are obtained by smoothing the experimental data.

The most reliable evidence of two-dimensional properties of carriers in the system is the results of magnetoresistance and Hall measurements performed on the samples in strong quantizing magnetic fields. The characteristic shape of the curves for one of samples in magnetic fields up to 10T at liquid helium temperature is shown in Figure 3. The formation of the narrow potential barrier in the vicinity of the upper heterojunction enhances the probability of interlayer transitions of charge carriers in structures with a high doping level in the upper layer. In some cases, it has a noticeable effect on the transport characteristics of the system [6]. It can be assumed that the considerable scatter of points on measured magnetotransport characteristics, observed in our experiments, is associated with the specific of potential distribution in the layers of investigated heterostructures, which makes possible an effective transfer of electrons between the parallel transport channels in the system.

The numerical analysis of the potential distribution in the layers of the structure was carried out on the basis of the Poisson equation and quasihydrodynamics equation for the better understanding the results obtained in samples with the different character of the elastic strain in the layers and the different levels of doping. Dependences of the shape of potential barrier, formed in the vicinity of the heterojunction, and the characteristics of the quantum well, formed in the silicon layer, versus different parameters of the system, such as the composition and doping level in the layers and the width of quantum well, were found. The characteristic shape of the energy band diagram for Sample 414 is shown in Figure 4. The width of the barrier at its half-maximum ($h_b/2$ = 73 meV) equals $W_b \approx$ 2 nm, as defined relatively to the bottom of conduction band in the upper doped SiGe layer (see Figure 4.b). The calculation shows a strong dependence of the potential characteristics, formed in the vicinity of the silicon transport channel, on a number of initial parameters of the system. The small width of the barrier layer and the built-in electric field in the potential well yield the deviation of its shape from the rectangular form.

The corresponding distribution of electrons in the layers of the same Sample 414 is shown in Figure 5. It is seen that electrons are localized near the upper edge of the quantum well,

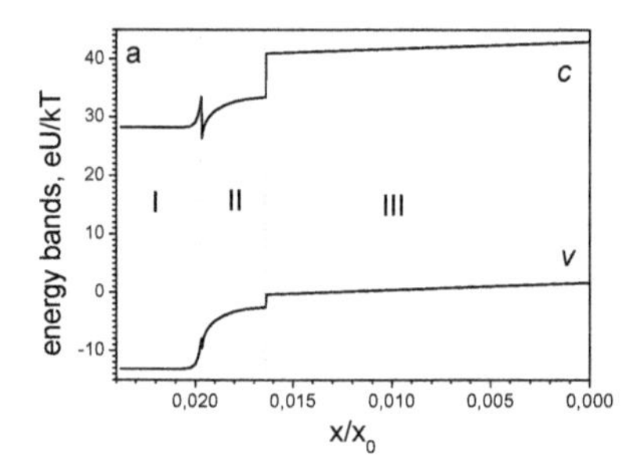

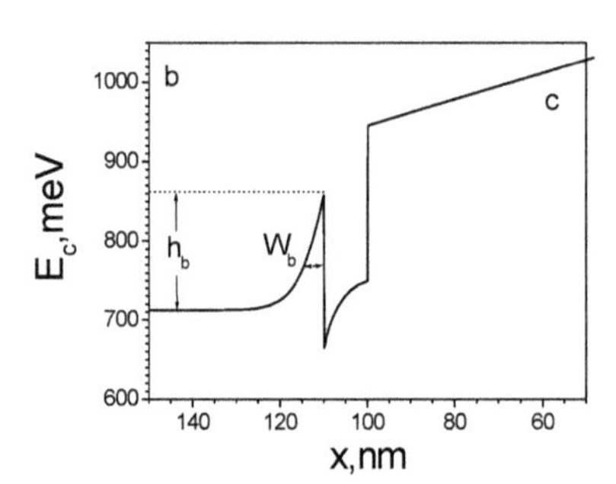

Figure 4. The potential distribution across the Si electron quantum well and the adjacent SiGe barrier layers for a donor concentration $N_D = 2*10^{18}$ cm^{-3} in Sample 414 with the maximum elastic-strained layer of Si (d_{Si} = 10 nm) with the parameters E_g(Si) = 0.93 eV, $E_g(Si_{0.75}\,Ge_{0.25})$ = 1.07 eV, ΔE_v = 0.056 eV [7].

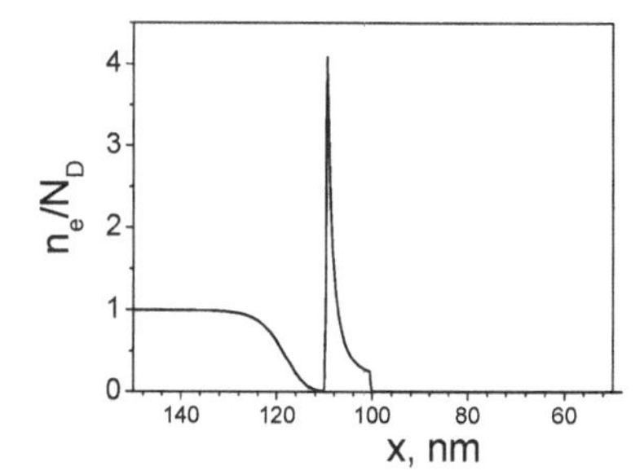

Figure 5. Distribution of the electron density n_e in the structure layers, referred to the donor concentration $N_D = 2*10^{18}$ cm^{-3} in the upper SiGe layer.

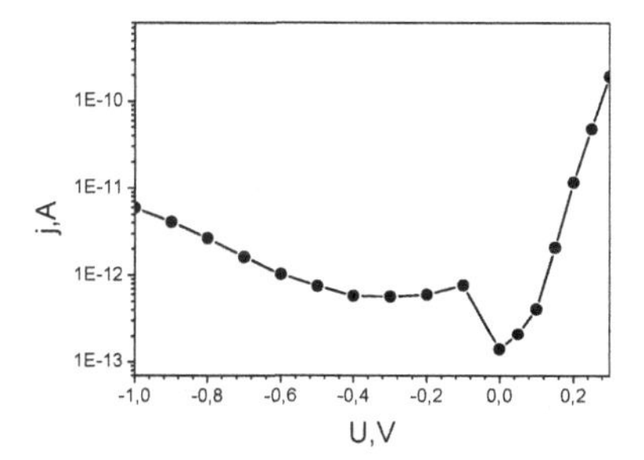

Figure 6. Transverse current-voltage characteristics of the n^+-SiGe/n-Si/p-SiGe heterostructure measured on a mesa structure with Al ohmic contacts at 100K.

and their concentration can significantly exceed the average level of doping of the top barrier layer. The higher positions of allowed electronic states in narrower energy barrier well up to the values, where tunneling is possible via the barrier, can provide a variation of measured transport properties even for the samples prepared at the same conditions. The instability observed during the transport measurements in high magnetic fields can be explained by interlayer tunneling transitions of 2D carriers from the quantum well to a three-dimensional state.

The small width of the barrier formed at the heterojunction leads to the possibility of interlayer tunneling, even in non-resonant conditions. Interlayer tunneling can occur even, if there are local states, which may additionally lead to a resonant tunneling via the layers at lower electric fields in the vicinity of the heterojunction resulting in a peak in the current at low applied bias and decreasing current with increasing the bias above the peak. Such a possibility was first discussed concerning the output characteristics of a InGaAs/InAlAs transistor structure in Ref. [6]. This mechanism may occur in silicon transistor structures as well, in particular studying the current flow across the plane of the structure [8]. The possibility of observation the falling down section on the current-voltage characteristic of diodes based on the n^+-SiGe/n-Si/p-SiGe heterostructure is shown in Figure 6. It was measured on a mesa structure with the Al ohmic contacts. The effect was observed only at temperatures below 120K at relatively low currents, corresponding mostly to the reverse branch of current-voltage characteristics.

CONCLUSIONS

The results of magnetoresistance and Hall measurements indicated the presence of a two dimensional electron gas in the studied n^+-SiGe/n-Si/p-SiGe heterostructures. However, it has been concluded that the features of magnetic field dependences and current-voltage characteristics can be explained by a variety of mechanisms, such as the existence of few groups of charge carriers in a system and the interlayer tunneling transitions of two-dimensional carriers from the quantum well to a three-dimensional state.

ACKNOWLEDGMENTS

The work was carried out in cooperation between the Russian and Hungarian Academies of Sciences (project number 17) and with the financial support from the Federal target program "Scientific and scientific-pedagogical personnel of innovative Russia" in 2009 -2013 years.

REFERENCES

1. J. Halstedt, M. von Haartman, P. E. Hellstrom, M. Ostling, H. H. Radamsson. *"Hole mobility in ultrathin body SOI pMOSFETs with SiGe or SiGeC channel"*, IEEE EDL, **27**, 466 (2006).
2. R. Hull, J. C. Bean, D. J. Werder, R.E. Leibenguth, *"In situ observations of misfit dislocation propagation in GeSi/Si(100) heterostructures"*, Appl. Phys. Lett., **52**, 1605 (1988).
3. E. B. Olshanetsky, V. Renard, Z. D. Kvon, J. C. Portal, N. J. Woods, J. Zhang, J. J. Harris, *"Conductivity of a two-dimensional electron gas in a Si/SiGe heterostructure near the metal-insulator transition: Role of the short- and long-range scattering potential"*, Phys. Rev. B., **68**, 085304 (2003).
4. T. G. Yugova, V. I. Vdovin, M. G. Milvidskii, L. K. Orlov, V. A. Tolomasov, A. V. Potapov, N. V. Abrosimov, *"Dislocation pattern formation in epitaxial structures based on SiGe alloys"*/ Thin Solid Films, **336**, 112 (1998).
5. L. K. Orlov, Z. J. Horvath, M. L. Orlov, A. T. Lonchakov, N. L. Ivina, L. Dobos, *"Anomalous electrical properties of $Si/Si_{1-x}Ge_x$ heterostructures with an electron transport channel in Si layers"*, Physics of the Solid State, **50**, 330, (2008).
6. M. L. Orlov, L. K. Orlov, *"Mechanisms of negative resistivity and generation of terahertz radiation in a short-channel InGaAs/InAlAs transistor"*, Semiconductors, **43**, 652 (2009).
7. M. M. Rieger, P. Vogl, *"Electronic-band parameters in strained $Si_{1-x}Ge_x$ alloys on $Si_{1-y}Ge_y$ substrates"*, Phys. Rev. B, **48**, 14276 (1993).
8. L. K. Orlov, Z. J. Horvath, A. V. Potapov, M. L. Orlov, S. V. Ivin, V. I. Vdovin, E. A. Steinman, V. M. Phomin, *"Electrical characteristics and the energy band diagram of the isotype n-SiGe/n-Si heterojunction in relaxed structures"* Physics of the Solid State, **46**, 2139 (2004).

III-V group semiconductors

Mater. Res. Soc. Symp. Proc. Vol. 1534 © 2013 Materials Research Society
DOI: 10.1557/opl.2013.298

Comparative study of Photoluminescence variation in InAs Quantum dots embedded in InAlGaAs Quantum wells

J.L. Casas Espinola[1], T.V. Torchynska[1], L. D. Cruz. Diosdado[2] and G. Polupan[3]
[1]ESFM– Instituto Politécnico Nacional, México D. F. 07738, México
[2]UPIITA– Instituto Politécnico Nacional, México D. F. 07720, México
[3]ESIME– Instituto Politécnico Nacional, México D. F. 07738, México

ABSTRACT

The photoluminescence (PL) and its temperature dependences have been studied in MBE grown InAs quantum dots (QDs) embedded in $Al_{0.3}Ga_{0.7}As/In_{0.15}Ga_{0.85}As/Al_xGa_{1-x}In_yAs$ quantum wells (QWs) in dependence on the composition of capping layers. Two types of capping layers ($Al_{0.3}Ga_{0.7}As$ and $Al_{0.40}Ga_{0.45}In_{0.15}As$) were investigated.

Temperature dependences of PL peak positions in QDs have been analyzed in the range of 10-300K and compared with the temperature shrinkage of the band gap in the bulk InAs crystal. This permits to investigate the efficiency of the Ga(Al)/In inter diffusion processes between QDs and capping layers in dependence on the capping layer compositions. The band gap fitting parameters obtained for InAs QDs have been compared with known ones for the bulk InAs crystal. It is shown that the efficiency of the Ga(Al)/In inter diffusion is high in the QD structures with $Al_{0.3}Ga_{0.7}As$ capping layer. Finally the reasons of higher thermal stability of the structure with $Al_{0.40}Ga_{0.45}In_{0.15}As$ capping layer have been analyzed and discussed.

INTRODUCTION

Over the past twenty years, zero-dimensional (0D) quantum dot (QD) systems with three-dimensional quantum confinement have attracted considerable interest from both the fundamental physics and potential device applications in: light – emitting and photo diodes, solar cells and memory devices [1-3]. The realization of the efficient light-emitting devices operated at room temperature requires understanding the QD PL temperature dependences and the study of the reasons of PL variation versus temperature. The PL intensity decay in InAs QDs, as a rule, is attributed to thermal escape of excitons from the QDs into a wetting layer (WL) or into the GaAs barrier [4-8], or to a thermally activated capture of excitons by the nonradiative defects in the GaAs barrier or at the GaAs/InAs interface [5, 8]. It was shown experimentally that the main reason for the PL thermal decay in QD structures is related to the thermal escape of the excitons or correlated electron-hole pairs from QDs [8-10].

In QD structures introducing the additional intermediate $Al_xGa_{1-x}As$ layers into InGaAs/GaAs QWs will increase the height of the potential barrier for the exciton thermal escape from QDs into the barrier and can permit the application of these QD structures at higher temperatures. The same effect can be achieved with the introduction of the intermediate $Al_xGa_{1-x}In_yAs$ QW layers in InGaAs/GaAs QWs. It is known that QD lasers fabricated with AlInGaAs QWs have shown superior performances at a higher temperature due to a larger conduction band offset [11, 12]. Even though the better device performance of QD-laser diodes in comparison with the QW laser diodes have been demonstrated, more reliable and reproducible manipulations during the formation of self-assembled QDs is essential for improving the QD optical properties. Improved understanding the operation and design peculiarities of InAs QDs embedded into InGaAs/InAlGaAs QWs could be obtained from the study of the variation of the PL spectra with temperatures.

EXPERIMENTAL DETAILS

The solid-source molecular beam epitaxy (MBE) in V80H reactor was used to grow the waveguide structures consisting of the layer of InAs self-organized QDs inserted into 9 nm $In_{0.15}Ga_{0.85}As$ QW layer. The thickness of the buffer $In_{0.15}Ga_{0.85}As$ layer was 1nm, which was grown on the 300.nm $Al_{0.30}Ga_{0.70}As$ buffer layer and the 2 inch (100) GaAs SI substrate. Then an equivalent coverage of 2.4 monolayers of InAs QDs were confined by the first capping (8 nm) $In_{0.15}Ga_{0.85}As$ layer, by the second 100 nm $In_yAl_xGa_{1-x}As$ capping layer, and by the 10 nm AlAs and 2 nm GaAs layers (Fig.1).

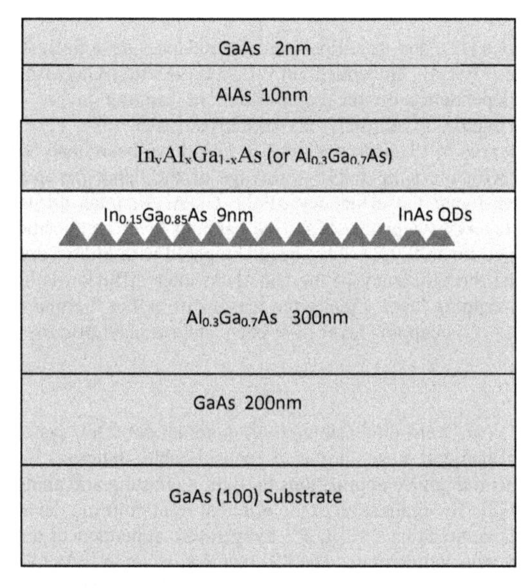

Figure 1. The design of studied QD structures.

Two groups of QD samples with the different compositions of the second $In_y Al_x Ga_{1-x}As$ (100nm) capping layers were investigated. In the first group the second capping layer is $Al_{0.30}Ga_{0.70}As$ (#1) and in the second group it is $Al_{0.40}Ga_{0.45}In_{0.15}As$ (#2). Investigated QD structures are grown under As-stabilized conditions at the temperature 510°C, during the deposition of the InAs active region and InGaAs layers. The in-plane density of QDs is estimated from previous AFM study was 4 x $10^{10}cm^{-2}$. The samples were mounted in a closed-cycle He cryostat where the temperature is varied in the range of 10 - 500 K. PL spectra were measured under the excitation of the 488 nm line of a cw Ar+-laser at an excitation power density of 500 W/cm^2 in the temperature range 10-500K. The setup used for PL study was presented earlier in [9,10].

EXPERIMENTAL RESULTS AND DISCUSSION

Typical PL spectra of the freshly prepared structures #1 and #2 measured at different temperatures in the range 10-500K are shown in Fig.2a,b.

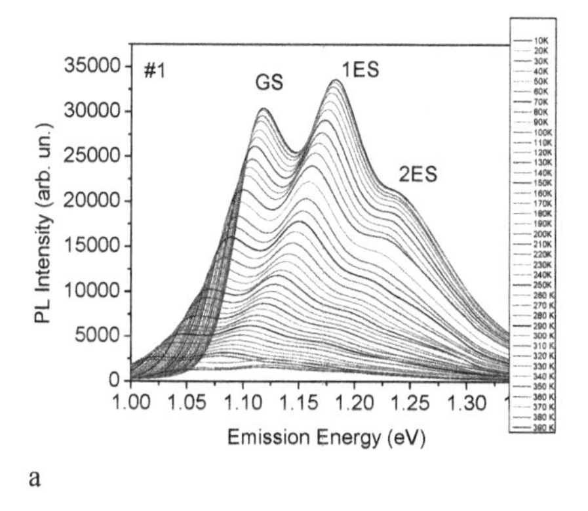

a

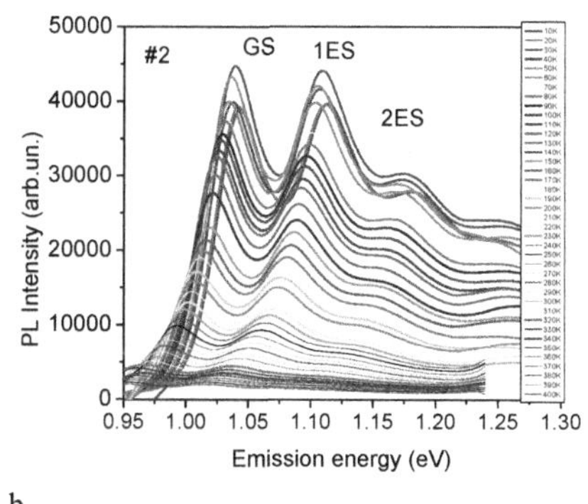

b

Figure 2a. PL spectra of the QD structure #1 measured at different temperatures from the range 10-300K.

Figure 2b. PL spectra of the QD structure #2 measured at different temperatures from the range 10-300K.

At least three overlapping PL bands appear due to the recombination of excitons localized at a ground state (GS) and at the first (1ES) and second (2ES) excited states in InAs QDs (Fig.2a,b). The peak positions of the GS at 10K are 1.118 eV (#1), 1.039 eV (#2) and the 1ES are 1.182 eV (#1), 1.109 eV (#2). As one can see the full width at half maximum (FWHM) of GS and 1ES (2ES) states and their overlapping in the structure #1 is higher in comparison with #2.

The composition of $In_{0.15}Ga_{0.85}As$/ $Al_{0.30}Ga_{0.70}As$ buffer layers and the growth conditions were the same for both studied QD structures; as a result, the original (uncovered) sizes of QDs are similar in #1 and #2. Thus it is possible to expect the same positions of GS emission in the QD structures #1 and #2. But as it is clear from figure 2a,b the GS PL peak in #2 locates at 1.039 eV and in #1 it is at1.118 eV at 10K. More "red" position of GS emission in #2 testifies that this QD structure, apparently, is characterized by the smaller level of strains and, as a result, the less efficient process of Ga/Al/In intermixing.

To analyze the process of Ga/Al/In intermixing the PL spectra of InAs QDs have been studied at different temperatures (Fig.2a,b). Then the dependences of GS PL peak positions versus temperatures have been obtained (Fig. 3a,b). It can be seen from Fig. 3 that the PL peaks shift to lower energies due to the shrinkage of the band gap with increasing temperature. Generally, due to the temperature induced lattice dilatation and the electron–lattice interaction, the band gap energy follows the well-known Varshni formula [13]:

$$E(T) = E_o - \frac{aT^2}{T+b} \quad , \quad (1)$$

where Eg(0) is the band gap at the absolute temperature T = 0 K; "a" and "b" are the Varshni thermal coefficients.

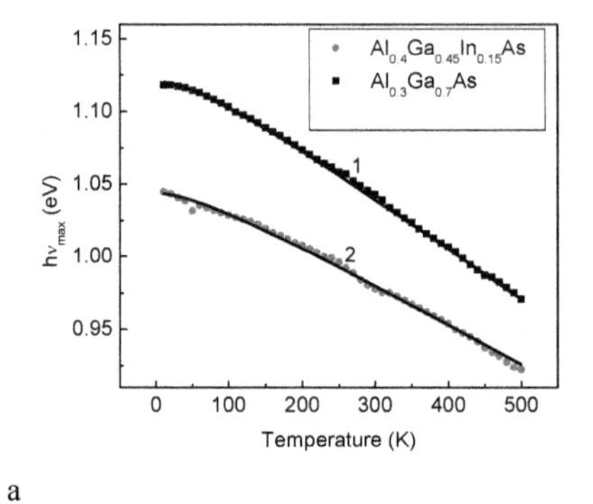

a

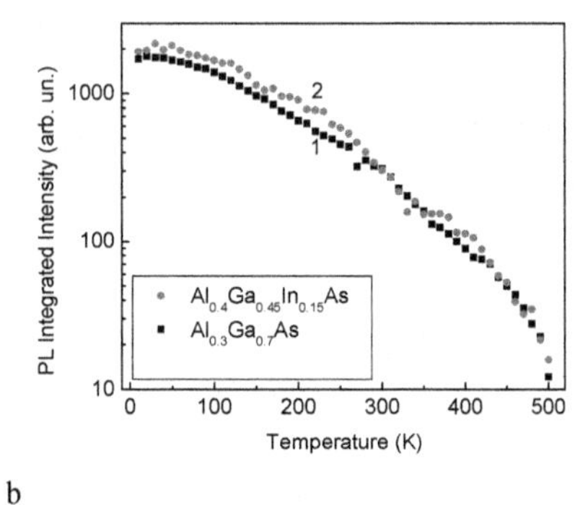

b

Figure 3. The variation of PL peak positions versus temperature in the structures #1 (1) and #2 (2)

Figure 4. The variation of integrated PL intensities versus temperature in the structures #1 (1) and #2 (2)

The lines in Fig.3 present the fitting results for the structures #1 and #2. The comparison of fitting parameters "a" and 'b' with the variation of the energy band gap versus temperature in the bulk InAs crystal (Table 1) has revealed that in QD structures #2 the fitting parameter "a" and 'b' are the same as their values in the bulk InAs crystal (Table 1). At the same time in the structure #1 the Varshni fitting parameters "a" and 'b' are little bit different than those in the bulk InAs. It is clear that a QD material composition in the structure #1 (AlGaAs capping) varies in comparison with the structure #2 (AlGaInAs capping) in the process of InAs QD capping at high temperatures due to efficient the Ga(Al)/In intermixing [15, 16].

Table 1. Fitting Varshni parameters in studied QD structures and in the bulk InAs

QD structure	E_o	α (meV/K)	β (K)
#1 ($Al_{0.3}Ga_{0.7}As$)	1.12	0.35	95
#2 ($Al_{0.4}Ga_{0.45}In_{0.15}As$)	1.04	0.28	83
InAs [14]	0.415	0.28	83

The variation of GS integrated PL intensities versus temperature in the QD structures #1 and #2 has been presented in figure 4. As it is clear the integrated PL intensities decay versus temperature by the same way in #1 and #2. It means that introducing the both types ($Al_{0.3}Ga_{0.7}As$ and $Al_{0.40}Ga_{0.45}In_{0.15}As$) of intermediate layers stimulate increasing the potential barrier in QWs for the exciton thermal escape from QDs into the barrier with the following nonradiative recombination there.

CONCLUSIONS

The photoluminescence (PL) and its temperature dependences have been studied in $GaAs/Al_{0.3}Ga_{0.7}As/In_{0.15}Ga_{0.85}As/In_yAl_xGa_{1-x}As/GaAs$ quantum wells (QWs) with embedded InAs QDs in dependence on the composition of capping ($Al_{0.3}Ga_{0.7}As$ and $Al_{0.40}Ga_{0.45}In_{0.15}As$). It is shown that a QD material composition in the structure #1 ($Al_{0.3}Ga_{0.7}As$ capping) varies in comparison with the structure #2 ($Al_{0.40}Ga_{0.45}In_{0.15}As$ capping) and the bulk InAs due to the Ga/Al/In intermixing. At the same time the both types of intermediate layers ($Al_{0.3}Ga_{0.7}As$ and $Al_{0.40}Ga_{0.45}In_{0.15}As$) stimulate increasing of the potential barriers in QWs for the exciton thermal escape from QDs into the barrier that manifest themselves by the same character of thermal decay of the integrated PL intensities in the structures #1 and #2.

ACKNOWLEDGEMENT

The author would like to thank the CONACYT (project 130387) and SIP-IPN, Mexico, for the financial support, as well as the Dr. Andreas Stinz from the Center of High Technology Materials at University New Mexico, USA, for growing the studied QD structures.

REFERENCES

1. D. Bimberg, M. Grundman, N. N. Ledentsov, Quantum Dot Heterostructures, Ed. Wiley & Sons (2001) 328.
2. A. Stintz, G. T. Liu, L. Gray, R. Spillers, S. M. Delgado, K. J. Malloy, J. Vac. Sci.Technol. B. **18(3),** 1496 (2000).
3. T. V. Torchynska, J. L. Casas Espínola, E. Velazquez Losada, P. G. Eliseev, A. Stintz, K. J. Malloy, R. Peña Sierra, Surface Science **532**, 848 (2003).
4. C. M. A. Kapteyn, M. Lion, R. Heitz, and D. Bimberg, P. N. Brunkov, B. V. Volovik, S. G. Konnikov, A. R. Kovsh, and V. M. Ustinov, Appl. Phys. Lett. **76,** 1573 (2000)
5. C. A. Duarte, E. C. F. da Silva, A. A. Quivy, M. J. da Silva, S. Martini, J. R. Leite E. A. Meneses and E. Lauretto, J. Appl. Phys., **93,** 6279 (2003).
6. X.Q. Meng, B. Xu, P. Jin, X.L. Ye, Z.Y. Zhang, C.M. Li, Z.G. Wang, Journal of Crystal Growth **243,** 432 (2002).
7. L. Seravalli, P. Frigeri, M. Minelli, P. Allegri, V. Avanzini, S. Franchi, Appl. Phys. Lett. **87,** 063101 (2005).
8. T. Torchynska, J. Appl. Phys., **104,** 074315, n.7 (2008).
9. T.V.Torchynska, A. Stintz, J. Appl. Phys. **108**, 2, 024316 (2010).
10. T. V. Torchynska, J. L. Casas Espinola, L. V. Borkovska, S. Ostapenko, M. Dybic, O. Polupan, N. O. Korsunska, A. Stintz, P. G. Eliseev, K. J. Malloy, J. Appl. Phys., **101,** 024323 (2007).

11. C.E. Zah, R. Bhat, B.N. Pathak, F. Favire, F., W. Lin, M.C. Wang, etc, IEEE J. Quantum Electron., **30(2),** 511–523 (1994).
12. K. Takemasa, M. Kubota, T. Munakata, and H. Wada, IEEE Photon. Technol. Lett., **11(8),** 949-951 (1999).
13. Y. P. Varshni, Physica **34**, 149 (1967).
14. A. Landolt-Boernstein, Numerical Data and Functional Reationship, in: Science and Technology, Vol. 22 (Springer, Berin, 1987), p. 118.
15. M. Dybiec, S. Ostapenko, T. V. Torchynska, E.Velasquez Losada, Appl. Phy. Lett.**84,** 5165-5167 (2004).
16. M. Dybiec, L. Borkovska, S. Ostapenko, T. V. Torchynska, J. L. Casas Espinola, A. Stintz and K. J. Malloy, Applid. Surface Science, **252**, 5542-5545(2006)

Mater. Res. Soc. Symp. Proc. Vol. 1534 © 2013 Materials Research Society
DOI: 10.1557/opl.2013.299

Effect of the Surface on Optical properties of AlGaAs/GaAs heterostructures with double 2-DEG

L. Zamora-Peredo[1*], I. Cortes-Mestizo[1], L. García-Gonzalez[1], J. Hernandez-Torres[1], D. Vázquez-Cortes[2], S. Shimomura[2],A. Cisneros[3], V. Méndez-García[3] and M. Lopez-Lopez[4].

[1] Centro de Investigación en Micro y Nanotecnología, Universidad Veracruzana, Calzada Adolfo Ruiz Cortines # 455, Fracc. Costa Verde, C.P. 94292, Boca del Río, Veracruz, México.
[2] Graduate School of Science and Engineering, Ehime University, 3 Bukyo-cho, Matsuyama, Ehime 790-8577, Japan.
[3]Coordinación para la Innovación y Aplicación de la Ciencia y Tecnología, Universidad Autónoma de San Luis Potosí, Av. Karakorum No. 1470, Lomas 4a Secc., San Luis Potosí, S.L.P. 78210, México.
[4] Departamento de Física, Centro de Investigaciones y de Estudios Avanzados - IPN, México D. F., México

ABSTRACT

In this work we studied a set of heterostructuresAlGaAs/GaAs by photoreflectance (PR) and photoluminescence (PL) spectroscopy in order to determine the effect of the surface on optical properties of a symmetric double two-dimensional electron gas(2DEG). The potential profiles of the edge of the conduction band were modeled by nextnano software. By Hall effect measurements at 77K we determined the electron mobility of the samples. Thesurface electric field was calculated by the analysis Franz-Keldysh oscillations observed in thePR spectra.Thecalculation of the band structure shows that the surface electric field prevents the creation of 2DEG closest to the surface. PL spectra shows a peak originated in thedeepest AlGaAs layer, close to the 2DEG region,associated with to the free exciton whose intensity increases as the electric field decrease.

INTRODUCTION

Usually, the electronic characteristics of the semiconductor devices can be affected by the properties of its surface. In order to eliminate chemical instability that may cause undesired effects it is well established that the surfaces of semiconductor-based devices have to be treated properly. Frequently, the surfaces of semiconductor devices are passivated in order to stabilize their chemical nature and to eliminate reactivity [1-4]. In the AlGaAs/GaAsheterostructures case, is possible to study the effect of the surface on the electronic properties of a 2DEG by electrical measurements [3, 4], inspecific the relationship between the surface electric field and the electron mobility.

Optical approaches like photoreflectance and photoluminescence spectroscopy are very useful techniques that have been widely used for the study and characterization of AlGaAs/GaAs system [5-11]. Some reports have been focused to determinate the origin of Franz-Keldyshoscillations (FKO) that usually are observed at the PR spectra, now we know that wide-period FKO are associated to surface electric field and short-period FKO are originated by internal AlGaAs/GaAs interfaces [9-11]. In this work, we studied a set of heterostructures by PR and PL measurements in order to determinate the effects of the surface electric fieldover the conduction band behavior.

EXPERIMENT

Fiveheterostructures were grown on semi-insulating GaAs (100) substrate by molecular beam epitaxy. Table 1 shows the thickness of different layers grown for each sample. All samples have a 1 μm-thick GaAs buffer layer (BL), a first spacer layer (1SL) of 7 nm undoped $Al_xGa_{1-x}As$, followed by a 80-nm-thick Si-doped AlGaAs barrier (M1 and M3), next a second spacer layer (2SL) of 7 nm undoped $Al_xGa_{1-x}As$, finally the structure was capped with 25 nm of undopedGaAs. The barrier layer was with a Si concentration of $1.4x10^{18}$ atoms/cm^3 and nominal Al concentration of 36%. Some samples have an intermediary layer of 10 nm-thick AlGaAs between AlGaAs:Si barrier (M2, M4 and M5), the other layers were similar to M1. Two sampleswere grown with an additional 2-nm-thick GaAs cap layer doped with $5x10^{17}$ cm^{-3} of Si (M3 and M4).

The mobility and concentration of electrons were determined from Hall measurements at 77K. The measurements were accomplished for use the Van der Pauw method with a current intensity of 100 mA and under a magnetic field of 0.5 T. Room temperature PR measurements were carried out employing and experimental setup similar to those described elsewhere [7]. With a 543 nm line of solid state laser as the modulation source with a maxima output power of 80 mW chopped at a frequency of 200 Hz. The 0.5 m focal distance Acton monochromator and a Spec 10 CCD camera system of Princeton Instruments have been used.PL measurements at 20 K were carried out with the same laser, monochromator and detector used in PR experiments.

Table 1. Thickness of the layers grown in the samples studied in this work.

	Sample					
Layer	**M1**	**M2**	**M3**	**M4**	**H5**	
GaAs : Si	0	0	2	2	0	nm
GaAs	25	25	25	25	25	
$Al_xGa_{1-x}As$	7	7	7	7	7	
$Al_xGa_{1-x}As$: Si	40	40	40	40	40	
$Al_xGa_{1-x}As$	0	10	0	10	10	
$Al_xGa_{1-x}As$: Si	40	40	40	40	40	
$Al_xGa_{1-x}As$	7	7	7	7	7	
BL	1 μm					
Substrate	GaAs (100)					

RESULTS AND DISCUSSION

With parameters listed at the table 1, if the surface electric field is not considered, it is possible to calculate the potential profile of the edge of the conduction band and we expect a symmetric double 2DEG (SD-2DEG) in the AlGaAS/GaAsheterojuntions. Figure 1 shows the conduction band behavior for M1 and M4samples where is clear that there is a SD-2DEG,similar situation was observed for the other samples. However, it is well established that the surface electric field produces a band bending from the surface to a distance *d* toward inside the heterostructure, which may cause that the triangular well potential nearest to the surface does not have electron confinement.

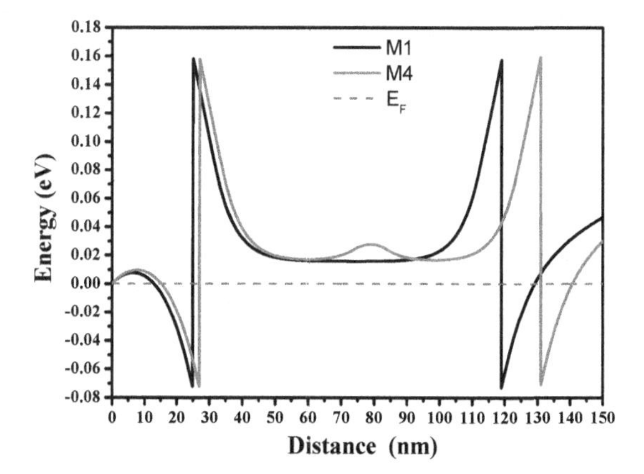

Figure 1. Potential profiles of the edge of the conduction band for M1 and M4 calculated by nextnano software (not considering the surface electric field). E_F is the Fermi´s energy.

In order to determine the surface electric field, PR measurements were carried out. Figure 2 shows the PR spectra of the samples. As we can see, there are two types of Franz-Keldysh oscillations (FKO) in PR spectra. Short-period oscillations (SP-FKO) near to the energy band gap of GaAs range from 1.42 to 1.44 eV and broad-period oscillations (BP-FKO) range from 1.42 to 1.68 eV. All PR spectra have very similar line shape despite that the samples have different thickness layers. In previous works, BP-FKO has been associated to the surface electric field [9-11]. By FKO analysis similar to those described elsewhere [12] we calculated the electric field strength (F_S), which range from 2.27 to 2.57 x 10^7 V/m (see table 2). Thereafter, using the Gauss's law and considering that on the sample surface there is a sheet charged with concentration of charge per unit areaN_S, we calculated this Ns by using the electric field F_S obtained by PR through the equation 1 [13]:

$$N_S = \varepsilon\varepsilon_0 F_S \tag{1}$$

Table 2 shows the concentration n_H and electron mobility μ_H obtained by Hall measurements at 77 K and the surface electric field F_S and the charge density N_S obtained by the PR analysis. There is not a clear relationship between the surface electric field and the electron mobility, which is an indicationthat perhaps there is no double-2DEG system.

In order to research the effect of the $F_S$over the profile of the conduction band, we used the charge density of M1 and M2 (with minimum and maximum electric field, respectively) in the calculations of the conduction band profile. Figure 3 shows the potential profile resulting. As we can see, the band bending provoked by the surface electric field affects deep inside the heterostructure down to approximately 60 nm, forbidding the formation of the 2DEG in the triangular well potential close to the surface. That is, there is only a single 2DEG in the AlGaAs/GaAsheterostructures due to the bending of the conduction band, which is originated by the surface electric field. The electron mobility specified in the table 2 is associated to the buffer/first-spacer-layer interface. Then, the variability observed is associated to the change on the electron concentration n_H and to the internal electric field originate by the ionized Si atom in the AlGaAs layer. This relationship can be studied by Franz-Keldysh analysis of theSP-FKO, presented in other reports [11].

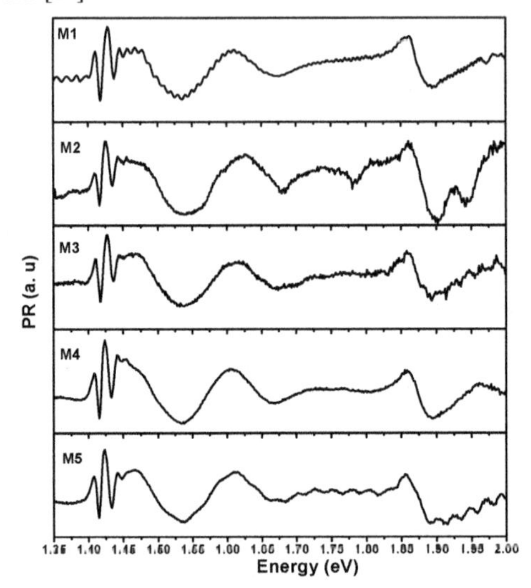

Figure 2. PR spectra of AlGaAs/GaAsheterostructures obtained with a 543 nm laser line as modulation source.

Table 2. Concentration n_H and electron mobility μ_H obtained by Hall measurements at 77 K. Surface electric field $F_S$and charge density N_S obtained by room temperature PR measurements.

Sample	n_H (x 10^{11} cm^{-2})	μ_H (cm^2/V.s)	F_S (x10^7 V/m)	N_S (10^{12} cm^{-2})
M1	9.32	88 413	2.27	1.59
M2	8.35	96 291	2.57	1.80
M3	9.23	84 388	2.36	1.65
M4	9.30	93 785	2.53	1.77
M5	10.13	83 695	2.31	1.62

The electron mobility Figure 4 shows PL spectra for all samples. One can resolve two parts of the spectra indicate by segments. The first one, below 1.54 eV, is attributed to the GaAs layers. Two peaks at 1.49 eV and 1.51 eV are observed, associated to the free exciton and band-to-band transitions, respectively. The inset 4(a) shows the comparison of PL spectra of the M1 and M3 samples. The signal associated to the free exciton decrease when the heterostructure is caped with the doped GaAs layer due to weak increment of the surface electric field.

The second part of the PL spectra is a wide peak close to 1.95 eVassociated with the AlGaAs layers. As we can see, the PL intensity from this peak increase when the undopedAlGaAs layer is grown between doped AlGaAs barrier layer (M2, M4 and M5). This part of the PL spectrum can be interpreted as the superposition of three peaks originated by the C acceptors (*C-A*), free exciton (*X*) and band-to-band (*B-B*) transitions. The inset 4(b) shows the behavior of the transitions for M1, M2, and M4 samples. The increase of the intensity associated to the free exciton transition is caused by the decrease of the electric field intensity.

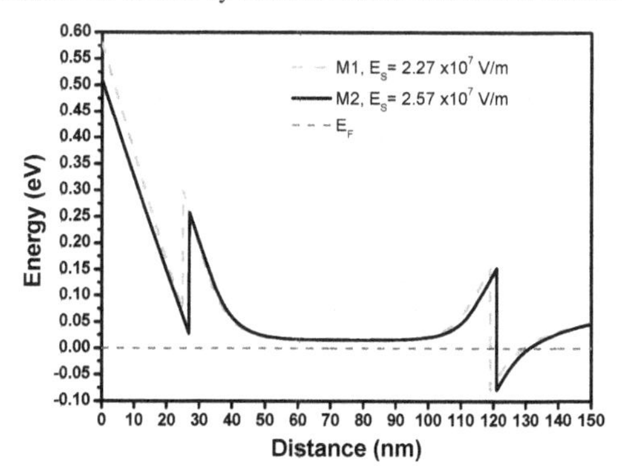

Figure 3. Potential profiles of the edge of the conduction band for M1 and M2 samples calculated by nextnano software considering the surface electric field. E_F is the Fermi´s energy.

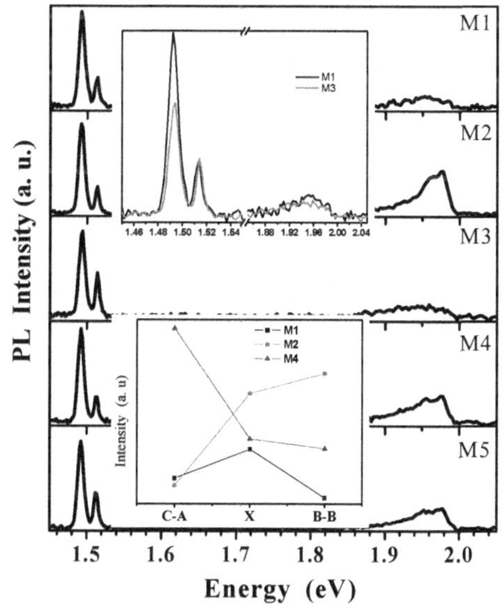

Figure 4.PL spectra at 20 K of AlGaAs/GaAsheterostructures. The inset 4(a) shows a close up of PL spectra of (M2) and without (M1) GaAs:Si cap layer. The inset 4(b) shows the behavior of the PL intensity of the free exciton in AlGaAs layer for M1, M2, and M4 samples.

CONCLUSIONS

In this work, the effect of the surface electric field on optical and electrical properties of AlGaAs/GaAsheterostructures is analyzed. By PR spectra analysis it is possible to determine experimentally the magnitude of the surface electric field. Calculations of the potential profile of the edge of the conduction band shows that the heterostructure have not the double-2DEG owing to the band bending originated by the surface electric field. The surface can be affecting the heterostructure until a distance of 60 nm toward inside of the device. By PL measurements we found that the exciton peak intensity (associated to theAlGaAs layers) increases as the influence of the surface decrease.

REFERENCES

1. H. Hasegawa and M. Akazawa, Appl. Surf. Sci. 255 (2008) 628.
2. H. Oigawa, J. Fan, Y. Nannichi, H. Sugahara and M. Oshima, Jpn. J. Appl. Phys. 30 (1991) L322-L325.
3. R. A. Khabibullin, I. S. Vasil'evskii, G. B. Gabliev, E. A. Klimov, D. S. Ponomarev, V. P. Gladkov, V. A. Kulbachinskii, A. N. Klochkov, and N. A. Uzeeva, Semiconductors 45 (2011) 657.
4. G. Kopnov, *V. Y. Umansky, H. Cohen, D. Shahar, and R. Naaman*, Phys. Rev. B 81 (2010) 045316.
5. A. Cerskus, J. Kundrotas, V. Nargeliene, A. Suziedelis, S. Asmontas, J. Gradauskas, A. Johannessen, and E. Johannessen, Lithuanian Journal of Physics Vol. 51, 4, (2011) 330.
6. N. Pan, X. L. Zheng, H. Hendriks, and J. Carter, J. Appl. Phys. 68 (1990) 2355.
7. J. A. N. T. Soares, R. Enderlein, D. Beliaev, J. R. Leite and M. Saitoz, Semicond. Sci. Technol. 13 (1998) 1418.
8. M. Sydor, N. Jahren, W.C. Mitchel, W.V. Lampert, T.W. Haas, M.Y. Yen, S.M. Mudare, D.H. Tomich, J. Appl. Phys. 67 (1990) 7423.
9. L. Zamora-Peredo, M. López-López, A. Lastras-Martínez, V. H. Méndez-García, Journal of Crystal Growth 278 (2005) 591.
10. J. Misiewicz, P. Sitarek, G. Sek, R. Kudrawiec, Materials Science, Vol. 21, No. 3, (2003) 263.
11. L. Zamora-Peredo, I. E. Cortes-Mestizo, L. Garcia-Gonzalez, J. Hernandez-Torres, D. Vázquez-Cortés,S. Shimomura, A. Cisneros-de la Rosa, and V. H. Méndez-García, Journal of Crystal Growth, in press.
12. H. Shen and F. H. Pollak, Phys. Rev. B 42, (1990) 7097.
13. B. JayantBaliga, Fundamentals of Power Semiconductor Devices, Springer, USA, 2008, pp. 306.

Mater. Res. Soc. Symp. Proc. Vol. 1534 © 2013 Materials Research Society
DOI: 10.1557/opl.2013.300

Emission and Strain in InGaAs/GaAs structures with Embedded InAs quantum dots

Ricardo Cisneros Tamayo, Georgiy Polupan and Leonardo G. Vega Macotela
ESIME– Instituto Politécnico Nacional, México D. F. 07738, México.

ABSTRACT

The photoluminescence (PL) and X ray diffraction (XRD) have been studied in the GaAs /$In_xGa_{1-x}As$ /$In_{0.15}Ga_{0.85}$ As/GaAs quantum wells with embedded InAs quantum dots (QDs) in dependence on the composition of the capping $In_xGa_{1-x}As$ layers. The parameter x in capping $In_xGa_{1-x}As$ layers varied from the range 0.10-0.25. In concentration (x) increasing in capping layers is accompanied by the variation non-monotonously of InAs QD emission: the PL intensity and peak positions. To understand the reasons of emission variation, the PL temperature dependences and XRD have been investigated in strained QD structures. It was revealed that the process of Ga/In inter diffusion at the $In_xGa_{1-x}As$/InAs QD interface and the level of elastic deformation are characterized by the dependence non monotonous versus parameter x in capping $In_xGa_{1-x}As$ layers. The physical reasons of the variation no monotonously of elastic strains in studied structures have been discussed.

INTRODUCTION

Self-assembled InAs/GaAs quantum dots (QDs) are the subject of great attention during last twenty years due to the fundamental scientific and application reasons. In this system the strong localization of the electronic wave function leads to an atomic-like electronic density of states and permits to realize the novel and improved photonic and electronic devices. InAs QDs embedded in the InAs/GaAs quantum wells (QWs) have been studied for the variety of applications such as: the low-threshold semiconductor lasers, infrared photo-detectors, one electron transistors and memory devices [1-5].

It was shown early that the InAs QD density can be enlarged significantly if the dots growth on the surface of $In_xGa_{1-x}As$ buffer layer within $In_xGa_{1-x}As$/GaAs QWs [6]. In these structures photoluminescence (PL) has been enhanced due to the better crystal quality of layers surrounding QDs [7-9] and owing the more effective exciton capture into QWs and QDs [10-14]. It was revealed as well that the emission intensity and PL peak positions vary versus InGaAs layer composition none monotonously [7]. In this paper we try to understand the physical reasons of the emission variation non-monotonously in InAs QD structures with the different In composition in capping $In_xGa_{1-x}As$ layers.

EXPERIMENTAL CONDITIONS

The experimental set of QD structures was created using the molecular beam epitaxial growth on the (100) oriented 2''diameter semi-insulating GaAs substrates. Each structure included a 300 nm GaAs buffer layer and a 70 nm GaAs upper capping layer. Between GaAs layers where located three self-organized InAs QD arrays (formed by depositing 2.4 ML of InAs at 490 °C) embedded into an external asymmetric In_xGa_{1-x} As/$In_{0.15}$ $Ga_{0.85}$ As/GaAs QWs separated by the GaAs spacer layers of 30 nm. The InAs QDs in all investigated structures were grown on a buffer layer of 1 nm with the composition $In_{0.15}Ga_{0.85}$ As under the InAs wetting layer (WL). The capping In_xGa_{1-x} As layers of 8.0 nm are characterized by the different In compositions with the parameter x: 0.10 (#1), 0.15 (#2), 0.20 (#3) and 0.25 (#4). PL spectra were measured at 80-300 K using the excitation by a 532 nm line of a solid state laser model V-5 COHERENT Verdi at an excitation power density 500 W/cm^2 [10,14]. PL spectra were dispersed by a SPEX 500M spectrometer with a Ge detector. The X-ray diffraction (XRD) experiments were done using the XRD equipment model of D-8 advanced (Bruker Co.) with $K_{\alpha 1}$ line from the Cu source (λ=1.5406 A).

EXPERIMENTAL RESULTS AND DISCUSSION

Typical PL spectra of the QD structures #1, #2, #3 and #4 are presented in figure 1. PL bands related to the recombination of excitons localized at the QD ground state (GS) and an excited state (1ES) are clearly seen in #1, #2 and #3.

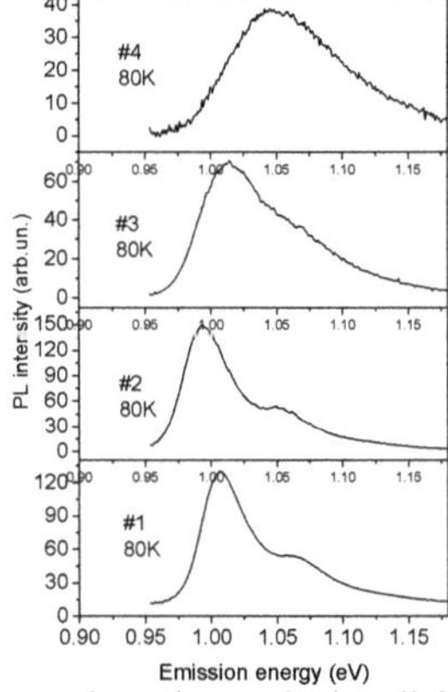

Figure 1. Typical PL spectra measured at the excitation light power 500 W/cm2 for the structures #1, #2, #3 and #4.

The "red" shift of GS PL bands is observed when x increases from 0.10 to 0.15. PL spectral red shift is accompanied by the enhancement of the PL band intensity (Fig.1). The GS PL band shifts into high energy spectral range ("blue" shift) when x is changed from 0.15 to 0.20 and to 0.25. Simultaneously, the full width at half maximum (FWHM) of PL bands increases and the QD PL intensity decreases (Fig.1). The variation no monotonously of the GS integrated PL intensity and PL peak positions have been detected in studied QD structures.

QD structures were prepared at the same technological conditions: with QD grown at 490 °C on the buffer layer of $In_{0.15}$ $Ga_{0.85}$ As and with the quantity of InAs material of 2.4 MLs. Thus we can expect that QDs in all structures are characterized by the same sizes and densities. In this case two reasons can explain the different GS PL peak positions in QDs (Fig. 1): (i) the Ga/In inter diffusion at the In_xGa_{1-x}As /InAs QD interface, and (ii) the different levels of elastic strains stimulated by the difference of the In concentration in the capping In_xGa_{1-x}As layers. To analyze these reasons PL spectra at different temperatures have been studied (Fig.2).

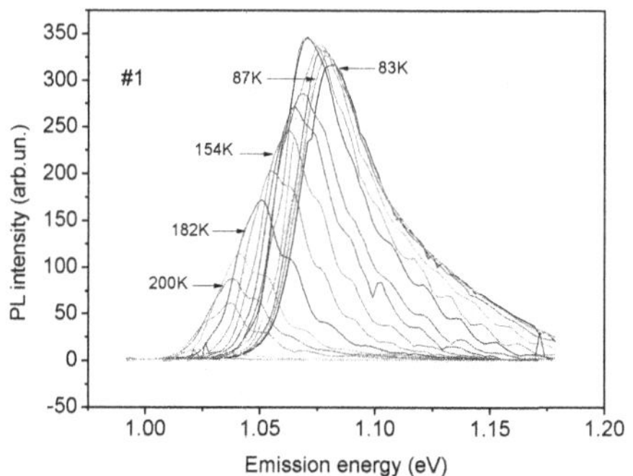

Figure 2. PL spectra of the structure #1 measured at different temperatures

Figure 3 presents the variation of GS PL peak positions versus temperature in the range of 100-300 K for all QD structures.

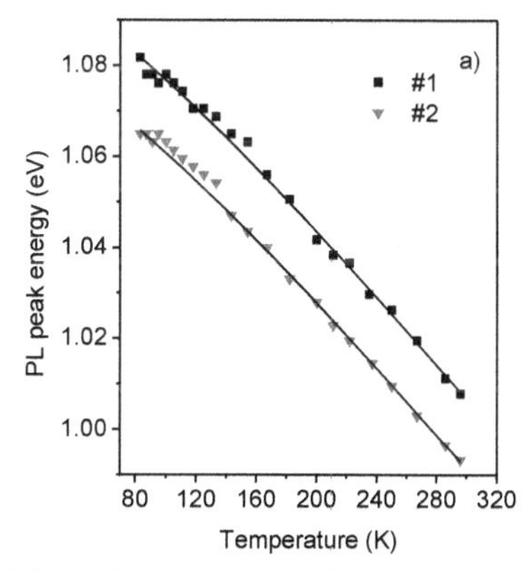

Figure 3a. Temperature dependences of PL peak positions in the structures #1 and #2. The lines are the fitting results.

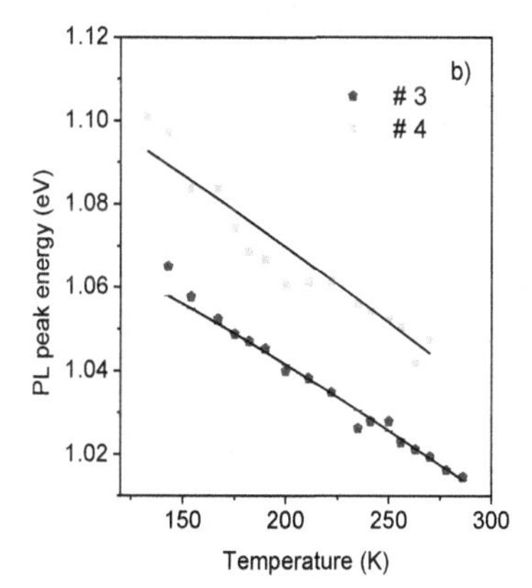

Figure 3b. Temperature dependences of PL peak positions in the structures #3 and #4. The lines are the fitting results.

PL peaks shift to low energy with increasing the temperature due to the QD optical gap shrinkage. The lines in Fig.3 present the fitting results obtained on the base of the Varshni relation that reflects the energy gap variation with temperature as [15]:

$$E(T) = E_o - \frac{aT^2}{T+b} \tag{1}$$

The comparison of fitting parameters with the variation of the energy band gap versus temperature in the bulk InAs crystal (Table 1) has revealed that the fitting parameters "α" and 'β' are closer to their values in InAs for the structure #2 (Table 1).

Table 1. Fitting results on the base of Varshni relation

In Composition	x = 10%	x = 15%	x = 20%	x = 25%	InAs [16]	GaAs [16]
E_0[eV]	1.10	1.08	1.10	1.22	0.40	1.519
α [meV/K]	0.399	0.363	0.395	0.401	0.276	0.540
β [K]	100	95	97	101	93	204

The fitting parameters "α" and 'β' distinguish essentially in the structures #1, #3 and #4 from those values in InAs (Table 1). Last fact testifies that the process of Ga/In intermixing takes place partially in the structure #2, and essentially in #1, #3 and #4. Note that the In content in InGaAs capping layers increases monotonically, but the process of Ga/In inter diffusion passes non monotonously. It means that some other factors are essential as well.

Figures 4 presents the XRD results obtained in the studied QD structures. XRD data are a superposition of XRD peaks that correspond to the diffraction of $K_{\alpha1}$, $K_{\alpha2}$ and K_{β} lines of a Cu X ray source from the (400) crystal planes in GaAs substrate, GaAs buffer and spacer layers, as well as in the InGaAs layers. It was surprise that XRD data testify concerning the $In_{0.60}Ga_{0.40}As$ alloy in QD structures (Fig.4a). The permanent position of XRD peaks, related to the $In_{0.60}Ga_{0.40}As$ alloy in all InAs QD structures permits assigning this $In_{0.60}Ga_{0.40}As$ alloy to the buffer layer obtained at the material mixture between the $In_{0.15}Ga_{0.85}As$ buffer and InAs wetting layers at QD growth temperatures.

The high intensity XRD peaks related to the diffraction of the $K_{\alpha1}$ and $K_{\alpha2}$ lines from the (400) crystal planes in GaAs and capping $In_xGa_{1-x}As$ layers are presented in more details in figure 4b.

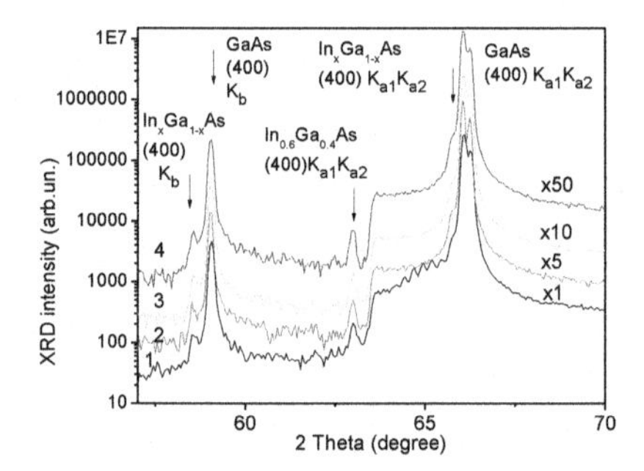

Figure 4a. XRD peaks related to the diffraction of $K_{\alpha1}$, $K_{\alpha2}$, K_{β} lines of a Cu source from the (400) crystal planes in the GaAs substrate, GaAs and InGaAs QW layers.

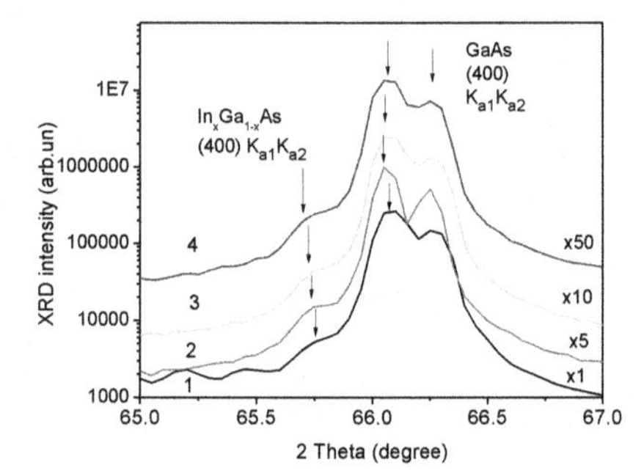

Figure 4b. XRD peaks related to the diffraction of the $K_{\alpha1}$ and K_{a2} lines from the (400) crystal planes in GaAs and InGaAs layers.

As follows from XRD data the variation of diffraction angles in GaAs spacer and buffer layers versus parameter x has the behavior non monotonous (Table 2). Actually the GaAs (400) XRD peak in the structure #2 locates closely to the (400) XRD peak (66.040 [17]) in the bulk cubic GaAs than those values in the structures #1, #3 and #4 (Fig.4b, Table 2).

Table 2. XRD angles of Kα1 beam diffraction from (400) GaAs crystal planes in studied QD structures

Samples	$2\Theta_{subst}$, GaAs substrate, [degree]	$2\Theta_{QW}$, GaAs QW [degree]	$2\Theta_{QW} - 2\Theta_{subst}$, [degree]	ε
#1 (x=0.10)	66.040	66.070	0.030	0.023
#2 (x=0.15)	66.040	66.050	0.010	0.007
#3 (x=0.20)	66.040	66.075	0.035	0.026
#4 (x=0.25)	66.040	66.100	0.060	0.046

It is known that the value of elastic deformation can be estimated by following formula [18]:

$$\varepsilon = -(\theta - \theta_o)\cot\theta_o \ , \tag{2}$$

where Θ and Θ_o are the diffraction angles in the strained layer (Θ) and in the reference layer without strain (Θ_o). The Θ_o value in present cases has been chosen as the diffraction angle meaning in the bulk cubic GaAs (GaAs substrate). Thus the elastic deformation in the InGaAs/GaAs QWs is proportional to the difference between the diffraction angles ($\Theta - \Theta_o$) and the elastic strain in the structure #2 is smaller than in the structures #1, #3 and #4 (Table 2). The lowest integrated PL intensities have been detected in the QD structures #3 and #4, apparently, due to the high concentration of nonradiative (NR) defects [19]. The high level of elastic strain enhances the partial stress relaxation in the QD structures #1, #3, #4 that accompanies by the Ga/In interdiffusion and the appearance of nonradiative recombination (NR) defects.

It is essential to note that the application of the buffer and capping In_xGa_{1-x} As layers coupled with InAs QDs stimulates the lattice mismatch and strain decreasing in the vicinity of InAs QDs. Simultaneously, the lattice mismatch and the strain increase at the In_xGa_{1-x} As/GaAs QW interface for the surface area between the QDs [7]. The competition of mentioned two processes, apparently, is a reason for the variation non-monotonously of elastic strains in studied QD structures versus parameter x. The last fact is necessary to take into account at the InAs QD structure design.

CONCLUSIONS

It is shown that in strained GaAs /$In_xGa_{1-x}As/In_{0.15}Ga_{0.85}As$/GaAs QW structures with embedded InAs QDs the monotonous increasing the In composition in $In_xGa_{1-x}As$ capping layers ($0.1 \leq x \leq 0.25$) has accompanied by the variation non monotonously of PL parameters and elastic strains, apparently, due to the different roles of the lattice mismatch at the In_x Ga_{1-x} As/InAs QD and In_x Ga_{1-x} As/GaAs QW interfaces in the process of elastic strain formation. The minimum of elastic deformation is detected in the structure with the $In_{0.15}Ga_{0.85}As$ capping layer. The high level of elastic strains stimulates: i) the process of Ga/In interdiffusion at the In_x Ga_{1-x} As/InAs QD interface, ii) the increase of the FWHM of GS PL bands, iii) the shift of PL peak positions in the high energy spectral range, iii) the partial strain relaxation with the formation of NR centers that manifests themselves by the variation of PL intensity non monotonously as well.

ACKNOWLEDGMENTS

The work was supported by CONACYT Mexico (project 130387) and by SIP-IPN, Mexico. The authors thank Dr. A. Stintz from Center of High Technology Materials at University of New Mexico, Albuquerque, USA for growing QD structures, the Dr. G. Gómez Gasga for XRD measurements and the Dr. T. Torchynska for fruitful discussions.

REFERENCES

1. M. Takahasi, T.Kaizu, Journal of Crystal Growth **311**, 1761–1763 (2009)
2. S. Dhamodaran , N. Sathish , A.P. Pathak, S.A. Khan, D.K. Avasthi, T. Srinivasan, R. Muralidharan, B.M. Arora, Nuclear Instruments and Methods in Physics Research **B 256,** 260–265 (2007)
3. Amtout, S. Raghavan, P. Rotella, G. von Winckel, A. Stinz and S. Krishna, J. Appl. Phys. **96**, pp. 3782 (2004)
4. T. V. Torchinskaya, Opto-electronics Review, **121-130**, 1998 (2).
5. D. Haft, R.J. Warburton, K. Karrai, S. Huant, G. Medeiros-Ribeiro, J. Garsia, W. Schoenfeld and P.M. Petroff, Appl. Phys. Lett. **78**, pp. 2946 (2001)
6. Stintz, G.T. Liu, L. Gray, R. Spillers, S.M. Delgado and K.J. Malloy, J. Vac. Sci. Technol. **B 18**(3), pp. 1496 (2000)
7. T.V. Torchynska, J. Appl. Phys. **104** (7), art.no. 074315 (2008)
8. T.V. Torchynska, J.L. Casas Espinola, L. Borkovska, S. Ostapenko, M. Dybiec, O. Polupan, N. Korsunska, A. Stintz, P.G. Eliseev and K.J. Malloy, J. Appl. Phys. **101** (2), art.no. 024323 (2007)
9. H. BenNaceur , I.Moussa, O.Tottereau, A.Rebey, B.El Jani, Physica **E 41**, 1779–1783 (2009)
10. T.V. Torchynska, S. Ostapenko, M. Dybiec, Phys. Rev. **B 72** (19), pp. 1-7 (2005)
11. M. Dybiec, S. Ostapenko, T. V. Torchynska and E. Velasquez Losada, Appl. Phys. Lett. **84** (25), pp. 5165-5167 (2004)
12. G.W. Shu, J.S. Wang, J.L. Shen, R.S. Hsiao, J.F. Chen, T.Y. Lin, C.H. Wu, Y.H. Huang, T.N. Yang, Mater. Scien. and Engin. B, **166**, 46–49 (2010)
13. M. Dybiec, L. Borkovska, S. Ostapenko, T.V. Torchynska, J.L. Casas Espinola, A. Stinz, K.J. Malloy, Appl. Surf. Scien. **252** (15) 5542-5545 (2006).
14. T.V. Torchynska, A. Diaz Cano, M. Dybic, S. Ostapenko, M. Mynbaeva, Physica B, Condensed Matter, **376-377** (1), 367-369 (2006).
15. Y. P. Varshni, Physica **34**, 149 (1967).
16. A. Landolt-Boernstein, Numerical Data and Functional Reationship, in: Science and Techology, Vol. **22** (Springer, Berin, 1987), p. 118.
17. http://www.ioffe.ru/SVA/NSM/Semicond.
18. T.V. Torchynska, A. Stintz, J. Applied Physics, **108**, 2, 024316 (2010).
19. T.V. Torchynska, E. Velazquez Lozada, J.L. Casas Espinola, J. Vacuum. Scien. And Techn. B, **27** (2) 919-922 (2009).

Mater. Res. Soc. Symp. Proc. Vol. 1534 © 2013 Materials Research Society
DOI: 10.1557/opl.2013.301

Photoluminescence spectrum Dependences on Temperature and Excitation power in InAs Quantum dots embedded in InGaAs/GaAs Quantum wells

Alejandro Vivas Hernandez[1], Erick Velázquez Lozada and Ingri J. Guerrero Moreno
ESIME– Instituto Politécnico Nacional, México D. F. 07738, México

ABSTRACT

The photoluminescence (PL), its temperature and excitation power dependences have been studied in the symmetric $In_{0.15}Ga_{1-0.15}As/GaAs$ quantum wells (QWs) with embedded InAs quantum dots (QDs), obtained with the variation of the QD growth temperature (470-535°C). The increase of QD growth temperatures is accompanied by the enlargement of the QD lateral sizes (from 12 up to 28 nm) and by the shift non monotonously of the PL peak positions. The fitting procedure has been applied for the analysis of the temperature dependences of PL peaks. The obtained fitting parameters testify that in studied QD structures the process of In/Ga inter diffusion between QDs and capping/buffer layers takes place. The intensity of the In/Ga inter diffusion depends on the density and the size of QDs in studied QD structures. Additionally the In/Ga intermixing influent essentially on the dependence of the integrated PL intensity versus excitation power.

INTRODUCTION

InAs quantum dots (QDs) embedded in InGaAs/GaAs quantum wells (QWs) have been the subject of great interest due to the variety of their applications in optoelectronics and microelectronics, such as: lasers for the IR optical range [1-3], infrared photo-detectors [4-6] and electronic memory devices [7, 8]. It was shown earlier [3] that the high density of InAs QD can be achieved if the QDs are grown inside of the GaAs/capping $In_{0.15}Ga_{0.85}As$/buffer $In_{0.15}Ga_{0.85}As/GaAs$ quantum wells [3]. But even for the optimal QD growth parameters and the capping/buffer layer compositions, the InAs QD structures are characterized by photoluminescence (PL) inhomogeneity along the wafers [9-19]. The technology of growth of InAs QD structures has become more reliable enabling the systematic studies of their physical properties. In this paper we try to understand the physical reasons of the emission variation inhomogeneously versus temperature and excitation power in InAs QDs coupled with the symmetric $In_{0.15}Ga_{0.85}As/GaAs$ QWs at the variation of InAs QD growth temperatures.

EXPERIMENT

A set of samples was prepared using molecular beam epitaxy on (100) oriented 2'' diameter semi-insulating GaAs substrates (Fig.1). InAs quantum dots (QDs) were grown inside of the symmetric structures: buffer GaAs (200nm)/ buffer $In_{0.15}Ga_{0.85}As$ (2nm)/ capping $In_{0.15}Ga_{0.85}As$ (10nm)/capping GaAs (100nm) quantum well layers, at the five QD growth temperatures 470 (#1), 490 (#2), 510(#3) 525 (#4) and 535 (#5)°C during the deposition of InAs QD active regions and InGaAs quantum well layers [12,14].

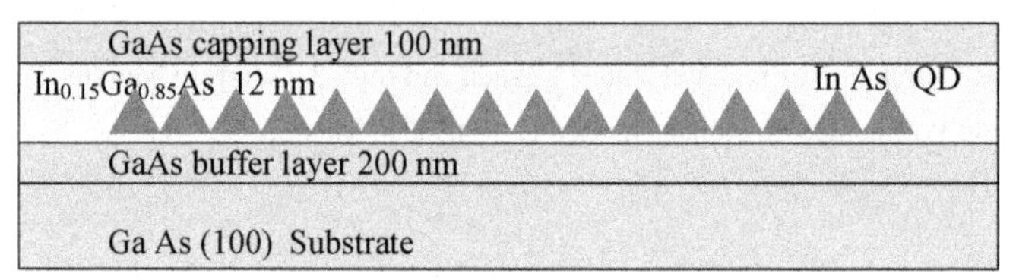

Figure 1. The design of QD structures

The QD size increases from 12 to 28 nm and the QD density decreases from 1.1 x 10^{11} down to $1.3x10^{10}$ cm^{-2} versus QD growth temperatures [19]. The photoluminescence spectra were measured in the temperature range of 10-300 K using the excitation by the 532 nm line of a solid state laser model V-5 COHERENT Verdi at an excitation power density from the range of 50-800 W/cm^2. PL spectra were dispersed by a SPEX 500M spectrometer with a Ge detector.

RESULTS AND DISCUSSION

Typical PL spectra of the structure #3, measured at different temperatures at the excitation light density of 300 W/cm^2, are shown in Fig.2. Two PL bands appear due to the recombination of excitons localized at a ground state (GS) and at an excited state (1ES) in QDs (Fig.2).

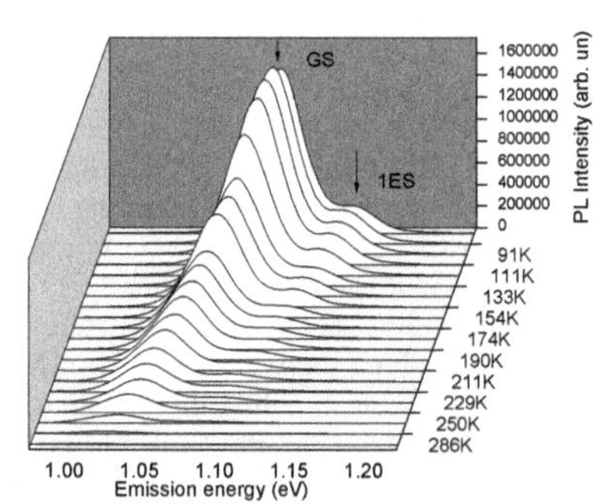

Figure 2. Typical PL spectra measured at different temperatures for the DWELL structure #3 at the excitation light density of 300 W/cm^2.

The excitations light density of 300 W/cm^2 is chosen for prevent the influence of 1ES emission on the GS PL intensity (Fig.2). QD diameters in studied structures increase monotonously from 12 up to 28 nm with the rise of QD growth temperatures from 470 up to 535 °C. It is possible to expect that the PL peak position in QDs has to shift monotonously to low energy as well. The PL peak positions and the integrated PL intensities in the structures (#1- #5) measured for the ground state PL bands versus temperatures have been presented in figures 3 and 4, respectively. As it is clear the variation of PL intensity and PL peak positions with the enlargement of QD growth temperatures and QD diameters is none monotonous. Note that the lower PL peak energy corresponds to the higher PL intensity (Fig.3, 4, curves 3).

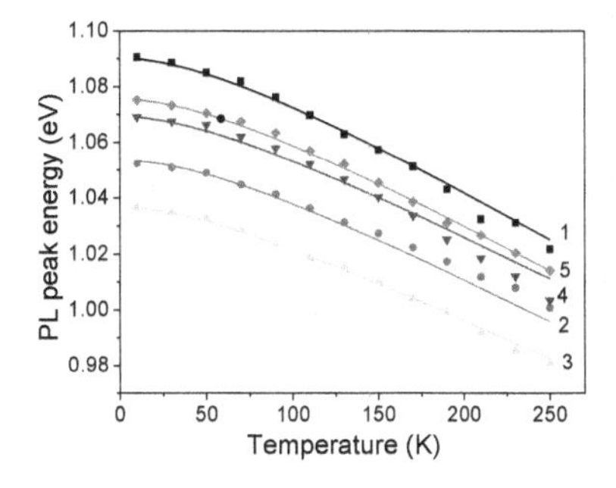

Figure 3. The variation of PL peak positions versus temperatures. The lines present the Varshni fitting results: 1- #1, 2- #2, 3 - #3, 4 - #4 and 5 - #5.

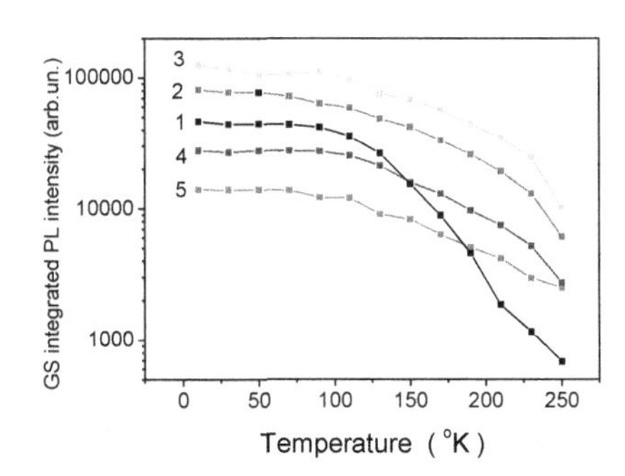

Figure 4. The variation of GS integrated PL peak intensities versus temperatures in the structures: 1- #1, 2- #2, 3 - #3, 4 - #4 and 5 - #5.

Two reasons can explain the variation non monotonously of PL peak positions and the PL intensity in studied QD structures: (i) the change of the QD composition due to the Ga/In interdiffusion between the InAs QDs and capping/buffer $In_{0.15}Ga_{0.85}As$ QW layers or (ii) the different level of elastic strains in QD structures due to the difference in QD densities and sizes. To distinguish between these two reasons PL spectra at different temperatures have been studied. The variation of PL peak positions of ground state PL bands in all studied QD structures is presented in Fig. 3. PL peaks shift to low energy with increasing temperature due to the optical gap shrinkage. The lines in Fig.3 present the fitting results analyzed on the base of Varshni relation that reflects the energy gap variation with temperature as [20]:

$$E(T) = \mathrm{E}_{o} - \frac{aT^2}{T+b} \quad (1)$$

The comparison of fitting parameters with the variation of the energy band gap versus temperature in the bulk InAs and GaAs crystals has revealed (Table 1) that in studied QD structures the fitting parameter "*a*" and '*b*' in the temperature range 10-250 K are very close to their values in the bulk InAs crystal for the QD structures #2 and #3.

Table 1. Varshni fitting parameters

	E_0	a	b
Structure numbers	eV	meV/ °K	°K
#1	1.082	0.355	110
#2	1.089	0.330	100
#3	1.010	0.300	95
#4	1.049	0.346	107
#5	1.079	0.335	108
InAs [21]	0.415	0.276	93
GaAs [21]	1.519	0.540	204

But in other QD structures the fitting parameters "*a*" and '*b*' are different from the values in InAs Actually the process of Ga/In interdiffusion takes place in these QD structures. Note that the process of Ga/In interdiffusion in studied structures passes none monotonously versus QD growth temperatures. It means that not only temperature but some other factors, apparently connected with elastic strain, are essential as well.

The power dependences of the integrated over entire spectrum ground state PL intensities measured at 120K are shown in Fig.5. The integrated intensity (I) of studied GS PL bands versus power (P) can be presented as $I \sim P^n$, where the parameter n is estimated as n=0.9-1.0 for QDs grown at 490 - 525 °C, as well as n=0.6-0.7 for QDs grown at lower (470 °C) and higher temperatures (535 °C) (Fig.5).

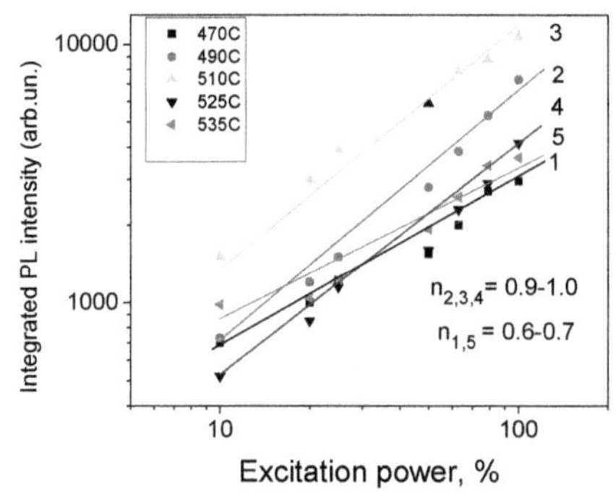

Figure 5. Dependences of the integrated GS PL intensities versus excitation light power. The highest power equals to 800W/cm^2 to prevent the saturation of GS PL intensity.

The last fact indicates that the Ga/In inter-diffusion process is accompanied by appearing the nonradiative (NR) recombination centers in QD structures. As it was shown earlier, on the base of investigation of exciton recombination parameters and thermal escape, these NR centers are localized in mentioned structures in the $In_{0.15}Ga_{1-0.15}As$ QW layers [22-24].

CONCLUSION

The influence of QD growth temperatures on the Ga/In interdiffusion has been investigated in GaAS/InGaAs/GaAs QW structures with InAs QDs. It was revealed that PL intensity and peak positions, as well as a level of elastic stain change none monotonously with the enlargement of QD growth temperature. The Ga/In interdiffusion is responsible for the shift of PL peaks into the high energy spectral range in #1, #2, #4 and #5 in comparison with #3.

ACKNOWLEDGEMENTS

The work was supported by CONACYT Mexico (project 130387) and by SIP-IPN, Mexico. The authors thank Dr. A. Stintz from Center of High Technology Materials at University of New Mexico, Albuquerque, USA for growing the studied QD structures.

REFERENCES

1. D. Bimberg, M. Grundman, N. N. Ledentsov, Quantum Dot Heterostructures, Ed. Wiley & Sons (2001) 328.
2. D. Bimberg, M. Grundmann, F. Heinrichsdorff, N.N. Ledentsov V.M. Ustinov, A.E. Zhukov, A.R. Kovsh, M.V. Maximov, Y.M. Shernyakov, B.V. Volovik, A.F. Tsatsul'nikov, P.S. Kop'ev, Zh.I. Alferov, Thin Solid Films, 367, 235 (2000).
3. G. T. Liu, A. Stintz, H. Li, K. J. Malloy and L. F. Lester, Electron Lett, 35, 1163 (1999).
4. A. Amtout, S. Raghavan, P. Rotella, G. von Winckel, A. Stinz, S. Krishna, J. Appl. Phys. 96, 3782 (2004).
5. R.S. Attaluri, J. Shao, K.T. Posani, S.J. Lee, J.S. Brown, A. Stintz, S. Krishna, J. Vac. Sci. Technol. B, 25, 1186 (2007).
6. T.V. Torchinskaya, Opto-Electronics Review, 2, 121-130 (1998).
7. D. Haft, R. J. Warburton, K. Karrai, S. Huant, G. Medeiros-Ribeiro, J. M. Garsia, W. Schoenfeld, P. M. Petroff, Appl. Phys. Lett., 78, 2946 (2001).
8. M. Geller, A. Marent, T. Nowozin, D. Feise, K. Potschke, N. Akcay, N. Oncana, D. Bimberg, Physica E, 40, 1811 (2008).
9. I. Kamiya, I. Tanaka, K. Tanaka, F. Yamada, Y. Shinozuka, and H. Sakaki: Physica E 13, 131 (2002).
10. K. Karrai, R. J. Warburton, C. Schulhauser, A. Hogele, B. Urbaszek, E. J. McGhee, A. O. Govorov, J. M. Garcia, B. D. Gerardot, and P. M. Petroff: Nature 427, 135 (2004).
11. R. M. Thompson, R. M. Stevenson, A. J. Shields, I. Farrer, C. J. Lovo, D. A. Ritchie, M. L. Leadbeater, and M. Pepper: Phys. Rev. B 64, 201302 (2001).
12. A. Stintz, G. T. Liu, L. Gray, R. Spillers, S. M. Delgado, K. J. Malloy, J. Vac. Sci. Tech. B. 18(3), 1496 (2000).
13. S.P. Ryu, N.K. Cho, J.Y. Lim, H.J. Lee, W.J. Choi, J.D. Song, J.I. Lee and Y.T. Lee Japan. J. Appl. Phys. 48, 095506 (2009).
14. T. V. Torchynska, J. Appl. Phys. 104 (7), 074315 (2008).
15. T. V. Torchynska, S. Ostapenko, M. Dybic, Phys.Rev.B, 72 (19) 1-7 (2005).
16. M. Dybiec, S. Ostapenko, T. V. Torchynska, E.Velasquez Losada, Appl.Phys.Lett. 84 (25), 5165-5167 (2004).
17. M. Dybiec, S. Ostapenko, T. V. Torchynska, E.Velasquez Lozada, P. G. Eliseev, A. Stintz, K. J. Malloy, phys. stat. sol. (c), 2 (8), 2951-2954 (2005).

18. M. Dybiec, L. Borkovska, S. Ostapenko, T. V. Torchynska, J. L. Casas Espinola, A. Stintz and K. J. Malloy, Applid. Surface Science, 252 (15), 5542-5545 (2006).
19. T.V. Torchynska, A. Stintz, J. Appl. Phys., 108, 2, 024316 (2010).
20. Y. P. Varshni, Physica 34, 149 (1967).
21. A. Landolt-Boernstein, Numerical Data and Functional Reationship, in: Science and Technology, Vol. 22 (Springer, Berin, 1987), p. 118.
22. T.V. Torchynska, G. Polupan, Superlattice and Microstructure, 45, 222-227 (2009).
23. T.V. Torchynska, E. Velazquez Lozada, J.L. Casas Espinola, Journal Vac. Sci. & Technol. B, 27(2), 919-922, (2009)
24. T. V. Torchynska, A.I. Diaz Cano, M. Dybic, S. Ostapenko, M. Mynbaeva, Physica B: Condensed Matter, 376-377 (1), 367-369 (2006).

Mater. Res. Soc. Symp. Proc. Vol. 1534 © 2013 Materials Research Society
DOI: 10.1557/opl.2013.302

Emission and HR-XRD study in InAs quantum dot structures prepared at different QD's growth temperatures

Leonardo G. Vega Macotela[1] and Tetyana V. Torchynska[2]
[1]E.S.I.M.E, National Polytechnic Institute, Mexico D. F. 07738, Mexico
[2]E.S.F.M, National Polytechnic Institute , Mexico D. F. 07738, Mexico

ABSTRACT

The structure of the symmetric GaAs/$In_{0.15}$ $Ga_{0.85}$GaAs/GaAs quantum wells (QWs) with embedded InAs quantum dots (QDs) has been studied using High resolution X-ray diffraction HR-XRD method. The QDs were grown at different temperatures from the range 470 - 535°C. The increase of growth temperature stimulates decreasing the QD surface density and the enlargement of QD lateral sizes. Simultaneously, the variation of PL intensity and PL peak position non monotonously have been detected. To understand the reason of PL variations the HR-XRD method with a resolution of 0.0001 degree has been applied. The high intensity peaks that correspond to the diffraction from the (400) crystal planes in GaAs QWs, InAs QDs and $In_xGa_{1-x}As$ QWs have been detected by HR-XRD. The simulation fitting of experimental HR XRD curves has been done on the base of the dynamic diffraction theory. This simulation fitting testifies the different level of intermixing of Ga/In atoms in QDs and QW layers in studied QD structures. The simulation fitting parameters for QD structures with the QDs grown at 490°C and 510°C are very close to the original technological parameters. The simulation fitting parameters in QD structures with QDs grown at the temperatures 470°C, 525°C and 535°C are very different from technological one. Actually in these QD structures the process of Ga/In intermixing is essential due to the higher level of elastic strains that stimulates the variation of PL parameters as well.

INTRODUCTION

InGaAs/GaAs quantum wells (QWs) with InAs quantum dots (QDs) have been the subject of great interest due to the variety of applications, such as: semiconductor lasers for the optical fiber communication [1-3], infrared photo-detectors [4-6] and electronic memory devices [7, 8]. Extensive efforts have been applied during the last two decades for the manipulation and control the position, size, shape and density of QDs, as well as a number of stacking QD layers for the high efficient devices [9-12]. But even for the optimal QD growth parameters and the capping/buffer layer compositions, the InAs QD structures are characterized by non homogeneous photoluminescence (PL) behavior [13-16]. The technology of growth of InAs QD structures has become more reliable enabling systematic studies their physical properties and the emission nonhomogeneity of QD ensembles. In this paper we try to understand the physical reasons of emission variation in InAs QDs coupled with symmetric $In_{0.15}Ga_{0.85}As$/GaAs QWs at the variation of the QD sizes and densities.

EXPERIMENT

A set of samples was prepared using molecular beam epitaxy on (100) oriented 2" diameter semi-insulating GaAs substrates. Each structure was grown on a 200nm GaAs buffer layer and capped by 100 nm GaAs layer [12]. An equivalent coverage of 2.4 ML of InAs QDs were grown inside the symmetric Ga/$In_{0.15}Ga_{0.85}As$/GaAs QWs (the same In concentration in buffer and capping layers), at five QD growth temperatures: 470, 490, 510°, 525 and 535°C (table 1) during the deposition of InAs active layers and InGaAs wells [17, 19]. The dot size and the density were determined by the AFM observation of parallel wafer that had not been covered by QWs

Table 1. Growth parameters of studied structures

Structure number	QD growth temperatures, °C	QD density, cm^{-2}
#1	470	$1.1x10^{11}$
#2	490	$7.0\ x10^{10}$
#3	510	$3.4\ x10^{10}$
#4	525	$1.8\ x10^{10}$
#5	535	$1.3\ x10^{10}$

The HR-X-ray diffraction (XRD) experiments were done using the XRD equipment Model XPERT MRD with the Pixel detector, three axis goniometry and parallel collimator with the resolution of 0.0001 degree. The X ray beam was from the Cu source with $K_{\alpha 1}$ line (λ=1.5406 A). Simulations were done by the variation of the sizes and compositions of capping, wetting and barrier layers. The X'Pert Epitaxy software based on the dynamic diffraction theory for X-ray diffraction in layered structures has been used [20-21].

RESULTS AND DISCUSSION

It was shown earlier that these samples are characterized by the shift non monotonous of GS PL peak positions versus QD growth temperatures (Fig.1) [12].

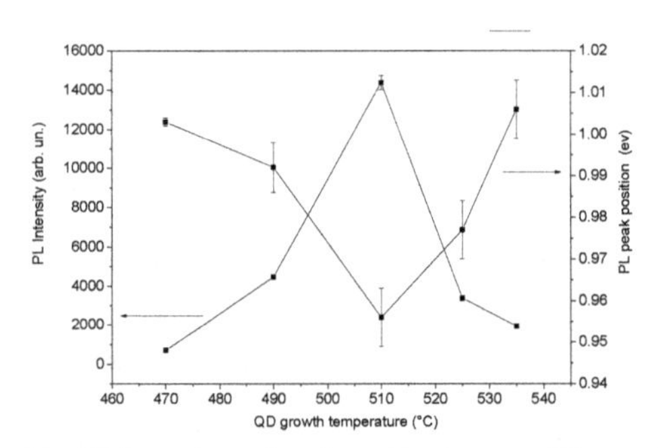

Figure 1. Variation of the PL intensity and peak positions versus QD growth temperatures

Figure 2 presents the typical HR-XRD results obtained in the samples #3 and #5. The different shoulders that have seen in the main HR XRD peaks related to the interference between the diffraction from the different QW layers.

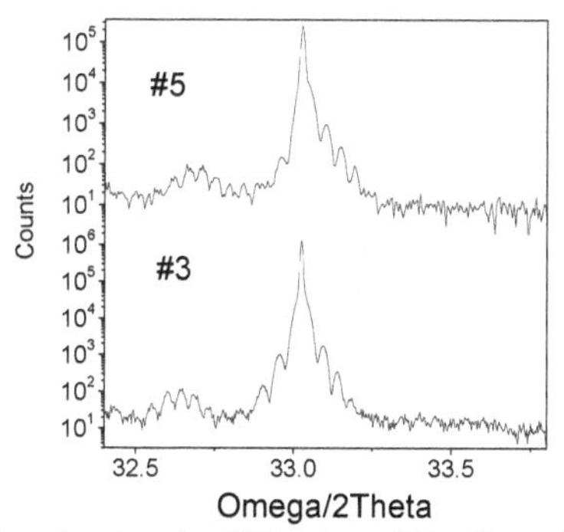

Figure 2. HR-XRD peaks related to the diffraction of $K_{\alpha 1}$ lines of the X-ray Cu source from the (400) crystal planes of GaAs and InGaAs layers in the structures #3 and #5.

As we can see there is the shape variation of HR-XRD for different samples. To explain this effect we supposed that a shape variation related to: a) the Ga/In atom inter diffusion between InAs QDs and InGaAs capping/buffer layers and/or b) the different levels of elastic strains in studied structures.

Figure 3 presents the results of simulation fitting and experimental curves for samples #3 and #5

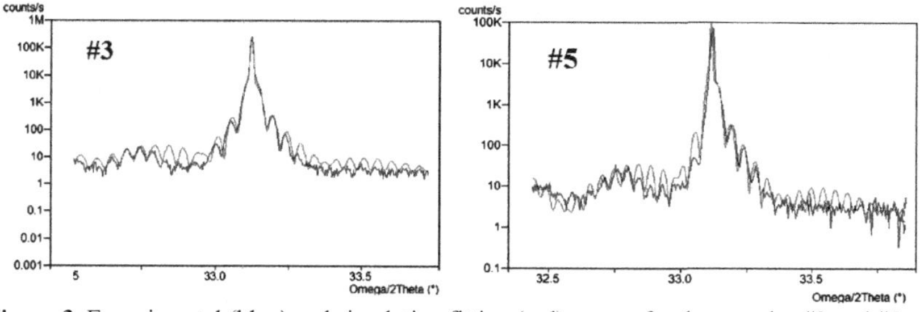

Figure 3. Experimental (blue) and simulation fitting (red) curves for the samples #3 and #5

The parameters used for the simulation fitting are shown in table 2. The comparison of fitting parameters (Table 2) has revealed that in studied QD structures the fitting parameters for the structure #3 are very close to their original technological compositions. From the other hand the fitting parameters for the structures #1,#2, #4 and #5 are a little bit different from the growth technological parameters due to the process of Ga/In interdiffusion. Note that the process of the

Ga/In inter diffusion in studied structures passes non monotonously versus QD growth temperatures.

Table 2. Simulation fitting parameters

Sample	Layer	Composition	Thickness (nm)
#1	Buffer layer	$In_{0.67}Ga_{0.33}As$	2.00
	Wetting layer	$In_{0.70}Ga_{0.30}As$	0.07
	Capping layer	$In_{0.25}Ga_{0.75}As$	10.65
#2	Buffer layer	$In_{0.30}Ga_{0.70}As$	2.00
	Wetting layer	$In_{0.80}Ga_{0.20}As$	0.09
	Capping layer	$In_{0.18}Ga_{0.82}As$	10.63
#3	Buffer layer	$In_{0.28}Ga_{0.72}As$	2.00
	Wetting layer	InAs	0.15
	Capping layer	$In_{0.17}Ga_{0.83}As$	10.57
#4	Buffer layer	$In_{0.30}Ga_{0.70}As$	2.00
	Wetting layer	$In_{0.85}Ga_{0.15}As$	0.24
	Capping layer	$In_{0.18}Ga_{0.82}As$	10.48
#5	Buffer layer	$In_{0.57}Ga_{0.43}As$	2.00
	Wetting layer	$In_{0.80}Ga_{0.20}As$	0.26
	Capping layer	$In_{0.21}Ga_{0.79}As$	10.46

In order to understand the relationship between the HR-XRD and the variation of elastic strains, the parameters of the perpendicular mismatches (M*) were calculated for the studied structures. The perpendicular layer mismatch (as a part per million) is given by formula (1):

$$M^* = \left[\frac{(\omega_L - \omega_S)}{\frac{\sin(2\phi_S)}{2} + \tan(\theta_S)\cos^2(\phi_S)}\right]\left[\frac{\pi}{180}\right]10^6 \quad (1)$$

where $(\omega_L - \omega_S)$ is the difference between the diffraction peaks for the two nearest layers that are considered at the numerical calculation, θ_S is the GaAs substrate Bragg diffraction angle and ϕ_S is the tilt of the substrate reflection plane in the case of asymmetric reflection. M* is multiplied then by $(1-\nu)/(1+\nu)$ to give the fully strained mismatch (where ν is the Poisson ratio for the layer). Results of those calculations for different layers are shown in figure 4.

Thus the simulation of HR-XRD results (table 2) and elastic strain calculation permit to show that the process of Ga/In interdiffusion is strain stimulated. The higher level of elastic strains enhanced the higher Ga/In intermixing between QDs and QW layers that influents on the PL intensity and PL peak positions in the studied QD structures as well.

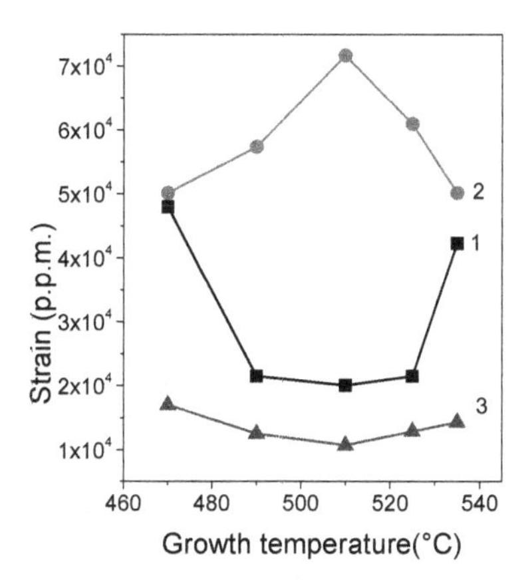

Figure 4. Calculated strains for the buffer (1), wetting (2) and capping (3) layers for the samples of: #1, #2, #3, #4, #5, versus QD growth temperatures

CONCLUSIONS

The HR-XRD has been studied in InAs QDs embedded in the symmetric $In_{0.15}Ga_{0.85}As/GaAs$ QWs with QDs grown at 470, 490, 510, 525 and 535°C with the different QD densities. It is shown that studied structures are characterized by the different level of Ga/In inter-diffusion between the QDs and QW layers, owing to the difference in elastic strains. The variation of PL parameters is the result of the process of Ga/In intermixing.

ACKNOWLEDGMENTS

The work was supported by CONACYT Mexico (project 130387) and by SIP-IPN, Mexico. The authors would like to thank Dr. A. Stintz from Center of High Technology Materials at University of New Mexico, Albuquerque, USA for growing the studied QD structures and Dr. Jose Alberto Andraca Adame from Center of Nanoscience Micro and Nanotechnologies at National Polytechnic Institute of Mexico for the HR-XRD measurements.

REFERENCES

1. D. Bimberg, M. Grundman, N. N. Ledentsov, Quantum Dot Heterostructures, Ed. Wiley & Sons (2001) 328.
2. J.F. Chen, P.Y. Wang, J.S. Wang, N.C. Chen, X.J. Guo, Y.F. Chen, J. Appl. Phys. **87**, 1251 (2000).
3. G. T. Liu, A. Stintz, H. Li, K. J. Malloy and L. F. Lester, Electron Lett, **35**, 1163 (1999).

4. A. Amtout, S. Raghavan, P. Rotella, G. von Winckel, A. Stinz, S. Krishna, J. Appl. Phys. **96**, 3782 (2004).
5. R.S. Attaluri, J. Shao, K.T. Posani, S.J. Lee, J.S. Brown, A. Stintz, S. Krishna, J. Vac. Sci. Technol. B, **25**, 1186 (2007).
6. T.V. Torchinskaya, Opto-Electronics Review, **6** (2), 121 (1998).
7. D. Haft, R. J. Warburton, K. Karrai, S. Huant, G. Medeiros-Ribeiro, J. M. Garsia, W. Schoenfeld, P. M. Petroff, Appl. Phys. Lett., **78,** 2946 (2001).
8. M. Geller, A. Marent, T. Nowozin, D. Feise, K. Potschke, N. Akcay, N. Oncana, D. Bimberg, Physica E, **40**, 1811 (2008).
9. I. Kamiya, I. Tanaka, K. Tanaka, F. Yamada, Y. Shinozuka, and H. Sakaki: Physica E **13,** 131 (2002).
10. T. V. Torchynska, J. L. Casas Espinola, L. V. Borkovska, S. Ostapenko, M. Dybic, O. Polupan, N. O. Korsunska, A. Stintz, P. G. Eliseev, K. J. Malloy, J. Appl. Phys., **101**, 024323 (2007).
11. K. Karrai, R. J. Warburton, C. Schulhauser, A. Hogele, B. Urbaszek, E. J. McGhee, A. O. Govorov, J. M. Garcia, B. D. Gerardot, and P. M. Petroff: Nature **427,** 135 (2004).
12. T. V. Torchynska, J. Appl. Phys. **104** (7), 074315 (2008).
13. M. Dybiec, S. Ostapenko, T. V. Torchynska, E.Velasquez Losada, Appl.Phys.Lett. **84**, 5165 (2004).
14. M. Dybiec, S. Ostapenko, T. V. Torchynska, E.Velasquez Lozada, P. G. Eliseev, A. Stintz, K. J. Malloy, phys. stat. sol. (c), **2**, 2951 (2005).
15. M. Dybiec, L. Borkovska, S. Ostapenko, T. V. Torchynska, J. L. Casas Espinola, A. Stintz and K. J. Malloy, Applid. Surface Science, **252,** 5542 (2006).
16. T. V. Torchynska, S. Ostapenko, M. Dybic, Phys. Rev. B, 72, 195341 (2005).
17. T. V. Torchynska, J. Appl. Phys. **104**, 074315 (2008).
18. R. M. Thompson, R. M. Stevenson, A. J. Shields, I. Farrer, C.
J. Lovo, D. A. Ritchie, M. L. Leadbeater, M. Pepper, Phys.
Rev. B **64,** 201302 (2001).
19. A. Stintz, G. T. Liu, L. Gray, R. Spillers, S. M. Delgado, K.
J. Malloy, J. Vac. Sci. Technol. B **18**, 1496 (2000).
20. H. Li, T. Mei, W. D.H. Zhang, S.F. Yoon and H.Yuan. J. Appl. Phys. **98**, 054905 (2005).
21. P. Mukhopadhyay, P Das, S Pathak, E. Y. Chang, D. Biswas. CS MANTECH Conference, Palm Springs, California, USA (2011)
22. B. Zhang, G. S. Solomon, M. Pelton, J. Plant, CH Santori, et. al. J. Appl. Phys. **97**, 073507 (2005).
23. M. Dybiec, S. Ostapenko, T. V. Torchynska, E.Velasquez Lozada, P. G. Eliseev, A. Stintz, K. J. Malloy, phys. stat. sol. (c), **2**, 2951 (2005).

Mater. Res. Soc. Symp. Proc. Vol. 1534 © 2013 Materials Research Society
DOI: 10.1557/opl.2013.303

Photoluminescence variation in InAs Quantum dots embedded in InGaAs/AlGaAs Quantum wells at thermal annealing

I. J. Guerrero Moreno[1], G. Polupan[1] and J.L. Casas Espinola[2]
[1]ESIME– Instituto Politécnico Nacional, México D. F. 07738, México
[2]ESFM– Instituto Politécnico Nacional, México D. F. 07738, México

ABSTRACT

The photoluminescence (PL) and its temperature dependence have been studied in MBE grown InAs quantum dots (QDs) embedded in $GaAs/Al_{0.3}Ga_{0.7}As/In_{0.15}Ga_{0.85}As/Al_xGa_{1-x}As/$ GaAs quantum wells (QWs) in dependence on the composition of the capping $Al_xGa_{1-x}As$ layers and after the thermal annealing at 640°C during 2 hours. Two types of capping layers (GaAs and $Al_{0.3}Ga_{0.7}As$) were investigated.It is shown that annealing initiates the shift of PL peak positions into the high energy spectral range and the value of this shift depends on the composition of capping layers. Temperature dependences of PL peak positions in QDs have been analyzed in the range of 10-300K and are compared with the temperature shrinkage of the band gap in the bulk InAs crystal. This permits to investigate the efficiency of the Ga(Al)/In inter diffusion processes in dependence on the capping layer compositions andthermal annealing. Experimental and fitting parameters obtained for the InAs QDs have been compared with known one for the bulk InAs crystal. It is revealed that the efficiency of the Ga(Al)/In inter diffusion depends essentially on the capping layer compositions.

INTRODUCTION

Self-assembled InAs/GaAs quantum dots (QDs) are the subject of great interest in the last two decades owing to the fundamental and application reasons [1-3]. It was shown that the QD density can be enlarged significantly by growing the dots within $In_xGa_{1-x}As/GaAs$ quantum wells (QWs) [3, 4]. In mentioned QD structures the photoluminescence (PL) has been enhanced due to the better crystal quality of QW layers and owing to the more effective exciton capture into the QWs and QDs [4]. But even for optimal capping/buffer layer compositions the PL of QDs is characterized by essential non homogeneity [5-8].

The realization of efficient light-emitting devices operated at room temperature requires understanding the QD PL temperature dependence and the study of the reasons of PL non homogeneity and its variation at thermal annealing. The PL intensity decay in InAs QDs, as a rule, attributed to thermal escape of excitons from the QDs into a wetting layer (WL) or into the GaAs barrier [9-15]. In QD structures introducing the additional $Al_xGa_{1-x}As$ layers into InGaAs/GaAs QWs, as it is expected, leads to the increase of the height of the potential barrier for the exciton thermal escape from QDs and can permit the application of these QD structures at higher temperatures. Improved understanding the operation and design peculiarities of InAs QDs embedded into InGaAs/AlGaAs QWs could be obtained from the study of the variation of the PL spectra at thermal annealing.

EXPERIMENT

The solid-source molecular beam epitaxy (MBE) in V80H reactor was used to grow the waveguide structures that contain the InAs self-organized QDs inserted into 9 nm $In_{0.15}Ga_{0.85}As/Al_{0.30}Ga_{0.70}As$ QWs. The thickness of the buffer $In_{0.15}Ga_{0.85}As$ layer was 1nm, which was grown on the 300.nm $Al_{0.30}Ga_{0.70}As$ buffer layer and the 2 inch (100) GaAs SI substrate. Then an equivalent coverage of 2.4 monolayers of InAs QDs were confined by the first capping (8 nm) $In_{0.15}Ga_{0.85}As$ layer, by the second 100nm $Al_xGa_{1-x}As$ capping layer, and by the 10 nm AlAs and 2 nm GaAs layers (Fig.1). Two groups of QD samples with the different compositionsof the second AlxGa1-xAs (100nm) capping layerswere investigated. In the first group the second capping layer is $Al_{0.30}Ga_{0.70}As$ (#1) and in the second group it is GaAs (#2).

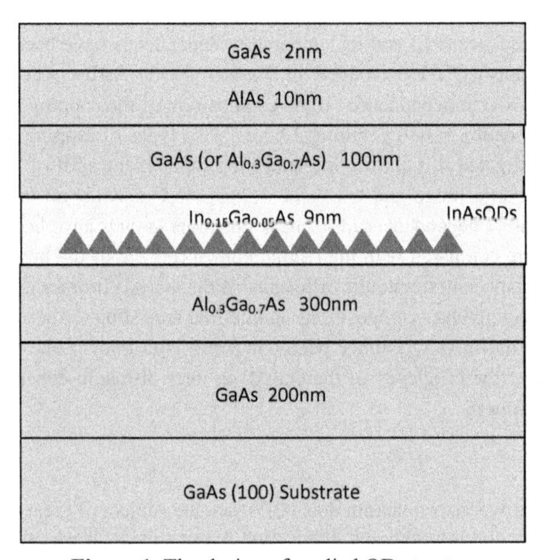

Figure 1. The design of studied QD structures

The in-plane density of QDs is estimated from previous AFM study was $4x10^{10}cm^{-2}$. PL spectra were measured under the excitation of the 488 nm line of a cwAr+-laser at an excitation power density of 500 W/cm^2 in the temperature range 10-500K. The setup used for PL study was presented earlier in [14, 15].The freshly prepared states are labeled by the letter A (#1A, #2A). The thermal annealing was carried out for some sets of the structures #1 and #2 at 640°C (state labeled by B, as #1B and #2B) during 2 hours.

RESULTS AND DISCUSSION

Typical PL spectra of the freshly prepared structures #1 and #2 measured at different temperatures are shown in Fig.2.Three overlapping PL bands appears due to the recombination of excitons localized at the ground state (GS) and at the first (1ES) and second (2ES) excited states in InAs QDs (Fig.2). The peak positions ofGS at 10K are 1.118 eV (#1), 1.139 eV (#2) and 1ES are 1.182 eV (#1), 1.187 eV (#2). As one can see the full width at half maximum (FWHM) of GS and 1ES (2ES) states and their overlapping in the structure #2 are higher in comparison with #1.

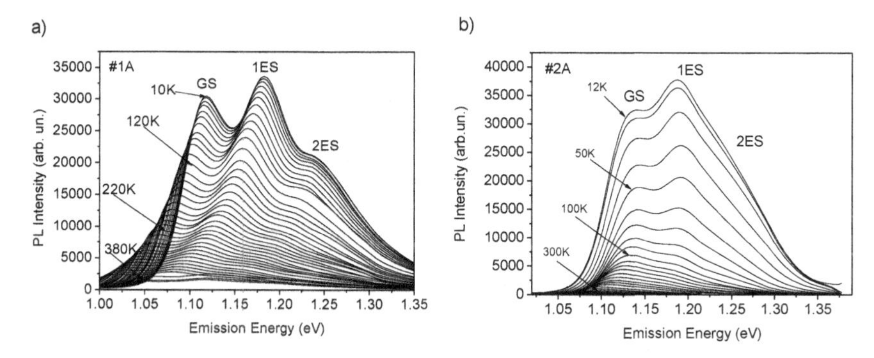

Figure 2a. PL spectra of the QD structure #1 measured at different temperature in a freshly prepared state

Figure 2b. PL spectra of the QD structure #2 measured at different temperature in a freshly prepared

The thermal annealing stimulates the shift of GS PL peaks into the high energy range and the increaseof a half width of PL bands and their overlapping (Fig.3).

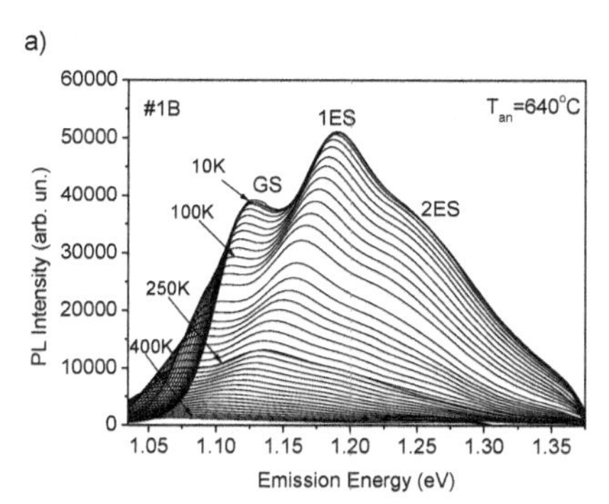

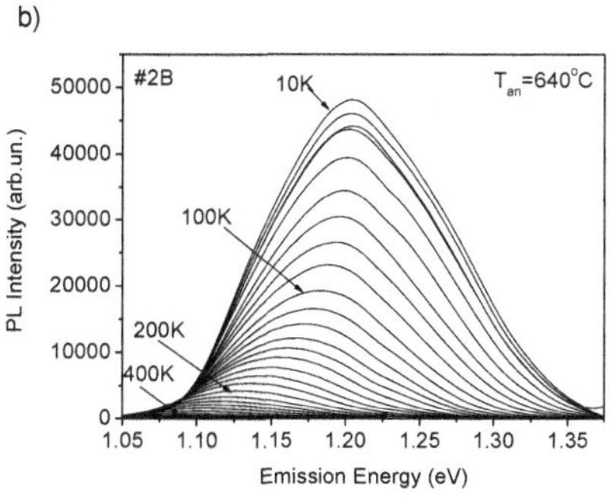

Figure 3a. PL spectra of the QD structure #1 after the thermal annealing at 640°C during 2 hours

Figure 3b. PL spectra of the QD structure #2 after the thermal annealing at 640°C during 2 hours

The $In_{0.15}Ga_{0.85}As$ buffer layer composition and the growth conditions were the same for both QD structures; as a result, the sizes of QDs are similar in #1 and #2. Moreover due to the close parameters of crystal lattices for the $Al_{0.30}Ga_{0.70}As$(5.6556 Å) and GaAs (5.6532Å)at 300K the level of elastic strain in #1 and #2, apparently, is similar as well. Thus it is possible to expect the same QD compositions in the studied structures owing to the similar conditions for the Ga/Al/In atom interdiffusion between the InAs QDs and InGaAs/AlGaAs QW layers.To analysis the process of Ga/Al/In inter diffusion the PL peak positions of InAs QDs have been studied at different temperatures in a freshly prepared state (A) and after the thermal annealing (states B, Fig.4).

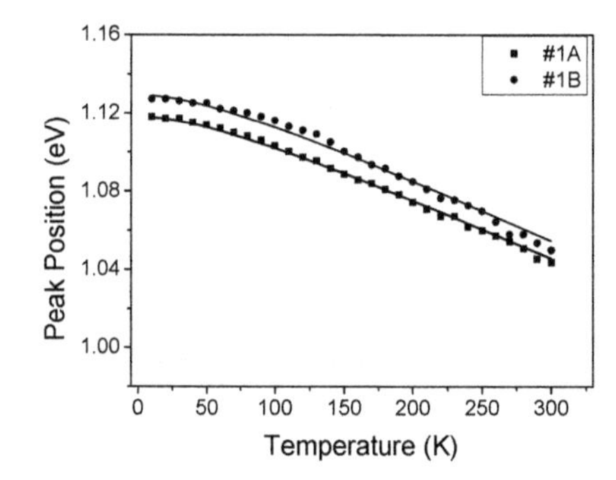

Figure 4a. The variation of PL peak positions versus temperatures in the structures #1 in freshly prepared states (A) and after thermal annealing at 640°C during 2 hours (B). The lines represent the Varshni fitting results.

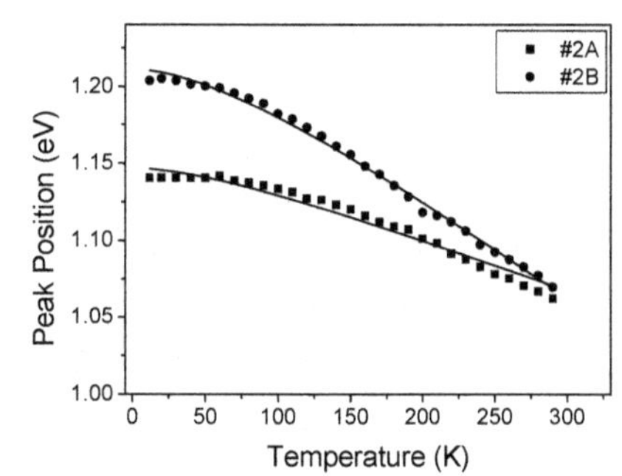

Figure 4b. The variation of PL peak positions versus temperature in the structures and #2 in freshly prepared states (A) and after thermal annealing at 640°C during 2 hours (B). The lines represent the Varshni fitting results.

Then the dependences of GS PL peak positions versus temperatures in all studied states have been obtained (Fig. 4). It can be seen from Fig. 4 that the PL peaks shift to lower energies with increasing temperature due to the shrinkage of the band gap with increasing temperature. Generally, due to the temperature induced lattice dilatation and electron–lattice interaction, the band gap energy follows the well-known Varshni formula [16]:

$$E(T) = \mathrm{E}_o - \frac{aT^2}{T+b} \quad , \qquad (1)$$

where $Eg(0)$ is the bandgap at the absolute temperature T = 0 K; "a" and "b" are the Varshni thermal coefficients.The linesin Fig.4a,bpresent the fitting results for the structures #1 and #2 in a freshly prepared state (A) and after thermal annealing (B).The comparison of fitting parameters "a" and 'b' with the variation of the energy band gap versus temperature in the bulk InAs crystal (Table 1) has revealed that in both QD structures (#1 and #2) in a freshly prepared statethe fitting parameter "a" and 'b' are close to their values in the bulk InAs crystal (Table 1).

The transformation of PL spectra of InAs QDs in the #1 and #2 structures after the thermal annealing (Fig.3 and Fig.4) has shown that the GS peak positions shift into a higher energy range owing the change of the QD sizes and material compositions. The fitting parameters obtainedforthe both QD structures after the thermal annealing's shown in Table 1. It is clear that a QD material composition in the structure #2 (GaAs capping) varies essentially in comparison with the structure #1 (AlGaAs capping) in the process of thermal annealing due to the more efficient the Ga/In inter diffusion.

Table 1, Vashni fitting parameters

Samples	A Freshly prepared	B T_{an}=640°C
2160 #1 AlGaAs/ InGaAs	E_o=1.12 eV α = 0.33 meV/K β = 93 K	E_o=1.13 eV α = 0.36 meV/K β = 97 K
2161 #2 GaAs/ InGaAs	E_o=1.14 eV α = 0.37 meV/K β = 95 K	E_o=1.20 eV α = 0.50 meV/K β = 120 K
InAs	Eo =0.415 eV α = 0.276 meV/K β = 83 K	

CONCLUSIONS

The photoluminescence (PL) and its temperature dependence have been studied in $GaAs/Al_{0.3}Ga_{0.7}As/In_{0.15}Ga_{0.85}As/Al_xGa_{1-x}As/GaAs$ quantum wells (QWs) with embedded InAs QDs in dependence on the composition of capping (GaAs and $Al_{0.3}$ $Ga_{0.7}$ As) layers and thermal annealing. It is shown that a QD material composition in the structure #2 (GaAs capping) varies essentially in comparison with the structure #1 (AlGaAs capping). It is concluded thatGa/Al/In inter diffusion is essential for the QDs in the structure with the InGaAs/GaAs capping layer, than in the structure with the InGaAs/AlGaAs capping layer. Our results have shown that the structures with AlGaAs capping layer can be used at higher temperatures up to 640 °C.

ACKNOWLEDGEMENT

The author would like to thank the CONACYT (project 130387) and SIP-IPN, Mexico, for the financial support, as well as the Dr. Andreas Stinz from the Center of High Technology Materials at University New Mexico, USA, for growing the studied QD structures.

REFERENCES

1. D. Bimberg, M. Grundman, N. N. Ledentsov, Quantum Dot Heterostructures, Ed. Wiley & Sons (2001) 328.

2. V. M. Ustinov, N. A. Maleev, A. E. Shukov, A. R. Kovsh, A. Yu .Egorov, A. V. Lunev, B. V. Volovik, I. L. Krestnikov, Yu. G. Musikhin, N. A. Bert, P. S. Kopev, Zh .I. Alferov, N. N. Ledentsov, D. Bimberg, Appl.Phys.Lett. **74**, 2815 (1999).
3. G. T. Liu, A. Stintz, H. Li, K. J. Malloy and L. F. Lester, Electron Lett, **35,** 1163 (1999).
4. A. Stintz, G. T. Liu, L. Gray, R. Spillers, S. M. Delgado, K. J. Malloy, J. Vac. Sci.Technol. B. **18(3**), 1496 (2000).
5. M. Dybiec, S. Ostapenko, T. V. Torchynska, E.Velasquez Losada, Appl. Phy. Lett.**84,** 5165-5167 (2004).
6. M. Dybiec, L. Borkovska, S. Ostapenko, T. V. Torchynska, J. L. Casas Espinola, A. Stintz and K. J. Malloy, Applid. Surface Science, **252**, 5542-5545(2006)
7. T.V.Torchynska, A. Diaz Cano, M.Dybic, S. Ostapenko, M.Mynbaeva, Physica B, Conden. Matter **376-377**, 367-369 (2006).
8. T.V. Torchynska, M. Dybiec, S. Ostapenko, Phys. Rev. B. (USA) **72**, 195341 (2005)
9. T.V.Torchinskaya, Opto-ElectronicsReview, **6(2),** 121-130(1998).
10. C. M. A. Kapteyn, M. Lion, R. Heitz, and D. Bimberg, P. N. Brunkov, B. V. Volovik, S. G. Konnikov, A. R. Kovsh, and V. M. Ustinov, Appl. Phys. Lett. 76, 1573 (2000)
11. C. A. Duarte, E. C. F. da Silva, A. A. Quivy, M. J. da Silva, S. Martini, J. R. Leite E. A. Meneses and E. Lauretto, J. Appl. Phys., 93, 6279 (2003).
12. X.Q. Meng, B. Xu, P. Jin, X.L. Ye, Z.Y. Zhang, C.M. Li, Z.G. Wang, Journal of Crystal Growth 243, 432 (2002).
13. L. Seravalli, P. Frigeri, M. Minelli, P. Allegri, V. Avanzini, S. Franchi, Appl. Phys. Lett. 87, 063101 (2005).
14. T. Torchynska, J. Appl. Phys., **104,** 074315, n.7 (2008).
15. T.V. Torchynska, A. Stintz, J. Appl. Phys. **108**, 2, 024316 (2010).
16. Y. P. Varshni, Physica 34, 149 (1967).

Mater. Res. Soc. Symp. Proc. Vol. 1534 © 2013 Materials Research Society
DOI: 10.1557/opl.2013.304

Nanoscale Structuration of Semiconductor Surface Induced by Cavitation Impact

Tetyana G. Kryshtab[1], Rada K. Savkina[2], Alexey B. Smirnov[2]

[1]Instituto Politécnico Nacional – ESFM, Department of Physics, Av. IPN, Ed. 9, U.P.A.L.M., 07738 Mexico D.F., Mexico
[2]V. Lashkaryov Institute of Semiconductor Physics at NAS of Ukraine, pr. Nauki 41, 03028 Kiev, Ukraine

ABSTRACT

The results of studies of the complex structures formed on the semiconductor substrates exposed to the acoustic cavitation (AC) near the liquid-solid interface are reported. Gallium arsenide and silicon substrates were exposed to the cavitation impact initiated by the focusing a high-frequency (MHz) acoustic wave into the liquid nitrogen. Optical, atomic force and scanning electron microscopy methods as well as energy dispersive X-ray spectroscopy (EDS), X-ray diffraction (XRD) and Raman spectroscopy were used for analysis of the surface morphology and chemical composition of semiconductor compounds. The formation of separated circular regions with the nanostructured surfaces inside was found. Electron micrograph images of the silicon surface show the creation of the dendritic objects inside of the ultrasonically structured region. The Raman spectroscopy and EDS data have confirmed the change of the chemical composition of the structured gallium arsenide surface and the Ga-N bond formation. The incorporation of nitrogen atoms into a silicon lattice has not been observed while XRD results have shown the formation of silicates of alkali metals on the silicon surface.

INTRODUCTION

Interactions of the energetic flows beams with the solid surfaces demonstrate unique phenomena and promise new applications for surface modification technology. Solid surfaces can develop a wide range of useful topological features upon bombardment with ions, clusters of atoms and molecules, as well as on laser processing as a result of the deposition of large amounts of energy in the form of heat and pressure or as a result of the plasma effect.

It is well known that ultrasonic cavitation also creates extreme energetic conditions with the local temperatures of about 5000K and the pressure of several hundred MPa [1]. It is necessary to notice also the possibility of the plasma generation in a cavitating liquid [2]. These extreme conditions are widely used in chemistry, as for example to synthesize nanoparticles [3], to enhance the electrochemical reactions and to modify the surface properties of electrodes [4], as well as to generate the novel materials in a liquid medium [5-7], etc.

Cavitation near extended liquid-solid interfaces is very different from cavitation in pure liquids. The impingement of microjets and shockwaves on the surface creates the localized erosion, which can generate newly exposed, highly heated surface and even eject matter from the surface. This is a grave disadvantage in many industrial systems. Despite corrosion problem, the controlled cavitation near a solid surface proves a powerful tool for modern technologies like the salmonella destruction [8], the treatment of kidney stones [9], producing the hydrogenated/ deuterated metallic powders by sonoimplantation [10], etc.

With respect to semiconductors, the main application of acoustic cavitation is the ultrasonic cleaning. At the same time, we think that the combination of the factors, which accompany bubbles collapse near the semiconductor surface, can results in semiconductor surface structuration up to a synthesis of a new phase. Possibility of the application of acoustic cavitation for the material deposition from a solution on the semiconductor surface was confirmed by the results obtained in [11].

Earlier we presented experimental results on the modification that occurs on the gallium arsenide surface subjected to the megasonic cavitation exposure [12]. The nitrogen atoms incorporation into the GaAs lattice and the Ga-N bond formation in the region of the maximal structural change due to the cavitation impact was found.

In the present paper, we report the effects of megasonic cavitation on the several semiconductor surfaces. Silicon and gallium arsenide samples were studied. Experiments employ the cavitation in an attempt to drive the chemical and structural transformations on a semiconductor surface, creating features with a number of remarkable and potentially useful properties.

EXPERIMENTAL DETAILS

We have undertaken studies of the cavitation impact in cryogenic liquid such as nitrogen (LN_2). The experimental setup consisted of a reactor vessel and the ultrasonic (US) equipment. A pumped, cylindrical stainless steel tank with the internal copper cell filled with technical nitrogen was used for the reactor vessel [12]. A ceramic US transducer (PZT-19) operating at 3 MHz (or 6 MHz) with a diameter of 12 mm was installed in the copper cell.

It is well known that at high sonic frequencies, on the order of the MHz, the production of cavitation bubbles becomes more difficult than at low sonic frequencies, of the order of the kHz. To achieve cavitation, the intensity of the applied sound must be increased. In our experiments, for the cavitation activation, a high frequency system (MHz) with focused energy resonator was used. Besides, the operating temperature of LN_2 (78 K) is near the critical temperature of this fluid, and the thermodynamic effect of cavitation can be easily reached.

The initial value of an acoustic intensity W_{US} did not exceed 1 W/cm^2. A cylindrical copper concentrator (lens) was used and the intensity gain of the acoustic system (PZT + copper lens) was about 58. The acoustic matching the PZT to copper lens is sufficient for satisfying the condition of transparent boundary (~98%). The impedance mismatch at the interface between the concentrator and LN_2 has led to the fact that the ratio of the emitted acoustic power to the dissipated power was about 55%. And the maximal value of pressure in the focus of the acoustic system was about 8 bars.

Two different types of materials, Si (001) and GaAs (001), were treated in a cryogenic liquid. We have used silicon wafers with diameter about 76.2-mm and semi-insulating gallium arsenide wafers with diameter about 40 mm. The samples were cut into the 5 mm$\times$ 5 mm squares and were cleaned for 10 minutes in ethanol and then in distilled water. The initial surface roughness of samples was found below 1nm.

A semiconductor sample was placed inside of the acoustically driven copper cell in the focus region. After sonication the semiconductor surface was investigated using the optical microscopy (NV2E, Carl Zeiss Jena), atomic force microscopy (AFM) (Digital Instruments NanoScope IIIa AFM operating in the tapping mode), and scanning electron microscopy (SEM) (JSM-6490). The energy dispersive X-ray spectroscopy (EDS), X-ray diffraction (XRD) and Raman spectroscopy were used as well.

RESULTS AND DISCUSSION

The exposure of semiconductor wafers to megasonic cavitation in LN_2 leads to the surface structuration. After a treatment of GaAs sample at 3 MHz in an acoustically driven copper cell, ripple-like patterns are developed [13]. Micron size rounded bumps and rings about 5-10 μm in diameter located in a random way are formed as well (see figure 1a). The modified region integrally is below of the original surface. A treatment of the Si sample at the same conditions leaded to the similar results.

It was revealed that the characteristic dimension of the structures on the semiconductor surface depends on the exposure parameters and can be controlled by the regulation of an acoustic frequency. The reduction of acoustic intensity at the same frequency results to a more random character of surface modifications.

The increase in a processing frequency changes both the species and the size of the structures formed on the surface. The formation of separated circular regions with nanostructured insides is observed (see figure 1b).

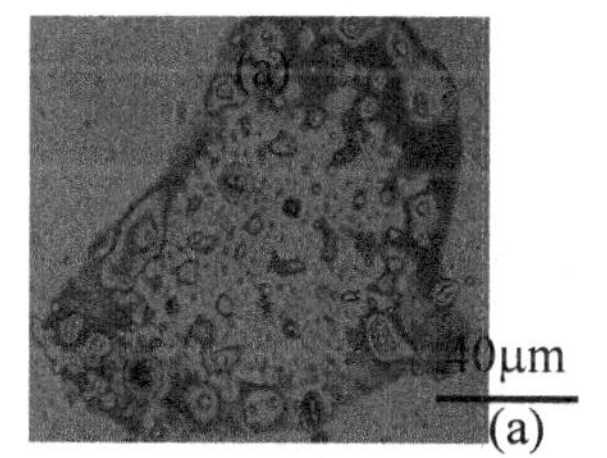

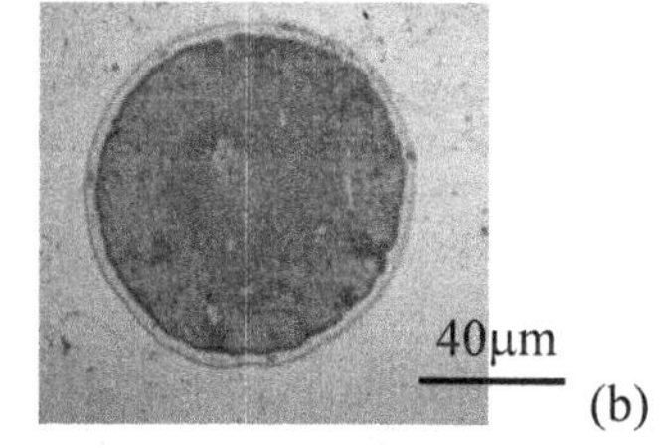

Figure 1.Optical micrographs of samples exposed to the acoustic cavitation in liquid nitrogen at 3MHz during 15 min (a) and at 6MHz during 15 min (b).

AFM studies revealed the presence of a rim around the structured region. The interior of circular region is located below of the original surface. Figures2a, 2b, and 2c show the AFM images of such region. The roughness of the above-mentioned surface regions are described by the surface height histograms (see figures 2d, 2e, and 2f). In these histograms, several local maxima are seen. One can separate the several groups of structures with the typical heights of h_i ~3-8 nm and~11-15 nm for GaAs as well as ~30 nm and ~ 70 nm for Si substrates.

Thus, it should be noted the similarity of the relief produced on GaAs and Si surfaces at the cavitation.

At the same time, there were found the distinctive features for each semiconductor studied. It was found that the prolonged cavitation treatment of silicon wafer resulted in the dendritic structure formation. Figure 3 presents the optical (Inset) and SEM images of the silicon surface that show the creation of dendritic objects inside of the ultrasonically structured region and a content of different atoms incorporated into the structured silicon surface exposed to the acoustic cavitation.

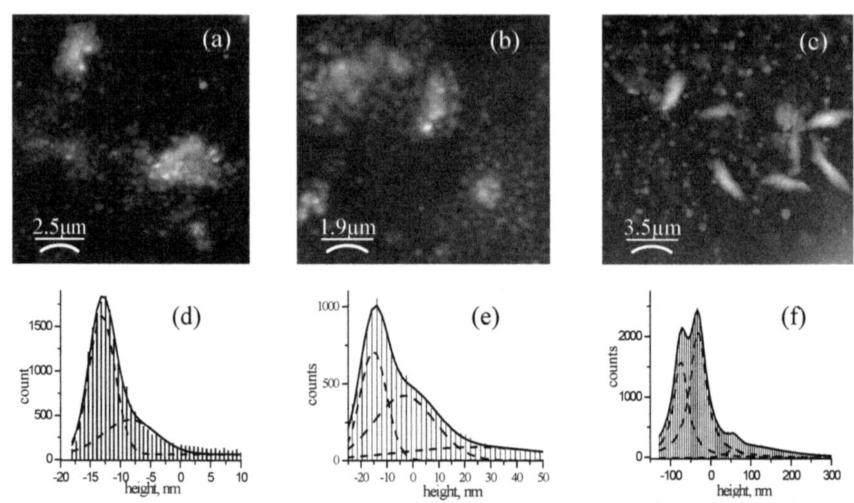

Figure 2.AFM images of the GaAs (a, b) and Si (c) surfaces exposed to the acoustic cavitation in the liquid nitrogen at 6MHz; (d), (e), (f) - corresponding histograms of a height distribution over the structures.

Chemical properties of nanostructured semiconductor surface

It was found that various gaseous atoms (including nitrogen) can be strongly implanted into the metal powders under ultrasonic cavitation [10]. In our experiment, one can expect that significant material intermixing in the impact sites takes place, and there is a probability for nitrogen atoms to be incorporated into a semiconductor wafer in the modificated region.

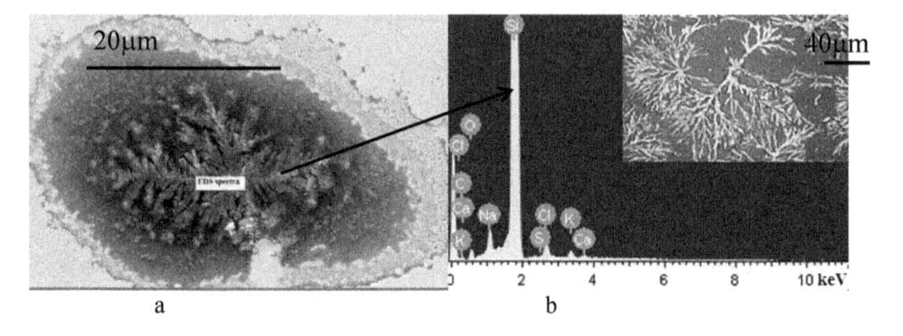

Figure 3. SEM micrograph (a) and the atomic composition of the structured silicon surface (b) exposed to the acoustic cavitation in liquid nitrogen at 6 MHz during 15 min. Inset: an optical image of silicon surface exposed to the acoustic cavitation during 1 hour.

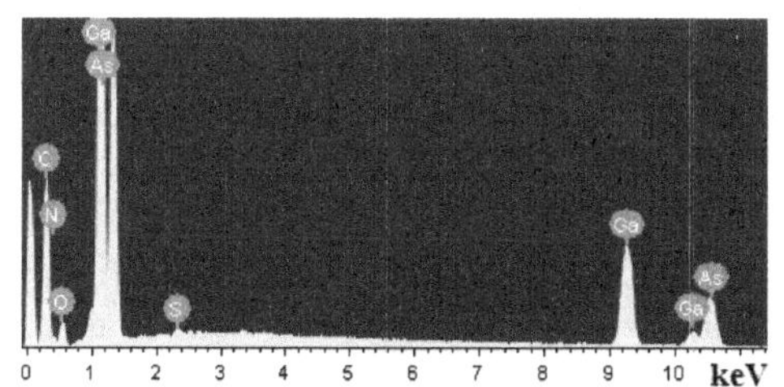

Figure 4. The atomic composition of structured gallium arsenide surface exposed to the acoustic cavitation in liquid nitrogen at 3 MHz.

The atomic composition of microstructured gallium arsenide surfaces was investigated using EDS. The chemical composition of samples was studied in numerous randomly selected areas of 5 × 5 μm. We found an inhomogeneous character of the nitrogen atom incorporation into the substrate lattice (from 5% to 7.5%) after the cavitation exposure. EDS analysis also indicated the presence of oxygen and carbon atoms in the nitrated areas (see figure 4).

Moreover, the Raman spectroscopy data has confirmed the nitrogen atom incorporation into the GaAs lattice and the Ga-N bond formation in the region of the maximal structural change at the cavitation impact [12]. At the same time, the nitrogen atom incorporation into the silicon lattice has not been observed while we identified the peaks corresponding to the following elements: Na, S, K, Cl, and Ca (see figure 3b). Figure 5 represents the XRD result for the Si(100) substrate before and after the cavitation treatment.

The XRD pattern for an initial state reveals only the Si 400 diffraction peak that indicates (001) orientation of the substrate.

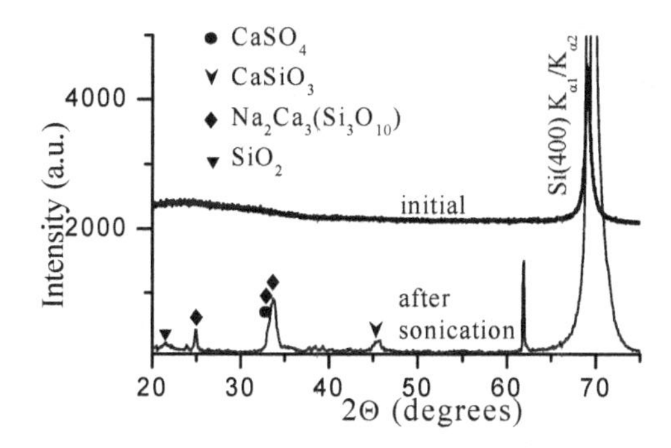

Figure 5.XRD patterns of silicon substrate before and after exposure to cavitation impacts. The crystallographic database WWW-MINCRYST [14] was used for peaks identification.

The presence of higher intensity background in the range of 2θ= 20-30 degrees in XRD pattern for the initial state denotes the existence of a small amount of amorphous phase on the Si substrate surface, obviously the amorphous Si oxides. A new peaks connected with the formation of some compounds are observed in XRD pattern obtained after the cavitation treatment. In particular, we found the formation of silicates and sulfates of alkali metals. The broad diffraction peak appeared at $2\theta = 21.5^{\circ}$ shows the formation of SiO_2 nanocrystals at the cavitation treatment.

CONCLUSIONS

Gallium arsenide and silicon substrates exposed to the cavitation impact, obtained by focusing a high frequency acoustic wave into liquid nitrogen, were investigated. Based on the experimental results, the following conclusions can be drawn.

The exposure of semiconductor substrate to megasonic cavitation in LN_2 leads to the surface structuration. It was revealed that the characteristic dimension of the peculiarities on the semiconductor surface depended on the exposure parameters and can be controlled (from micron- to nano-scale dimension) by the regulation of an acoustic frequency. The change of the chemical composition of the semiconductor surface was also found. EDS analysis indicated on the nitrogen atom incorporation into the gallium arsenide lattice. XRD results showed the formation of silicates of alkali metals on the silicon surface as well.

REFERENCES

1. K. S. Suslickand D. J. Flannigan, *Annu. Rev. Phys. Chem.* **59**, 659 (2008).
2. D. J. Flannigan and K.S. Suslick, *Phys. Rev. Lett.* **95**,044301 (2005).
3. Y. Mastai and A. Gedanken,"Sonochemistry and other novel methods developed for synthesis of nanoparticles", *The Chemistry of Nanomaterials: Synthesis, Properties and Applications,*edited by C. N. R. Rao, A. Müller, and A. K. Cheetham (WILEY-VCH, 2004) pp.113-207.
4. T. J. Mason, J. P. Lorimer and D. J. Walton, *Ultrasonics* **28**, 333 (1990).
5. M. Ashokkumar and T. Mason,*SonochemistryinKirk-Othmer Encyclopedia of Chemical Technology* (John Wiley and Sons, 2007).
6. A. Gedanken, *Ultrasonics Sonochemistry* **11**, 47 (2004).
7. N. Perkas, Y. Wang, Y. Koltypin, A. Gedanken, and S. Chandrasekaran, *Chem. Comm.* **11**, 988 (2001).
8. D. M. Wrigley and N. G. Llorca, *Journal of Food Protection* **55**, 678 (1992).
9. Y. A. Pishchalnikov, O. A. Sapozhnikov, M. R. Bailey, J. C. Williams Jr, R. O. Cleveland, T. Colonius, L. A. Crum, A. P. Evan, and J. A. McAteer, *J.Endourol.* **17**, 435 (2003).
10. Y. Arata and Y.-C. Zhang, *Appl. Phys. Letters* **80**, 2416 (2002).
11. S. Nomura and H. Toyota, *Appl. Phys. Letters* **83**, 4503 (2003).
12. R. K. Savkina, A. B.Smirnov, *J. Phys. D: Appl. Phys.* **43**, 425301 (2010).
13. R. K. Savkina, *Functional material,* **19**, 38(2012).
14. http://database.iem.ac.ru/mincryst.

Mater. Res. Soc. Symp. Proc. Vol. 1534 © 2013 Materials Research Society
DOI: 10.1557/opl.2013.305

Micro-Raman Cross-Section Study of Ordered Porous III-V Semiconductor Layers

Tetyana R. Barlas[1], Nicolas L. Dmitruk[1], Nataliya V. Kotova[1], Denys O. Naumenko[1,2], Valentinas Snitka[2]
[1]Institute for Physics of Semiconductors, National Academy of Sciences of Ukraine, 45 Nauki Prospect, 03028 Kyiv, Ukraine ,
[2]Research Centre for Microsystems and Nanotechnology, Kaunas University of Technology, Studentu 65, LT-51369 Kaunas, Lithuania

ABSTRACT

Porous layers have been obtained by electrochemical etching of *n*-type III-V semiconductor single crystals (GaAs, InP, and GaP) in water or the ethanol solutions of different acids. The InP porous layers have been shown a more ordered and perfect structure than those of GaP and GaAs. Metal inclusions have been incorporated into the porous layers in an electrochemical cell from an aqueous solution of the Au salt. As found, phonon band intensities are significantly increased in the porous region. The homogeneity of the porous layers prepared has been reliably proved by the Raman micro-spectroscopic mapping of their fresh cleavages. The incorporation of gold into pores leads to a stronger Raman signal of the TO and LO modes, especially in the two-phonon absorption region, and also significantly enhances the photoluminescence of porous layers.

INTRODUCTION

Porous polar III-V semiconductor compounds having a direct band gap and a large refractive index that is promising for photonics and optoelectronics applications, and their large surface-to-volume ratio offers a great potential for sensors. Moreover, such materials can be employed in the photoelectric devices and solar cells. Surface plasmon or surface plasmon polariton excitation in the metal nanoparticles or on a corrugated metal surface leads to localization, concentration, and local enhancement of electromagnetic fields in their vicinity thus causing an enhancement of many photophysical phenomena such as photoluminescence (PL), infrared absorption (SEIRA), Raman scattering (SERS), photocurrent in barrier structures (plasmonic photovoltaics), electromagnetic forces, electroluminescence etc. Here we propose a facile and cheap method for the fabrication of a new class of nanocomposite materials, viz., ordered porous III-V semiconductor layers with metal nanoparticles incorporated into the pores. Optical properties of nanocomposites have been investigated by micro-Raman spectroscopy, a non-destructive technique giving information on the chemical composition and crystallinity of a sample with a microscopic spatial resolution [1]. Operating with visible-light photons this technique has a high sensitivity for the investigation of vibrational properties of semiconductors and particularly of the Brillouin zone center phonons. Thus, quantum confinement effects or intrinsic stresses in a semiconductor lead to significant changes in phonon frequencies and in

Raman spectra, respectively [2-4], making the technique attractive for studying a wide range of phenomena.

EXPERIMENT

Porous layers were obtained by electrochemical etching of n-type GaAs, InP, and GaP single crystals with the free carrier concentration N of 10^{16} – 10^{18} cm^{-3} and the (111) and (100) surface orientations. The etching was performed in the galvanostatic regime with a platinum counter electrode in water or ethanol solutions of the HF, HCl and H_2SO_4 acids for GaAs, InP, and GaP samples, respectively, at room temperature [5, 6]. In particular, 5% HCl was used as an etchant for InP. The current density was varied from 2 to 100 mA/cm^2 and the etching time from 1 to 30 min. By selecting appropriate etching regimes, we aimed to form reproducible, structurally homogeneous, and sufficiently thick porous layers with well-defined boundaries.

A metal was incorporated into the thus prepared porous layers. We have chosen gold as a inclusion metal taking advantage of its resistance to aging and oxidation. Also, it is experimentally convenient that the plasmon of gold nanoparticles is found in the visible region and has been much studied. Metal nanoparticles were deposited on pore surfaces by passing direct electric current at a voltage of 1 V through an aqueous $AuCl_3$ solution (in the same electrolytic cell where porous layers were fabricated). The Au^{+3} ion concentration was varied from 0.01 to 0.5 gr cm^{-3} and the deposition time from 2 to 60 min. This method can be easily extended to incorporating various metals into porous layers.

The obtained porous layers and resulting nanocomposites were structurally analyzed by scanning electron microscopy (SEM) using a JEOL 6700 instrument (with a resolution of 1 to 2.2 nm). We investigated both the surfaces and, even more informatively, the fresh cleavages of the porous layers. The elemental analysis of the samples under study was carried out with a microanalyzer for Energy dispersive X-ray spectroscopy (EDAX) attached to the microscope used.

The Raman measurements were undertaken using NTEGRA Spectra system (NT-MDT Inc.) in an upright configuration. NTEGRA Spectra instrument is equipped with high performance optics (100x 0.7 NA objective) that provides scanning of the sample area with the resolution of 0.4 μm. Its optical measuring head serves for illuminating the surface with visible light at wide angle and for collecting the scattered radiation. A 20-mW DPSS laser (LCM-S-111-20-NP25) has been a source of p-polarized light of 532 nm wavelength with a controlling laser power at the sample surface in the range of 0.005-5 mW. The laser beam is focused on the sample, which rests on a controlled XYZ piezostage capable of scanning samples over a 130x130x7 μm. The scattered signals are collected by the same objective and then focused through a 100 μm pinhole. A 532-nm ultrasteep long-pass edge filter rejects residual backscattered laser light and the signal is directed into a spectrometer (Solar TII, MS5004i) equipped with a TE-cooled (down to -60°C) CCD camera (1024x128 pixels, DV401-BV, Andor Technology). The diffraction grating of 1800 lines/mm has been chosen for the experiments, resulting in a spectral resolution of 0.79 cm^{-1}. Another CCD camera is used to collect white light microscope images.

Fresh cuts of both porous InP with gold nanoparticles inside the pores and without them were prepared immediately before the measurements. The samples were then mounted perpendicularly to the incident light and characterized by Raman mapping procedure. The Raman maps of the main characteristic peaks, corresponding to LO, TO phonons and their combinations had been plotted as follow. A full spectrum map is recorded for a scan area of 68x17 μm with a step of 0.532 μm between two points, resulting in a map of 128x32 pixels (spectra). The acquisition time was chosen 8 s for each pixel. Then the peak's position and intensity were calculated using the standard NT-MDT software (Nova 1.0.26). To avoid a heating of the samples, a 1 mW laser power was used.

RESULTS AND DISCUSSION

Morphology of the porous GaAs, InP, and GaP semiconductors layers

SEM investigations have confirmed that we can fabricate both ordered and disordered porous layers of essentially different morphologies, with the layer parameters varied in a very wide range: the porosity from 1% to 70% (as judged by SEM images), the pore diameter from 50 to 200 nm, and the porous layer thickness from 5 to 90 μm. Figure 1 presents the porous layers typical for each semiconductor concerned. As easily seen, InP forms porous layers of a more ordered and perfect structure compared to GaP and GaAs. It is with InP that we succeeded to fabricate up to 100- μm-thick layers, whereas this proved impossible with GaP and GaAs (their thick porous layers are too brittle and friable, and tend to flake from the surface). The InP porous layers appear as a system of cylindrical cavities (which is typical of the (100) surface and at relatively large currents of a few mA) or embedded tetrahedrons demonstrating, in the latter case, a horizontal-plane correlation between neigbouring pores (see Figure.1a). The pore diameter is in the range 50-100 nm, while the wall thickness varies from 50 to 300 nm and the pore length from 10 to 80 μm; an average period is estimated to be 100 – 350 nm. These properties of InP porous layers motivated us to select them for the further design of nanocomposites.

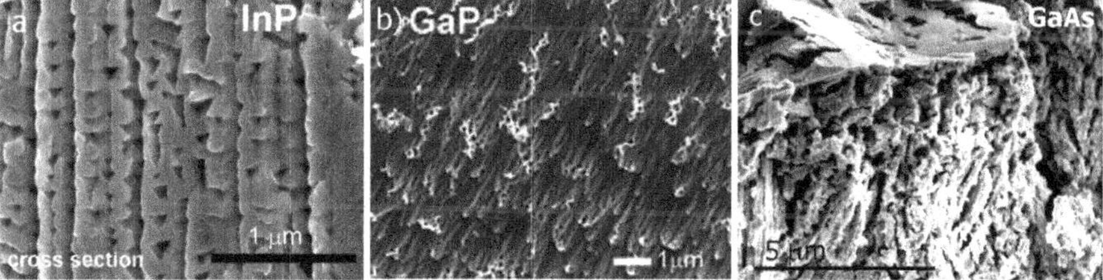

Figure 1. SEM images of the InP(a), GaP (b) and GaAs (c) porous layers.

Metal-semiconductor nanocomposite

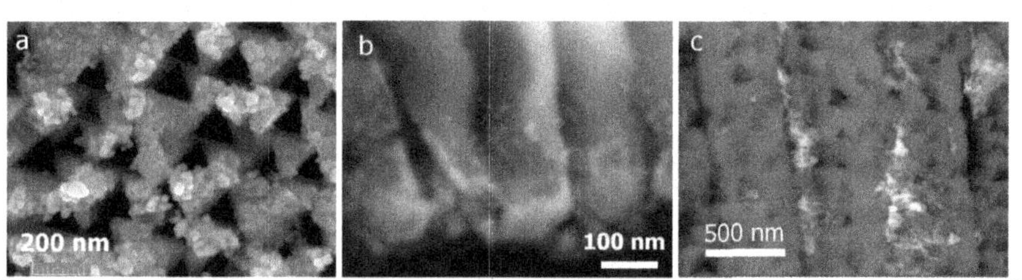

Figure 2. The SEM images of InP porous layers with Au inclusions: a surface image (a) and different cross sections (b, c). The metal was deposited from 0.1 mg/cm^3 aqueous $AuCl_3$ solution at 1 V DC voltage for 10 min.

SEM images of the porous layers with gold particles incorporated into them are shown in Figure 2. The surface of the sample (see Figure 2a) is "decorated" with 10-40 nm sized metal particles, whereas the cleavage image (see Figure 2b) exhibits that the particles (of a somewhat smaller size, 10-20 nm) also fill pores. Inclusions of this kind causing the surface plasmon enhancement effect are very attractive for applications in sensorics, optoelectronics, and photovoltaics. As evidenced by EDAX point elemental analysis of porous layers, gold particles are indeed present on the layer surface as well as inside pores. The measurements have confirmed the unchanged stoichiometry of semiconductor pore walls, which, along with SEM images, suggests that the walls represent a homogeneous perfect monocrystal.

Micro-Raman spectra and mapping of Au-InP porous metal-semiconductor nanocomposite

Figure 3 depicts the Raman spectra of bulk InP and its porous layers both with and without gold nanoparticles incorporated. Sharp bands at 305 and 345 cm^{-1} correspond to the TO and LO phonons of the InP crystal lattice [7, 8]. The bands in the region 600-700 cm^{-1} arise from the TO and LO phonon combinations with the characteristic frequencies of the 2TO, TO+LO, and 2LO vibrations [7].

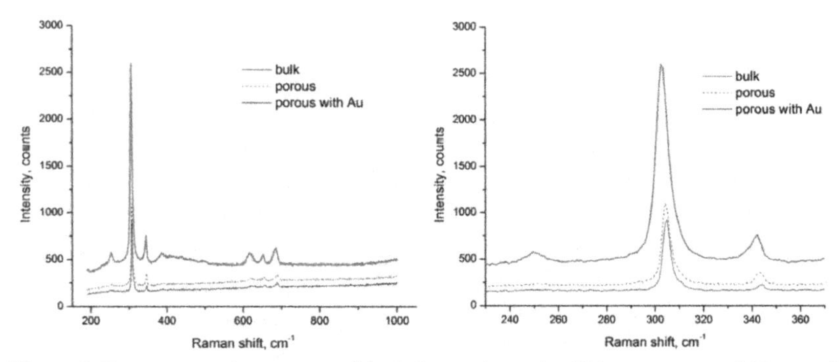

Figure 3. Raman scattering spectra of the InP samples at 1 mW laser power and 30s acquisition time in the range of 200-1000 cm^{-1} (a) and 230-370 cm^{-1} (b).

To establish the effects caused by pore formation and gold incorporation into pores on the lattice vibration region of the InP spectrum, the peaks of TO phonons were fitted by the Gaussian curve using formula (1) (for the fitted curve parameters see Table 1).

$$y = y_0 + A/(w\sqrt{\pi/2}) * e^{-2\frac{(x-x_c)^2}{w^2}} \quad (1),$$

where y_0 is a baseline, x_c is a curve center, w – its width, and A – its area. Then a Full Width at Half Maximum and Height of the peak can be calculated as:

$$FWHM = \sqrt{2ln2} \cdot w, \quad Height = A/(w\sqrt{\pi/2}) \quad (2)$$

Table I. Parameters of Gaussian curves (formula 1) of the fitted TO phonon Raman spectra.

	y_0	Height	FWHM	x_c
Bulk	183±5	712	3.97	304.9±0.1
Porous	259±7	779	4.66	304.7±0.1
Porous with Au	571±14	1913	6.72	303.1±0.1

Noteworthy, going from bulk InP to its porous layers and further to the gold-incorporated material is accompanied by a number of spectral changes, namely, the background signal (i), the band intensity (ii) and half-width (iii) are increased, whereas the band itself is red-shifted (iv). The first effect is attributed to enhanced photoluminescence in the region 530–550 nm [9]; indeed, a Raman shift of 300 cm^{-1} at an excitation wavelength of 532 nm corresponds to the signal at 541 nm. A significant increase in band intensity on gold incorporation is due to the formation of metal nanoparticles in the layer; as a result, nanoparticle local plasmons are excited, with their absorption maximum at 530 nm [10, 11], and the TO signal is enhanced threefold (a so-called surface enhanced Raman scattering (SERS) effect) [12]. A red shift of the TO peak, along with its increased half-width, suggests the influence of quantum-dimensional effects [2, 8].

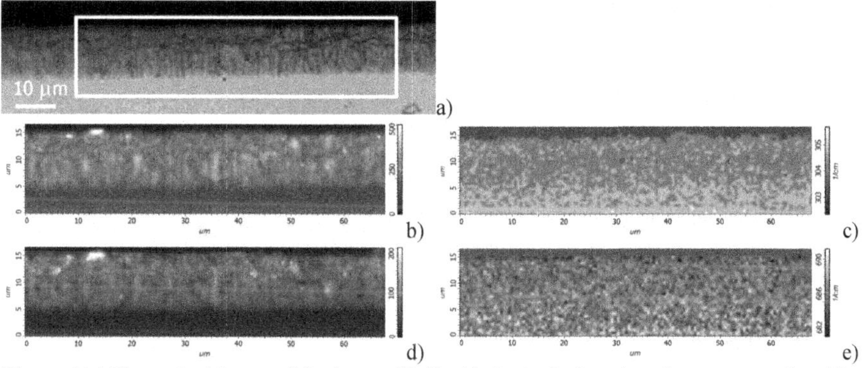

Figure 4(a) The optical image of fresh cut of InP with Au inclusions into the porous region. The marked area shows a region, when Raman mapping was performed. Intensity **(b)** and peak position mapping **(c)** of TO peak (304 cm^{-1}) and 2LO one (686 cm^{-1}, d, e).

A detailed analysis of the LO phonon region (about 345 cm^{-1}) in the spectra of gold-incorporated porous layers has revealed the band asymmetry thus implying the existence of a surface mode [13, 14] which is likely to also be SERS-enhanced (see Figure 3b).

Table II. Normalized parameters of the phonon modes and PL in bulk and porous InP.

	InP porous sample		InP porous sample with Au inclusions	
	Peak position	Peak value normalized	Peak position	Peak value normalized
Flat region	304,8±0,1	1	304,6±0,3	1
Porous region	304,5±0,4	1,12	304,2±0,3	1,53
Flat region	344±0,5	0,08	343,6±0,3	0,09
Porous region	343,5±0,4	0,19	343,2±0,5	0,23
Flat region	687±1	0,09	686±1.5	0,14
Porous region		0,12		0,2
Flat region	PL (705cm^{-1}, λ=550 nm)	0,14	PL (705 cm^{-1}, λ=550 nm)	0,16
Porous region		0,18		0,53

To characterize the samples prepared and to verify the conjectures made above, the Raman mapping of the porous layer cross sections was performed (for details see Experiment).

Figure 5a displays an optical microscopic picture of a fresh cleavage of the InP sample, with gold nanoparticles incorporated into its porous layer; the TO and 2LO phonon band intensities, with the luminescence background subtracted, are presented in Figures 5b, d, and the corresponding band positions in Figures 5c, e. It is seen from Figures 5b, d, that the porous layer with gold inclusions is fairly homogeneous; its statistical parameters as well as those of its gold-free counterpart are given in Table 2 (mapping data are not presented). The already mentioned red-shifted TO peak is indicated in Figure 5c. A similar 2LO shift is less noticeable due to the peak shape (see Fig. 4), since the exact position of a broad maximum is difficult to determine. The data in Table II also illustrate that the two main phonon modes are red-shifted in InP porous layers as compared to the bulk material. In addition, the TO and LO peak intensities strongly increase, especially on gold incorporation thus showing the high potentiality of the proposed method for material characterization.

CONCLUSIONS

Metal-semiconductor nanocomposites based on well-ordered InP porous layers have been electrochemically created. Raman micro-spectroscopic mapping of the porous layer fresh cleavages has reliably proved the homogeneity of the porous layers. The intensity of phonon bands is significantly increased in the porous region. Incorporation of gold nanoparticles into InP layer pores leads to an enhancement of the Raman signal from TO and LO modes, especially in the two-phonon band region. Also, gold inclusion results in a significantly increased intensity of the photoluminescence signal of the porous region.

REFERENCES

1. E. Smith, G. Dent, *Modern Raman Spectroscopy: A Practical Approach* (John Wiley & Sons, Chichester, 2005) p. 222.
2. G. Gouadec, P. Colomban, Prog. Cryst. Growth. Ch. 53 (1), 1 (2007).
3. D.Naumenko, V.Snitka, M.Duch, N.Torras, J.Esteve, Microelectron Eng. 98, 488 (2012).
4. Falkovsky L A, *Phys. Usp.* **47**, 249 (2004).
5. L. Santinacci, T. Djenizian, C. R. Chimie 11, 964 (2008).
6. N. Dmitruk, T. Barlas, V. Serdyuk, Physics and Chemistry of Solid State 11, 13 (2010).
7. M. Rojas-Lopez, J. Nieto-Navarro, E. Rosendo, H. Navarro-Contreras, M.A. Vidal, Thin Solid Films. 379, 1 (2000).
8. N Dmitruk, T. Barlas, I. Dmitruk, S. Kutovyi, N. Berezovska, J. Sabataityte, I. Simkiene, phys. stat. sol. (b) 247 (4), 955 (2010).
9. U. Schlierf, D.J. Lockwood, M.J. Graham, P. Schmuki, Electrochimica Acta. 49, 1743 (2004).
10. N.L. Dmitruk, A.V. Goncharenko, E.F. Venger, *Optics of small particles and composite media*, (Naukova Dumka , Kyiv, 2009) p. 386.
11. Susie Eustis and Mostafa A. El-Sayed, Chem. Soc. Rev. 35, 209 (2006).
12. M. Moskovits, J. Raman Spectrosc. 36, 485 (2005).
13. M. Dmitruk, V. Litovchenko, V. Strizshevsky, *Surface Polaritons in Semiconductors and Dielectrics*, (Naukova Dumka, Kyiv, 1989), p. 376.
14. A. Sarua, J. Monecke, G. Irmer, I.M. Tiginyanu, G. Gartner, H L Hartnagel, Phys. Condens. Matter **13**(31), 6687 (2001).

Mater. Res. Soc. Symp. Proc. Vol. 1534 © 2013 Materials Research Society
DOI: 10.1557/opl.2013.316

Structural and Optical Properties of Porous III-V Semiconductors GaAs, InP Prepared by Electrochemical Etching

Nicholas L. Dmitruk[1], Natalia I. Berezovska[2], Igor M. Dmitruk[2], Denis O. Naumenko[1,4], Irene Simkiene[3] and Valentinas Snitka[4]
[1]Institute for Physics of Semiconductors, National Academy of Sciences of Ukraine, 45 Nauki Prospect, Kyiv, 03650, Ukraine
[2]Taras Shevchenko National University of Kyiv, 64 Volodymyrs'ka, Kyiv, 01601, Ukraine
[3]Semiconductor Physics Institute, 11 Gostauto, Vilnius, 01108, Lithuania
[4]The Research Centre for Microsystems and Nanotechnology, Kaunas University of Technology, 65 Studentu, 51369, Kaunas, Lithuania

ABSTRACT

Properties of the electrochemically prepared porous III-V semiconductors, GaAs and InP, have been studied using scanning electron microscopy (SEM), atomic force microscopy (AFM), monochromatic multi-angle-of-incidence (MAI) ellipsometry, Raman scattering (RS), including confocal micro-Raman measurements. Two-layer oxide/porous structures have been observed for porous samples. The optical constants and filling factors of porous layers have been calculated in the frame of the effective medium approximation. The peculiarities of Raman spectra of porous GaAs and InP have been analyzed using the critical point analysis of the phonon dispersion.

INTRODUCTION

For the last years the researchers actively study different composite materials, in particular porous semiconductors, due to their unusual or unique properties, namely the capability to reduce the optical losses and to increase the efficiency/sensitivity of devices with large specific surface area. Engineering the optical properties of given materials by the modification of their structure is highly motivated due to the perspective possible applications in optoelectronics, quantum electronics, photovoltaics, etc. Besides, the porous semiconductors with rather large specific surface are very useful for sensorics. The porous semiconductor is an appropriate template for a wide variation of effective optical parameters by the incorporation of metal or other semiconductor nanoparticles, and in particular for the design of so-called metamaterials with negative value of the dielectric permittivity or even the refractive index. Therefore determination of porosity of these semiconductors and their effective optical parameters is an important task.

In present paper the morphology and optical properties of the porous layers of III-V semiconductors (GaAs and InP) have been studied by atomic force microscopy (AFM) and scanning electron microscopy (SEM), monochromatic multi-angle-of-incidence (MAI) ellipsometry, Raman scattering (RS), including confocal micro-Raman measurements.

EXPERIMENTAL DETAILS

The porous layers were prepared by electrochemical etching under following technological conditions: 1) n-GaAs, (100), 1M HCl or $HF:C_2H_5OH:H_2O$ (2:1:1), the current density, j= 5 - 20 mA/cm^2, the etching time, t= 10 - 20 min; 2) n-InP, (100), 1M HCl, j= 1 - 5 mA/cm^2, t= 3 - 10 min. At the anodization process the wafers were illuminated by an incandescent lamp providing the light flux of 600–1200 lx from the top or from the bottom of the etched substrates to generate additional charge carriers (holes) at semiconductor surfaces.

The surface morphology was monitored by scanning electron microscopy (SEM) and atomic force microscopy (AFM). SEM images have been obtained using field emission SEM JEOL JSM-6700F and SEM TESLA BS300, and AFM Dimension 3100 (Digital Instruments) has been used for AFM images.

RS measurements were carried out using an optical setup made on the basis of a DFS-24 (LOMO) double-grating monochromator. Raman spectra were measured at room temperature with s-polarized Ar-laser light with the wavelengths of 488 or 514.5 nm at φ=45° angle of incidence. The scattered light was collected at the normal to the sample surface, i.e. in quasi-backscattering geometry. Micro-Raman measurements were undertaken using NTEGRA Spectra system (NT-MDT Inc.) in an upright configuration. The excitation wavelength was 532 nm (20-mW DPSS laser). To avoid the sample heating, the laser power of 1 mW was used. The instrument was equipped with an objective resulting in a spot diameter of 0.5 μm at the laser focus. The scattered signals were analyzed using spectrometer (Solar TII, MS5004i) equipped with a cooled CCD camera (DV401-BV, Andor Technology). The peak position and intensity were calculated using the standard NT-MDT software (Nova 1.0.26).

DISCUSSION

SEM and AFM investigations demonstrate that studied materials are microporous with the pores of average diameter up to 2 μm and length of 0.4 – 10 μm. The surface concentration of pores varies from ~10^8 to 10^{10} cm^{-2} (see figure 1).

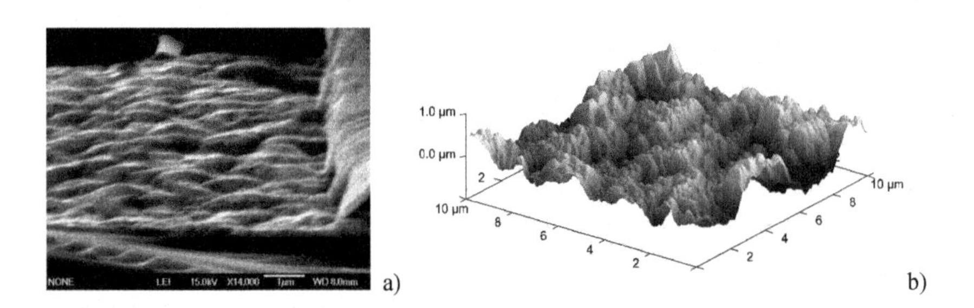

Figure 1. SEM image (a) of GaAs sample (j=20 mA/cm^2, t=10 min) and AFM image (b) of InP sample (j=5 mA/cm^2, t=3 min).

The surface structure of studied materials consists of two-layers as was revealed from the SEM and AFM investigations. In particular, the top layer of GaAs sample primarily consists of As_2O_3, Ga_2O_3 and GaAs nanocrystals (see, for example [1-4]). The bottom layer with pores of different sizes depending on the substrate resistance and etching conditions (current, duration of anodization) was formed on the surface of the monocrystalline substrate. The surface morphology strongly depends on the current density and illumination during the etching procedure [4]. The two-layer structure of porous InP contains native oxides and reaction products. We have discussed previously [5] that porous InP layers are much less oxidized by etching than GaAs layers. In common with other studied III-V compounds, the tendency of the formation of pores with lager diameter under the lager current density has been also observed for InP samples (the pore width varying from 1 μm under the treatment conditions j=1 mA/cm^2, t=5 min to 4 μm under the treatment conditions j=5 mA/cm^2, t=5 min).

The optical parameters, the reflection index (n) and extinction coefficient (κ) have been determined by laser ellipsometer at wavelength λ = 632.8 nm. At first, the polarization angles ψ and Δ were measured using a double-zone method. Then, n, κ have been determined from the dependencies of ψ and Δ on the incidence angle φ (55° ÷ 80°) (see figure 2).

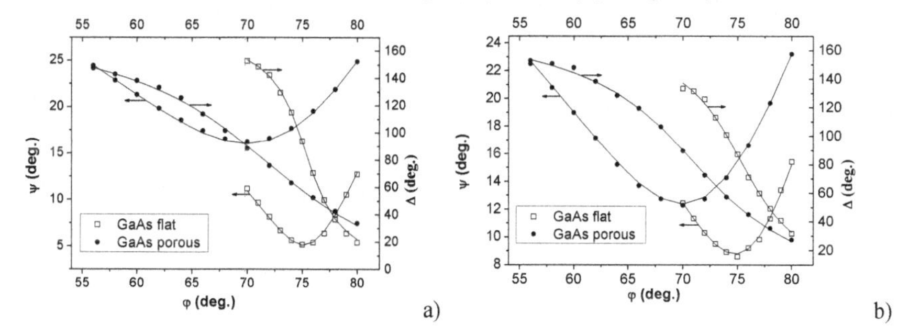

Figure 2. MAI ellipsometry data for GaAs samples (j=20 mA/cm^2, t=10 min) (a) and (j=5 mA/cm^2, t=10 min) (b).

The characteristic feature of the $\psi(\varphi)$ curve is a minimum at the Brewster's angle φ_B, at which the $\Delta(\varphi)$ curve shows the value Δ = 90°. This means that at the wavelength of 632.8 nm the porous GaAs layer on single crystalline GaAs can be presented as a semi-infinite medium with optical parameters (n, κ) of the porous layer, with some oxide overlayer (n_{0x}, κ_{0x}, d_{0x}). The fact is that the light penetration depth, $D = \lambda/4\pi\kappa \approx 0.2$ μm, is much less than the length of pore. Thus, by using equation (1) applied to the model of semi-infinite medium (see, for example [6]) we could obtain the "zero" approximation for optical constants:

$$n_0 \cong \mathrm{tg}\varphi_B, \qquad \kappa_0 \cong \mathrm{tg}\psi(\varphi_B)\frac{(1+\sin\varphi_B)^2}{2\cos\varphi_B} \qquad (1)$$

Further, by fitting the experimental curves of ellipsometric angles $\psi(\varphi)$ and $\Delta(\varphi)$ to experimental ones using an optical model for "ambient/oxide film/porous", the effective optical parameters and thickness of oxide film (d_{0x}) were estimated: $n_{0x} \approx 1.7 - 1.8$, $\kappa_{0x} \approx 0.02$. The thickness d_{0x} of the surface oxide layer before the anodization is $d_{0x} \approx 5 - 12$ nm (in various samples), and after anodization the value of d_{0x} increases up to 20 – 30 nm. As one can see from table I, the porous layer has lower refractive index and extinction coefficient than the flat surface. Thus, the porous layer can be considered as an effective medium composed of GaAs and voids because the light wavelength is larger than the pore dimensions. Therefore, so-called symmetrical Bruggeman effective medium approximation (EMA) (see, [7]) can be used for describing of optical constants.

$$f\frac{\varepsilon-\tilde{\varepsilon}}{\varepsilon+2\tilde{\varepsilon}}+(1-f)\frac{1-\tilde{\varepsilon}}{1+2\tilde{\varepsilon}}=0, \quad (2)$$

where f is the filling fraction of GaAs, ε and $\tilde{\varepsilon}$ are the dielectric permittivity's of the GaAs and effective medium, respectively. The estimated values of the filling fraction of voids in porous layer are listed in table I. The consistency of the obtained values indicates the applicability of the previously described model for porous layers and demonstrates the possibility of determining the optical constants and filling factors of porous layer.

Table I. Optical parameters of model layers for flat and porous GaAs surfaces.

Technological specification of sample	**Surface**	**Initial values (“zero” approximation)**		**Final results of fitting**		
		n_0	κ_0	n	κ	$1-f$
j=20 mA/cm^2, t=10 min	flat	3.732	0.68	3.823	0.225	–
	porous	2.747	1.58	3.040	0.209	0.22
j=5 mA/cm^2, t=10 min	flat	3.732	1.13	3.839	0.229	–
	porous	2.747	1.20	3.080	0.200	0.27

Besides, the porous layer can be considered as an intermediate layer between monocrystalline semiconductor and ambience. The porous layer with reduced optical parameters (n, κ) facilitate the luminescence or RS to leave the material. The consequent enhancement of photoluminescence and RS signals for porous surface has been detected by numerous studies (see for example [1, 8, 9]).

First-order Raman spectra of the porous n-GaAs (100) and n-InP (100) surfaces contain bands which can be related to the Γ-point transversal (TO) and the longitudinal (LO) optical phonon modes which are the result of splitting of optical phonons due to mixing with electric field (see figures 3, 4). According to the selection rules for RS and for our geometry of experiment only the LO-mode should be observable at the (100) surface orientation. In studied porous samples as compared with the flat ones, the TO and LO modes often are of the same order of intensity. The appearance of forbidden TO-mode in the first-order RS spectra for the quasi-backscattering geometry of the experiment is caused by the violation of the selection rules due to several reasons, namely the deviation from back-scattering geometry due to the complicated path of light in the pores, the existence of pores of different orientations, disorder of crystallographic orientations in the walls of the porous layer (skeleton), the influence of the surface electric field in the lateral surfaces of pores on the selection rules for the macroscopic symmetry of porous crystal.

In the second-order Raman spectra of GaAs sample the wide structural band between 480 cm^{-1} and 600 cm^{-1} for flat surface and the bands at 533, 561 and 578 cm^{-1} for porous surface have been observed in micro-Raman measurements (see figure 3, b). Critical-point analysis can be used for explanation of these features. Γ, X, L, W points are the critical points on the phonon dispersion curves of zincblende structures. The selection rules for two-phonon processes [10]

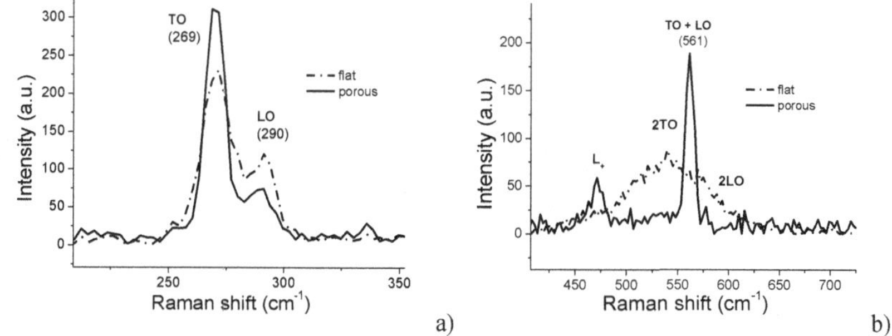

a) b)

Figure 3. First-order (a) and second-order (b) Raman spectra for GaAs sample (j=15 mA/cm^2, t=15 min) at room temperature, λ = 532 nm.

predict the overtone scattering in Γ point and (TO + LO) combination scattering in X and L points. In [11] it was reported that the overtone scattering is predominant in the second-order Raman spectra of GaAs. But in our case the band with the frequency of 561 cm^{-1} likely is caused by scattering by (TO + LO) two-photon combination states with wave vectors at the Γ critical point of Brillouin zone. In [12] the phonon dispersion curves of GaAs in four directions of high symmetry have been calculated using the eleven-parameter rigid-ion model. The calculated frequencies were in a reasonable agreement with neutron scattering data throughout most of the Brillouin zone. And in this model we find the conformation of our assignment of the band at 561 cm^{-1} in the second-order Raman spectrum of porous GaAs. The band at 471 cm^{-1} is attributed to the coupled LO-phonon-plasmon mode L_+.

The second-order Raman spectrum of porous InP at the energy region from 618 to 688 cm^{-1} (figure 4, b) is caused by two-phonon overtone and combination states.

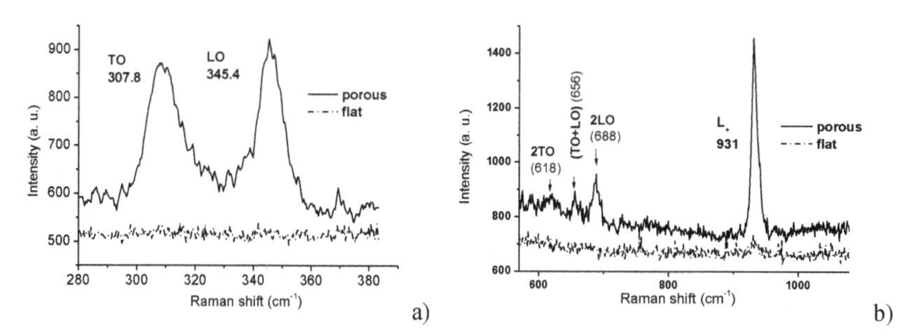

a) b)

Figure 4. First-order (a) and second-order (b) Raman spectra for InP sample (j=5 mA/cm^2, t=3 min) at room temperature, λ = 514.5 nm.

Another prominent band at 931 cm^{-1} in this spectrum can be attributed to the coupled LO-phonon-plasmon mode L+. However the nature of its small bandwidth cannot be explained in a simple way. For this sample with the wafer electron concentration of n=7.1 x 10^{18} cm^{-3} the position of L_+ mode (figure 4) indicates that carrier density decreases approximately to 6.9 x 10^{18} cm^{-3} in the porous layer. For InP substrate only weak band in the region of L_+ mode has been observed. Such behavior is in accordance with the study of the distribution of free carriers in porous layers of some III-V semiconductors presented in [13]. But at the same time we observe the pure LO-mode in the first-order Raman spectrum (see figure 4, a). The morphology of the sample demonstrates the inhomogeneous surface with a system of pores of different dimensions. Thus there is no sufficient decreasing the free carries concentration that would be resulted in the decrease of frequency of L_+ mode.

CONCLUSIONS

The consistency of results obtained by MAI ellipsometry, SEM, AFM indicates the applicability of the model of "ambient/oxide film/porous" and demonstrates the possibility of determination of optical constants and a filling factor of porous layer. The first-order and second-order Raman spectra, and a behavior of the coupled LO-phonon-plasmon modes in porous GaAs and InP have been analyzed. For porous III-V semiconductors, as a rule, the TO- and LO-modes are of the same order of their intensity due to the light scattering enhancement at the pores inner surface and the violation of the RS selection rules. The critical point analysis of zincblende structures has been used for the explanation of peculiarities of second-order Raman spectra of porous III-V semiconductors. In particular, the intense band with the frequency of 561 cm^{-1} in Raman spectrum of porous InP is caused by scattering by (TO + LO) two-photon combination states with wave vectors at the Γ critical point of Brillouin zone. The charge carrier depletion of the skeleton is monitored by Raman scattering experiments.

REFERENCES

1. D.J. Lockwood, P. Schmuki, H.J. Labbe, and J.W. Fraser, *Physica* E **4**, 102 (1999).
2. C.M. Finnie, P.W. Bohn, *J. Appl. Phys.* **86**, 4997 (1999).
3. D.J. Lockwood, *J. Solution Chem.* **29** 1039 (2000).
4. I.·Simkiene, J. Sabataityte, A. Kindurys, and M. Treideris, *Acta Physica Polonica* A. **113**, 1085 (2008).
5. N. Dmitruk, T Barlas, I. Dmitruk, S. Kutovyi, N. Berezovska, J. Sabataityte, and I. Simkiene, *Phys. Status Solidi* B **247**, 955 (2010).
6. V.N. Antonyuk, N.L. Dmitruk, and M.F. Medvedeva, *Ellipsometry in science and technique*, (Novosibirsk, 1987) pp. 66-71 (in Russian).
7. N.L. Dmitruk, A.V. Goncharenko, and E.F. Venger, *Optics of Small Particles and Disperse Media*, (Naukova Dumka, Kyiv, 2009) 386 p.
8. L. Beji, L. Sfaxi, B. Ismail, S. Zghal, F. Hassen, and H. Maaref, *Microelectron. J.* **34**, 969 (2003).
9. A. Liu, C. Duan, *Physica* E **9**, 723 (2001).
10. J. Birman, *Phys. Rev.* **131**, 1489 (1963).
11. T. Sekine, K. Uchnokura, and E. Matsuura, *J. Phys. Chem. Solids* **38**, 1091 (1977).
12. C. Patel, T.J. Parker, H. Jamshidi, and W.F. Sherman, *Phys. Status Solidi* B **122**, 461 (1984).
13. G. Irmer, *J. Raman Spectrosc.* **38**, 634 (2007).

II-VI group semiconductors

Mater. Res. Soc. Symp. Proc. Vol. 1534 © 2013 Materials Research Society
DOI: 10.1557/opl.2013.306

Double core infrared (CdSeTe) / ZnS quantum dots Conjugated to IgG antibodies

Tetyana V. Torchynska[1], Jose L. Casas Espinola[1], Chetzyl Ballardo Rodriguez[2], Janna Douda[2] and Karlen Gazaryan[3]
[1]ESFM– Instituto Politécnico Nacional, México D. F. 07738, México
[2]UPIITA– Instituto Politécnico Nacional, México D. F. 07320, México
[3]Instituto de Investigaciones Biomedicas, UNAM, Mexico DF, Mexico

ABSTRACT

Double core CdSeTe/ZnS quantum dots (QDs) with emission at 800 nm have been studied by photoluminescence (PL) and Raman scattering methods in the non-conjugated state and after the conjugation to the IgG antibodies. The PL energy shift into the high energy range has been detected in PL spectra of bioconjugated QDs. Raman scattering spectra are studied with the aim to reveal the CdSeTe core compositions, as well as to design the QD energy diagrams. The QD energy diagrams permit to analysis PL spectra and their transformation at the bioconjugation of QDs. It is revealed that the interface in double core QDs has the type II quantum well character that permits to explain the IR optical transition (800nm) in CdSeTe QD core materials with relatively wide band gaps. It is shown that essential PL energy shift is promising for the study of QD bioconjugation with specific antibodies and can be a powerful technique in biology and early medical diagnostics.

INTRODUCTION

Core/shell quantum dots (QDs) of II-VI group semiconductors (CdSe, CdS, CdSeTe, etc) with the size of 2-8 nm are of great interest owing their bright size-tunable emission, with relatively narrow bandwidth of 30–45 nm [1, 2], and dimensional compatibility with biological macromolecules [3-5]. The QDs have a broad absorption spectrum that allows simultaneous excitation of particles of several sizes with a single excitation wavelength. Additionally, II-VI core/shell QDs are usually characterized by high PL quantum yield (up to 75%) and high resistance to photo- and metabolic degradation.

One of the high important scientific areas in biology is the new luminescent markers needed for the better assessment of treatments or the antibody detection at the many types of important human diseases. Core/shell QDs of II-VI semiconductors as the luminescent markers have been investigated intensively during the last decade. QD luminescence intensity depends on concentration of attached bio-molecules, enabling QD application as protein sensors [4, 6]. Note, the full impact of bioconjugation on the optical properties of double core/shell CdSeTe/ZnS QDs and the bioconjugation mechanisms are not yet completely understood.

EXPERIMENTAL DETAILS

Double core/shell CdSeTe/ZnS QDs (Fig.1) commercially available [7] covered by the amine (NH_2)-derivatized polyethylene glycol (PEG) polymer were used in a form of colloidal particles diluted in a phosphate buffer saline (PBS) with a 1:200 volumetric ratio. CdSeTe/ZnS QDs are characterized by IR emission with the PL maximum at 800 nm (1.60 eV). At the first the PL and Raman scattering spectra of QDs are studied in nonconjugated states (named 800N).

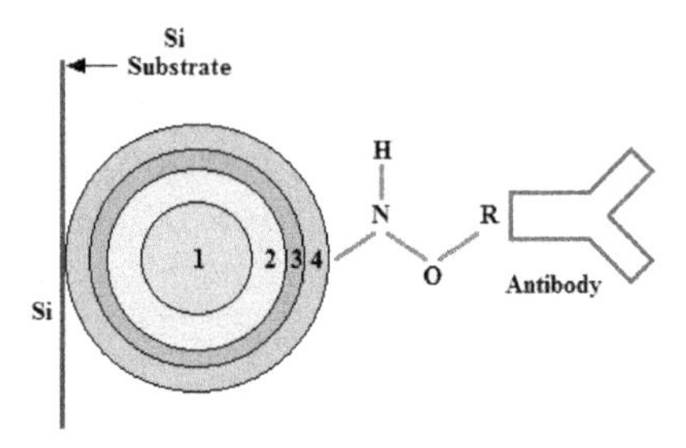

Figure 1. The scheme of QD and its bioconjugation to antibodies: 1- $CdSe_{0.6}Te_{0.4}$, 2 – $CdSe_{0.8}Te_{0.2}$, 3 – ZnS, 4 – PEG polymer.

Then one part of QDs has been conjugated to the IgG antibodies (Ab - immunoglobulin G (IgG) affinity purified with Protein G-Cepharose from rabbit antiserum to Pseudo rabies virus (PRV), stock concentration of 1mg/ml in PBS) using the 800 nm QD conjugation kits [8]. The samples of QDs (bioconjugated and nonconjugated) in the form of a 5 mm size spot (Fig.1) were dried on a surface of crystalline Si substrates as described earlier in [9].

PL spectra were measured at the excitation by a He-Cd laser with a wavelength of 325 nm and a beam power of 20 mW at 300K using a PL setup on a base of spectrometer SPEX500 described in [4,10]. Raman scattering spectra were measured at 300 K using a Raman spectrometer of model Lab-Raman HR800 Horiba Jovin-Yvon in the range of Raman shift of 100-500 cm^{-1} at the excitation by a solid state laser with a wavelength of 532.1 nm and a beam power of 20 mW [11,12].

EXPERIMENTAL RESULTS

Figure 2 presents the PL spectra of nonconjugated (800N) and bioconjugated (800P) QDs. In nonconjugated state the PL spectra are characterized by one Gaussian shape PL band with the maximum at 1.60 eV (Fig. 2, curves 1,2) related to exciton emission in CdSeTe (1.60 eV) cores. PL spectra have varied essentially at the QD bioconjugation. In bioconjugated QDs the PL peaks shift to high energies up to 1.90-1.91eV (Fig. 2, curves 3,4) and the shape of PL bands becomes asymmetric with the essential low energy shoulders at 1.75-1.78 eV (Fig. 2, curves 3, 4). At the same time the PL intensity in some cases does not change and for other cases increases a little in 800P QDs bioconjugated to IgG antibodies (Fig. 2, curves 3, 4).

The energy shift of PL spectra at the QD bioconjugation to antibodies can be assigned to: i) the Stark effect stimulated by the charge of antibodies, ii) a compressive strain applied to bioconjugated QDs [7, 13, 14], iii) the quantum confinement effect owing to the shift of QD energy levels, stimulated by the change of electric potential at the surface of bioconjugated QDs [15] or iv) the domination of the emission of excitons localized at the excited states of the QD core. To distinguish these reasons the Raman scattering spectra are studied as well. It is essential to note that the compressive strains, as a rule, do not change the symmetry of PL bands, but has to change the Raman peak positions of QDs [16].

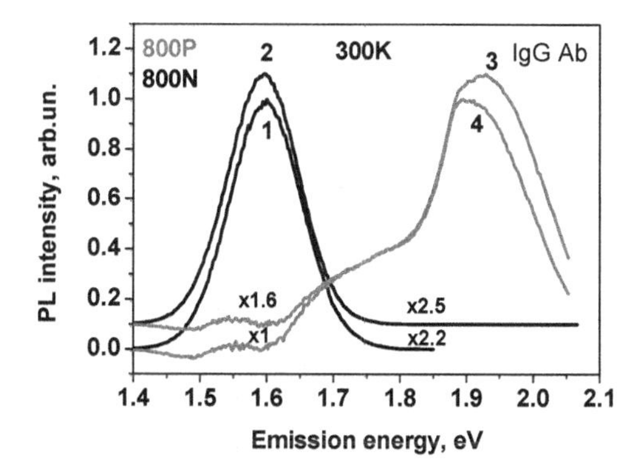

Figure 2. Normalized PL spectra of two nonconjugated 800N QD (curves 1, 2) and two bioconjugated 800P QD (curves 3, 4) samples measured at 300 K. The corresponding PL spectra are shifted with the aim to prevent overlapping. Numbers at the curves (x1.6, x2.2, x2.5) indicate on the multiplication coefficients used at the normalization of experimental PL spectra.

The Raman scattering spectrum of nonconjugated 800nm CdSeTe/ZnS QDs presents a complex peak at 203.3 cm-1, which is the superposition of two Raman peaks at 193.1 and 203.3 cm^{-1} (Fig.3a). The Raman peaks at 193.1 and 203.3 cm^{-1} correspond to the LO phonons in the $CdSe_{0.6}Te_{0.4}$ core and the $CdSe_{0.8}Te_{0.2}$ cover layer of 800nm QD core, respectively [16-18]. Additionally, the Raman spectrum demonstrates the peaks at 177 and 280 cm^{-1}. The peak at 177 cm^{-1} can be assigned to a surface phonon, SP, in the QD core [16]. The Raman peak at 280 cm-1 can be assigned to the LO phonon in the CdSeS intermediate alloy layer at the CdSeTe/ZnS interface. The Raman scattering study shows (Fig.3a,b) that the position of Raman peaks related to a LO phonon in CdSeTe core has not varied at the QD bioconjugation. This fact testifies that the PL energy shift in bioconjugated QDs (Fig.2). has been not connected with the compressive strains in QDs.

DISCUSSION

In present work an ensemble of IR double core/shell CdSeTe/ZnS QDs has been investigated. At the first let us consider the energy diagram of the CdSeTe and ZnS bulk materials which composing the studied QDs. Figure 4a presents the values of band gaps and the electronic affinities for the $CdSe_{0.6}Te_{0.4}$ and $CdSe_{0.8}Te_{0.2}$ alloys and ZnS bulk crystals [17-19]. To design the energy diagram of nonconjugated QDs the variation of optical band gaps in the $CdSe_{0.6}Te_{0.4}$ core and the $CdSe_{0.8}Te_{0.2}$ cover layer versus their sizes has been estimated on the base of the effective mass approximation model [20]. Note, that the big size of IR double core/shell CdSeTe/ZnS QDs permits to use the effective mass approximation model in our case [15].

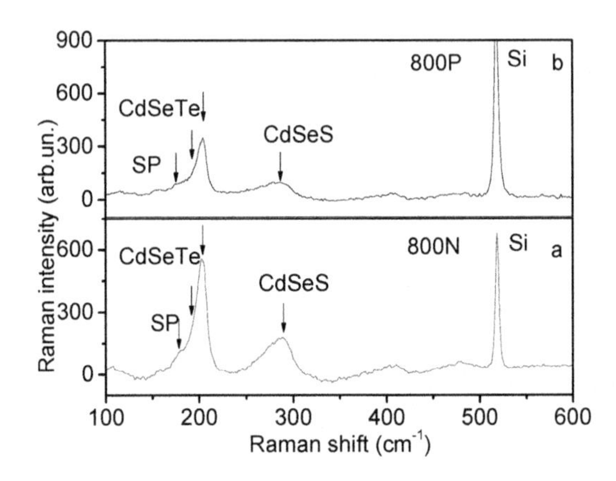

Figure 3. Raman scattering spectra of nonconjugated 800N (a) and bioconjugated 800P (b) QDs.

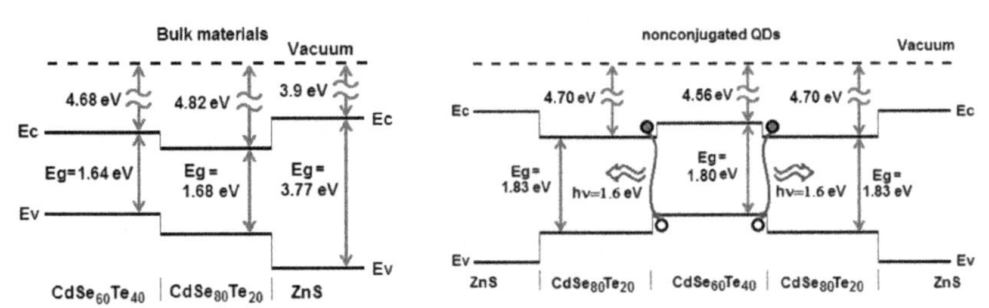

Figure 4a. Energy band diagrams of the CdSeTe and ZnS bulk materials

Figure 4b. Energy band diagrams of the nonconjugated CdSeTe/ZnS QDs

To analysis the optical band gap the next parameters are used for the bulk CdSe: Eg = 1.730 eV at 300K, me* = 0.13 mo, mh*= 0.45mo, the high frequency dielectric constant $\varepsilon = 8.2$ [17, 18]. Taking into account that in the bulk CdTe: Eg = 1.50 eV at 300K, $m_e^* = 0.11m_o$, $m_h^* = 0.35m_o$ [17, 18], the values of optical band gaps (Fig.4a) and corresponding effective masses for the bulk $CdSe_{0.6}Te_{0.4}$ and $CdSe_{0.8}Te_{0.2}$ alloys have been obtained by the linear interpolations. Calculated data of the ground electron-hole pair states are used for the design of energy diagram of nonconjugated QDs (Fig. 4b). The energy diagram of nonconjugated QDs, presented in figure 4b, explains the type of optical transitions and the energy of emission quanta of 1.60 eV detected in nonconjugated QDs. As one can see in figure 4b, due to the type-II quantum well ($CdSe_{80}Te_{20}/CdSe_{60}Te_{40}/CdSe_{80}Te_{20}$) in the core of QDs, the wide band gap core and covered layer with E_g = 1.80 and 1.83 eV, respectively, permits to obtain IR emission with peak at 1.60 eV in nonconjugated QDs.

As we mentioned above the PL spectrum transformation in bioconjugated QDs related to the electronic reasons and the one of them is the Stark effect. The quantum-confined Stark effect stimulated by an applied electric field has been studied in CdS, CdSSe and CdSe QDs theoretically [2] and experimentally [21] in the early 90th. Only the "red" Stark energy shift was predicted theoretically for the CdS and CdSSe QDs [2]. Nevertheless, the both type ("red" and "blue") Stark energy shifts were detected experimentally in CdSe QDs [21,22]. The quantum-confined Stark shifts in QD ensemble emission were found early to be purely quadratic in the applied electric field, increasing with the QD sizes and related to the QD polarizability mainly [21].

The other reason of the "blue" shift of emission energy in PL spectra of bioconjugated QDs can be related to the change of electric potential (or potential barrier) at the surface of bioconjugated QDs. It is known that the position of energy levels in QDs for the strong quantum confinement regime depends on the value of potential barrier. This question was discussed in [23] and it was shown that the influence of a finite potential barrier on the electron-hole energy states increases with decreasing dot radius. As expected, the lower potential barrier reduces and the higher potential barrier increases the quantum confinement energy levels and these changes are more essential for high energy states [23].

Additionally, in our experiment the external electric field created by charged antibodies has transformed the energy diagram of bioconjugated QDs. Two types of optical transitions manifest themselves in the PL spectra of bioconjugated QDs. The PL bands with the peaks at 1.90 and 1.75 eV can be assigned, as well, to the recombination of excitons localized at the excited and ground states in the $CdSe_{60}T4_{50}$ core and/or the $CdSe_{80}Te_{20}$ covered layer (Fig.2, curves 3,4).

CONCLUSIONS

Photoluminescence and Raman scattering spectra of IR double core/shell CdSeTe/ZnS QDs in nonconjugated states and after the conjugation to the Pseudo rabies virus antibodies have been studied. It is shown that the bioconjugation to antibodies stimulates the shift of Pl spectra into high energy spectral range. The energy diagrams of QDs have been designed, which permit to explain the types of optical transitions in QDs and its transformation at the bioconjugation. The electronic effects stimulated the PL spectrum transformations in bioconjugated QDs have been discussed. Better understanding the QD bioconjugation with specific antibodies is expected to produce the major advances in biology and medicine and can be a powerful technique in medical early diagnostics.

The work was partially supported by CONACYT Mexico (project 130387) and by SIP-IPN, Mexico.

REFERENCES

1. D.J. Norris, M.G. Bawendi, Phys. Rev. B, **53**, 16338 (1996).

2. G.W. Wen, J.Y. Lin, H.X. Jiang, Z. Chen, Phys. Rev. B, **52**, 5913 (1995).
3. M. Kuno, D.P. Fromm, H.F. Hamann, A. Gallagher, D.J. Nesbitt, J. Chem. Phys.**115**, 1028 (2001).
4. T.V. Torchynska, Nanotechnology, **20**, 095401 (2009).
5. N. Tessler, V. Medvedev, M. Kazes, S.H. Kan, U. Banin, Science, **295**, 1506 (2002).
6. A.R. Clapp, I. L. Medintz, J. M, Mauro, Br. R. Fisher, M. G. Bawendi, and H. Mattoussi, J. Am. Chem. Soc. **126**, 301-310 (2004).
7. www.invitrogen.com
8. http://probes.invitrogen.com/media/pis/mp19010.pdf
9. T. V. Torchynska, A. Diaz Cano, M. Dybic, S. Ostapenko, M. Morales Rodrigez, S. Jimenes Sandoval, Y. Vorobiev, C. Phelan, A. Zajac, T. Zhukov, T. Sellers, phys. stat. sol. (c), 4, 241 (2007).
10. T. V. Tor chynska, J. Douda, and R. Peña Sierra, phys. stat. sol. (c), 6, S143 (2009).
11. A. Diaz Cano, S. Jiménez Sandoval, Y. Vorobiev, F. Rodriguez Melgarejo and T. V. Torchynska, Nanotechnology, 21, 134016 (2010).
12. L. G. Vega Macotela, T. V. Torchynska, J. Douda, and R. Peña Sierra, phys.stat.solid. (c), 7, 1192 (2010).
13. R.W. Meulenberg, T. Jennings, G. F. Strouse, Phys. Rev. B, 70, 235311 (2004).
14. X. Gao, W. C.W. Chan, Sh. Nie, J. of Biomedical Optics, 7(4), 532-537 (2002).
15. Torchynska, T.V. "Nanocrystals and quantum dots. Some physical aspects" in the book "Nanocrystals and quantum dots of group IV semiconductors", Editors: T. V. Torchynska and Yu. Vorobiev, American Scientific Publisher, 1- 42 (2010).
16. T. Torchynska and Yu. Vorobiev, Semiconductor II-VI Quantum Dots with Interface States and Their Biomedical Applications, in book : Advanced Biomedical Engineering, InTech Publisher, Croatia , Editors A. McEwan and G. D. Gargiulo, 143-183 (2011).
17. Physics of II-VI compounds, Editors A.N. Georgobiany and M.K. Sheinkman, Publisher „Nauka", Moscow, Russia, 1986, 300 p.
18. Physics and chemistry of II – VI compounds, Editors M. Aven and J.S. Prener, North-Holland, Amsterdam, 1967, 625p.
19. A.G.Milnes, D.L. Feucht, Heterojunctions and Metal-Semiconductor junctions, Academic Press, New York and London, 1972.
20. Y. Kayanuma, Phys. Rev. B., 38, 9797 (1988)
21. S.A. Empedocles, M.G. Bawendi, Science, 278, 2114 (1997).
22. D.J. Norris, A. Sacra, C.B. Murray, M.G. Bawendi, Phys. Rev. Lett., 72, 2612 (1994).
23. D.B. Tran Thoai, Y.Z. Hu, S.W. Koch, Phys. Rev. B, 41, 6079 (1990).

Mater. Res. Soc. Symp. Proc. Vol. 1534 © 2013 Materials Research Society
DOI: 10.1557/opl.2013.307

Effects of Bio-conjugation and Annealing on the Photoluminescence and Raman Spectra of CdSe/ZnS Quantum Dots

L. Borkovska[1], N. Korsunska[1], T. Stara[1], O. Kolomys[1], V. Strelchuk[1], O. Rachkov[2] and T. Kryshtab[3]
[1]V. Lashkaryov Institute of Semiconductor Physics of NASU, pr. Nauky 41, 03028 Kyiv, Ukraine
[2]The Institute of Molecular Biology and Genetics of NASU, Zabolotnogo Str. 150, 03680 Kyiv, Ukraine
[3]Instituto Politécnico Nacional – ESFM, Av. IPN, Ed.9 U.P.A.L.M., 07738 Mexico D.F., Mexico
E-mail:bork@isp.kiev.ua; kryshtab@gmail.com

ABSTRACT

The effect of thermal annealing at ambient conditions on the photoluminescence (PL) and Raman scattering spectra of non-conjugated and conjugated with S6K2 antibody CdSe/ZnS core-shell quantum dots (QDs) has been studied. In the PL spectra, the annealing is found to produce the blue shift of the PL band spectral position and degradation of the PL intensity. These changes are found to be much more intense in conjugated QDs and are accompanied by the increase of a PL band half-width. In the Raman scattering spectra, the spectral shift and a decrease of the intensity of CdSe LO phonon peak as well as a increase of the intensity and narrowing of the peak ascribed to a mixed interface layer are found. The effect of annealing is supposed to be the result of partial core-shell intermixing that is accompanied by the QD oxidation in bio-conjugated QDs.

Keywords: Quantum dots; Bio-conjugation; Photoluminescence

INTRODUCTION

Recently, the application of colloidal II-VI quantum dots (QDs) in fluorescence bio-sensing and bio-imaging is getting increased attention [1]. We have found earlier that the samples of bio-conjugated CdSe(Te)/ZnS core-shell quantum dots (QDs) dried on a Si substrate can be used for detecting of bio-molecules utilizing the effect of a photoluminescence (PL) spectral shift [2]. We have shown that if a liquid solution of bio-conjugated QDs is deposited on a crystalline Si substrate and then dried, a spectral position of the QD photoluminescence band shifts to shorter wavelengths (a "blue" shift effect) in comparison with the identical non-conjugated QDs [3, 4]. The shift magnitude increased gradually during the sample storage at atmospheric ambience for several days. This effect was observed in CdSe/ZnS and CdSeTe/ZnS core-shell QDs conjugated to different monoclonal antibodies [4]. It has also been found that the process exhibited as the PL spectral shift is facilitated noticeably and the shift magnitude is increased for the samples annealed at above room temperature for hours [3]. For instance, thermal annealing of conjugated QD sample at 190 °C for 1 hour in air stimulated a blue shift of the PL band of about 15 nm, while thermal annealing at 250 °C under the same conditions resulted in a blue shift of about 60 nm [3]. Therefore, the thermal annealing can be used as an effective processing to change relatively quickly the emission color in conjugated QDs in such a way that this change will be clearly seen by the naked eye.

The "blue" shift effect observed in the samples stored at room temperature in air was supposed to be caused by a change in elastic strain in the QDs as well as by oxidation of the QD core [5]. An oxidation was observed in the samples stored in an atmospheric ambience for a long period of time (up to 2 years) [5]. An oxidation can occur in the samples subjected to thermal annealing too. Here we present the results of PL and Raman scattering study of the influence of bio-conjugation on the processes occurred in the core-shell QDs subjected to thermal annealing.

EXPERIMENTAL DETAILS

Commercial water-soluble CdSe/ZnS core-shell QDs covered with an amphiphilic polymer and polyethylene glycol (PEG) purchased from Invitrogen Inc. [6] were used. The QDs were ellipsoids with minimum and maximum axis of 6 and 12 nm, respectively [6]. The QDs were conjugated to mouse monoclonal antibodies against S6K2 (ribosomal protein S6 kinase 2). S6K2 is known to participate in various cellular processes, including mRNA processing and protein synthesis, as well as in some pathologies, e.g., diabetes and cancer [7]. The conjugation procedure was performed using a commercially available 655 nm QD conjugation kit from Invitrogen Inc. [6]. The bio-conjugation was confirmed with immune-enzyme analysis (ELISA).

Both QDs dispersed in water and QDs dried on a solid substrate were investigated. The dried QD samples of non-conjugated QDs (NC-QDs) as well as of conjugated QDs (C-QDs) were produced by depositing about 10 μL of a water mixture of corresponding QDs on a polished crystalline (100) Si wafer and stored in the atmospheric ambience at room temperature for 1 hour. This process produced dried droplets of about 3 mm in diameter. The samples were annealed in the atmospheric ambience at 190 °C for 2 hours under day light illumination.

The PL and Raman scattering spectra were measured in a backscattering geometry at room temperature using triple Raman spectrometer T-64000 Horiba Jobin-Yvon, equipped with a cooled charge-coupled device (CCD) detector. The 514.5 nm line of an Ar-Kr ion laser was used for excitation. Laser beam with power up to 2 mW was focused at the 1 μm in diameter spot.

RESULTS

The Raman and PL spectra of the QDs in aqueous media are shown in figure 1. In both NC-QDs and C-QDs, the spectral position of the Raman peak related to scattering by the CdSe longitudinal optical (LO) phonon is at about 212 cm^{-1} which is due to phonon confinement in the QDs [8] as well as the compressive strain in CdSe core resulted from the ZnS shell having a smaller lattice parameter [9, 10]. The LO peak from CdSe shows a low frequency shoulder at about 197 cm^{-1} usually ascribed to the contribution of the surface optical (SO) vibrations [11]. In addition, a peak at about 280 cm^{-1} can be resolved in the Raman spectra of the QDs studied. The band at 280 cm^{-1} has been reported to appear in the Raman spectra of CdSe QDs covered by CdS or ZnS shells and ascribed to the LO phonon mode from mixed interface layer of Cd(Zn)S(Se) caused by partial interdiffusion between the CdSe core and CdS or ZnS shells [12]. A satisfactory curve fitting of the spectra was achieved by using three Lorentzian profiles, representing LO and SO peaks from the CdSe core and the LO peak from the interface layer. A fitting of the Raman spectra showed that the LO peak from CdSe has nearly the same spectral position as well as the same full width at a half maximum (FWHM) for NC-QDs and C-QDs (see table I).

In the same manner, the PL spectra of QD liquid solutions (figure 1b) show the same spectral position and the same FWHM of the PL peak from NC-QDs and C-QDs (table I).

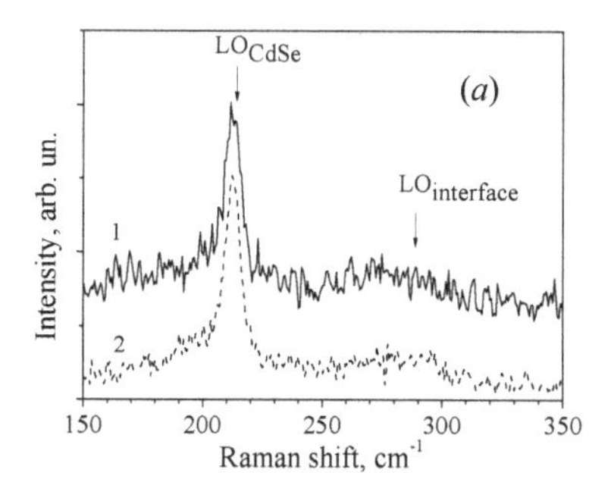

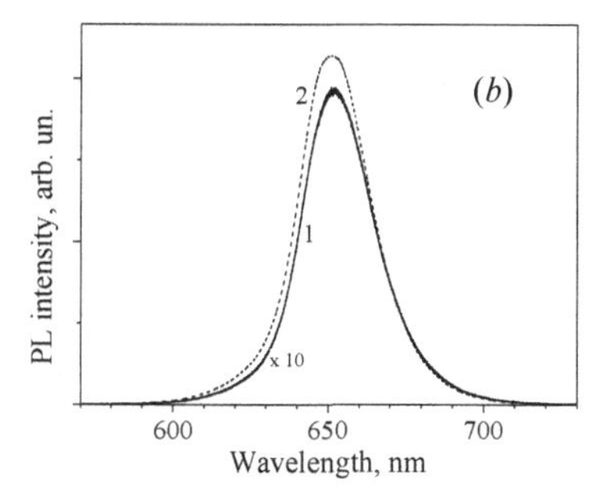

Figure 1. Raman scattering spectra in the range of CdSe longitudinal optical (LO) phonon mode (a) and photoluminescence spectra (b) of non-conjugated (1) and conjugated (2) CdSe/ZnS QDs in aqueous media, T=300 K, λ_{exc}=514.5 nm.

Table I. Some optical and PL characteristics of non-conjugated and conjugated CdSe/ZnS QDs in aqueous media and dried on the Si substrate, T=300 K, λ_{exc}=514.5 nm.

The sample	PL peak position, nm	PL peak FWHM, nm	CdSe LO phonon position, cm^{-1}	FWHM of CdSe LO phonon, cm^{-1}
NC-QDs in aqueous media	650,8	26,7	212.4	7.8
C-QDs in aqueous media	651,5	27,0	212.3	7.8
NC-QDs dried on Si wafer	650,8	25,6	210.6	8.0
C-QDs dried on Si wafer	652,3	26,3	210.8	7.7
annealed NC-QDs	641,1	25,4	212,2	6.5
annealed C-QDs	630,8	34	209,5	8,3

In the Raman spectra of the dried QDs, the CdSe LO phonon frequency is found to be slightly decreased (table I) apparently owing to tensile strains produced in consolidated solution of a droplet as a result of interface tension at the interface between Si substrate and the medium of a consolidated solution. In the PL spectra of dried QDs, both spectral position and FWHM of the PL peak remain close to that observed for the QDs diluted in aqueous media and vary within ±1 nm in different points of the sample.

Thermal annealing of dried QD samples affects distinctly both the Raman and the PL spectra (figure 2). In the Raman spectra, the intensity of CdSe LO peak decreases, while the intensity of the interface-related peak increases. The spectral position of CdSe LO peak from NC-QDs shifts towards a high energy region, but the peak from the C-QDs shifts to a low energy region (see table I). In the PL spectra, the PL intensity falls down and the band shifts to a blue spectral region upon annealing. These changes in the PL spectra are found to be larger for C-

QDs (table I). Besides, the FWHM of the PL band of the C-QDs increases in ~ 1.3-1.4 times, while that of the NC-QDs does not change.

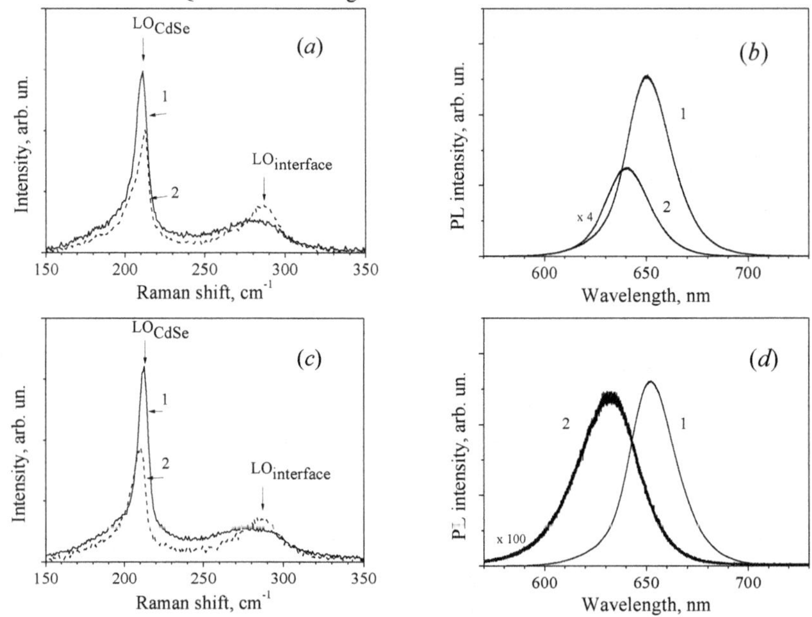

Figure 2. Raman scattering (a, c) and photoluminescence spectra (b, d) of non-conjugated (a, b) and conjugated (c, d) CdSe/ZnS QDs dried on the Si substrate, before (1) and after (2) thermal annealing in the atmospheric ambience at 190 °C for 2 hours; T=300 K, λ_{exc}=514.5 nm.

DISCUSSION

A "blue" spectral shift of QD luminescence band occurred upon thermal annealing and resulted from the increase of QD core band gap can be stimulated by different processes: (i) changing of a mechanical stress in QDs; (ii) formation of the ZnCdSe alloy in the result of Zn diffusion from a ZnS shell; (iii) decreasing of a QD core size owing to QD oxidation.

The spectral position of CdSe LO peak in CdSe bare and CdSe/ZnS core-shell QDs is known to be influenced strongly by mechanical stress [9, **Error! Bookmark not defined.**]. The Raman spectra of the QDs studied clearly demonstrate that both the NC-QDs and C-QDs are under compressive strains caused by the lattice mismatch between CdSe and ZnS that are compensated by tensile strains caused by the effect of substrate. Thermal annealing can stimulate the strain relaxation in the dried QDs samples. However, tensile strains caused by the substrate effect on dried QDs have negligibly small influence on the PL peak position. This is proved by the fact that the PL peak position in dried QDs is nearly the same as in diluted QDs (see table I). Therefore, the change of mechanical stress in the QDs under annealing cannot be the reason of a blue shift of the PL peak position, but can influence a spectral position of CdSe LO peak in the

Raman spectra.

In addition to strain relaxation the thermal annealing can stimulate a core-shell intermixing in the QDs. The effect of alloying of CdSe core and ZnS or ZnSe shell upon thermal annealing has been observed in CdSe/ZnS QDs [13] as well as in CdSe/ZnSe QDs [14] and nanorods [15]. It has been found that both CdSe/ZnS and CdSe/ZnSe core-shell structures evolve into an alloy at temperatures higher than 270 °C as it is evidenced by noticeable blue shifts in both the absorption and emission maxima [**Error! Bookmark not defined.**, **Error! Bookmark not defined.**4], as well as by corresponding shifts of CdSe diffraction peaks in powder X-ray diffraction patterns and of the CdSe single phonon mode in Raman LO phonon spectra [**Error! Bookmark not defined.**]. However, the slight diffusion of Zn into CdSe has already been observed at 240 and 220 °C by the blue shift of the peaks in the absorption spectra [**Error! Bookmark not defined.**]. Moreover, in CdSe/ZnSe nanorods, the formation of ZnCdSe alloy at the core-shell interface was supposed to occur already at 180 °C during ZnSe shell growth [**Error! Bookmark not defined.**]. A blue shift of PL and optical absorption peaks upon progressive thermal treatment was found to be accompanied by decrease a quantum yield from alloyed ZnCdSe versus CdSe/ZnSe core/shell nanorods that was ascribed to the lack of surface passivation on the ZnCdSe nanorods [**Error! Bookmark not defined.**]. The changes occurred in the PL spectra of annealed QDs under investigation are similar to that described above and can be the result of a core-shell intermixing. This assumption is confirmed by the transformations observed in the Raman spectra. The increase in intensity and the narrowing of CdS-like interface phonon found in the Raman spectra of both NC- and C-QDs can be the evidence of the changes in composition and thickness of intermixed interface region in thermally treated QDs. The blue shift of CdSe LO peak in annealed NC-QDs can also be the result of a core-shell intermixing. Therefore, blue shift of PL peak position and the decrease in the PL intensity observed in annealed NC- and C-QDs can be caused by partial core-shell intermixing.

However, in the PL spectra of thermally treated C-QDs, a larger blue shift of PL band position as well as a pronounced increase in PL band FWHM and disastrous degradation of the PL intensity are observed. These results indicate additional process that contributes to a blue shift effect as compared with NC-QDs. Since the samples were annealed in the ambient atmosphere under day light illumination, an oxidation of the QDs can happen. Oxidation has been observed in the QDs placed in the environments containing O_2 and occurred both under continuous irradiation with above-the-band gap light (photo-oxidation) and in the dark [16-19]. It has been shown that oxidation of CdSe QDs results in the formation of SeO_2 or $CdSeO_x$ ($x = 2$ and 3) oxides on the surface of the CdSe as well as in the decrease of the QD core diameter [17, **Error! Bookmark not defined.**]. It has been shown that coating of CdSe QDs with ZnS shell slows down, but not completely terminates the oxidation of CdSe core [**Error! Bookmark not defined.**]. An oxidation of CdSe and CdSe/ZnS QDs has been found to result in a blue shift of PL peak position, increased FWHM and decreased PL intensity [19]. These changes are similar to those observed in our thermally annealed C-QDs. The decrease of QD core dimensions owing to oxidation is confirmed by the red shift of CdSe LO peak position in the Raman spectra. This process obviously is more pronounced for C-QDs. Thus, we can conclude that conjugation promotes the QD oxidation.

CONCLUSIONS

The effect of the thermal annealing on the PL and optical properties of non-conjugated and conjugated CdSe/ZnS core-shell QDs dried on a Si substrate is studied. It is found that the thermal annealing results in a blue shift of PL peak position and a decrease in the PL intensity. In C-QDs these changes are found to be larger and accompanied by the increase of PL band FWHM. It is supposed that the thermal annealing stimulates interdiffusion at core-shell interface in both NC-QDs and C-QDs. The latter is proved by the increase in the intensity and the narrowing of the interface-related peak in the Raman spectra. The larger blue shift observed in the PL spectra of C-QDs is assumed to be caused by QDs oxidation. Though the role of bio-molecules in QD oxidation is not clear now, it is proposed that bio-conjugation promote QD oxidation during thermal annealing.

ACKNOWLEDGMENTS

This work was partially supported by the National Academy of Sciences of Ukraine through the program "Nanotechnologies and nanomaterials" (Grant No 2.2.1.14/26) and the project "Physical and Physical-Technological Aspects of Fabrication and Characterization of Semiconductor Materials and Functional Structures for Modern Electronics" (Grant No III-41-12).

REFERENCES

1. H. Mattoussi, G. Palui, and H. B. Na, *Advanced Drug Delivery Reviews* **64**, 138 (2012).
2. G. Chornokur, S. Ostapenko, E. Oleynik, C. Phelan, N. Korsunska, T. Kryshtab, J. Zhang, A. Wolcott, and T. Sellers, *Superlattices and Microstructures* **45**, 240 (2009).
3. M. Dybiec, G. Chornokur, S. Ostapenko A. Wolcott, J. Z. Zhang A. Zajac, C. Phelan, T. Sellers, and D. Gerion, *Appl. Phys. Lett.* **90**, 263112 (2007).
4. G. Chornokur, S. Ostapenko, Yu. Emirov, N. E. Korsunska, T. Sellers, and C. Phelan, *Semicond. Sci. Tech.* **23**, 075045 (2008).
5. T. G. Kryshtab, L. V. Borkovska, O. F. Kolomys, N. O. Korsunska, V. V. Strelchuk, L. P. Germash, R.Yu. Pechers'ka, G. Chornokur, S. S. Ostapenko, C. M. Phelan, and O. L. Stroyuk, *Superlattices and Macrostructures* **51**, 353 (2012).
6. http://www.invitrogen.com
7. T. R. Fenton, and I. T. Gout, *Int. J. Biochem. Cell. Biol.* **43**, 47 (2011).
8. A. Tanaka, S. Onari, and T. Arai, *Phys. Rev.* **B 45**, 6587 (1992).
9. A. V. Baranov, Yu. P. Rakovich, J. F. Donegan, T. S. Perova, R. A. Moore, D. V. Talapin, A. L. Rogach, Y. Masumoto, and I. Nabiev, *Phys. Rev.* **B 68**, 165306 (2003).
10. R. W. Meulenberg, T. Jennings, and G. F. Strouse, *Phys. Rev.* **B 70**, 235311 (2004).
11. A. Roy, and A. K. Sood, *Phys. Rev.* **B 53**, 12127 (1996).
12. V. M. Dzhagan, M. Ya. Valakh, A. E. Raevskaya, A. L. Stroyuk, S. Ya. Kuchmiy, and D. R. T. Zahn, *Nanotechnology* **18**, 285701 (2007).
13. X. Xia, Z. Liu, G. Du, Y. Li, and M. Ma, *J. Phys. Chem. C* **114**, 13414 (2010).
14. X. Zhong, M. Han, Z. Dong, T. J. White, and W. Knoll, *J. Am. Chem. Soc.* **125**, 8589 (2003).
15. H. Lee, P. H. Holloway, and H. Yang, *J. Chem. Phys.* **125**, 164711 (2006).
16. W. G. J. H. M. van Sark, P. L. T. M. Frederix, A. A. Bol, H. C. Gerritsen, and A. Meijerink, *ChemPhysChem* **3**, 871 (2002).
17. J. E. B. Katari, V. L. Colvin, and A. P. Alivisatos, *J. Phys. Chem.* **98**, 4109 (1994).

18. H. Asami, Y. Abe, T. Ohtsu, I. Kamiya, and M. Hara, *J. Phys. Chem. B* **107**, 12566 (2003).
19. A. Y. Nazzal, X. Wang, L. Qu, W. Yu, Y. Wang, X. Peng, and M. Xiao, *J. Phys. Chem.* **B** **108**, 5507 (2004).

Mater. Res. Soc. Symp. Proc. Vol. 1534 © 2013 Materials Research Society
DOI: 10.1557/opl.2013.308

Raman Scattering and Photoluminescence transformation in core/shell CdSe/ZnS Quantum dots with Interface state at the bioconjugation

I. Ch. Ballardo Rodríguez[1] and A.Vivas Hernandez[2]
[1]UPIITA- Instituto Politécnico Nacional, México D. F. 07340, México.
[2]ESIME – Instituto Politécnico Nacional, México D. F. 07730, México.

ABSTRACT

This paper presents the results of photoluminescence (PL) and Raman scattering study in CdSe/ZnS quantum dots (QDs) covered by polyethylene glycol (PEG) polymer with and without bioconjugation to Osteopontin antibodies. Commercial CdSe/ZnS QDs used in the study are characterized by the color emission with a maximum at 640 nm (1.96 eV) at 300K. It is shown that the PL spectra of nonconjugated QDs can be presented as a superposition of PL bands related to exciton emission in the CdSe core (1.96 eV) and some high energy PL bands located between 2.37 – 3.00 eV. It is shown that the QD bioconjugation to Osteopontin antibodies is accompanied by the change of PL spectra and the variation in the intensity of all type Raman lines. The explanation of the mentioned effects has been proposed on the base of re-charging of interface related states and the surface enhanced Raman scattering (SERS) effect.

INTRODUCTION

Semiconductor crystallites, known as nanocrystals (NCs) or quantum dots (QDs), can essentially improve an application of fluorescent markers in biology and medicine [1] by using the ability to cover them by the biomolecules and producing the efficient bio-luminescent markers. The most popular QDs are the core/shell CdSe/ZnS QDs due to their specific and unique optical properties. Emission of CdSe/ZnS QDs is characterized by the high photoluminescence quantum yield *and can be changed* with the variation of CdSe core sizes due to the quantum confinement effect [2]. Overcoating the CdSe core with the higher band gap ZnS material has increased a PL quantum yield and improved QD chemical and photo- stabilities. Last effects are owing to the passivation of nonradiative surface recombination states and due to the creation of potential barriers on the QD surface [3]. As result the core/shell CdSe/ZnS QDs have been widely used in optoelectronic and biological applications [4] serving as luminescence tags in antigen-antibody biochemical reactions and can be used in early cancer diagnostics [1]. The confirmation of QD bioconjugation using the detection of the essential transformations of PL and Raman scattering spectra is highly important.

EXPERIMENT DETAILS

Commercially available core-shell CdSe/ZnS QDs, covered with PEG polymer, are used as colloidal particles diluted in a phosphate buffer saline (PBS) with a 1:200 volumetric ratio. Studied QDs are characterized by the size of 6.3 nm and color emission with the maxima at *640 nm (1.96 eV)*. The part of CdSe/ZnS QDs has been bioconjugated (named 640P) to the *anti* Osteopontin antibodies using the commercially available 640 nm QD conjugation kit [5]. This kit contains amine-derivatized polymer coated QDs and the amine-thiol crosslinker (SMCC) [5].

The conjugation reaction is based on the efficient coupling of thiols, that present in reduced antibodies, to reactive maleimide groups which exist on the QD surface after the SMCC activation. Some part of CdSe/ZnS QDs (named 640N) has been left nonconjugated and serves as a reference object. Samples of QDs (bioconjugated and nonconjugated) in the form of a 5 mm size spot were dried on a polished surface of crystalline Si substrates as described earlier in [6-10]. PL spectra were measured at 300K at the excitation by a He-Cd laser with a wavelength of 325 nm and a beam power of 20 mW using a PL setup described in [8, 9]. Raman scattering spectra were measured at 300 K using a Raman spectrometer of model Lab-Raman HR800 Horiba JobinYvon and a solid state laser for the excitation with a wavelength of 785 nm and a power of 20 mW in backscattering configuration.

RESULTS AND DISCUSSION

Normalized PL spectra of different nonconjugated CdSe/ZnS QD samples measured at 300K demonstrate the PL band related to the CdSe core (1.96 eV) and the broad PL bands in the spectral range of 2.00-3.50 eV with a main maximum at 2.50 -2.75 eV (Fig.1 a). Normalized PL spectra of bioconjugated QDs are characterized only one PL band related to exciton emission in the CdSe core (Fig.1b).

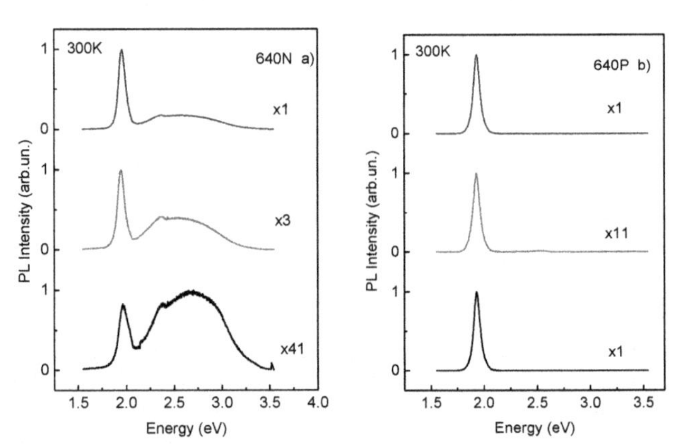

Figure 1. Normalized PL spectra of nonconjugated 640N (a) and bioconjugated 640P (b) QDs.

These broad PL bands are a superposition of a set of elementary PL bands. The deconvolution procedure has been applied to PL spectra permitting to represent them as a superposition of four elementary PL bands with the peaks at 2.20, 2.37, 2.73 and 3.06 eV (Fig.2). The nature of PL bands with the peaks at 2.20, 2.37, 2.73 and 3.06 eV needs to be discussed. These bands can be assigned to the recombination via: i) defects at the CdSe/ZnS interface; ii) excited states in the CdSe core or iii) defect states at the ZnS/polymer interface.

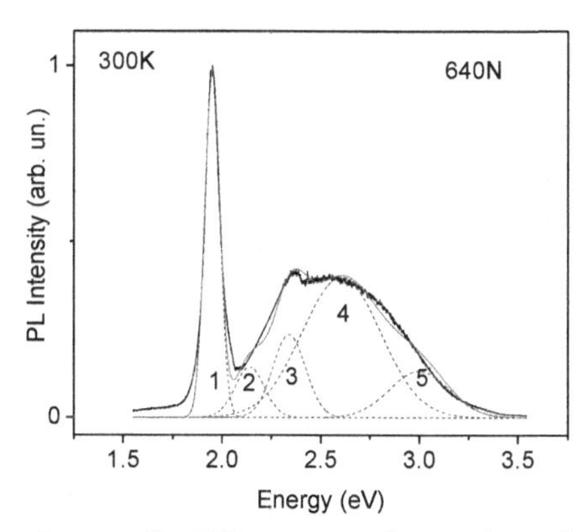

Figure 2. Deconvolution of a normalized PL spectrum of nonconjugated 640N QDs.

Earlier it was revealed the very fast (subpicosecond) electron and hole relaxation dynamics in the CdSe QDs with the rate exceeding that due to phonon emission in bulk semiconductors [11]. This fast initial decay of the electron population was strongly dependent on the degree of QD surface passivation. Taking into account very fast depopulation dynamics of electron and hole quantized states in the CdSe QDs it is possible to suppose that the radiative states responsible for the 2.20, 2.37, 2.73 and 3.06 eV PL bands are localized at the CdSe/ZnS interface (Fig.3). Note the thickness of ZnS shell is small enough (1-2 nm) and it can be permitted the carrier tunneling to the ZnS/polymer interface with the subsequent recombination there (Fig.3).

Presented PL results for bioconjugated QDs (Fig.1b) were explained on the base of the bioconjugation model proposed early in [10]. It was supposed in [10] that the interface states (IS), responsible for the hole trapping in nonconjugated QDs, are negatively charged acceptor-like defects (IS^-) [12-14]. Simultaneously, the interface states, responsible for the electron trapping in non-conjugated QDs, are positively charged donor-like defects (IS^+) [12-14]. These assumptions follow from the electrical neutrality of nonconjugated QDs (Fig.3). The negative charge of acceptor-like interface states is due to their compensation by electrons from donor-like interface states in nonconjugated CdSe/ZnS QDs (Fig.3). Thus the localization of defect states at the QD interface and their attractive potentials permit them to compete in the recombination process with exciton (inside the CdSe core) and nonradiative recombinations (Fig.3) [10, 15-18].

PL spectra of CdSe/ZnS QDs bioconjugated with the anti Osteopontin antibodies have been modified dramatically in comparison with those of nonconjugated QDs (Fig.1). PL spectrum of bio - conjugated QDs is characterized by the one Gaussian shape PL band with a narrow full width at a half maximum (90 meV) and the peak position of 1.96 eV at 300K (Fig.1 b).

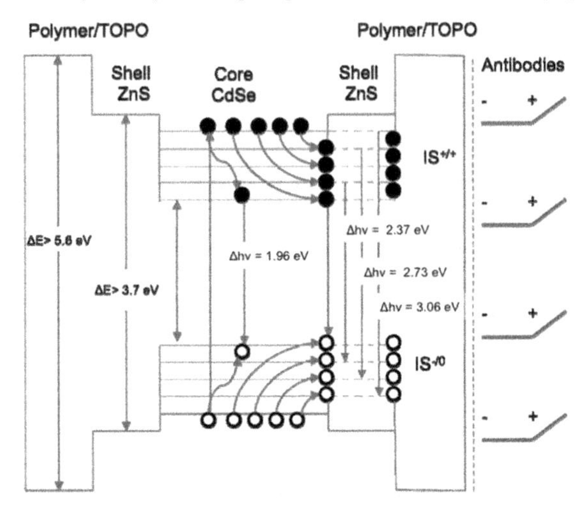

Figure 3. The energy diagram of CdSe/ZnS core/shell QDs covered by polymer and bioconjugated with Osteopontin antibodies. Symbols $IS^{+/+}$ and $IS^{-/0}$ present the charge of donor-like and acceptor- like interface states (IS), respectively, in nonconjugated (IS^{+} and IS^{-}) and bioconjugated (IS^{+} and IS^{0}) CdSe/ZnS QDs [10]. Dashed lines below IS symbols show the ways of carrier tunneling from CdSe/ZnS interface states to the ZnS/polymer interface.

Thus the hot electron-hole recombination via interface defects becomes ineffective in the bio-conjugated CdSe/ZnS QDs. The interface state re-charging was proposed in [10] for the explanation of disappearing the PL bands related to the interface states. As follows from the re-charging model, the acceptor-like interface states are neutral (IS^{o}) in bioconjugated QDs (Fig.3). Thus the carrier recombination via these defects cannot compete with exciton recombination inside the QDs. As a result the recombination flow via interface states decreases dramatically (Fig.1b).

Thus the proposed bioconjugation model assumes that the QDs have the electric charge (or dipole moments). To confirm the existance of the electric charge in antibodies the Raman scattering spectra of bio-conjugated (640P) and nonconjugated (640N) CdSe/ZnS QDs have been compared (Fig.4). Detected Raman peaks can be divided on two groups related to: i) a Si substrate (235, 302, 452, 521, 621, 662, 900, 935, 981 cm^{-1}) [19,20] and ii) a CdSe core (211.5 cm^{-1}). In the non-conjugated QD samples a set of Raman peaks at 1126, 1355, 1587, 1829, 2145, 2582, 2897 cm^{-1} have been detected as well. The intensity of Raman peaks is very low in nonconjugated QD states and it increases in bioconjugated QDs (Figure 4). All mentioned Raman peaks can be assigned to the different vibration modes in organic amine (NH_2)-

derivatized PEG polymer $[OH-(CH_2-CH_2-O)_n-H]$ cover of QDs. There are: 1126-1268-1340 and 1447 cm^{-1}stretching and bending vibrations of COOH, C-H and CH_2 groups, C-C, C=C, C-O, C=O skeleton vibrations (1657, 1829 and 2145 cm^{-1}), symmetric and anti-symmetric stretching vibrations of CH, CH_2 or CH_3 groups (2582, 2872, 2932 and 3064 cm^{-1}) [11].

For all measured Raman peaks their intensities in nonconjugated QDs (640N) were essentially smaller (twelve-fold) than those in bioconjugated (640P) QDs (Fig.4). The stimulation of optical field near the Si surface with illuminated bioconjugated QDs and increasing the intensity of all detected Raman peaks (Fig.4) can be attributed to the surface enhanced Raman scattering (SERS) effect [6,7]. The enhancement of Raman signals related to a CdSe core and a Si substrate in bioconjugated QDs indicates that the anti Osteopontin antibodies are characterized by the electrical charge that permits an electrostatic interaction of biomolecules with electric field of excitation light (785 nm) and induces the surface enhanced Raman scattering (SERS) effect.

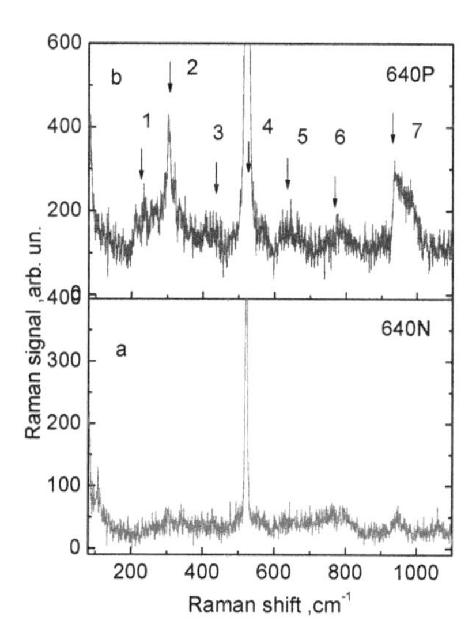

FIGURE 4. The Raman spectra of the nonconjugated QD (a) and bioconjugated QD (b) QDs on the Si substrate measured in the range 80 - 1100 cm^{-1}. The high intensity Raman peak at 521 cm^{-1} related to optic phonons of the Si substrate.

CONCLUSIONS

It is shown that the bioconjugation process of the CdSe/ZnS QDs to anti Osteopontin antibodies is complex and includes both the covalent and the electrostatic interactions. The variation of PL spectra at the bioconjugation is explained on the base of electrostatic interaction between the charged QDs and antibody electrical dipoles that lead to re-charging the QD luminescent interface states. The Raman scattering study has confirmed the existence of electrical charges for anti Osteopontin antibodies, which leads to the SERS effect in bioconjugated QDs as well.

ACKNOWLEDGMENTS

This work was partially supported by CONACYT (projects 130387) as well as by the SIP-IPN, Mexico

REFERENCES

1. M. Dybiec, G. Chomokur, S. Ostapenko, A. Wolcott, J. Z. Zhang, A. Zajac, C. Phelan, T. Sellers, G. Gerion, *Appl. Phys. Lett.* **90**, 263112 (2007).
2. M. Kuno, D.P. Fromm, H.F. Hamann, A. Gallagher, D.J. Nesbitt, *J. Chem. Phys.* **115,** 1028 (2001).
3. A. Mews, A. Eychmuller, M. Giersig, D. Schoos, H. Weller, *J. Phys. Chem.* **98**, 934 (1994).
4. Y. Ebenstein, T. Mokari, U. Banin, *J. Phys. Chem.* B **108**, 93 (2004).
5. www.invitrogen.com
6. T. V. Torchynska, J. Douda, S. S. Ostapenko, S. Jimenez-Sandoval, C. Phelan, A. Zajac, T. Zhukov, T. Sellers, *J. of Non-Crystal. Solid.* **354**, 2885 (2008).
7. T. V. Torchynska, A. Diaz Cano, M. Dybic, S. Ostapenko, M. Morales Rodrigez, S. Jimenes Sandoval, Y. Vorobiev, C. Phelan, A. Zajac, T. Zhukov, T. Sellers, *phys. stat. sol. (c),* **4**, 241 (2007).
8. T. V. Torchynska, J. Douda, P. A. Calva, S. S. Ostapenko and R. Peña Sierra. *J. Vac. Sci. &Technol.* **27(2)**, 836 (2009).
9. T. V. Torchynska, J. Douda, and R. Peña Sierra, *phys. stat. sol.* (c), **6,** S143 (2009).
10. T.V. Torchynska, *Nanotechnology,* **20**, 095401 (2009).
11. V. I. Klimov, D. W. McBranch, C. A. Leatherdale, M. G. Bawendi, *Phys. Rev. B*, **60**, 13740 (1999).
12. N.E. Korsunskaya, I.V. Markevich, T.V. Torchinskaya and M.K. Sheinkman, *phys. stat. sol* (a), **60**, 565 (1980).
13. N.E. Korsunskaya, I.V. Markevich, T.V. Torchinskaya and M.K. Sheinkman, *J. Physics C: Solid Sttate Physics*, **13**, 2975 (1980).
14. N.E. Korsunskaya, I.V. Markevich, T.V. Torchinskaya and M.K. Sheinkman, *J. Phys. Chem. Solids*, **43,** 475 (1982).
15. T.V. Torchynska, A. Diaz Cano, M. Morales Rodriguez, L. Yu. Khomenkova, *Physica B, Conden, Matter.* **340-342**, 1113 (2003).
16. L. Khomenkova, N. Korsunska, M. Sheinkman, T. Stara, T.V. Torchynska, A. Vivas Hernandez, *J. Lumines.* **115**, 117 (2005).
17. T.V. Torchynska, N. Korsunskaya, B.R. Dzumaev, L.Yu. Khomenkova, *J. Phys. Chem. Solids*, **61**, 937 (2000).
18. T.V. Torchynska, M.K. Sheinkman, N.E. Korsunskaya, L.Yu Khomenkova, B.M. Bulakh, B.R. Dzhumaev, A. Many, E. Savie, Y. Godstein, *Physica B, Conden. Matter.* **273-274**, 955 (1999).
19. F.A. Johnson and R. Loudon, Proc. Roy. Soc. A, **281,** 274 (1964).
20. P.A. Temple and C. E. Hathaway, Phys. Rev. B, **7**, 3685 (1973).

Mater. Res. Soc. Symp. Proc. Vol. 1534 © 2013 Materials Research Society
DOI: 10.1557/opl.2013.309

Double core Infrared (CdSeTe) / ZnS quantum dots conjugated to Papiloma virus antibodies

J.L. Casas Espinola[1], T. V. Torchynska[1], J. A. Jaramillo Gómez[2], J. Douda[2] and K. Gazarian[3]
[1]ESFM– Instituto Politécnico Nacional, México D. F. 07738, México
[2]UPIITA – Instituto Politécnico Nacional, México D. F. 07320, México
[3]Instituto de Investigaciones Biomédicas, UNAM, México D.F. México

ABSTRACT

The paper presents the photoluminescence (PL) study of the double core/shell infrared CdSeTe/ZnS quantum dots (QDs) in nonconjugated states and after the conjugation to the anti papiloma virus (i) mouse anti-HPV 16-E7 or ii) mouse monoclonal [C1P5] to HPV16 E6 + HPV18 E6) antibodies. CdSeTe/ZnS QDs with infrared emission at nearly 800 nm (1.6 eV), have been investigated. PL spectra of nonconjugated QDs are characterized by one Gaussian shape PL band related to the exciton emission in CdSeTe cores. Raman scattering spectra have been studied with the aim to reveal the CdSeTe double core composition. The Raman scattering study has shown that the central part of the core in QDs has the composition $CdSe_{0.5}Te_{0.5}$ and the periphery part of the core has the composition $CdSe_{0.7}Te_{0.3}$.

PL spectra of bioconjugated QDs have changed: PL bands shift into the high energy and become asymmetric. The energy diagram of the double core/shell CdSeTe/ZnS QDs have been created for the nonconjugated QDs, which permits to explain the PL spectrum of nonconjugated QDs and its transformation at the bioconjugation to papiloma virus antibodies. It is shown that the PL spectrum transformation in bioconjugated QDs is promising for the study of the bioconjugation with specific antibodies and can be a powerful technique in biology and medicine.

INTRODUCTION

The application of II-VI core/shell QDs in biology and medicine is expected to produce major advances in molecular diagnostic [1], gene technology [2] and toxin detections [3]. Additionally, II-VI core/shell QDs are usually characterized by high PL quantum yield (up to 75%) and high resistance to photo- and metabolic degradation.

Core/shell QDs of II-VI semiconductors as the luminescent markers have been investigated intensively during the last decade. The PL intensity in bioconjugated QDs decreased [4] or increased [5] owing to the energy exchange between QDs and biomolecules. The QD luminescence intensity depends on the concentration of attached bio-molecules, enabling QD application as a protein sensor [4, 5]. Unfortunately, the limits of protein detection are not sensitive enough. Thus, it is desirable to have additional spectroscopic confirmation of bioconjugation with a spectral shift of the QD emission or changing the PL band half width.

Double core /shell CdSeTe QDs with IR emission (800 nm) are manufactured specially. This type of QDs is very interesting for the biological applications owing to the possibility of in vivo animal imaging, due to the lower absorption of IR light by animal tissues.

EXPERIMENTAL DETAILS

Commercially available double core/shell CdSeTe/ZnS QDs [6] covered by the amine (NH_2)-derivatized polyethylene glycol (PEG) polymer (Fig.1) were used in a form of colloidal particles diluted in a phosphate buffer saline (PBS) with a 1:200 volumetric ratio. CdSeTe/ZnS QDs were characterized by IR emission with the PL maximum at 800 nm (1.60 eV).

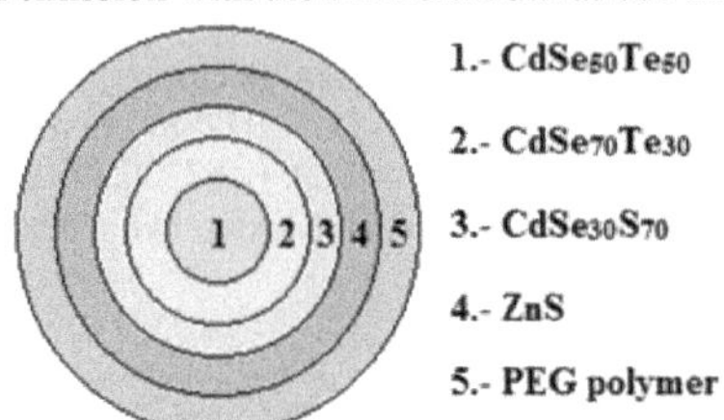

Figure1. The structure of QDs

At first the PL and Raman scattering spectra of QDs were studied in nonconjugated states (named 800N) with the aim to reveal the CdSeTe double core composition. Then some part of the QDs has been conjugated to the anti papiloma virus antibodies: (i) mouse anti-HPV 16-E7 or ii) mouse monoclonal [C1P5] to HPV16 E6 + HPV18 E6) using the 800 nm QD conjugation kits [7]. The samples of QDs (bioconjugated and nonconjugated) in the form of a 5 mm size spot were dried on the surfaces of crystalline Si substrates as described earlier in [8]

The PL spectra of QDs were measured using an excitation wavelengths of 325 nm from a He-Cd laser with a beam power of 45 mW at 300K, using a PL setup based on a spectrometer SPEX500 and described in [5,9]. Raman scattering spectra were measured at 300 K using a Raman spectrometer of a model Lab-Raman HR800 Horiba Jovin-Yvon in the range of Raman shifts of 100-500 cm^{-1} excited by a solid state laser with a wavelength of 532 nm and a beam power of 20 mW [10,11].

EXPERIMENTAL RESULTS

Figures 2 and 3 represent the PL spectra of nonconjugated (800N) and bioconjugated (800P) QDs. In nonconjugated state the PL spectra are characterized by one Gaussian shape PL band with the maximum at 1.60 eV (Fig. 2 and 3, curves 1) related to the exciton emission in CdSeTe cores. The small PL peak at 2.5 eV in nonconjugated QDs belong to the emission in the CdSeS intermediate alloy layer at the CdSeTe/ZnS interface.

The PL spectra changed upon QD bioconjugation. In figures 2 and 3 the PL spectra of bioconjugted QDs for two different concentrations of antibodies have been presented. PL spectra for the small and high antibody concentrations are presented by the curves 2 and 3, respectively (figures 2 and 3). In QDs bioconjugated to the high concentration of anti-HPV 16-E7 antibodies (Fig.2) the PL spectrum is characterized by the wide and complex PL band which can be presented as a superposition of the original 1.6 eV PL band and the additional high energy PL band with the peak at 1.90eV (Fig.2).

In QDs bioconjugated to anti HPV16 E6 + HPV18 E6 antibodies the PL peaks shift to higher energies, up to 1.80 eV for low concentration or to 1.90 eV for high antibody concentration and the shape of PL band becomes asymmetric with pronounced low energy shoulders (Fig. 3, curves 2, 3). Additionally, the PL intensity of the band peaked at 2.5 eV decreases and the new high energy PL bands appear in the spectral range of 2.6-3.0 eV.

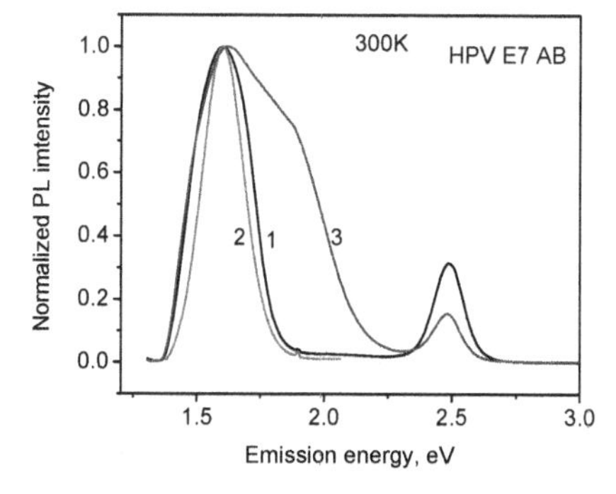

Figure 2. Normalized PL spectra of nonconjugated 800N QDs (curves 1) and two bioconjugated 800P QDs (curves 2, 3) to anti-HPV 16-E7 antibodies measured at 300 K.

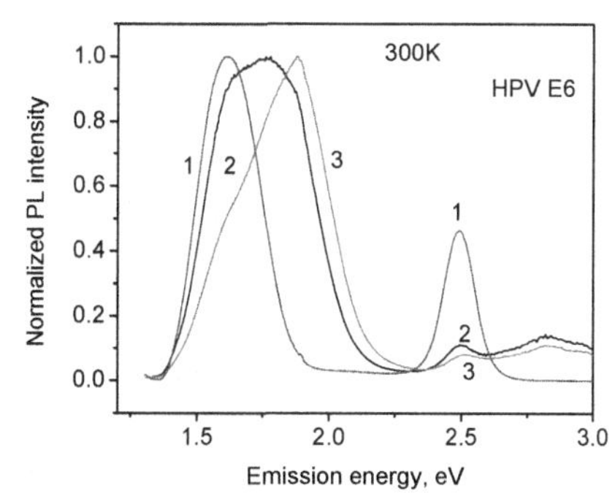

Figure 3. Normalized PL spectra of nonconjugated 800N QDs (curves 1) and two bioconjugated 800P QDs (curves 2, 3) to anti HPV16 E6 + HPV18 E6 antibodies measured at 300 K.

At the PL study using the UV excitation light (325nm) it is possible to excite the QD emission together with emission of the biomolecules or the PBS buffer. To distinguish between these possibilities the PL spectra of antibodies and PBS buffer without QDs have been investigated as well (Fig.4). The wide small intensity PL band has been detected in the high energy spectral range of 2.2-3.2 eV for three studied objects. Note that antibodies were dissolved in PBS buffer as well. Thus the PL spectrum of antibodies includes the emission of PBS buffer. It is essential that the intensities of the antibody and PBS buffer emissions are smaller ten-fold in comparison with the PL intensity of bioconjugated QDs and this emission located in the high energy spectral range. Thus the emission of PBS buffer or antibodies cannot influent on the PL spectra of bioconjugated QDs excited by UV light (325nm).

To understand the structure of the double core/shell CdSeTe/ZnS QDs the Raman scattering spectra were studied as well. The Raman scattering spectrum of nonconjugated 800 nm CdSeTe/ZnS QDs presents a complex peak at 200.7 cm^{-1}, which is the superposition of two Raman peaks at 190.7 and 200.7 cm^{-1} (Fig.5). The Raman peaks at 190.7 and 200.7 cm-1 correspond to the LO phonon in the $CdSe_{0.5}Te_{0.5}$ core and the $CdSe_{0.7}Te_{0.3}$ core cover layer in 800nm QDs, respectively [12, 13]. Additionally, the Raman spectrum demonstrates the peaks at 177.0 and 280 cm-1. The peak at 177.0 cm-1 deals with a surface phonon, SP (177.0 cm-1), in the CdSeTe core [12]. The Raman peak at 280 cm-1 can be assigned to the LO phonon in the $CdSe_{30}S_{70}$ intermediate alloy layer at the CdSeTe/ZnS interface mentioned above [12, 13].

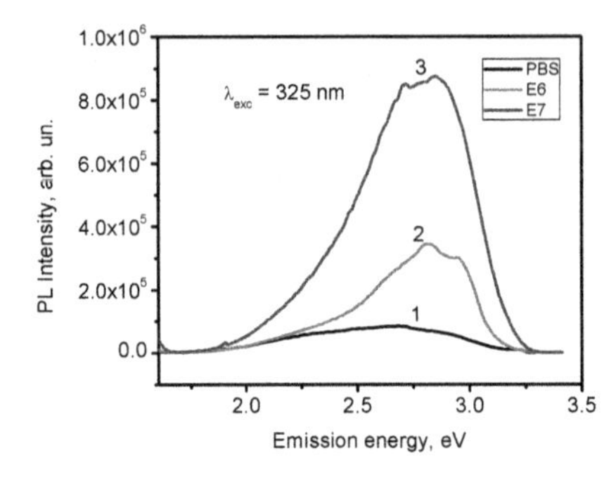

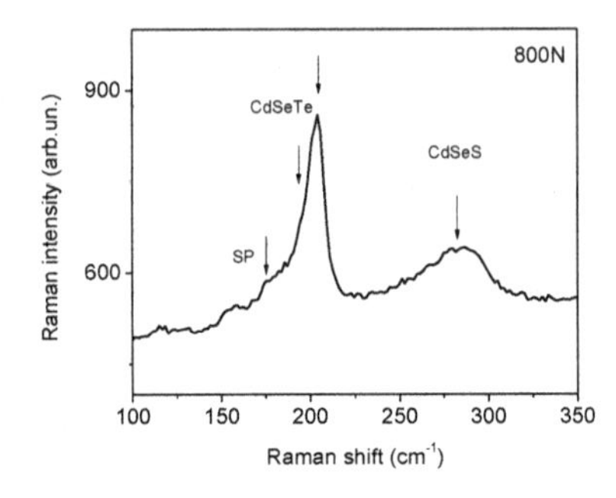

Figure 4. PL spectra of PBS buffer (1), anti HPV16 E6 + HPV18 E6 (2) and anti-HPV 16-E7 (3) antibodies

Figure 5. Raman spectrum of nonconjugated 800N QDs

DISCUSSION

In present work an ensemble of double core/shell CdSeTe/ZnS QDs has been investigated. At the first let us consider the energy diagram of the CdSeTe/ZnS QDs (Fig.6).

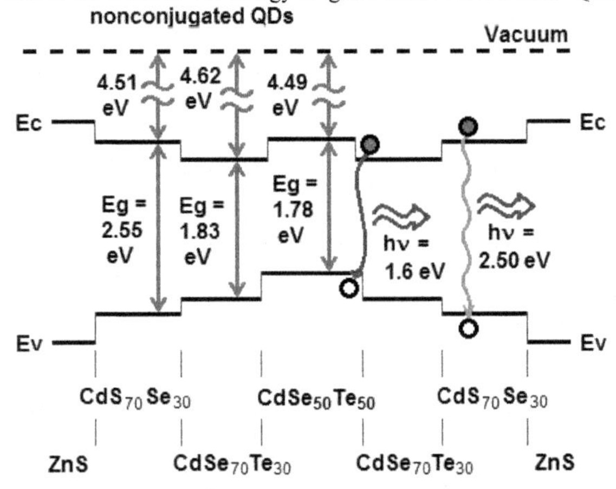

Figure 6. Energy diagram of nonconjugated QDs

To design the energy diagram of nonconjugated QDs the variation of optical band gaps in the $CdSe_{50}Te_{50}$ core and the $CdSe_{70}Te_{30}$ cover layer versus their sizes has been estimated using the effective mass approximation model [14]. Note, that the big size of double core/shell CdSeTe/ZnS QDs permits us to use the effective mass approximation model in our case [14]. To analyse the optical band gap the following parameters are used for the bulk CdSe: Eg = 1.730 eV at 300K, $m_e^* = 0.13\ m_o$, $m_h^* = 0.45m_o$, and the high frequency dielectric constant $\varepsilon = 8.2$ [12, 13]. Taking into account that in the bulk CdTe: Eg = 1.50 eV at 300K, $m_e^* = 0.11m_o$, $m_h^* = 0.35m_o$ [12,13], the values of the optical band gaps and corresponding effective masses for bulk $CdSe_{0.7}Te_{0.3}$ and $CdSe_{0.5}Te_{0.5}$ alloys have been obtained by linear interpolations. The values of electronic affinities of CdSe, CdTe and ZnS were taken from [15]. To estimate the PL excitation peak positions related to the electron-hole pair ground states in QDs with emission at 800 nm, the value of Stokes shift has been taken in account equals to 50-60 meV [6,7]. Obtained data of electron-hole pair ground states are used for the energy diagram design of the double core/shell QDs (Fig. 6).

The energy diagram of nonconjugated QDs, presented in figure 6, explains the type of optical transitions and the energy of emission quanta of 1.60 eV detected in nonconjugated double core/shell QDs. As one can see in figure 6, due to the type-II quantum well ($CdSe_{70}Te_{30}/CdSe_{50}Te_{50}/CdSe_{70}Te_{30}$) in the double core of the QDs, the wide band gap core and wide band gap covered quantum well layer with the band gaps of E_g = 1.78 and 1.83 eV, respectively, enable IR emission with a peak at 1.60 eV in nonconjugated QDs (Fig.2, 3 and 6).

As we mentioned above the PL spectrum transformation in bioconjugated QDs is due to the application of electric charges of antibodies to QDs. It was shown earlier that antibody molecules are characterized by electric charges (dipoles) that enable them to stimulate the transformation of PL spectra and to initiate the surface enhanced Raman scattering (SERS) effects in the bioconjugated QDs [5, 9, 10, 16]. The application of electric charges to QDs can change the profile of energy diagram that will be accompanied by the concentration of carriers in the core of bioconjugated QDs. The last effect can make possible fulfilling by excitons both the ground and the excited states in QD cores and, as a result, appearing the QD excited state emission in the PL spectrum (Fig.2 and Fig.3) [17-19].

The "blue" shift of the emission energy in PL spectra of bioconjugated QDs (Fig.2 and Fig.3) can be related, as well, to the change of the electric potential (potential barrier) at the surface of the bioconjugated QDs. It is known that the position of energy levels in QDs for the strong quantum confinement regime depends on the value of the potential barrier. This question was discussed in [20] and it was shown that the influence of a finite potential barrier on the electron-hole energy states increases with decreasing dot radius. As expected, the lower potential barrier reduces and the higher potential barrier increases the quantum confinement energy levels [20].

CONCLUSION

Photoluminescence of IR double core/shell CdSeTe/ZnS QDs in nonconjugated states and after the conjugation to the anti papiloma virus antibodies have been studied. It is shown that the bioconjugation to antibodies stimulates a shift of PL spectra to the higher energy spectral ranges. The energy diagram of QDs has been designed for the nonconjugated QDs that permits to explain the types of optical transitions in QDs and their transformation with the bioconjugation. A better understanding of the QD bioconjugation with specific antibodies is expected to produce major advances in biology and medicine and can be a powerful technique in early medical diagnostics.

ACKNOWLEDGEMENT

The work was partially supported by CONACYT Mexico (project 130387) and by SIP-IPN, Mexico.

REFERENCES

1. T. Jamieson, R. Bakhshi, D. Petrova, R. Pocock, M. Imani, A. M. Seifalian, Biomaterials **28**, 4717 (2007).
2. D. Gerion, W.J. Parak, S.C. Williams, D. Zanchet, C.M. Micheel, A.P. Alivisatos, J Am Chem Soc **124**, 7070 (2002).
3. L. J. Yang, Y.B.Li, Analyst **131**, 394 (2006).
4. A.R. Clapp, I. L. Medintz, J. M, Mauro, Br. R. Fisher, M. G. Bawendi, and H. Mattoussi, J. Am. Chem. Soc. 126, 301-310 (2004).
5. T.V. Torchynska, Nanotechnology, **20**, 095401 (2009).
6. www.invitrogen.com
7. http://probes.invitrogen.com/media/pis/mp19010.pdf
8. T. V. Torchynska, A. Diaz Cano, M. Dybic, S. Ostapenko, M. Morales Rodrigez, S. Jimenes Sandoval, Y. Vorobiev, C. Phelan, A. Zajac, T. Zhukov, T. Sellers, phys. stat. sol. (c), 4, 241 (2007).
9. T. V. Torchynska, J. Douda, and R. Peña Sierra, phys. stat. sol. (c), 6, S143 (2009).
10. A. Diaz Cano, S. Jiménez Sandoval, Y. Vorobiev, F. Rodriguez Melgarejo and T. V. Torchynska, Nanotechnology, 21, 134016 (2010).
11. L. G. Vega Macotela, T. V. Torchynska, J. Douda, and R. Peña Sierra, phys.stat.solid. (c), 7, 1192 (2010).
12. Physics of II-VI compounds, Editors A.N. Georgobiany and M.K. Sheinkman, Publisher „Nauka", Moscow, Russia, 1986, 300 p.
13. Physics and chemistry of II – VI compounds, Editors M. Aven and J.S. Prener, North-Holland, Amsterdam, 1967, 625p.
14. Torchynska, T.V. "Nanocrystals and quantum dots. Some physical aspects" in the book "Nanocrystals and quantum dots of group IV semiconductors", Editors: T. V. Torchynska and Yu. Vorobiev, American Scientific Publisher, 1- 42 (2010).
15. A.G. Milnes, D.L. Feucht, Heterojunctions and Metal-Semiconductor junctions, Academic Press, New York and London, 1972.
16. T. V. Torchynska, J. Douda, S. S. Ostapenko, S. Jimenez-Sandoval, C. Phelan, A. Zajac, T. Zhukov, T. Sellers, J. of Non-Crystal. Solid. **354**, 2885 (2008).
17. N.E. Korsunskaya, I.V. Markevich, T.V. Torchinskaya and M.K. Sheinkman, phys. status solidi (a), **60**, 565 (1980).
18. N.E. Korsunskaya, I.V. Markevich, T.V. Torchinskaya and M.K. Sheinkman, J. Phys. Chem. Solids, **43,** 475 (1982).
19. N.E. Korsunskaya, I.V. Markevich, T.V. Torchinskaya and M.K. Sheinkman, J. Phys. C. Solid St. Phys. **13**, 2975 -2982 (1980).
20. D.B. Tran Thoai, Y.Z. Hu, S.W. Koch, Phys. Rev. B, **41**, 6079 (1990).

Mater. Res. Soc. Symp. Proc. Vol. 1534 © 2013 Materials Research Society
DOI: 10.1557/opl.2013.310

Electronic effects in Emission of core/shell CdSe/ZnS Quantum dots conjugated to anti-Interleukin 10 antibodies

Janna Douda, Oscar S. Lopez de la Luz and Aaron I. Díaz Cano
UPIITA, Instituto Politécnico Nacional, av. IPN, 2580, México D. F. 07320, México

ABSTRACT

The paper presents the study of photoluminescence (PL) and Raman scattering spectra of nonconjugated and bioconjugated CdSe/ZnS core-shell quantum dots (QDs). Commercial CdSe/ZnS QDs used are characterized by color emission with the maxima at 605 nm (2.05eV) and 655 nm (1.89 eV). The QD conjugation has been performed with biomolecules – anti Interleukin-10 antibodies (anti IL10 mAb). PL spectra of nonconjugated QDs are characterized by only one symmetric PL band related to the exciton emission in the CdSe core. PL spectra of bioconjugated QDs have changed: PL band shifts into the high energy side and becomes asymmetric. To explain these effects the model has been proposed which assumes that the PL spectrum transformation in bioconjugated QDs, apparently, is connected with the quantum confinement and/or the Stark effects.

To confirm the existence of electric charges (or dipoles) in the anti-Interleukin 10 antibodies, which can provoke the Stark effect, the Raman scattering spectra have been investigated as well. The enhancement of Raman scattering is observed in bioconjugated CdSe/ZnS QDs. The last effect testifies that the IL10 antibodies are characterized, actually, by the electric charges (or dipoles) that permit them to interact with an electric field of excitation light and to provoke the surface enhanced Raman scattering (SERS) and the Stark effects in QDs.

INTRODUCTION

The integration of nanotechnology with biomedicine is expected to produce the major advances in bioengineering, early cancer diagnostic, biosensors, therapeutic etc [1, 2]. Semiconductor core/shell CdSe/ZnS quantum dots (QDs) due to their unique optical properties and dimensional similarities to biomolecules have attracted the great attention in biomedicine during a last decade [3-9]. These QDs used as bio-luminescent markers can vary the photoluminescence (PL) intensity when coupled to different biomolecules. The process of QD bioconjugation, as a rule, is accompanied by the variation of QD PL intensity mainly. The confirmation of the QD bioconjugation and the study its details using the detection of PL spectrum transformation are highly important as well. Additionally it is important to look for other optical effects such as Raman scattering in QD bioconjugation, which could offer the important information on the structure of bioconjugated QDs.

This paper presents the results of PL and Raman scattering study of the CdSe/ZnS core/shell QDs with color emission at 605 and 655 nm in nonconjugated state and bioconjugated to the anti-Interleukin 10 antibodies (anti IL10 mAb).

EXPERIMENTAL DETAILS

The commercially available core-shell CdSe/ZnS QDs covered with a polyethylene glycol (PEG) polymer were used in a form of colloidal particles diluted in a phosphate buffer saline (PBS) with a 1:200 volumetric ratio. Studied QDs are characterized by the size of 5.2 and 6.4 nm and by color emission with the maxima at 605 nm (2.05eV) and 655 nm (1.89 eV), respectively. The part of CdSe/ZnS QDs (605P, 655P) has been conjugated to anti-Interleukin (IL10) antibodies (antihuman IL10, Rt IgG1, stock concentration of 1mg/ml, clone JES3-9D7, code RHCIL1000) using the commercially available 605 and 655 nm QD conjugation kits [10,11]. Some part of CdSe/ZnS QDs (605N, 655N) has been left nonconjugated and serves as a reference object. Samples of QDs (bioconjugated and nonconjugated) in the form of a 5 mm size spot were dried on a polished surface of the crystalline Si substrate (Fig. 1) as described earlier in [12-15]. PL spectra were measured at 300 K at the excitation by a He-Cd laser with a wavelength of 325 nm and a beam power of 50 mW using a PL setup on a base of spectrometer SPEX500. Raman scattering spectra were measured at 300K using a Raman spectrometer of model Lab-Raman HR800 Horiba Jovin-Yvon and a solid state laser with a light wavelength of 532 nm and a power of 20 mW in backscattering configuration.

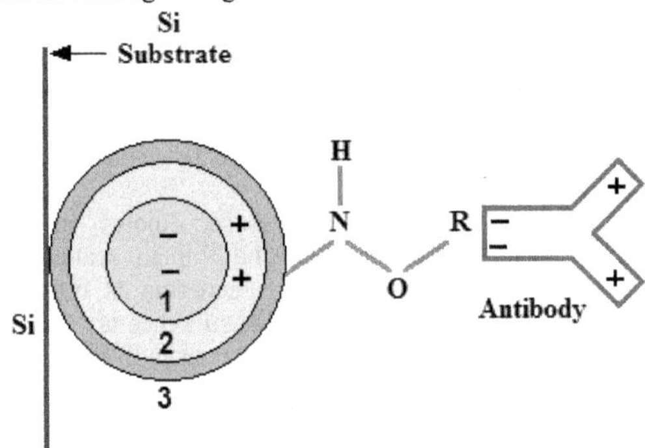

Figure 1. Core/shell QDs bioconjugated to antibodies.

RESULTS AND DISCUSSION

Typical PL spectra for nonconjugated and bioconjugated CdSe/ZnS QDs have been presented in figures 2 and 3. In nonconjugated state the PL spectra of QDs are characterized by one PL band with the symmetric Gaussian shape and the maxima at 2.05 eV (Fig.2) and 1.89 eV (Fig.3), respectively, related to the exciton recombination in CdSe cores. At the bioconjugation of QDs with emission of 2.05 eV (605P) the PL spectra have changed a little: the PL band peak shifts into higher energy (10meV) and a half width increases from 120 meV in nonconjugated QDs up to 140 meV in bioconjugated QDs (Fig.2). At the bioconjugation of QDs with emission of 1.89 eV (655P) the PL spectra have changed as well (Fig.3). The PL intensity decreases and a PL peak position shifts (48meV) to higher energy. At the same time the shape of PL bands become asymmetric with an essential high energy tails (Fig.3).

A set of possible explanations of the PL spectral shift at the QD bioconjugation to antibodies can be related to varying the QD energy levels owing: i) the Stark effect [2, 12], ii) a compressive strain applied to bioconjugated QDs [2,13], iii) core/shell material intermixing [14], iv) core size decreasing owing the oxidation [14] or v) quantum confinement effect related to the change of QD surface potential barrier at the bioconjugation [15].

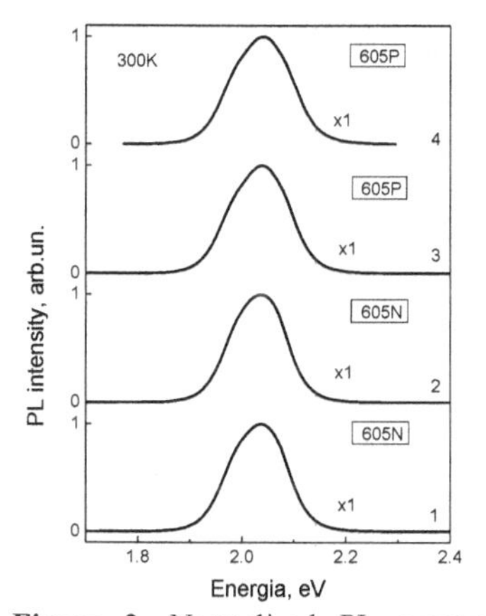

Figure 2. Normalized PL spectra of non-conjugated 605N and bioconjugated 605P QDs measured at 300 K. Numbers at the curves (x 1, x 1, ...) indicate the multiplication coefficient used at the normalization of experimental PL spectra

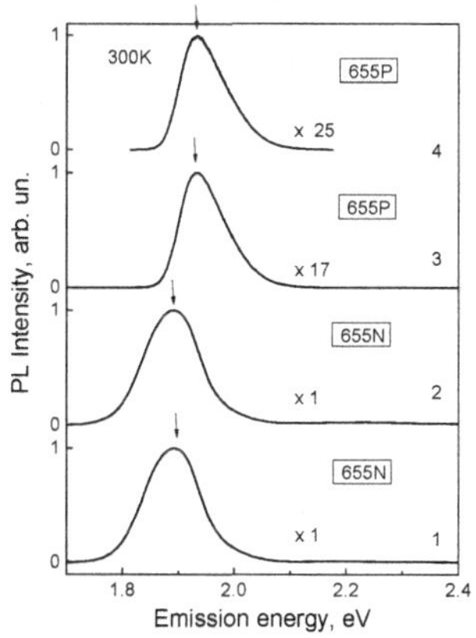

Figure 3. Normalized PL spectra of non-conjugated 655N and bioconjugated 655P QDs measured at 300 K. Numbers at the curves (x 1, x 1, x 1.7 and x 2.5) indicate the multiplication coefficient used at the normalization of experimental PL spectra.

To distinguish between the different reasons it is essential to note that the compressive strain has not change the symmetry of PL bands. Moreover, as Raman scattering study testifies the position of Raman peaks related to the LO phonon of CdSe core has not changed at the bioconjugation. That means that the compressive strains, the core/shell intermixing or oxidation processes are not relevant at the bioconjugation. To investigate the possible influence of the Il-10 mAb and PBS emissions, the PL spectra of pure Il-10 mAb and PBS have been investigated (Fig.4). The PL intensities of emission of the pure Il-10 mAb and PBS are fifteen fold less than the PL intensity of QDs and their PL spectra located very far from the spectral range of QDs. Additionally it is important to note that the impact of changing the QD surface potential barrier has to be more essential in smaller QDs (emitted at 605nm), than in bigger QDs (emitted at 655nm). However the higher PL shift is detected in the bigger QDs (Fig.3). These facts, as well as the asymmetric shape of PL band appeared in bioconjugated QDs (Fig.2,3), testify that the Stark effect can be essential in the PL spectrum transformation at the bioconjugation.

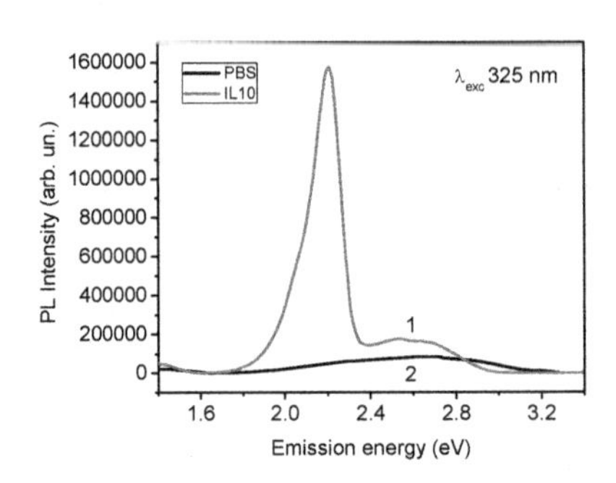

Figure 4. PL spectra of anti-IL-10 mAb (1) and PBS buffer (2) measured at 300K. Excitation wavelength is 325 nm.

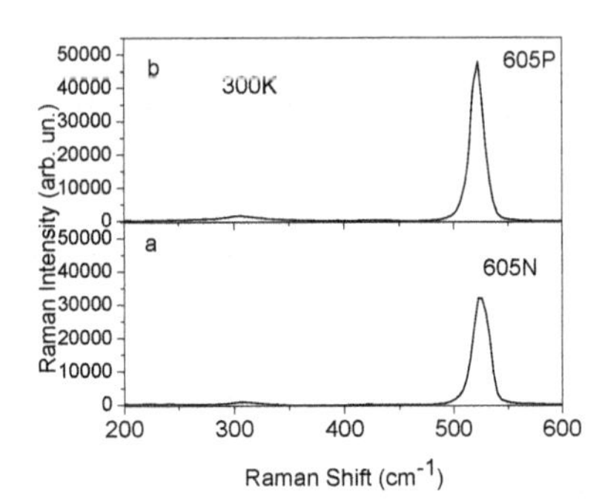

Figure 5. High intensity Raman peaks in Raman spectra of nonconjugated 605N (a) and bioconjugated 605P (b) CdSe/ZnS QDs.

The Stark shift of energy levels, ΔE, as a function of electric field applied, ξ, can be presented as a sum of a linear (first order Shark effect) and quadratic (second order Stark effect) functions of field [16]: $\Delta E=\mu_{QD}\xi + 0.5\alpha_{QD}\xi^2$, where E is the energy of optical transition, μ_{QD} and α_{QD} are the projections of the excited-state dipole and polarizability, respectively, along the applied electric field. Note the Stark shifts in QD ensemble emission were found earlier increased with the size of QDs and to be purely quadratic in the applied field [16]. The last fact is the reason, apparently, why the energy shift, ΔE, in our experiment, is higher for 655 nm QDs (d=6.4 nm) in comparison with one in 605nm QDs (d=5.2nm) (Fig.2, 3).

Note in the present model we suppose that an electric field applied to bioconjugated QDs is created by electric charges of antibodies (Fig.1). To confirm the above mentioned model it is important to show that anti IL-10 antibodies have an electric charge. For this aim the Raman scattering spectra have been studied. Raman scattering spectra of nonconjugated QDs show a high intensity peak at 520 cm^{-1}, corresponding to the optic phonon Raman line from the Si substrate (Fig.5). The intensity of this peak in the Raman spectrum of the noncojugated (605N) QD sample is smaller in comparison with its intensity in the bioconjugated (605P) QD sample (Fig.5).

Additionally a set of low intensity peaks at: 211.9, 306.8, 424.3, 619.8 and 667.9 cm^{-1} has been detected (Fig.6). The Raman peak at 211.9 cm^{-1} (and its two LO phonon overtone at 424.3 cm^{-1}) corresponds to the LO phonon in CdSe core of QDs. The small shift of LO phonon Raman line (211.9 cm^{-1}) from its position in the bulk CdSe (213 cm^{-1}) has to do, apparently, with the phonon confinement effect in small size QDs [17-20]. Raman scattering in the region of 0-700 cm^{-1} is related to the Si substrate as well. The peak at 306.8 cm^{-1} can be considered as two TA phonon overtones at Raman scattering at the X point of the silicon Brillioun zone [19,20]. The Raman peaks at 619.8 and 667.9 cm^{-1} in silicon were assigned to the two phonon peaks, which, as assumed, are the combinations of the acoustic and optic phonons in the X and Σ directions of

Brillioun zone [19, 20]. In the Raman spectrum of bioconjugated QDs the intensity of all Raman peaks increases essentially and they have seen clearly (Fig.6).

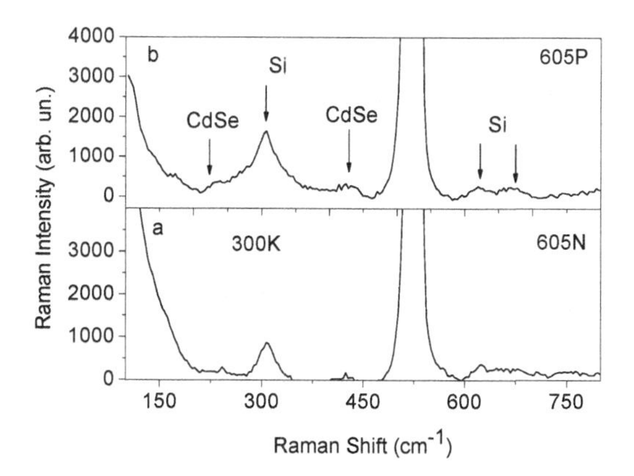

Figure 6. Low intensity Raman peaks in Raman spectra of nonconjugated 605N (a) and bioconjugated 605P (b) CdSe/ZnS QDs.

The stimulation of optical field near the surface of illuminated bioconjugated QDs located on the Si substrate and the appearance, as a result, of low intensity Raman peaks at 211.9, 306.8, 424.3, 619.8 and 667.9 cm^{-1} can be attributed to the surface enhanced Raman scattering (SERS) effect [17,21,22]. The last fact indicates that the anti-Interleukin 10 antibody molecules are characterized by the electric charges which can provoke the SERS effect in Raman scattering and can be the reason of the shift of QD energy levels and emission of the bioconjugated QDs.

CONCLUSIONS

It is shown that the PL bands shift into high energy side and becomes asymmetric in QDs bioconjugated to the anti-Interleukin 10 antibodies. To explain this effect the model has been proposed which assumes that the PL spectrum transformation in bioconjugated QDs is connected with the Stark effect. The enhancement of Raman scattering is observed in CdSe/ZnS QDs bioconjugated with ant-Interleukin 10 antibodies as well. The last fact testifies that the antibody biomolecules are characterized by the electric charge that permits them to interact with an electric field of excitation light at the SERS effect. Additionally, the electric charge of anti IL-10 antibodies creates the electric field, which stimulates the shift of QD energy levels and emission of the bioconjugated QDs.

The work was partially supported by CONACYT Mexico (project 130387) and by SIP-IPN, Mexico. The authors thank of the Dr. J.L. Casas Espinola for the photoluminescence measurement.

REFERENCES

1. R.E. Bailey, A.M. Smith, Sh. Nie, (2004). Physica E, **25**, 1-12 (2004).
2. M. Dybiec, G. Chomokur, S. Ostapenko, A. Wolcott, J. Z. Zhang, A. Zajac, C. Phelan, T. Sellers, G. Gerion, Appl. Phys. Lett. **90**, 263112 (2007).
3. W. U. Huynh, J. J. Dittmer, A. P. Alivisatos, Science **290**, 2425 (2002).
4. M.P. Bruchez, M. Moronne, P. Gin, S. Weiss, A.P. Alivisatos, Science, **281**, 2013 (1998).
5. D. Gerion, F. Pinaud, S. C. Williams, W. J. Parak, D. Zanchet, S. Weiss, and A. P. Alivisatos, J. Phys. Chem. B **105,** 8861 (2001).
6. T. V. Torchynska, Nanotechnology, **20,** 095401 (2009),
7. L. G. Vega Macotela, T. V. Torchynska, J. Douda, and R. Peña Sierra, phys.stat.solid. (c), **7,** 1192-1195 (2010).
8. L. G. Vega Macotela, J. Douda, T. V. Torchynska, R. Peña Sierra and L. Shcherbyna, phys.stat.solid. (c), **7**, 724-727, 2010.
9. T.V.Torchynska, A.L. Quintos Vazquez, R. Pena Sierra, K. Gazarian, L. Shcherbyna, J. of Physics, Conference Ser. **245**, 012013 (2010).
10. www.invitrogen.com
11. http://www.invitrogen.com/site/us/en/home.html
12. T. V. Torchynska, J. Douda, S. S. Ostapenko, S. Jimenez-Sandoval, C. Phelan, A. Zajac, T. Zhukov, T. Sellers, J. of Non-Crystal. Solid. **354**, 2885 (2008).
13. T. V. Torchynska, A. Diaz Cano, M. Dybic, S. Ostapenko, M. Morales Rodrigez, S. Jimenes Sandoval, Y. Vorobiev, C. Phelan, A. Zajac, T. Zhukov, T. Sellers, phys. stat. sol. (c), **4**, 241 (2007).
14. T. V. Torchynska, J. Douda, P. A. Calva, S. S. Ostapenko and R. Peña Sierra. J. Vacuum Scien. &Technol. **27(2),** 836-838 (2009).
15. T. V. Torchynska, J. Douda, and R. Peña Sierra, phys. stat. sol. (c), **6,** 143-145 (2009).
16. S.A. Empedocles, M.G. Bawendi, Science, **278,** 2114 (1997).
17. A. Diaz Cano, S. Jiménez Sandoval, Y. Vorobiev, F. Rodriguez Melgarejo and T. V. Torchynska, Nanotechnology, **21**, 134016 (2010)
18. N.E. Korsunskaya, I.V. Markevich, T.V. Torchinskaya and M.K. Sheinkman, phys. status solidi (a), **60**, 565 (1980).
19. P.A. Temple and C. E. Hathaway, Phys. Rev. B, **7**, 3685 (1973).
20. F.A. Johnson and R. Loudon, 1964 Proc. R. Soc. **A 281**, 274 (1964).
21. N.E. Korsunskaya, I.V. Markevich, T.V. Torchinskaya and M.K. Sheinkman, J. Phys. Chem. Solids, **43,** 475 (1982).
22. N.E. Korsunskaya, I.V. Markevich, T.V. Torchinskaya and M.K. Sheinkman, J. Phys. C. Solid St. Phys. **13**, 2975 -2982 (1980).

Mater. Res. Soc. Symp. Proc. Vol. 1534 © 2013 Materials Research Society
DOI: 10.1557/opl.2013.311

Size Control of Cadmium Sulfide Nanoparticles in Polyvinyl Alcohol and Gelatin by Polyethyleneimine Addition

Oleksandra E. Rayevska[1], Galyna Ya. Grodzyuk[1], Oleksandr L. Stroyuk[1],
Stepan Ya. Kuchmiy[1], Victor F. Plyusnin[2]
[1]L. V. Pysarzhevsky Institute of Physical Chemistry of National Academy of Sciences of Ukraine, 31 prosp. Nauky, 03028, Kyiv, Ukraine, stroyuk@inphyschem-nas.kiev.ua
[2]Institute of Chemical Kinetics and Combustion of Siberian Branch of Russian Academy of Sciences, 3 Institutskaya str., 630090, Novosibirsk, Russian Federation,

ABSTRACT

A simple method of controlling the size and photoluminescence of CdS nanoparticles stabilized by gelatin and polyvinyl alcohol in water by introducing small amounts of polyethyleneimine at the moment of nanoparticle formation is reported. It was found that by adding 0.01–0.10 w.% polyethyleneimine the size of CdS nanoparticles can be changed in the range of 3.8–6.6 nm. The luminescence efficiency and radiative life time of polyvinyl alcohol-stabilized CdS nanoparticles were found to increase four-, fivefold with increasing the polyethyleneimine content at the moment of CdS nanoparticles synthesis. A more complex behavior was observed for gelatin-based media, where the interaction between the polyethyleneimine and gelatin can have a detrimental effect of on the size distribution and luminescence properties of CdS nanoparticles.

INTRODUCTION

Cadmium chalcogenide nanoparticles (NPs) stabilized in aqueous solutions by water-soluble polymers, for example, gelatin and polyvinyl alcohol (PVA) attract considerable attention as promising materials for the luminescent biodiagnostics, photocatalysis, non-linear optics, light-emitting devices, etc. [1-5]. The main challenges at the synthesis of aqueous CdS colloids are low photoluminescence (PL) efficiency, and the limited means of exerting control over the size and the size distribution of CdS NPs [1-3]. Recently we have reported on the synthesis of ultra-small, ~2 nm (with ~10% size dispersion), colloidal CdS nanoparticles (NPs) stabilized by polyethyleneimine (PEI) in water and ethanol [6]. The CdS-PEI NPs exhibit excellent chemical and photo-chemical stability and PL emitted in a broad band at λ = 400-700 nm with the quantum yields up to 20-30% and a life-time of 60-70 ns [6]. However, though producing CdS NPs with superior PL efficiency and size distribution to those stabilized by gelatin or PVA, this approach does not allow varying the NP size without concomitant deterioration of PL properties [6]. Here we report on a simple method of controlling the average size and PL properties of CdS NPs stabilized by gelatin and PVA in water by introducing small amounts of PEI at the moment of NP formation.

EXPERIMENTAL DETAILS

Polyethyleneimine (50 wt% aqueous solution, molecular weight 50.000 g/mole), $CdSO_4$, Na_2S, gelatin, polyvinyl alcohol of reagent grade quality were supplied by Sigma-Aldrich.

Aqueous colloidal CdS solutions were prepared by interaction between $CdSO_4$ and Na_2S in the presence of gelatin [7, 8], PVA [7, 9] and PEI [6] alone or a mixture of PEI with gelatin or PVA. In a typical procedure, 0.8 mL 0.1 M aqueous $CdSO_4$ solution was added to 9.2 mL of aqueous solution containing gelatin (or PVA) and PEI at vigorous stirring. The concentrations of CdS and gelatin (PVA) were 8×10^{-3} M and 2.0 wt%, respectively, while the PEI content was varied from 0.01 to 0.1 wt%. To this mixture 0.8 mL of 0.1 M aqueous Na_2S solution was added drop-wise at vigorous stirring resulting in formation of colloidal CdS NPs. Then an additional PEI amount was added to the colloidal solutions so that the final PEI concentration was 0.10 w.% in all the synthesized samples. The colloidal solutions were used to prepare polymer films on the microscope glass substrates. The glass plates were cleaned by into the piranha solution and keeping in an ultrasonic bath. The polymer films were prepared by depositing of 1.0 mL colloidal solution onto a 2.5×2.5 cm glass plate followed by drying at ambient temperature till complete evaporation of the solvent. The final films had a thickness of 0.2-0.3 mm and contained 3.5 wt% CdS.

The absorbance spectra were registered with an HP Agilent 8453 spectrophotometer. The photoluminescence (PL) spectra and PL decay profiles were taken on an Edinburgh Instruments FLS920 luminescence spectrometer equipped with a photon counting system based on a EPL-375 picosecond diode laser emitting 60 ps pulses with $\lambda = 375$ nm. The average PL life time $<\tau>$ was determined after deconvoluting the decay curves into four monoexponential functions with the characteristic time constants τ_i and amplitudes A_i as $<\tau>=\sum_i \frac{A_i\tau_i^2}{\sum_j A_j\tau_j}$ [10].

DISCUSSION

Absorbance spectra of CdS NPs incorporated into the gelatin- and PVA films reveal broad bands with non-distinct edges at around 455 and 470 nm, respectively (Fig. 1a,b, curves 1), shifted to shorter wavelengths as compared to the bulk CdS [1, 3], indicating the presence of an ensemble of quantum-sized NP with a broad size distribution. The size of CdS NPs, d, can be estimated using well-known empirical correlations between d and NP band gap, E_g [1, 3, 6]. As the absorbance band edge is formed by a fraction of larger NPs in the ensemble, the band gap energy determined by plotting the tangent to the absorption band slope (absorption threshold energy, see inset in Fig. 1a), E_t, can be used to find the upper NP size limit, d_{max}, while the energy corresponding to the minimum of the first derivative of the absorbance spectrum, E_d (see the inset in Fig. 1b), characterizes the average size of NP in the ensemble, $<d>$ [1]. A difference between d_{max} and $<d>$ may be used as a measure of the NP size distribution, Δd (Table I).

As can be seen from Fig. 1 and Table I addition of PEI to polymer solutions before CdS NPs formation results in a "blue" shift of the absorption threshold of CdS NPs formed in such solutions and appearance of a distinct maximum at 360-370 nm. The changes are the more pronounced the higher is the PEI concentration. At the same time, an additional amount of PEI added to solutions after formation of CdS NPs in order to achieve equal PEI concentration in all the samples, 0.10 w.%, does not influence the optical properties of NPs.

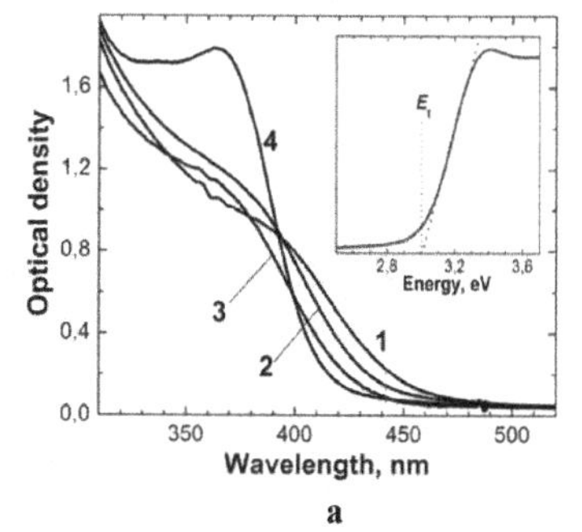

a

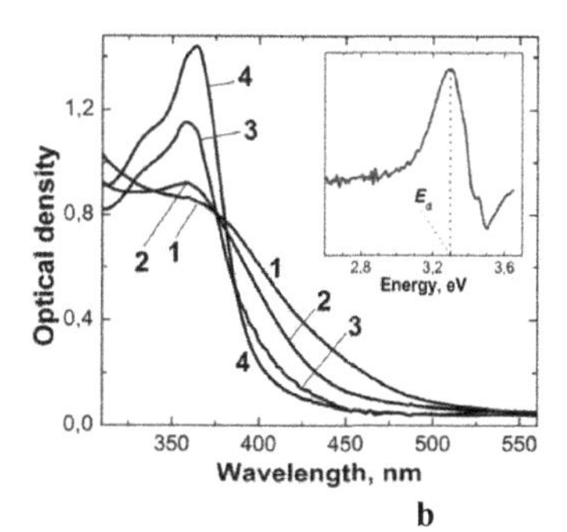

b

Figure 1. Absorbance spectra of CdS NPs in polymer films composed of PEI and gelatin (a) and PEI and PVA (b) at various content of PEI during the NP formation in original colloidal solutions – no PEI (curves 1), 0.02 w.% PEI (curves 2), 0.04 w.% PEI (curves 3), and 0.10 w.% PEI (curves 4). Insets: examples of determination of E_t and E_d from the absorption spectra.

Estimations based on spectral data show introduction of PEI and increase of its concentration from 0.02 to 0.10 w.% result in decreasing d_{max} from 6.6 nm to 3.8 nm in case of PVA and from 2.9 to 2.0 nm in case of gelatin (Table I). As can be seen from the table, PEI addition allows focusing considerably the size distribution of CdS NPs – from 70% to 10% in case of PVA and from 30% to 15% in case of gelatin. It should be noted that in the latter case Δd first grows when 0.02 w.% PEI is introduced into solution, then, at higher PEI content, it begins to decrease.

Table I. Energy of the absorption threshold (E_t) and the minimum of first derivative of the absorption spectrum (E_d), size determined from E_t (d_{max}) and E_d ($\langle d \rangle$), size distribution Δd, integral PL intensity I_{PL}, PL band maximum position (λ_m) and the average PL life time of CdS incorporated into polymer films of gelatin, PVA and PEI of different composition.

Polymer	PEI, w.%	E_t, eV	E_d, eV	d_{max}, nm	$\langle d \rangle$, nm	Δd, %	I_{PL}, 10^{-7}, a.u.	λ_m, nm	$\langle \tau \rangle$, ns
PVA	–	2.65	3.10	6.6	3.8	70	0.88	495	6
	0.02	2.87	3.27	5.0	3.7	40	1.44	475	13
	0.04	3.07	3.34	4.0	3.6	20	1.58	480	30
	0.10	3.14	3.30	3.8	3.5	10	3.67	490	43
Gelatin	–	2.74	2.98	5.7	4.7	20	2.64	560	50
	0.02	2.83	3.07	5.0	4.0	30	1.49	530	45
	0.04	2.90	3.13	4.8	3.8	25	1.80	505	40
	0.10	3.03	3.20	4.2	3.8	15	4.78	500	50
PEI	–	3.40	3.45	2.0	1.8	10	7.16	475	70

Notes: the experimental errors are 0.01 eV for E_d and E_t, 1×10^4 a.u. for I_{PL}, 1 ns for <τ>.

PL spectra of gelatin- or PVA-stabilized CdS NPs reveal broad PL bands in the range of 400–700 nm centred at around 550–560 nm (Fig. 2a) typical for the surface-trap-state-related radiative recombination [1, 2, 10]. The two polymer stabilizers react differently on the PEI addition (Fig. 2b). While in case of PVA introduction of PEI results in an increase in PL intensity (Fig. 2b, curve 1) at an almost constant maximum position (Table I) and FWHM of the PL band, in the gelatin-based systems PEI addition at low concentrations inhibits the radiative recombination efficiency (Fig. 2b, curve 2). At the same time, CdS NPs synthesized at PEI content higher than 0.05–0.06 w.% are superior by PL intensity as compared to those prepared using the gelatin alone.

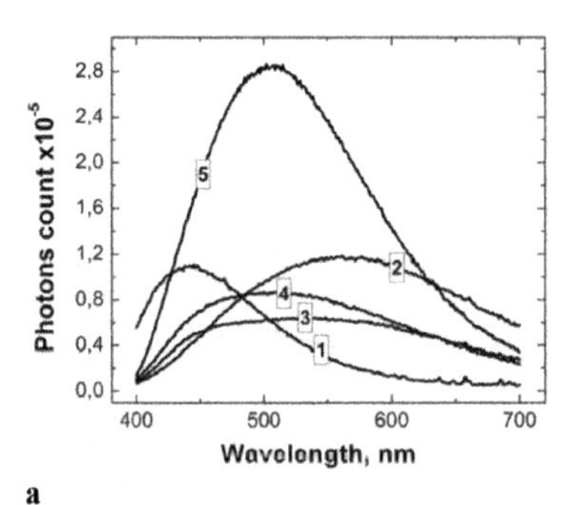

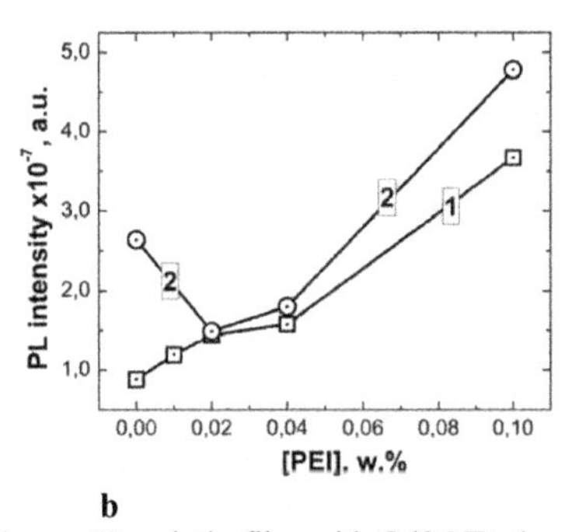

a **b**

Figure 2. (a) PL spectra of pure gelatin film (curve 1), gelatin film with CdS NPs (curve 2), and mixed gelatin–PEI films with CdS NPs containing different amount of PEI: 0.02 w.% (curve 3), 0.04 w.% (curve 4), and 0.10 w.% (curve 5). (b) Integral intensity of PL bands of CdS NP in pure PVA and gelatin films, and in mixed PVA–PEI films (curve 1) and gelatin–PEI films (curve 2) as a function of PEI content.

The PL kinetic profiles of gelatin- and PVA-stabilized CdS NPs have prominently non-exponential character (Fig. 3a, curve 1) typical for the trap-state-related emission [1, 2, 10] with the average time constants of 50 and 6 ns, respectively (Table I). A much higher radiative life time of gelatin-stabilized CdS NPs indicates the more efficient passivation of the surface states responsible for the radiationless recombinative processes, as compared with PVA. This fact as well as a much higher PL efficiency of the stationary PL, a lower NP size and a narrower size distribution evidence a superior ability of gelatin to stabilize and passivate colloidal CdS NPs in water. The difference can be accounted for by a different nature of the interaction between the two polymers and the surface of CdS NPs. In contrast to PVA, the gelatin molecules contain terminal carboxyl and amine groups and a lot of peptide –CONH– fragments [11] capable of efficient bounding to the under-coordinated Cd atoms on the surface of CdS NPs and passivating the sites of the radiationless processes.

Similarly to the case of stationary PL, increase in PEI content has a different influence on the PL decay kinetics of CdS NPs stabilized by PVA and gelatin (Fig. 3b, Table I). In the first case a monotonous increase in the radiative life time is observed (Fig. 3a, curves 2,3, Fig. 3b, curve 1) while in the gelatin-based systems < τ> first decreases when PEI is introduced to the system and then, at higher PEI content, increases almost to the original value (Fig. 3b, curve 2).

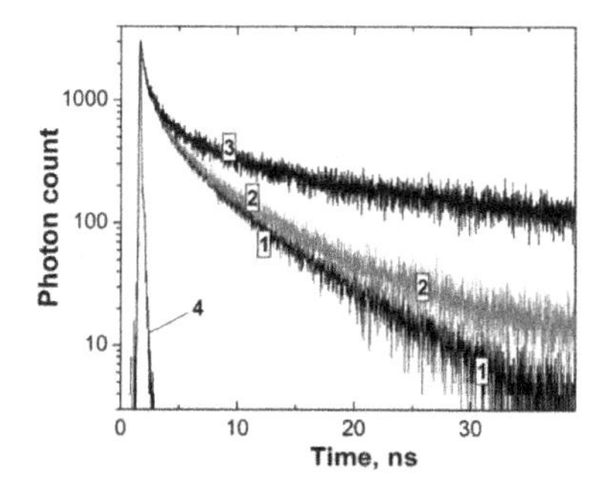

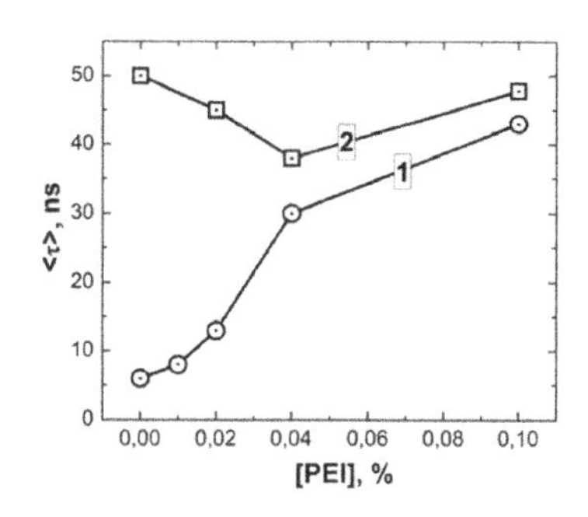

a b

Figure 3. (a) PL decay profiles for CdS NPs incorporated in PVA film (curve 1), and PVA–PEI films with 0.02 w.% PEI (curve 2) and 0.10 w.% PEI (curve 3). Curve 4 – instrumental response function. (b) Average PL life time for CdS NPs incorporated in pure PVA and gelatin films, and mixed PVA–PEI films (curve 1) and gelatin–PEI films (curve 2) as a function of PEI content.

In summary, the CdS NPs stabilized by PVA alone are characterized by the largest size, the most broad size distribution, the lowest PL efficiency and the shortest life time. As the PEI is introduced to the system, the former containing a lot of amine groups capable of formation of comparatively strong coordination bonds with the Cd atoms on the surface of forming CdS NPs, the size and its distribution decrease, while the PL efficiency and time constants of the charge recombination grow monotonously with increasing the PEI content. At the same time, similar gelatin-based systems reveal quite a different behavior. In particular, addition of PEI affects the NP size and its distribution for the gelatin-stabilized CdS NPs to a lesser extent as compared to the PVA-based systems. The changes of the size distribution, the PL intensity and the life time with increasing PEI content are of a non-monotonous character and reach extremes at 0.02-0.04 w.% PEI. From the facts one can conclude that in this concentration range some interaction between the gelatin and PEI occurs that prevents them from effectively bounding to the surface of CdS NPs. A possible reason for the observed irregular dependences can be related to the formation of the well-known polyionic complexes between the gelatin and PEI molecules. Polyethyleneimine is a well-known polyelectrolyte due to the protonation of the amine nitrogen of PEI at the expense of water dissociation. The gelatin is reported to reveal mild polyanionic properties as a result of dissociation of terminal carboxyl groups [11]. When both polymers are present in the system a polyionic complex can form between the positively charged PEI chains and negatively charged gelatin molecules prohibiting them from stabilizing and passivating CdS NPs. At higher PEI concentrations only a fraction of PEI gets involved into the formation of such polycationic complexes while the rest of PEI molecules can interact with the surface of forming CdS NPs and stabilizing them in water.

CONCLUSIONS

A simple method of controlling the size and photoluminescence of CdS NPs stabilized by gelatin and PVA in water by introducing small amounts of PEI at the moment of NP formation has been reported. It was found that by adding 0.01–0.10 wt% PEI, the size of PVA-stabilized CdS NPs can be tuned from 6.6 nm to 3.8 nm, while the dispersion Δd narrowed from 70% to 10%. In the case of gelatin a size decrease from 5.7 nm to 4.2 nm but only a modest size distribution narrowing from 30 to 15 % can be achieved. The PL efficiency of PVA-stabilized CdS NPs increases more than 4-fold upon introduction of 0.10 wt% PEI. The averaged PL life-time $<\tau>$ was found to increase from 6 ns to 43 ns when 0.10 wt% PEI is added to PVA solution before the formation of CdS NPs, indicating efficient elimination of the radiationless recombination sites by complexation of amino-groups of PEI with the Cd(II) on the surface of CdS NPs. A more complex behavior was observed for gelatin-based media, where PEI introduction results in a decrease of $<\tau>$ at low PEI content and an increase of $<\tau>$ upon further increase of PEI content to 0.10%. The detrimental effect of small PEI additives upon the size distribution and PL properties of gelatin-stabilized CdS NPs is tentatively ascribed to the formation of polyionic complexes between the dissociated carboxyl-groups of gelatin and protonated amino-groups of PEI.

ACKNOWLEDGEMENTS

This work was supported by the Ukrainian State Fund for Fundamental Research (project # F40.3./040).

REFERENCES

1. *Semiconductor Nanocrystal Quantum Dots – Synthesis, Assembly, Sperctroscopy and Applications*, Ed. A.L. Rogach (Springer Verlag, 2008).
2. D. V. Talapin, J. S. Lee, M. V. Kovalenko, E. V. Shevchenko, *Chem. Rev.* **110**, 389-458 (2010).
3. O. L. Stroyuk, S. Ya. Kuchmiy, A. I. Kryukov, V. D. Pokhodenko, *Semiconductor Catalysis and Photocatalysis on the Nanoscale* (Nova Science Publishers, 2010).
4. W. R. Algar, A. J. Tavares, U. J. Krull, *Anal. Chim. Acta* **673**, 1-25 (2010).
5. M. Zorn, W. K. Bae, J. Kwak, H. Lee, C. Lee, R. Zentel, K. Char, *ACS Nano* **5**, 1063-1068 (2009).
6. O. Ye. Rayevska, G. Ya. Grodzyuk, V. M. Dzhagan, O. L. Stroyuk, S. Ya. Kuchmiy, V. F. Plyusnin, V. P. Grivin, M. Ya. Valakh, *J. Phys. Chem. C* **114**, 22478-22486 (2010).
7. A. E. Raevskaya, A. L. Stroyuk, S. Ya. Kuchmiy *J. Nanoparticle Res.* **6**, 149-158 (2004).
8. V. M. Dzhagan, O. L. Stroyuk, O. E. Rayevska, S. Ya. Kuchmiy, M. Ya. Valakh, Y. M. Azhniuk, C. von Borczyskowski, D. R. T. Zahn, *J. Colloid Interface Sci.* **345**, 515-523 (2010).
9. Yu. M. Azhniuk, V. M. Dzhagan, Yu. I. Hutych, A. E. Rayevskaya, A. L. Stroyuk, S. Ya. Kuchmiy, M. Ya. Valakh, D. R. T. Zahn, *J. Optoelectron. Adv. Mater.* **11**, 257-263 (2009).
10. M. Jones, S. S. Lo, G. D. Scholes, *J. Phys. Chem.* C **113**, 18632-18642 (2009).
11. R. Schrieber, H. Gareis, *Gelatin Handbook: Theory and Industrial Practice* (Wiley VCH, 2007).

Mater. Res. Soc. Symp. Proc. Vol. 1534 © 2013 Materials Research Society
DOI: 10.1557/opl.2013.312

Photoinduced Photoluminescence Enhancement in CdSe Quantum Dot – Polyvinyl Alcohol Composites

L. Borkovska[1], N. Korsunska[1], T. Stara[1], V. Bondarenko[1], O. Gudymenko[1], O. Stroyuk[2], O. Raevska[2] and T. Kryshtab[3]
[1] V. Lashkaryov Institute of Semiconductor Physics, NASU, pr. Nauky 41, 03028 Kyiv, Ukraine
[2] L. Pysarzhevsky Institute of Physical Chemistry, NASU, pr. Nauky 31, 03028 Kyiv, Ukraine
[3] Instituto Politécnico Nacional – ESFM, Av. IPN, Ed.9 U.P.A.L.M., 07738 Mexico D.F., Mexico

ABSTRACT

The effect of thermal annealing at 100 °C under irradiation by LED's light of 409 nm on the photoluminescence (PL) of polymer films of polyvinyl alcohol (PVA) and PVA films with embedded CdSe quantum dots (QDs) was investigated. X-ray diffraction study revealed that annealing increases a degree of crystallinity of the PVA matrix in the both films studied. In the PL spectra of the annealed QD-PVA composite, both an increase (up to thirteen- folds at room temperature) of the QD-related PL band and a small spectral shift toward the red spectrum region were found. The PL intensity at low temperatures and the activation energy of the PL thermal quenching also increased. In the PL spectra of pure PVA, a small decrease in the PL intensity and a red shift of the PL band were observed. The color of the composite film changed from yellow to slightly reddish with no changes in the PVA film transparency. The effect of PL enhancement in CdSe QD-PVA composite is ascribed to the improvement of QD surface defect passivation by the functional groups or the polymer chain fragments of PVA matrix caused by photochemical transformations in the polymer.

INTRODUCTION

An interest to polymers with embedded metal or semiconductors nanoparticles in device application is motivated by novel and distinctive properties of such composites resulted from the combination of intrinsic characteristics of polymers and the unique properties of nanoparticles [1]. Polyvinyl alcohol (PVA) is a well accepted human- and environmental-friendly polymer that is used in various pharmaceutical, medical, cosmetic, food and agricultural products [2]. It is also characterized by good thermo-stability, chemical resistance, and film forming ability [3]. Thus, PVA is considered as a good host material for metal and semiconductor nanoparticles. PVA-protected Ag [4], Pt [5] and Au [6] nanoparticles as well as ZnO [7], CdS[8] and CdSe [9, 10] quantum dots (QDs) have been reported. PVA is known to undergo photochemical transformations under γ-quanta or UV-light irradiation that change its mechanical, optical and luminescent characteristics [3]. However, the photoinduced evolution of the properties of PVA-based nanocomposites is less studied [8].

Here, we present the results of our investigations of the effect of visible light irradiation at elevated temperatures on the photoluminescence (PL) and structural properties of the nano-composite PVA films with embedded CdSe QDs.

EXPERIMENTAL DETAILS

The samples studied were the films of pure PVA as well as of CdSe QDs in PVA matrix deposited on a glass slide. For synthesis of QDs-polymer composite the reagents purchased from Aldrich were used without additional purification. CdSe QDs were produced by the reaction of Na_2SeSO_3 and $CdCl_2$ in aqueous solution of gelatin [11]. After the dialysis the QDs-gelatin solution was mixed with 20 % solution of PVA. To improve homogeneity of the polymer mixture as well as to maintain optical transparency of the films an HCl acid was added in the QDs-polymer solution. The average diameter of CdSe QDs obtained by this method is about 2.3 nm as estimated from spectral position of the absorption maximum (at about 2.5 eV at 300 K) in the optical absorption spectrum of QDs-PVA composite.

The films of pure PVA and of QDs-polymer composite were annealed at 100 °C for 5 minutes upon irradiation by light of 409 nm from a light-emitting diode (LED) in an atmospheric ambience. Before annealing the films were removed from the glass. The PL spectra were measured in the temperature range of 77-400 K. The PL was excited with a LED's light of 470 nm and recorded using a prism monochromator equipped with a photomultiplier and an amplifier with a synchronous detector. X-ray diffraction (XRD) study was realized using a D-8 ADVANCE (Bruker) one-crystal X-ray diffractometer operating with the Cu K_α - radiation.

RESULTS

The XRD patterns of the films of pure PVA and QDs-PVA composite are shown in figure 1. The patterns show the same diffraction peaks in both cases, with one dominating around $2\theta = 19.5°$ accompanied by the signals of much lower intensity. The peaks are evidently caused by PVA matrix of partially crystalline nature [3]. The peaks from crystalline CdSe phase are not observed apparently because of low QD concentration in the composite (less than 1 % w/v). In both the PVA and QDs-PVA films, the intensity of PVA-related peaks increases after annealing, indicating the increase of a degree of crystallinity of PVA matrix.

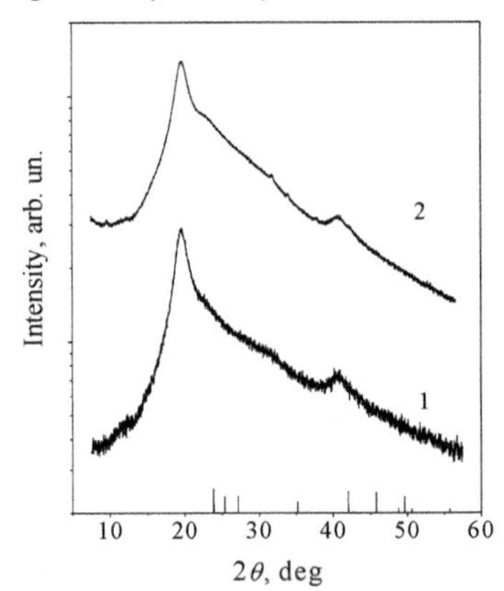

Figure 1. X-ray diffraction pattern of pure PVA film (1) and the composite film of CdSe QDs in PVA matrix (2). The line spectrum indicate the reflections for bulk hexagonal CdSe.

Room temperature PL spectra of an initial and annealed films of pure PVA and of QDs-PVA composite are presented in figure 2. The PL spectra of pure PVA film consist of a broad band centered near 2.5 eV originated from the functional groups of PVA. The PL spectra of the QDs-PVA composite typically show two bands. The high-energy PL band (at about 2.26 eV) can be ascribed to exciton radiative recombination or band-edge PL in the QDs while the low-energy PL band (at 1.79 eV) originates from the radiative recombination of carriers via deep levels of surface defects in the QDs [12]. A large contribution of deep trap emission in the PL spectrum is not surprising owing to small size of the QDs. This is often observed in bare CdSe QDs of small size passivated by organic capping groups and is explained by large surface-to-volume ratio and insufficient surface defect passivation by organic ligands [13]. The PL spectra similar to those presented in figure 2, but with a somewhat higher contribution of the exciton PL, have also been observed by us in CdSe QDs 2.4 nm in diameter embedded in gelatin matrix [14].

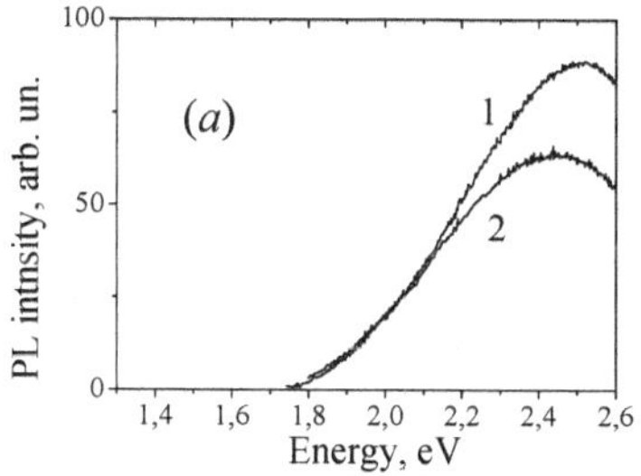

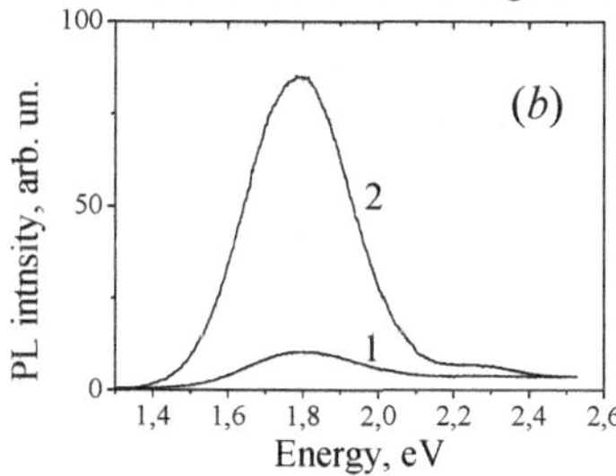

Figure 2. Photoluminescence spectra of pure PVA (a) and the composite of CdSe QDs in PVA matrix (b) before (curves 1) and after (curves 2) annealing at 100 °C under irradiation by LED's light of 409 nm, T=300 K, λ_{exc}=470 nm.

Figure 2 shows that annealing affects the PL spectra of pure PVA and of QDs-PVA composite in a different way. In the PL spectrum of pure PVA a red shift on about 70 meV of the PL band spectral position as well as a decrease in the PL intensity are found. In the PL spectrum of the QDs-PVA composite, the PL bands also shift a little towards red spectrum region, but the intensity of both band-edge and defect-related PL band increases considerably (up to thirteen-folds at room temperature).

The transparency of PVA film in the visible region does not change noticeably under annealing, while the color of the composite film changed from yellow to slightly reddish. The changes both in the PL spectra and in a color of the composite film are found to be irreversible. The changes occur only at the thermal annealing carried out under irradiation and not in the dark.

In order to have a deeper insight into the nature of the effects observed in the PL spectra of composite film, the temperature dependence of the QD photoluminescence intensity in the range of 77-400 K was studied. The intensity of defect-related PL band as a function of the film temperature in the Arrhenius plot measured before and after the thermal treatment is shown in figure 3. When the temperature increases from 77 up to about 300 K the PL intensity decreases owing to the thermal quenching. However, in the initial sample further temperature increase causes the increase of the PL intensity. This indicates that the effect of photoinduced PL enhancement in the QDs-PVA composite starts already at 300 K and occurs under irradiation by light of 470 nm LED too. After the film being annealed, only the decrease in the PL intensity is found as the temperature is elevated from 77 to 400 K.

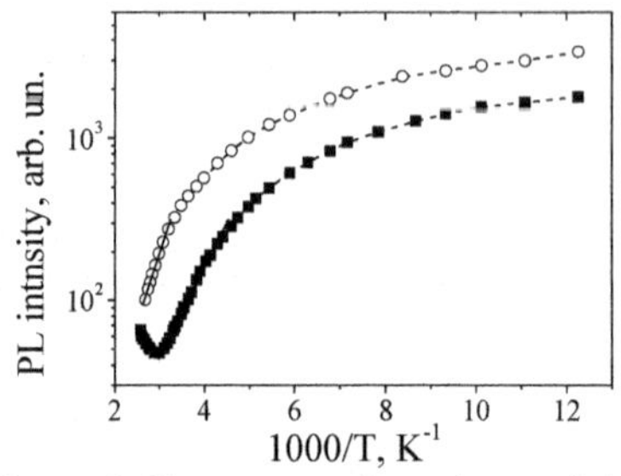

Figure 3. Temperature dependence of the PL intensity in QDs-PVA composite before (solid symbols) and after (open symbols) thermal annealing at 100 °C under irradiation by LED's light of 409 nm and their approximation by exponential dependence (dashed lines), λ_{exc}=470 nm.

Figure 3 shows that the composite annealing results in the increase of the PL intensity measured both at room and at liquid nitrogen temperatures. The PL intensity increase at room temperature is larger. In addition, a slope of the dependence in the temperature range of 250-350 K increases. In this range the Arrhenius plots can be approximated by a linear dependence indicating that the PL intensity W(T) varies exponentially with temperature T and can be fitted by a single exponential dependence:

$$W(T) = A + B\cdot\exp(-E_A/kT), \quad (1)$$

where A and B are the constants and E_A is the activation energy of PL thermal quenching. It is found that annealing increases the activation energy of thermal quenching of the intensity of defect-related PL band in 1.5-2 times in different samples. For the dependencies presented in figure 3 the E_A magnitude increases from 110 to 170 meV. These values are close to the E_A magnitudes found by us for CdSe QDs embedded in gelatin matrix, as well as to those observed by other authors for colloidal CdS or CdSe QDs embedded in inert matrices [14-16]. We are inclined to ascribe this activation energy to the height of potential barrier for carrier escape from the QD to the level of nonradiative defect on the QD surface.

DISCUSSION

The effect of the PL enhancement under continuous irradiation by visible light has been observed in colloidal QDs exposed to air, in QDs embedded in polymer and sol-gel glasses as well as in a single isolated QD on a glass substrate [17]. The effect was tentatively explained by different models: (i) suppression of a blinking effect from individual QDs as well as activation of non-emissive QDs caused by adsorption of oxygen and/or water molecules on the surface of QDs [17]; (ii) photo-induced recharging of the trapping centers on the QD surface that results in the change of electrical field affected exciton PL intensity [18]; (iii) passivation of non-radiative defects on the surface of the QDs caused by photo-induced rearrangement of surfactant molecules or other ligands [17]. In the first two models the effect of the PL enhancement is semi-reversible, but the effects found by us are irreversible. At the same time, the last model explains irreversible character of the PL enhancement. This model seems to be more relevant for the QDs embedded in solid films, and so we will discuss our results within this model.

In the QDs-polymer composite, the individual functional groups of polymer as well as the fragments of the polymer chains can passivate under-coordinated surface atoms of the QD [1]. The passivation of non-radiative defects on the QD surface should increase the QD photoluminescence intensity both at low and room temperatures. Moreover, charged organic surface ligands can increase the height of potential confining barrier for charge carriers in the QDs [14, 19] and increase in such a way the PL intensity at higher temperatures (where PL thermal quenching occurs). Therefore, the effects of the PL enhancement in the QDs-PVA composite films studied can be caused by enhancement of QD surface defect passivation by the functional groups or the whole fragments of PVA polymer chains stimulated by thermal annealing upon irradiation.

It is known that UV-light irradiation of PVA results in photochemical transformations in the polymer [3]. The initial stage of the photochemichal process in PVA is assumed to be an elimination of hydrogen atoms and formation of free radicals. In air, an UV-light irradiation results in oxidation of hydroxyl- groups to carbonyl- ones followed by splitting of polymer chains. At elevated temperatures various transformations of free radicals give rise to partial cross-linking of polymer chains as well as to the formation of polyene units. Both carbonyl- and polyene-groups contribute to coloration of the polymer. Therefore, a color change of the QDs-PVA composite can be due to formation of new chromophores in PVA matrix. However, the transparency of pure PVA film does not change upon the annealing. The possible reasons are:

(i) the energy of quanta of exciting light is enough to produce new chromophore groups in QDs-PVA composite, but not enough for that in pure PVA matrix. In fact, the photochemical transformations resulted in formation of new chromophores in pure PVA are usually observed upon UV-irradiation. At the same time, in the QDs-PVA composite an exciting light is absorbed not only in PVA molecules, but also in CdSe QDs. The photocarriers generated in the QDs can initiate the reaction of radical displacement in polymer near the QD/PVA interface [8]. In a similar way, in [20] an enhancement of the photooxidation process in PVA in the presence of collagen was explained by absorption of UV- light by chromophores in collagen and transfer of the energy to PVA molecules. Besides, the incorporation of CdSe nanocrystals into PVA can lead to the formation of vacancy defects at the polymer chain ends. This can promote detachment of hydrogen atoms and formation of radicals. The similar processes have been observed, for example, in PVA compositions doped with inorganic salts [21].

(ii) the addition of HCl to the QDs-polymer solution during fabrication of the composite can accelerate the reaction dehydrogenation and dehydration in PVA. In the places, where hydrogen atoms and hydroxyl groups are detached from polymer, the double bonds of C=C can be formed and coloration of polymer can occur.

Hence, we can suppose the next model of photoinduced PL enhancement in CdSe QDs-PVA composite. In the initial film, the QDs surface defects are partially passivated by functional groups of PVA and gelatin molecules. Dissociation of these bonds upon heating occurs. At the same time annealing of the composite under irradiation stimulates the formation of free radicals in PVA matrix, most likely near the QD/PVA interface, as well as their secondary reactions. This can result in partial cross-linking of polymer molecules and the formation of double bonds in polymer chains. Such a treatment can stimulate also the formation of stronger bonds of non-coordinated QD surface atoms with functional groups or the fragments of the polymer chains of PVA matrix and the increase of the height of barrier for carrier escape from the QDs.

CONCLUSIONS

The irreversible effect of photoinduced photoluminescence enhancement in the QDs-PVA composite is revealed. The effect is stimulated by the thermal annealing of composite at 100 °C under irradiation by light of 409 or 470 nm LED. The effect consists in the increase of intensity of QD-related PL (up to thirteen-folds at room temperature) and is accompanied by the increase of the activation energy of the PL thermal quenching as well as by a slight darkening of the film. The effect is not observed in the films of pure PVA subjected to the same treatment. The effect is supposed to arise from photochemical transformations occurred in a polymer matrix of composite and is assumed to result in improvement of QD surface defect passivation by the functional groups or the fragments of PVA chains.

ACKNOWLEDGMENTS

This work was partially supported by the National Academy of Sciences of Ukraine through the program "Nanotechnologies and Nanomaterials" (Grant N 2.2.1.14/26) and the project "Physical and Physical-Technological Aspects of Fabrication and Characterization of Semiconductor Materials and Functional Structures for Modern Electronics" (Grant N III-41-12).

REFERENCES

1. N. Tomczak, *et. al.*, *Progress in Polymer Science* **34**, 393 (2009).
2. C. C. DeMerlis, and D. R. Schoneker, *Food Chem. Toxicol.* **41**, 319 (2003).
3. J.G. Pritchad, in *Polyvinyl Alcohol. Basic Properties and Uses* (Gordon and Breach, 1970).
4. S. Porel, *et. al.*, *Chem. Mater.* **17**, 9 (2005).
5. Y. Luo, and X. Sun, *Mater. Lett.* **61**, 2015 (2007).
6. C. Sun, *et. al.*, *J. Nanopart. Res.* **11**, 1005 (2009).
7. X. M. Sui, C. L. Shao, and Y. C. Liu, *Appl. Phys. Lett.* **87**, 113115 (2005).
8. G. Yu. Rudko, *et. al.*, *Adv. Sci. Eng. Med.* **4**, 394 (2012).
9. H. S. Mansur, and A. A.P. Mansur, *Mat. Chem. Phys.* **125**, 709 (2011).
10. Y. Azizian-Kalandaragh, and A. Khodayari, *Mat. Sci. Semicond. Proc.* **13**, 225 (2010).
11. E. Raevskaya, A. L. Stroyuk, and S. Ya. Kuchmiy, *J. Colloid Interface Sci.* **302**, 133 (2006).
12. A.R. Kortan, R. Hull, R.L. Opila, M.G. Bawendi, M.L. Steigerwald, P.J. Carroll, and L.E. Brus, *J. Am. Chem. Soc.* **112**, 1327 (1990).
13. B. O. Dabbousi, *et. al.*, *J. Phys. Chem. B* **101**, 9463 (1997).
14. L. V. Borkovska, N. O. Korsunska, T. R. Stara, V. M. Dzhagan, O. L. Stroyuk, O. Ye. Raevska, and T. G. Kryshtab, *Phys. Status Solidi C* **9**, 1779 (2012).
15. J. Zhao, *et. al.*, *J. Lumin.* **66/67,** 332 (1996).
16. D. Valerini, A. Creti, M. Lomascolo, L. Manna, R. Cingolani, and M. Anni, *Phys. Rev. B* **71**, 235409 (2005).
17. C. Carrillo-Carrion, *et. al.*, Chem. Commun. 5214 (2009).
18. N. E. Korsunska, M. Dybiec, L. Zhukov, S. Ostapenko, and T. Zhukov, *Semicond. Sci. Technol.* **20**, 876 (2005).
19. S. Sarhar, N. Chandrasekharan, S. Gorer, and G. Hodes, *Appl. Phys. Lett.* **81**, 5045 (2002).
20. A. Sionkowska, A. Planecka, J. Kozlowska, J.Skopinska-Wisniewska, *Polymer Degradation and Stability* **94**, 383 (2009).
21. E. N. Suchkova, and A. B. Pagubko, *Russian Physics Journal*, **51**, 633 (2008).

Mater. Res. Soc. Symp. Proc. Vol. 1534 © 2013 Materials Research Society
DOI: 10.1557/opl.2013.313

Emission modification in ZnO nanosheets at thermal annealing

Aaron I. Diaz Cano*[1], Brahim El Filali[1], Tetyana V. Torchynska[2] and Jose L. Casas Espinola[2]
[1]UPIITA – Instituto Politécnico Nacional, México D. F. 07738, México.
[2]ESFM- Instituto Politécnico Nacional, México D. F. 07738, México.

ABSTRACT

Photoluminescence (PL) and its temperature dependences, as well as the X ray diffraction (XRD), have been studied in the freshly prepared amorphous phase ZnO nanosheets, obtained by the electrochemical (anodization) method, and in the crystalline annealed ZnO nanosheets. The freshly prepared samples have been divided in two groups. One of these groups has been annealed at 400 °C for 2 hours in ambient air. Defect related PL bands with the peaks at 2.10-2.13, 2.42-2.46 and 2.65-2.69 eV are detected in amorphous state. Appreciable changes in the size of nanosheets as a function of thermal treatments have been revealed. XRD study has shown that annealing stimulates the Zn oxidation and the creation of ZnO with a wurtzite crystal lattice. In crystalline ZnO seven PL bands appeared with the PL peaks 1.46, 1.58, 2.02, 2.43, 2.70, 2.93 and 3.16 eV at 10K. The reasons of emission transformation and the nature of optical transitions related to the studied PL bands have been discussed. It is shown that the anodization method permits by a controllable way to obtain the wide range ZnO emission that is interesting for the future applications in room temperature "white" light-emitting diodes.

INTRODUCTION

ZnO nanostructures have attracted considerable attention due to their promising applications in electronic, photoelectronic and sensing devices, mainly due to the potential for engineering the properties not obtained in the bulk materials. The past few years have witnessed a much progress in the synthesis of ZnO nanostructures that promises their applications in optical and electronic devices [1-4], ZnO nanowall and nanowire arrays can be used as room temperature white light-emitting diodes and ultraviolet nanolasers [5], ZnO nanoneedles possess excellent field emission performance [6], ZnO nanorods were used for high performance field–effect transistors [7]. ZnO nanostructures have attracted attention as well owing their possible applications in the low voltage and short-wavelength (368 nm) electro-optical devices, transparent ultraviolet (UV) protection films, gas sensors, and even spintronic devices [8].

Various methods to synthesis of ZnO nanostructures have been employed such as sol–gel methods [9, 10], spray pyrolysis [11, 12], metal–organic chemical vapor deposition [13, 14], rf sputtering [15, 16], thermal evaporation [17], pulsed laser deposition [18], chemical vapor deposition [19], catalysis driven molecular beam epitaxy [20] and sonochemical method [21]. Electrochemical (anodization) method has been recognized as one of the most simple, economic, non-vacuum and low temperature effective method to prepare the nanomaterials. ZnO nanosheets prepared by this method are of particular interest because of their high porosity and large surface area, which are crucial for optimized performance of dye-sensitized solar cells, sensors and hydrogen devices [8].

Even high-quality ZnO samples prepared by this method contain a variety of point defects which contribute to the visible luminescence. A typical photoluminescence (PL) spectrum of undoped crystalline ZnO contains sharp and intense exciton related lines in the UV region with one or more broad defect related PL bands in the visible range. The visible luminescence in ZnO has been studied intensively during the last two decades [22]. However, the majority of these defects remain unidentified and optical transitions responsible for the PL bands need to be verified. Actually it is not clear the relations between the defects appeared and the structural properties of ZnO nanosystems (nanosheets, nanorods etc.). In this work ZnO nanosheets were prepared by the anodization method and their optical properties have been studied before and after annealing.

EXPERIMENTAL DETAILS

The Zn foils were electrochemically anodized in a two electrode system with both Zn electrodes exposed in the electrolyte. The distance between them was 10 mm. The pieces of Zn foil (Aldrich 99.99% purity, 0.25 mm thickness) with a radius of 6 mm were degreased in acetone and ethanol for 15 min with ultrasonic cleaning and washed with deionized water. All electrolytes were prepared from chemical reagents (Aldrich) and deionized water. The aqueous solution used as electrolyte was a 1:10 volume mixture of HF and deionized water without any surfactant. The reaction time used was 1, 6 to 10 min at the voltage of 5 V. After that, white ZnO nanosheets were produced on the Zn foil. Then ZnO nanosheets were washed in deionized water to eliminate remainders and dried in air at room temperature. The part of samples was annealed at 400 °C for 2 hours in ambient air and other part of these samples without annealing has been left as a reference object).

The crystal structure of prepared ZnO nanosheets was investigated by X-ray diffraction (XRD) using the XRD equipment model XPERT MRD with the Pixel detector, three axis goniometry and parallel collimator with the resolution of 0.0001 degree. XRD beam was from the Cu source, $K_{\alpha 1}$ line λ=1.5406 A, and the angles used were from 20° to 80° with a step size of 0.05° and step time of 10 s. PL spectra were measured at the excitation by a He-Cd laser with a wavelength of 325 nm and a beam power of 80 mW using a spectrometer SPEX500 and a photomultiplier described in [23, 24].

DISCUSSION

The XRD study has shown that the freshly prepared ZnO nanosheets are characterized by an amorphous phase. Simultaneously very small Zn substrate related peaks at the angles 2θ equal to 38.993, 43.233 and 70.058 degrees have been detected as well (Fig.1a). These peaks correspond to the X- ray diffraction from the (100), (101) and (103) crystal planes, respectively, in the hexagonal Zn crystal lattice with the lattice parameters of a=2.6650Å and c=4.9470Å [25].

The annealing at 400 °C for 2 hours in ambient air stimulates the process of ZnO oxidation and crystallization. As it is clear from figure 1b, the conversion from the amorphous to the polycrystalline structure takes place at annealing. A set of XRD peaks appears after the annealing at the angles 2θ equal to 31.770, 34.422, 36.253, 47.540, 56.604 and 62.865 degrees (Fig.1b). The comparison of XRD peak positions with the available data base [25] has shown that these peaks correspond to the X- ray diffraction from the (100), (002), (101), (102), (110) and (103) crystal planes, respectively, in the wurtzite ZnO crystal structure with the hexagonal lattice parameters of a=3.2498Å and c=5.2066Å. In the samples obtained with the anodization duration of 6 and 10 min the volume of ZnO phase enlarges that manifests itself in increasing the XRD peak intensities.

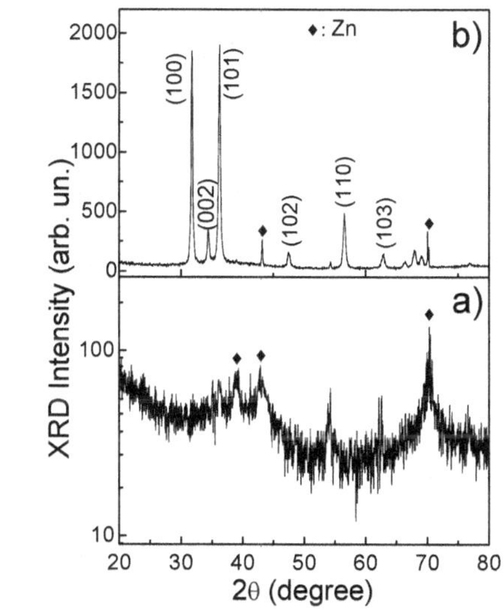

Figure 1. XRD results for the freshly prepared amorphous (a) and annealed crystalline (b) ZnO nanosheets obtained at the anodization duration of 10 min.

PL spectra of the freshly prepared ZnO nanosheets are shown in figure 2 for samples obtained at different anodization durations. It is clear that the PL spectra are complex and can be presented as a superposition of, at least, three elementary PL bands with the peaks at 2.10-2.13, 2.42-2.46 and 2.85-2.93 eV (Fig.2, curves a,b,c). However, the mechanism of optical transitions and the nature of defects for these PL bands are still not fully understood. Several authors assigned the green and yellow PL bands to oxygen vacancies V_O [26], zinc vacancies V_{Zn} [26,27] and oxygen interstitial O_i [28]. With increasing the anodization duration up to 6 min the PL intensity of mentioned peaks enlarges due to increasing the ZnO phase volume. Also the production of various defects, which were considered as an origin of green PL band, increases versus anodization time. The PL intensity in the ZnO samples obtained at the anodization time of 10 min falls down because decreasing the ZnO volume of nanosheets.

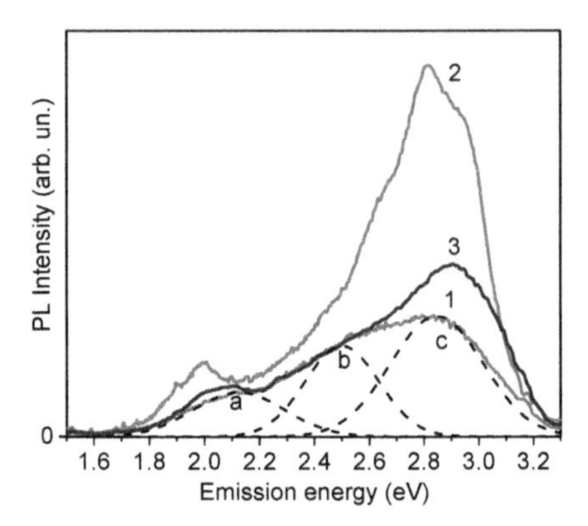

Figure 2. PL spectra of fresh ZnO nanosheets obtained at the anodization durations of 1 min (1), 6 min (2) and 10 min (3), the voltage is 5V. The curves (a), (b) and (c) correspond to PL bands obtained by the deconvolution of the curve 1.

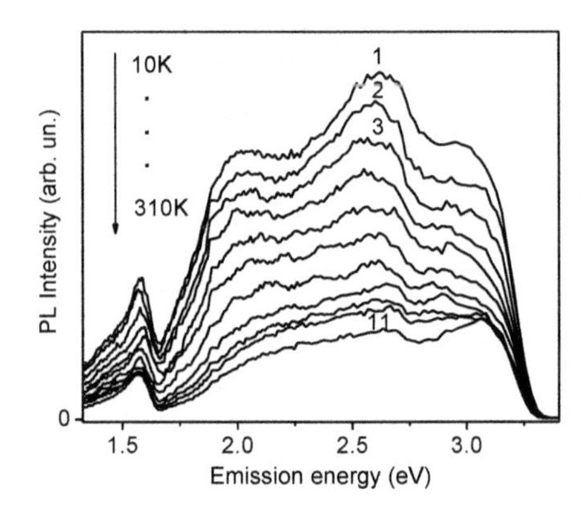

Figure 3. PL spectra measured at different temperatures for the annealed ZnO sample obtained at the anodization time 6 min and the voltage 15V

The amorphous ZnO phase disappeared after the annealing at 400 °C for 2 hours in ambient air in the process of ZnO crystallization (Fig.1b). The last process is accompanied by the transformation of PL spectra as well (Fig.3). PL spectra for four definite temperatures in more details are presented in figure 4. It is clear from figure 4 that the PL spectra could be decomposed on seven elementary PL bands with the peaks at 1.46, 1.58, 2.02, 2.43, 2.70, 2.93

and 3.16 eV at 10K. In crystalline ZnO nanosheets two new PL bands at 1.46 and 1.58 eV have appeared (Fig.3 and Fig.4). The temperature dependences of PL peak intensities for all elementary PL bands are shown in figure 5.

The 2.02 eV and 2.43 PL band have been attributed earlier to the oxygen defects such as oxygen vacancies (2.02 eV) and interstitial oxygen ion centers (band located around 2.4 eV) respectively [6]. Although there have been controversies for a long time about the origin of the peak at 2.70 eV, it is usually assigned to the interstitial zinc (Zn_i) [17]. However other authors attributed this PL band to the recombination in donor acceptor pairs including the shallow donor and oxygen vacancy [29]. The band centered at 2.93 eV disappeared completely at higher temperatures and it is related, apparently, to the V_{Zn}^- center [17]. The UV band at 3.16 eV is attributed to exciton related emission [30].

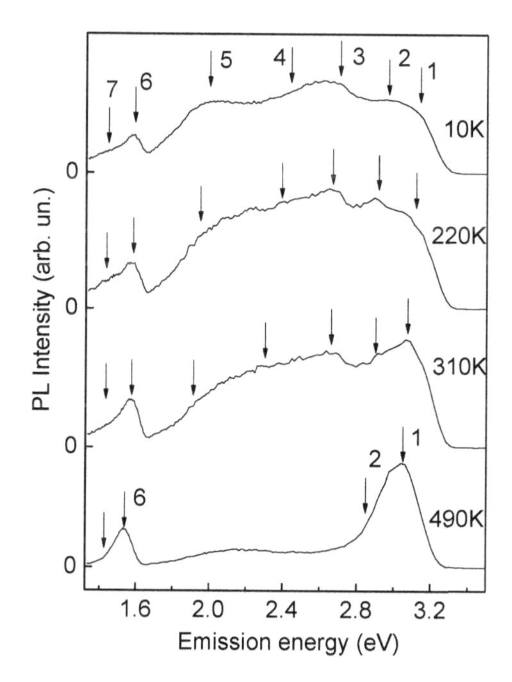

Figure 4. PL spectra for 4 definite temperatures of the sample obtained at the anodization of 6 min at 15V in the temperature range from 10 to 490 K.

To understand the nature of the PL bands centered at 1.46 and 1.58 eV, let us analysis figure 5. In this figure we observed that the temperature behavior of the PL bands is the same for the PL bands with the peaks centered at 1.58 and 3.16 eV as well as at 1.46 and 2.93 eV. This fact could be interpreted that the PL bands centered at 1.46 and 1.58 eV are the second order diffraction peaks of the 2.93 and 3.16 eV PL bands, respectively [31].

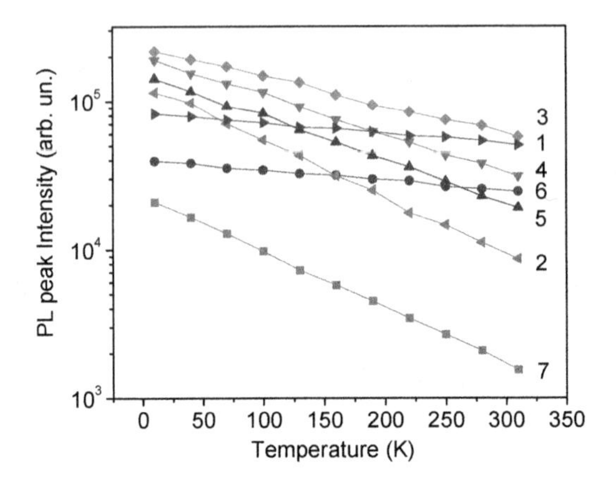

Figure 5. PL peak intensity dependences versus temperature for the PL bands centered at: 1-3.16eV, 2-2.93eV, 3-2.70eV, 4-2.43eV, 5-2.02eV, 6-1.58eV and 7-1.46eV at 10K for the ZnO sample presented in figure 3.

CONCLUSIONS

XRD study shows that the annealing stimulates the process of ZnO oxidation and crystallization. The transformation of defect-related PL bands has been studied in as grown and annealed ZnO samples. Seven elementary PL bands with the peaks at 1.46, 1.58, 2.02, 2.43, 2.70, 2.93 and 3.16 eV at 10K have been detected. The nature of elementary PL bands has been discussed. PL bands with the peaks centered at 1.58 eV and 1.46 eV could be interpreted as the second order diffraction peaks of the 2.93 and 3.16 eV PL bands, respectively. The anodization method is one of the simplest technological methods to prepare ZnO nanosheets and this method permits to obtain the wide range of emission in comparison with another technologies.

ACKNOWLEDGMENTS

The authors would like to thank the CONACYT (project 130387) and SIP-IPN, Mexico, for the financial support, as well as the CNMN- IPN for XRD measurements.

REFERENCES

1. RaminYousefi, A.K.Zak, Materials Science in Semiconductor Processing, **14**, 170–174 (2011).
2. Jianling Zhao, Xixin Wang, Jijing Liu,Yanchao Meng, Xuewen Xu, Cheng chunTang, Materials Chemistry and Physics, **126**, 555–559 (2011).
3. Zhi feng Liu, Lei E, Jing Ya,Ying Xin, Applied Surface Science, **255**, 6415–6420 (2009).

4. P.G. Li, W.H. Tanga, X. Wang, J. of Alloys and Compounds, **479**, 634-637 (2009).
5. A. B. Djurisic, A. M. C. Ng, X. Y. Chen, Progress in Quantum Electronics, **34**, 191–259 (2010).
6. C. E. Secu, Mariana Sima, Optical Materials, **31**, 876–880 (2009).
7. Zhiyong Fan and Jia G. Lu, Applied Physics Letters, **86**, 032111 (2005).
8. Chandra Sekhar Rout, S. Hari Krishna, S. R. C. Vivekchand, A. Govindaraj and C. N. R. Rao, Chemical Physics Letters, **418**, 586–590 (2006).
9. P. Hoyer, H. Weller, Chem. Phys. Lett., **221**, 379 (1994).
10. L. Spanhel, M. A. Anderson, J. Am. Chem. Soc.,**113**, 2826 (1991).
11. S. A. Studenikin, N. Golego, M. Cocivera, J. Appl. Phys., **84**, 2287 (1998).
12. A. Ortiz, C. Falcony, J. Hernandez, A. M. Garcia, J. C. Alonso, Thin Solid Films, **293**, 103 (1997).
13. T. Maruyama, J. Shionoya, J. Mater. Sci. Lett., **11**, 170 (1992).
14. S. Bethe, H. Pan, B. W. Wessels, Appl. Phys. Lett., **52**, 138 (1988).
15. L. Vasanelli, A. Valentini, A. Losacco, Sol. Energy Mater., **16**, 91 (1987).
16. Y. Sato, S. Sato, Thin Solid Films, **281**, 445 (1996).
17. J. Ma, F. Ji, H.-L. Ma, S. Li, Thin Solid Films, **279**, 213 (1996).
18. T. Premkumar, Y. S. Zhou, Y. F. Lu, K. Baskar, Appl. Mater. Interfaces, **2**, 2863–2869 (2010).
19. T. Tan, X. W. Sun, X. H. Zhang, B. J. Chen, S. J. Chua, A. Yong, Z. L. Dong, X. Hu, J. Cryst. Growth, **290**, 518-522 (2006).
20. Y. W. Heo, V. Varadarajan, M. Kaufman, K. Kim, D. P. Norton, F. Ren, P. H. Fleming, Appl. Phys. Lett., **81**, 3046 (2002).
21. S.-H. Jung, E. Oh, K.-H. Lee, Y. Yang, C. G. Park, W. Park, S.-H. Jeong, Cryst. Growth Des., **8**, 265 (2008).
22. Sung-Sik Chang, Sang Ok Yoon, Hye Jeong Park, Akira Sakai, Materials Letters, **53**, 432-436 (2002).
23. M. Dybic, S. Ostapenko, T.V. Torchynska, E. Velazquez Lozada, Appl. Phys. Lett. **84** (25), 5165-5167 (2004).
24. T. V. Torchynska, A.I. Diaz Cano, M. Dybic, S. Ostapenko, M. Mynbaeva, Physica B, Condensed Matter, **376-377**, 367-369 (2006).
25. PDF2 XRD database, card no. 36-1451.
26. Han N. S., Shim H. S., Seo J. H., Kim S. Y., Park S. M. and Song J. K., J. Appl. Phys., **107**, 084306 (2010).
27. Zhang S. B., Wei S. H. and Zunger A., Phys. Rev. B, **63**, 075205 (2001).
28. Vanheusden K., Seager C. H., Warren W. L., Tallant D. R. and Voigt J. A., Appl. Phys. Lett., **68**, 403 (1996).
29. D. H. Zang, Z. Y. Xue, Q. P. Wang, J. Phys. D : Appl. Phys. **35** (21) 2837-2840 (2002).
30. Srimala Sreekantan, Lee Ren Gee, Zainovia Lockman, J. of. Alloys and Compounds. **476**, 513-518 (2009).
31. J. Lv, Ch. Liu, W. Gong, Zh. Zi, X. Chen, K. Huang,T. Wang, G. He, Sh. Shi, X. Song, Zh. Sun, Optical Materials, **34**, 1917 (2012).

Mater. Res. Soc. Symp. Proc. Vol. 1534 © 2013 Materials Research Society
DOI: 10.1557/opl.2013.314

Emission Variety in ZnO Nanocrystals Obtained At Different Growth Temperatures

Erick Velázquez Lozada[1] and Luis Castañeda[2]
[1]ESIME – Instituto Politécnico Nacional, México D. F., 07738, MEXICO
[2]Instituto de Física, Benemérita Universidad Autónoma de Puebla, 72570, MEXICO

ABSTRACT

Scanning electronic microscopy, X ray diffraction and photoluminescence have been applied to the study of structural and optical properties of ZnO nanocrystals prepared by the ultrasonic spray pyrolysis at different temperatures. The variation of temperatures and times at the growth of ZnO films permits modifying the ZnO phase from the amorphous to crystalline, to change the size of ZnO nanocrystals, as well as to vary their photoluminescence spectra. The study has revealed three types of PL bands in ZnO NCs: defect related emission in visible spectral range, the near-band-edge PL, related to the LO phonon replica of free exciton recombination, and IR emission that, apparently, is the second-order diffraction peaks of near band edge PL. The PL bands, related to the LO phonon replica of free exciton, in room temperature Pl spectrum testify on the high quality of ZnO films prepared by the ultrasonic spray pyrolysis technology.

INTRODUCTION

Nanocrystalline Zinc oxide (ZnO) with wide band gap energy nearly 3.37 eV, high exciton binding energy (60 meV at 300K) and easy way of nanostructure preparation has attracted great attention during the last two decades [1]. In addition to exceptional exciton properties, ZnO possesses a number of deep defects that emit in whole visible range and, hence, can provide intrinsic "white" light emission. ZnO nanocrystals (NCs) are being investigated as promising candidates for different optoelectronic applications, such as: the non-linear optical devices [2], light-emitting devices [3-6], transparent electrodes for solar cells [7] and laser diodes [8], as well as for the excellent field emitters [9], electrochemical sensors and toxic gas sensors [10]. The control of the ZnO defect structure in these nanostructures is a necessary step in order to improve the device quality. Since the structural imperfection and defects generally deteriorate the exciton related recombination process, it is necessary to grow the high quality films for efficient light-emitting applications. The ultrasonic spray pyrolysis (USP) offers many advantages such as easy compositional modifications; easy introducing the various functional groups, relatively low annealing temperatures and the possibility of coating deposition on a large area substrate. It will be interesting to study the optical emission of USP produced ZnO NCs doped by Ag in order to identify the best regimes for obtaining the bright emitting nanosystems.

EXPERIMEMTAL DETAILS

ZnO:Ag thin solid films were prepared by the USP technique (Fig. 1) on the surface of soda-lime glass substrate for the two substrate temperatures (400 and 450 °C) and different deposition times (Table 1).

Table 1. Technological regimes and NC parameters from SEM images

Sample numbers	Growth temperature (°C)	Deposition time (min)	NC width (nm)
1	400	10	150-200
2	400	5	100-150
3	400	3	50-70
4	450	10	200-250
5	450	5	120-180
6	450	3	80-110

The deposition system presented in figure 1 includes a piezoelectric transducer operating at variable frequencies up to 1.2 MHz and the ultrasonic power of 120 W. ZnO:Ag thin solid films were deposited from a 0.4 M solution of zinc (II) acetate [$Zn(O_2CCH_3)_2$] (Alfa), dissolved in a mix of deionized water, acetic acid [CH_3CO_2H] (Baker), and methanol [CH_3OH] (Baker) (100:100:800 volume proportion). Separately, a 0.2 M solution of silver nitrate [$Ag(NO_3)$] (Baker) dissolved in a mix of deionized water and acetic acid [CH_3CO_2H] (Baker) (1:1 volume proportion) was prepared, in order to be used as doping source. A constant [Ag]/[Zn] ratio of 2 at. % was applied at the ZnO Ag film preparation.

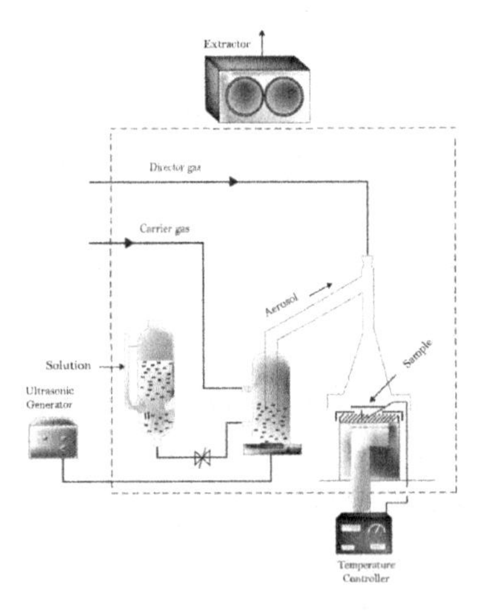

Figure 1. Schematic diagram of the experimental setup used for the deposition of the ZnO:Ag films by the ultrasonic spray pyrolysis method.

The morphology of ZnO:Ag films has been studied by secondary electrons signals using the scanning electron microscopy (SEM) Dual Beam, FEI brand, model Quanta 3D FEG with field emission gun. PL spectra were at the excitation by a He-Cd laser with a wavelength of 325 nm and a beam power of 20 mW at 300K using a PL setup on a base of spectrometer SPEX500 described in [11, 12]. The crystal structure of ZnO:Ag films was investigated by the X-ray diffraction (XRD) using the diffractometer XPERT MRD with the Pixel detector, three axis goniometry and parallel collimator with the resolution of 0.0001 degree. XRD beam was from the Cu source, Kα1 line λ=1.5418 Å, 45 kV, 40 mA and the angles used were from 20° to 80° with a step size of 0.05° and step time of 100 s.

EXPERIMENTAL RESULTS AND DISCUSSION

SEM images of the typical ZnO:Ag NCs obtained at the deposition times of 3 and 10 min for two substrate temperatures are presented in figure 2. It is clear that the ZnO NCs have the hexagonal cross section and the road orientation along the c axis. The cross section of ZnO NCs increases with the temperature and durations of the UPS process (Table 1).

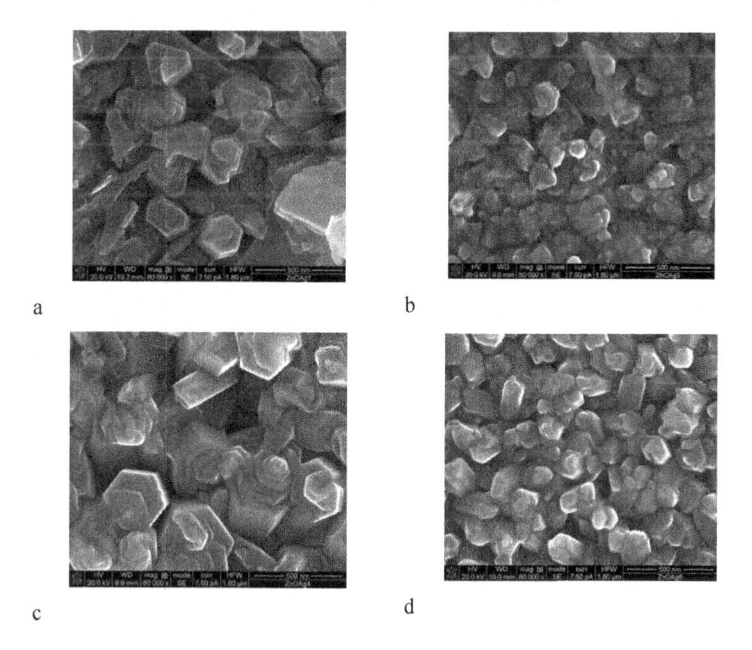

Figure 2. SEM images of the samples prepared at the substrate temperatures 400°C (a,b) and 450°C (c,d) and the durations of 10 (a, c) and 3 (b,d) min.

X-ray diffraction patterns of ZnO:Ag NCs obtained on the substrate with the temperature of 400°C and different deposition durations are shown in Fig. 3. At low deposition time (3min) the ZnO films showed an amorphous phase mainly with very small (1 0 0), (0 0 2) and (1 0 1) peaks. These peaks are the evidence of starting the conversion of an amorphous phase into a polycrystalline one. It was observed that with increasing the deposition duration from 3min to 10min, a set of new peaks appears, which correspond to the X- ray diffraction from the (100), (002), (101), (102), (110), (103) and (200) crystal planes (the angles 2θ equal to 31.770, 34.422, 36.253, 47.540, 56.604, 62.865 and 68.709 degrees), respectively, in the wurzite ZnO crystal structure with the hexagonal lattice parameters of a=3.2498A and c=5.2066A [13]. The (0 0 2) reflection peak is more intense, as compared to the other peaks, indicating a preferential c-axis orientation of ZnO:Ag NCs.

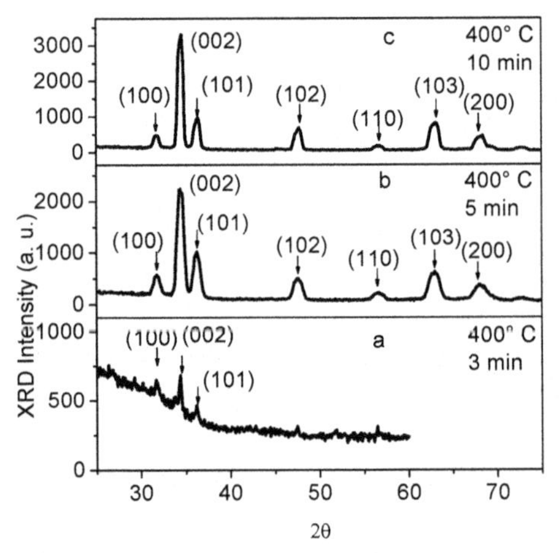

Figure 3. XRD diagrams of studied samples prepared on the substrate with T = 400° C at the deposition times of 3(a), 5(b) and 10 (c) min.

PL spectra of ZnO:Ag NCs are shown in figure 4. It is clear that the PL spectra are complex and can be represented as a superposition of elementary PL bands with the peaks in the spectral ranges: 3.14 eV (I), 2.00-2.70 eV (II) and 1.57 eV (III) (Fig.4). It is known that the UV-visible PL bands in ZnO owing to the near-band-edge (NBE) or exciton (I) and defect-related (II) recombination [14-17]. With increasing the substrate temperature NBE related PL bands in the range I enlarged mainly, in comparison with defect related (II) PL bands (Fig.4). At the same time the PL peak position of defect related PL bands shifts into the high energy (to 2.5 eV). With increasing the USP durations the intensity of defect related PL bands (II) raises mainly in comparison with those of NBE PL bands (Fig.4).

A great variety of luminescence bands in the UV and visible spectral ranges have been detected in the ZnO crystals [14]. The origin of these emissions has not been conclusively established. The NBE emission at 3.0-3.37 eV is attributed to the free (FE) or bound (BE) excitons, their LO phonon replicas, such as FE-1LO or FE-2LO, to optical transition between the free to bound states, such as the shallow donor and valence band, or to donor-acceptor pairs [14].

The blue PL band with the peak at 2.75-2.80 eV is attributed to Cu related defects [18], to Zn interstitials [19] or to donor-acceptor pairs including the shallow donor and oxygen vacancy [20]. The defect related green PL band in the spectral range 2.20-2.50 eV in ZnO is assigned ordinary to oxygen vacancies [19], Cu impurities [18] or surface defects [21]. The orange PL band with the peak at 2.02-2.10 eV was attributed earlier to oxygen interstitial atoms (2.02 eV) [22] or to the hydroxyl group (2.10eV) [23]. Taking into account that the PL intensity of 2.25 eV PL band increased with raising the USP duration the assumption that the corresponding defects are related to oxygen vacancies looks very reliable.

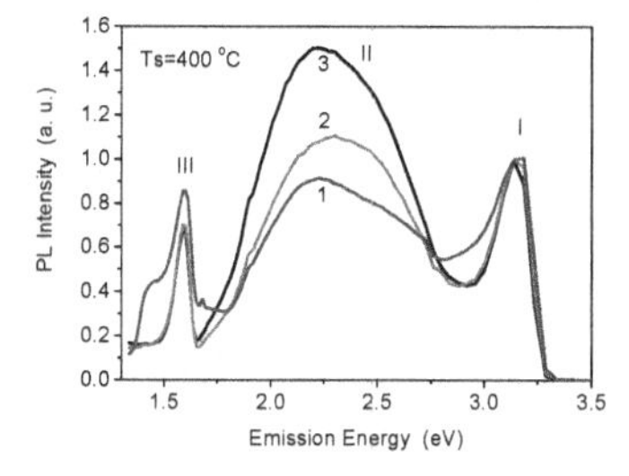

Figure 4a. PL spectra of samples prepared on the substrate with T = 400° C at the deposition times: 1-3 min, 2-5min, 3-10min.

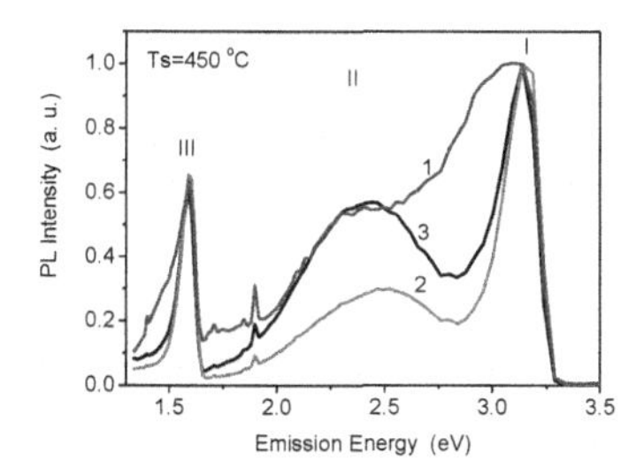

Figure 4b. PL spectra of samples prepared on the substrate with T = 450° C at the deposition times: 1-3 min, 2-5min, 3-10min.

Studied ZnO films were doped by Ag and, therefore, have the acceptor type defects, Ag_{Zn}, which were formed when the Ag atoms substitute of Zn atoms in the ZnO crystal lattice. Emission of the acceptor type defects in ZnO has been studied intensively during the last two decades as well. Note, the broad PL bands with the peaks at 3.238 and 3.315 eV at 4.3K earlier have been observed in acceptor N doped MBE ZnO crystals [24]. These peaks were assigned to acceptor (N_o) bound exciton (3.315 eV), to donor-acceptor pair emission, involving N_o acceptor, or to the LO-phonon replica (3.238eV) of the donor bound exciton line [24]. With temperature increasing the NBE intensity of acceptor bound exciton falls down due to the dissociation of bound excitons and the FE band with it's LO replicas dominates in the PL spectrum. Thus in our case the 3.14 eV PL band in the room temperature PL spectrum can be attributed to the LO phonon replica of FE emission. Note that the variation of PL intensity of the 1.57 eV PL band correlate with the intensity variation of the 3.14 eV PL band that permits assigning the 1.57 eV PL band to the second-order diffraction peak of 3.14 eV PL band.

CONCLUSIONS

ZnO:Ag NCs with hexagonal structures have been successfully synthesized by the USP method. With increasing the substrate temperature at USP up to 450°C the PL intensity of NBE related emission bands has enlarged. The study has revealed three types of PL bands in the room temperature PL spectra related to the LO replicas of FE and its second-order diffraction peak, as well as to the defect-related PL band, apparently, connected with oxygen vacancies. The PL band related to the LO phonon replicas of free exciton in PL spectra at room temperature testify on the high quality of the ZnO:Ag films prepared by USP.

The authors would like to thank the CONACYT (project 130387) and SIP-IPN, Mexico, for the financial support, as well as the CNMN-IPN for SEM and XRD measurements and the Dr. J.L. Casas Espinola for the PL measurements.

REFERENCES

1. S.J. Pearton, D.P. Norton, K. Ip, Y.W. Heo, T. Steiner, Prog. Mater. Sci. **50**, 293 (2005).
2. M.H. Koch, P.Y. Timbrel1, R.N. Lamb, Semicond. Sci. Technol. **10**, 1523 (1995).
3. K. Vanheusden, C.H. Seager, W.L. Wareen, D.R. Tallant, J. Caruso, M.J. Hampden-Smith, T.T. Kodas, J. Lumin.**75** , 11 (1997)
4. Z.K. Yang, P. Yu, G.L. Wong, M. Kawasaki, A. Ohtomo, H. Koinuma, Y. Segawa, Solid State Commun. **103**, 459 (1997).
5. N.H. Alvi, S.M. Usman Ali, S. Hussain, O. Nur and M. Willander Scripta Materialia **64**, 697 (2011).
6. M.H. Huang, S. Mao, H. Feick, Science **292,** 1897 (2001).
7] R. Scheer, T. Walter, H.W. Schock, M.L. Fearheiley, H.J. Lewerenz, Appl. Phys. Lett. **63,** 3294 (1993).
8. Y. Chen, D.M. Baghall, H. Koh, K. Park, K. Hiraga, Z. Zhu, T. Yao, J. Appl. Phys**. 84,** 3912 (1998).
9. Y.B. Li, Y. Bando, D. Golberg, Appl. Phys. Lett. **84** , 3603 (2004).
10. J. Ding, T.J. McAvoy, R.E. Cavicchi, S. Semancik, Sens. Actuat. B **77** , 597 (2001).
11. M. Dybic, S. Ostapenko, T.V. Torchynska, E. Velazquez Lozada, Appl. Phys. Lett. **84** (25), 5165 (2004)
12] T. V. Torchynska, A.I. Diaz Cano, M. Dybic, S. Ostapenko, M. Mynbaeva, Physica B, Condensed Matter, **376-377**, 367 (2006)
13. PDF2 XRD database, card no. 36-1451.
14. A. B. Djuris, A.M.C. Ng, X.Y. Chen. Progress in Quantum Electronics 34. 191-259 (2010).
15. N.E. Korsunskaya, I.V. Markevich, T.V. Torchinskaya and M.K. Sheinkman, J. Phys. Chem. Solid. **43,** 475-479 (1982).
16. N.E. Korsunskaya, I.V. Markevich, T.V. Torchinskaya and M.K. Sheinkman, J. Phys. C. Solid St.Phys. **13**, 2975 -2982 (1980).
17. N.E. Korsunskaya, I.V. Markevich, T.V. Torchinskaya and M.K. Sheinkman, phys. stat. sol (a), **60**, 565 -572 (1980).
18. M.A. Reshchikova, H. Morkoc, B. Nemeth, J. Nause, J. Xie, B. Hertog, A. Osinsky, Physica B, Condensed Matter, 401–402, 358–361 (2007).
19. M.K. Patra, K. Manzoor, M. Manoth, S.P. Vadera, N. Kumar, J. Lumin. 128 (2) 267–272 (2008).
20. D.H. Zhang, Z.Y. Xue, Q.P. Wang, J. Phys. D: Appl. Phys. 35 (21) 2837–2840 (2002).
21. A.B. Djurišic, W.C.H. Choy, V.A.L. Roy, Y.H. Leung, C.Y. Kwong. K.W. Cheah, T.K. Gundu Rao, W.K. Chan, H.F. Lui, C. Surya, Adv. Funct. Mater. 14 856-864 (2004).
22. X. Liu, X. Wu, H. Cao, R.P.H. Chang, J. Appl. Phys. 95 (6) 3141–3147 (2004).
23. J. Qiu, X. Li, W. He, S.-J. Park, H.-K. Kim, Y.-H. Hwang, J.-H. Lee, Y.-D. Kim, Nanotechnology 20 155603 (2009).
24. D.C. Look, D.C. Reynolds, C.W. Litton, R.L. Jones, D.B.Eason, G. Cantwell, Appl. Phys. Lett. **81**, 1830 (2002).

Mater. Res. Soc. Symp. Proc. Vol. 1534 © 2013 Materials Research Society
DOI: 10.1557/opl.2013.315

Thin Film Hybrid Ceramic-Polymeric Low Cost Solar Absorber

Edgar A. Chávez-Urbiola and Juan F. Pérez-Robles
CINVESTAV-Querétaro, C.P. 76230 Querétaro Qro., México

ABSTRACT

Different hybrid Ceramic-Polymeric coatings were prepared, from a suspension consisting of a mixture of tetraethyl orthosilicate (TEOS) as a silicon dioxide precursor, polyvinyl acetate (PVA) and colorant to obtain sol-gel SiO_2-PVA thin films. The films were prepared using Sol-Gel technology, applied by dip-coating technique. In order to determine the optimal formulation, different samples varying the proportion of PVA were prepared and evaluated. Both optical and mechanical properties were tested, finding an optimal value of 30 percent of PVA for the mechanical properties, and a value of 50 percent for the optimal optical properties. In both cases, the coatings made can be considered as a reasonable alternative for use as a Solar Absorber for Low-Mid range temperature, with a smaller thickness than the comparable commercial coating but with similar performance and lower cost.

INTRODUCTION

In the last decades, the use of solar energy has shown a remarkable increase due to the necessity of alternative power sources, as a result of the scarcity of fossil fuels and their constantly increasing costs [1]. The use of solar absorbers to convert sunlight to thermal power is one of the more affordable and economical way to use solar radiation [2]. The solar absorbers in systems like solar water heaters have been used for decades and now form a solid part of the market of renewable energy devices. It is also important to mention that there are many different techniques, configurations and materials that are suitable for the fabrication of solar absorbers [3]. The most common materials used in the mid-temperature range ($T<400^{\circ}C$) are NiCrOx [4] (Sputtering), black nickel, black chrome and black copper (Electrodeposition) [5-9], Graphitic films (PVD/PECVD), Ge, Si and PbS in silicon binder (Paint), Ag dielectric and CuFeMnO4/silica (Sol-Gel) among others [10-12], used over different substrates. The sol-gel method is a versatile and reproducible technique for the preparation of coatings having advantages over melting techniques; it was used in this work to prepare SiO_2-PVA films.

EXPERIMENT

In order to obtain an appropriated film that can be used as solar thermal absorber, several formulations were prepared using sol-gel process and varying the content of PVA of the sol. The films obtained were characterized mechanically (hardness [13] and adhesion [14]) and optically, before and after degradation process (250°C for 24 hours). The thickness of each film was also measured before and after degradation. Those results were used to determine the formulation with better performance. The experiments were carried out as described below.

All the chemicals used were of analytical grade, and used without further purification and all solutions were made up with distilled water. First, the stock solution (in order to form the SiO_2 part) was prepared by a mixture of Ethanol, H_2O, TEOS and 2ml HNO_3 (as catalyzer) in a small breaker at room temperature. As colorant carbon black die # 20 was added, and incorporated by first vigorously mixing for 5 minutes and then dispersed in the solution by an ultrasonic bath for half hour for the starting solution. Then, PVA was added to the sol; a homogenous solution was prepared within 30 min by vigorously mixing the stock solution with the PVA. Many solutions at different proportions were prepared to obtain different SiO_2-PVA ratios. Then, the substrates (Corning glass) were painted by "Dip Coating" technique, and then dried in quiet air at room temperature for 30min. Finally they were sintered in air at 150°C for 1 hour. The amount of PVA in the sol-gel/PVA was labeled as follows: F1 for 10% of PVA by weight, F2 for 20% , F3 for 30% etc. A commercial paint solar absorber is labeled like REF for performance comparison.

Hardness measurements were performed using TQC pencil hardness tester (ISO: 3900-E19) scratching the surface coatings in an area free of blemishes and minor surface imperfections, and then making visual assessment. Adhesion was measured by the Tape test (ASTM: D3359) also in an area free of blemishes. The sample was placed on a firm base and parallel cuts were made, and then cleaned with a soft brush to remove possible debris in order to place the tape over the clean scratched surface. After that, the tape was removed and the surface was imaged using an optical microscope. Also a profilometer analysis of the samples was performed. Finally, the optical absorbance of the films was determined using Photo spectrometer UV-VIS QE65000 Ocean Optics. All the measurements were made before and after the degradation process, that consist of a heat treatment in hot air at 250°C for 24 hours.

DISCUSSION

The mechanical properties as a function of the PVA content are shown in Table I. The column Hardness 1 shows the values of hardness after degradation and column Hardness 2 the values after degradation. In a similar way adherence is tabulated. As can be seen, the mechanical quality decreased after degradation, especially the adherence to the substrate, except for F3.

Table I. Mechanical properties measured before (1) and after (2) degradation.

Formulation	% PVA	Hardness (1)	Hardness (2)	Adherence (1)	Adherence (2)
F0	0	9H	------	0B	0B
F10	100	8B	------	5B	0B
REF	-----	2H	9B	4B	3B
F1	10	>9H	>9H	5B	0B
F2	20	5H	4H	5B	1B
F3	30	9H	9H	5B	5B
F4	40	9H	3H	4B	1B
F5	50	9H	9H	4B	2B

The adherence was more affected by degradation process than the hardness. The Figure 1 shows the images for the different films after the adherence test in two columns. Left side of each of the two columns presents the sample before degradation and after degradation in the right side.

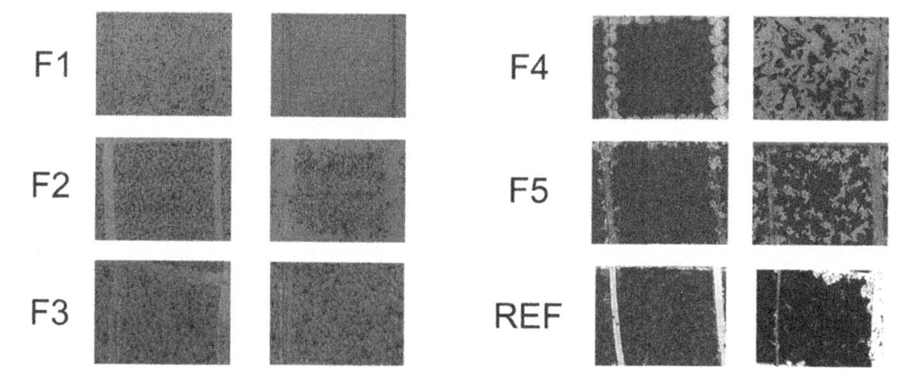

Figure 1. Different samples after the adherence test, undegraded and degraded.

What is expected is that a lower content of PVA leads to a more ceramic film, thus more fragile and also harder, but it is not clear in Table I nor Figure 1. Figure 2 shows different micro photographs, revealing that there exist cracks on the surface for some of the formulations that are the cause of loosing of the mechanical properties, which agrees with the results of the Table I.

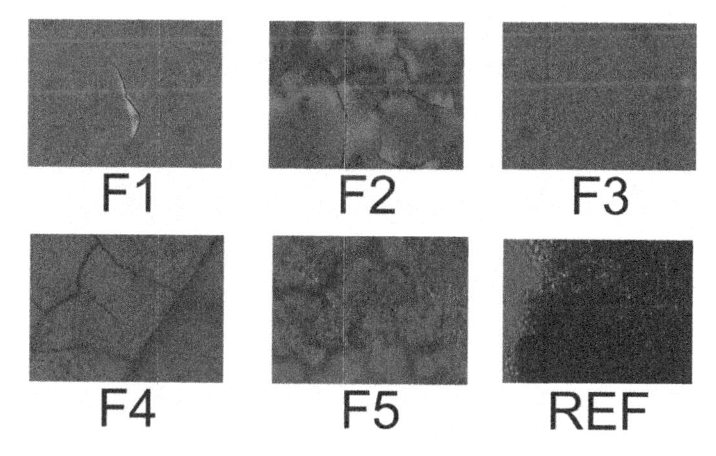

Figure 2. Micro photographs of surface cracking of the different coating formulations (1000X).

Also the changes in optical absorbance were analyzed before and after degradation process. In Figure 3 the obtained data is shown. Before degradation (Left), F5 and REF are overlapped, but after it the reference material absorbs less than the REF one. F3 shows little decrease in absorbance as well as F1, but the effect is much clearer in F1.

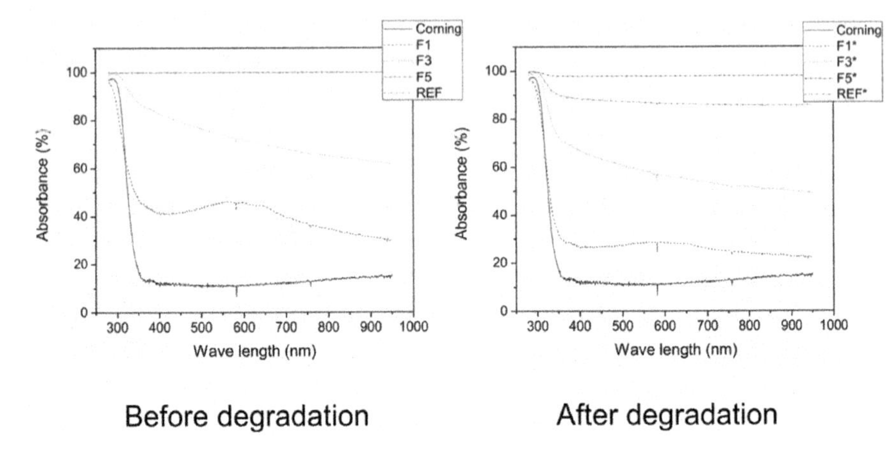

Figure 3. Optical absorbance of the different films, before and after degradation.

A profile analysis performed shows the changes in thickness due to degradation process. In Figure 4 the results are shown. In the left side, the REF material and the different hybrid coatings as a function of the PVA content are presented, before and after the degradation. On the right side, REF material is not present, in order to magnify the effect of PVA content in the prepared samples.

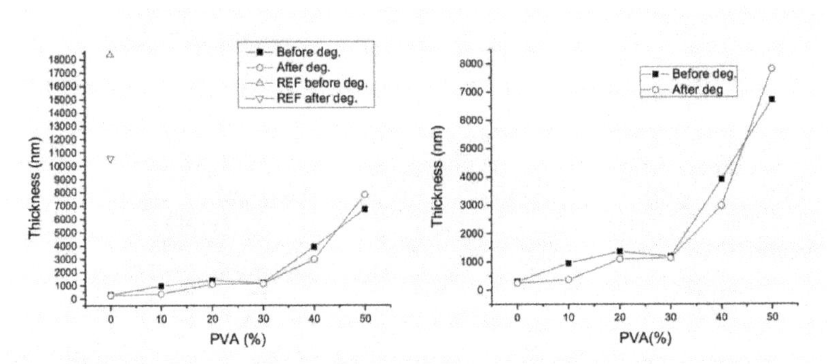

Figure 4. Thickness of the prepared samples as a function of the PVA content.

For all the films the degradation affects the thickness, being the more affected the REF material, and F3 almost unchanged, which agrees with the absence of cracking in the surface. In all the cases, the our hybrid films were thinner than the commercial coating. The thickness of the samples increased as the PVA content increases; this agrees with the increase of viscosity of the solution as the PVA is added to the stock solution: at higher PVA content, higher is the viscosity that leads to a thicker film.

CONCLUSIONS

The mechanical properties of the different films change with the composition, but their deterioration is more marked when cracking of the surface is present, especially in adherence.

F3 formulation is the more similar to the REF material, having much better mechanical properties, caused for its ceramic-polymeric composition. On the other hand, F5 have better absorbance and also better mechanical properties than REF material. That makes those materials good alternatives to use like a solar absorber for mid-range temperature solar heaters. Also all of the coatings made have a smaller thickness than commercial ones, using less material to make the films. The lower use of material and the simplicity of preparation of those thin films make them reproducible and will lead to cheaper surface coatings.

ACKNOWLEDGMENTS

The authors are grateful to CONACYT, for the PhD Scholarship.

REFERENCES

1. M. V. d. Hoven, *CO2 emissions from fuel combustion highlights*, 1 ed. vol. 1. France, 2011.
2. J. A. S. Jaisankar, "A comprehensive review on solar water water heaters," *Renewable and Sustainable Energy Reviews,* vol. 15, pp. 3045-3050, 2011.
3. K. C.E., "Review of Mid-to-High Temperature Solar Selective Absorber Materials," *NREL/TP,* 2002.
4. X. H. Yunzhen Cao, "Absrbing Film on Metal for Solar Selective Surface," *Thin Solid Films,* vol. 15, pp. 155-158, 1999.
5. N. V. S. N. K. Srinivasan, M. Selvam, S. John and B. A. Shenoi, "Nickel-Black Solar Absorber Coatings," *Energy Convers. Mgmt,* vol. 24, pp. 255-258, 1983.
6. J. S. A. Wazwaz, R. Bes, "The effects of nickel-pegmented aluminum oxide selective coating over aluminium alloy on the optical properties and thermal efficiency of the selective absorber prepared by alternate and periodic plating technique.," *Energy Conversion and Management,* vol. 51, pp. 1679-1683, 2012.
7. R. U. R. Vishal Saxena, A. K. Sharma, "Studies on ultra high solar absorber black electrodes nickel coatings on aluminum alloys for space application," *Surface and Coatings Technology,* vol. 201, pp. 855-862, 2006.
8. G. E. McDonald, "Spectral Reflectance Properties of Black Chrome for use as a Solar Selective Coating," *Solar Energy,* vol. 17, pp. 119-122, 1974.
9. J. S. A. Wazwaz, H. Hallak, R. Bes, "Solar thermal performance of a nickel-pigmented aluminum oxide selective absorber," *Renewable Energy,* vol. 27, pp. 277-292, 2002.

10. N. L. Z. Crnjak Orel, B. Orel, M. G. Hutchins, "Spectrally selective silicon paint coatings: Influence of pigment volume concentration ratio on their optical properties," *Solar Energy Materials and Solar Cells,* vol. 40, pp. 197-204, 1995.
11. Z. C. Orel, "Characterization of high-temperature-resistant spectrally selective paints for solar absorbers," *Solar Energy Materials and Solar Cells,* vol. 57, pp. 291-301, 1999.
12. M. K. G. Zorica Crnjak Orel, "Spectrally selective paint coatings: Preparation and characterization," *Solar Energy Materials and Solar Cells,* vol. 68, pp. 337-353, 2000.
13. ISO, "Methods of test for paints. Determination of film hardness by pencil test," vol. BS 3900-E19:1999, ISO 15184:1998, ed, 1998.
14. ASTM, "Standard Test Methods for Measuring Adhesion by Tape Test," vol. ASTM D3359 B.

AUTHOR INDEX

SUBJECT INDEX

Printed in the United States
by Baker & Taylor Publisher Services